Volume 136

First European Conference on
Optics Applied to Metrology

October 26-28, 1977, Strasbourg (France)

Honorary Chairman

D. Gabor (Nobel Prize)

Organized By

European Photonics Association
Comite International D'Optique
The Society of Photo-Optical Instrumentation Engineers

With the Assistance of

Université Louis Pasteur, Council of Europe, European Physical Society, D.G.R.S.T., D.R.M.E., S.E.U., C.N.R.S., D.G.A.O., A.D.R.E.R.U.S., C.E.T.I.M., B.N.M., B.I.P.M., Agfa-Gevaert, Coherent Radiation, Kodak, Lexel Optilas, Masson, Spectra Physics, Soptel.

Editors

Michel Grosmann
Patrick Meyrueis

ISBN 0-89252-163-5

Copyright © 1978 by the Society of Photo-Optical Instrumentation Engineers,
405 Fieldston Road, Bellingham, Washington 98225 USA.
All rights reserved. No part of this book may be reprinted, or
reproduced or utilized in any form or by any electronic, mechanical
or other means, now known or hereafter invented, including photo-
copying and recording, or in any information storage or retrieval
system, without permission in writing from the publisher.

Printed in the United States of America..

1st EUROPEAN CONGRESS ON OPTICS APPLIED TO METROLOGY
Volume 136

CONTENTS

Preface vii

SESSION 1.1. MEASUREMENT THROUGH OPTICS 1

136-76 **Optics in Europe** 2
Prof. J.-Ch. Viénot

136-01 **Space-Time Optics in Shape and Surface Metrology** 3
Jean-Pierre Goedgebuer and Jean-Charles Viénot

136-02 **Interferometric Measurement of Heterogeneities in Semiconductors** 8
Waldemar Kowalik

136-03 **Application of Interferential Correlation of Spectra to the Detection of Pollutants in the Atmosphere** 14
G. Fortunato, A. Maréchal, Melle Wolfer

136-04 **Measurement of Small Rotations** 19
Aline Huard and Christian Imbert

SESSION 1.2. MEASUREMENT THROUGH OPTICS 25

136-06 **Determination of the Index Profile of a Dielectric Plate by Optical Methods** 26
André Roger and Daniel Maystre

136-08 **Optical Analyser of Vibrations** 29
Alain Constans

136-09 **Modex Opto-Numerical Measurer of External Dimensions** 32
J. P. Elissalde, F. Pointeau, B. Turlier

SESSION 2. INTERFEROMETRY 37

136-10 **Comparison of Wavelengths to the Primary Standard at the French National Institute of Metrology** 38
P. Bouchareine and A. Janest

136-11 **Interferometer for Testing Infrared Materials and Optical Systems** 43
Michel Lamare

136-13 **Displacement and Vibration Measurement by Laser Interferometry** 52
L. E. Drain, J. H. Speake and B. C. Moss

136-14 **Applications of Phase Modulation Interferometry to the Characterization of Materials and to Dimensional Metrology** 58
G. Roblin

136-15 **Image Spectrograms of Three-Dimensional Objects: Metrological Applications** 65
Alain Lacourt

136-16 **Interferometrical Setup for the Study of Thermic Turbulence in a Plane Airstream** . . 69
Alain Gagnaire and Albert Tailland

SESSION 3. PHOTOGRAMMETRY AND HOLOGRAMMETRY 75

136-17   Hologrammetric Plotter (Apparatus for Three-Dimensional Measurement on an Image Reconstructed by Holographic Process) 76
  Gerard Bocquemo, Dominique Laroche, Bernard Turlier

SESSION 4. PHOTOELASTICIMETRY 81

136-20   Photoelasticimetry and Holographic Interferometry: Applications to the Study of Stresses and Deformations 82
  J. Monneret, P. Rastogi, M. Spajer

SESSION 5.1. HOLOGRAPHIC INTERFEROMETRY 91

136-72   Dimensional Metrology of Length Standards by Holographic Interferometry with Phase Heterodynage 92
  M. Grosmann, P. Meyrueis

136-24   Double Exposure Holographic Interferometry: Application to Nondestructive Testing and to Breaking Point Mechanics 101
  Nicole Jolly, Jacques Poirier

136-25   Holographic Interferometry Applied to Minimal Wear Measurement 107
  J. T. Atkinson and M. J. Lalor

136-26   Applications of Holography to the Study of Structures and Materials 114
  Gaël Cadoret

136-74   Holographical Disdrometry 127
  V. M. Zakharov, L. N. Rasumov, I. N. Sisakyan

136-27   Study by Holographic Interferometry of Dimensional Variability in Precision-Moulding Materials Used in Odontology 130
  M. Blandin, C. Durou, H. Soulet

SESSION 5.2. HOLOGRAPHIC INTERFEROMETRY 135

136-28   Application of Holographic Interferometry to the Study of Structural Deformations in Civil Engineering 136
  J. M. Caussignac

136-29   Study by Holographic Interferometry of Mass Transfer During Electrochemical Processes at Solid-Liquid Interfaces 143
  Michael Clifton, Victor Sanchez and Christian Durou

136-30   Holographic Analysis of Oscillations in Squealing Disk Brakes 148
  Armin Felske

136-31   Some Considerations on the Quantitative Interpretation of Holographic Interferograms 156
  R. F. C. Kriens

136-32   Video-Electronic Analysis of Holographic Interferograms 166
  Franz Lanzl and Michael Schlüter

SESSION 5.3. HOLOGRAPHIC INTERFEROMETRY 173

136-34   Holographic Interferometry with the Possibility of Modifying the Fringes During Reconstruction. 174
  W. Schumann and M. Dubas

136-35   Testing by Holographic Interferometry of Solid Propergol Engines 181
  Paul Smigielski, Daniel Cesario, Claude Patanchon

136-36	Application of Holographic Interferometry to Testing of Spun Structures .... 186 J. D. Dubourg	
136-37	Holographic Interferometry Applied to the Metrology of Gaseous Flows .... 192 Jean Surget, Jean Délery and Jean-Paul Lacharme	
136-38	Holographic Interferometry in Osteosynthesis.. 202 P. Meyrueis, M. Pharok, J. Fontaine	

SESSION 5.4. HOLOGRAPHIC INTERFEROMETRY 207

136-41	Holographic Testing of Aspherical Surfaces .... 208 R. Mercier	
136-42	Heterodyne Holographic Interferometry: A Review.................................. 215 René Dändliker	

SESSION 6.1. SPECKLE 217

136-43	Determining the Inclination of a Diffusing Surface with Regard to Viewing Direction by Speckle Photography.. 218 J. Montilla, R. Hernández	
136-44	Holographic Methods Made Useful by Phase Modulated ESPI......................... 222 Ole J. Løkberg and Kåre Høgmoen	
136-75	New Possibilities of Real-Time Interferometry with Photoconductive Electro-Optic Crystals $Bi_{12}SiO_{20}$ .... 226 J. P. Huignard, J. P. Herriau	
136-77	Autoprocessor Materials for the Recording of Phase Holograms: Photopolymers and Organo-Metallic Semiconductors .... 229 M. Jeudy	
136-45	Improvement of the Signal/Noise Ratio in the Method of Subtraction of Images by Speckle Interferometry .... 237 S. Debrus and V. Sokolov	
136-46	Detection of Axial Displacements of a Diffusing Object .... 245 Yves Dzialowski and Marie May	
136-07	The Diffracting Gauge in Extensometry.. 251 Jean P. L. Ebbeni	
136-47	Speckle Photography for Strain Measurement—A Critical Assessment .... 258 E. Archbold, A. E. Ennos and M. S. Virdee	

SESSION 6.2. SPECKLE 265

136-49	Study of the Distribution of Velocities in a Fluid by Speckle Photography.............. 266 R. Grousson and S. Mallick	
136-50	Deformation Measurements on Connecting Rods by Speckle Photography .... 270 Alfons Happe	
136-51	Analysis of the Diffraction Spectrum of a Population of Particles .... 277 J. Fleuret, H. Maitre, J. F. Thery	
136-52	Decorrelation Produced in the Image of a Diffusing Object by a Rotation of the Incident Wave— Application to the Study of Surface States.. 282 Michel Menu and Marie-Louise Roblin	
136-53	Two Wavelength Speckle Images Applied to Contouring .... 286 Gilbert Tribillon	

136-55	Application in Civil Engineering of a Sandwich Speckle Method	291
	Jean P. L. Ebbeni	

SESSION 7. MOIRE ... 295

136-56	Optical Differentiation of Moire-Holographic Fringes by Wavefront Reconstruction with White Light Sources	296
	Carlo Brutti, Giuseppe Di Chirico, Umberto Pighini	
136-58	The Measurement of Residual Stress by a Moire Fringe Method	302
	C. A. Walker, J. McKelvie	
136-59	Electronic Processing of Moire Fringes. Application to Moire Topography and Elements of Comparison with Photogrammetry	311
	J. C. Perrin and A. Thomas	
136-60	Dimensional Metrology of Large Objects by Projection Moire Techniques	318
	Luciano Pirodda	
136-61	Analysis of Grating Imaging and its Application to Displacement Metrology	325
	R. M. Pettigrew	
136-62	Measurement in Real Time of Transversal Micro-Vibrations (Down to 1 Å) on Diffusing Objects: Random Moire Captors	333
	D. Joyeux	

SESSION 8. ACOUSTO-OPTICS ... 339

136-63	Contributions of Acoustical Holography to Mechanical Metrology	340
	J. Pasteur and Y. Seyzeriat	

POSTER SESSION ... 347

136-64	Investigation of Cavitation Bubble Dynamics by High Speed Ruby Laser and Argon Ion Laser Holocinematography	348
	Karl Joachim Ebeling	
136-66	Minutes of the Round Table on Photo-Sensitive Surfaces	355
	J. P. Christy, J. Sagaut and J. L. Tribillon	
136-68	Principle of the Holographic Cinematography	358
	Victor G. Komar	
136-73	Holographic Art with Recording in Three-Dimensional Media on the Basis of Lippman Photographic Plates	365
	Yu. N. Denisyuk	

Author Index ... 369
Subject Index ... 371

1st European Conference on
Optics Applied to Metrology
Volume 136

PREFACE

Laboratoire de Spectroscopie et d'Optique du Corps Solide
(associé au C.N.R.S. n° 232)
Université Louis Pasteur
5, rue de l'Université
67000 Strasbourg (France)

The 1st European Congress on Optics Applied to Metrology was held in Strasbourg from 26th to 28th October 1977.

Previous European conferences have taken place, for example, in Brussels, 1959. The Strasbourg Congress is, however, the first of what will be a series of regularly scheduled meetings on this important subject.

About 300 participants of 35 countries were registered at this Congress.

Optics Applied to Metrology has become a rather wide domain of research. Numerous papers were submitted for the Strasbourg Congress, but the number of papers presented had to be reduced in order to keep the Congress within reasonable limits. The Program Committee had to accomplish the delicate task of eliminating papers in spite of their quality. It was decided not to have invited conferences, parallel sessions, or poster sessions. Instead, we opted for the presentation of a limited number of high quality papers. It was decided that these papers should be written and published after the Congress in order to enable their authors to take into account the comments and reflections made during the congressional period.

The scope was limited to the following topics:
- Measurements through optics,
- Interferometry,
- Photoelasticimetry,
- Photogrammetry and hologrammetry,
- Holographic interferometry,
- Speckle interferometry,
- Moire measurements,
- Acousto-optics,
- Advances in three-dimensional imagery.

Three panel discussions were also held:
- Emulsions and photosentive materials,
- Uses of optics and photonics in quality testings,
- Amelioration of coordination in optics and photonics in Europe.

Therefore, we feel that a comprehensive review of the most outstanding results in each area was presented at the Strasbourg Congress. The organization of one session at a time allowed for stimulating general discussion.

The sessions were held in the Assembly Room of the Council of Europe which kindly offered its hospitality. The Council of Europe also made possible the use of its smaller conference rooms for parallel sessions, as well as its offices and facilities. Particularly useful were the interpreters whose work was highly commendable.

In the Josephine Palace adjacent to the Council of Europe was held an exhibition of Scientific Apparatus as well as an International Exhibition of Holograms. More than 25,000 people visited and profited from the occasion to better understand the new possibilities in optics.

In the opening session Professor P. Karli, President of University of Strasbourg, welcomed the participants of the Congress to Strasbourg and welcomed the conference to the Council of Europe. He gave a short review of the activities of the Council in general and in particular its promotion of scientific cooperation. He emphasized the paramount importance of such cooperation for the promotion of mutual understanding and peace.

On behalf of the local Organization Committee, its Secretary, Dr. Meyrueis, expressed his thanks to previous speakers as well as to all organizations which helped, by financial or moral support, in the organization of the Congress. He thanked as well all members of the Organization Committee and participants. These Proceedings are published as part of SPIE's book series. We would like to express our gratitude to SPIE for its support in this venture.

We are confident that this Congress and its Proceedings will help develop a fascinating and deep understanding of the great possibilities in Optics Applied to Metrology. The great vitality of this branch of applied physics will become evident to the reader. It further would appear to me that this branch merits the acclaim it will, in all justification, receive at the following important events:

- The All Union Meeting of Holography in August 1978 in USSR

- The two meetings planned in 1978 by the European Photonic Association : one in Belgium on "Possible Economical and Industrial Impacts of New Developments in Optics and Photonics"; the other in Hungary on "Biomedical Applications of Optics and Photonics".

- The Meeting of the Federation of European National Committees of Optics which should occur in April 1979 in Germany.

On behalf of the Organization Committee of the European Optics Congress 1977, I should like to express our best wishes to the organizers of the Conference.

M. Grosmann,
Secretary of the Organizing Committee
of E.O.C. 77.

1st EUROPEAN CONGRESS ON OPTICS APPLIED TO METROLOGY

Volume 136

SESSION 1.1

MEASUREMENT THROUGH OPTICS

Session Chairmen
Francon
Hopkins

OPTICS IN EUROPE

Prof. J.-Ch. Viénot
Secretary General of the International Commission for Optics
Laboratoire de Physique Générale et Optique
Faculté des Sciences — Université de Franche-Comté
Route de Gray — 25030 Besancon cedex

On behalf of the International Commission for Optics who sponsors this European Conference, I want to express my best wishes to the organizers as I believe they need a lot of encouragement. Next I wish that those who have come to participate to the meeting first of all may find here what they expect they will find - I mean the opportunity to discuss their research work and that of exchanging ideas inside a truly European Community. By the end of this week, shall we all feel like members of European Optics, nay of Optical Europe ? Wouldn't this constitute the greatest aim we all here might wish to achieve, since in this European Hall are sitting together in particular, with all our colleagues from the West, those who have also been willing to start on such a long journey, I mention the scientists from the USSR, only to refer to the geographically remotest European nation.

One may wonder why, as French as well as English is an official ICO Language, a Frenchman should use an idiom from overseas to express his ideas, all the more so as a lot of subtlety can be conveyed through our French language, especially since we have imported a few valuable expressions from over the Channel.... A reason for this is that after reading the programme of this Conference in the English translation I thought I might as well spare the technicians some extra work by delivering this address directly in English. Indeed it is my impression that some transpositions of titles do not fit exactly the thoughts their authors would have liked to express. Who could, for example, feel enthusiastic about the title and summary of the first paper of this morning session ? I can assure you that I have no responsibility in their manufacturing and I guarantee the genuine scientific quality of their French wording.

The initiative of holding the Strasburg meeting was taken in the Spring 1976. The aim was to collect grey matter and compare techniques in optical metrology, but just among the French. They wanted to examine the state of the art in surface inspection and rugosity assessment, in stress and strain analysis, and other engineering topics. They, generally speaking, intended to make a nation wide estimation of the technological capacities that are often scattered among various laboratories and a few too scanty industries in the field of optics.

In recent weeks one has heard the Strasburg physicists say that they do not specialize in optics but in molecular physics. Yet they have been bold enough to get involved in the organization of a Conference in Optics. They had been wise to ask for other people's advice beforehand, since after all no one knew exactly which mood would best suit this meeting : should it be some sort of workshop ; a multi-session seminar ; a highly specialized symposium ; or a large conference ? Towards the fall of 1976 it became known that the French Physical Society was ready to help, while the National Committee of the CNRS has accepted to support such a meeting in the form of an Associated Colloqium provided that such a formula should imply the participation of both French and European physicists.

It is necessary now to step a few years back. If to-day I am speaking to you in the name of the ICO, it is also to remind you that our organization, which gathers 20 Member Countries throughout the world, has for many years pointed out the need to constitute some kind of body after the manner of the powerful OSA, the aim being to weave closer links between all the European physicists working in Optics, arrange regular optical meetings, and, at the dimensions of our Continent, try to comply with the objective defined by the article 1 of the ICO statutes. In tokyo in 1974, in Prague in 1975, then in Jerusalem in 1976 important ICO meetings took place, and, either during general Assemblies or during Bureau sessions, everybody called for an Optical European Union. Among the several participants from Europe last year in Jerusalem, there was a strong feeling that the time was ripe to seize the opportunity that the Strasburg Conference offered. The decision of the ICO Bureau to consent to sponsoring the present occasion shows undoubtedly the will that a first step be made towards an actual European cooperation. A few weeks ago a series of tangible proposals was circulated by the Belgian and Netherlands' Optical Committees. They are to be discussed here these days. Can we expect decisions to be taken about the real start of a European Optical Committee ?

Still a last word in the name of the ICO : the wish I have so often made public on the occasion of contacts with representatives of the USSR is that their great country join the ICO. It would bring the General Secretary utter joy if the first European Optical Conference could constitute a further element of weight for that matter. Finally I am happy to announce that the German Democratic Republic has officially applied for ICO membership. I hope other countries will still do so before the ICO 11th Conference to be held in Madrid next September.

So much so for to-morrow. Our leading star to-day is Strasburg, the capital city of Europe, who rejoices in giving you all here a most heartly welcome.

SPACE-TIME OPTICS IN SHAPE AND SURFACE METROLOGY

Jean-Pierre Goedgebuer and Jean-Charles Viénot

Laboratoire de Physique Générale et Optique, associé au C.N.R.S.
Université de Franche-Comté — Faculté des Sciences et des Techniques
25030 Besancon Cedex, France

Abstract

Basic considerations of similarity between the spatial and temporal frequencies in image formation have been developed along applications for which two classes of examples are given.

a) Absolute measurements of optical paths by interferometry in white light ; applications to the determination of surface mapping :

A channelled spectrum [1], observable at the output of a spectroscope set in cascade with a two-beam interferometer illuminated in white light, displays the local path-differences introduced along one arm of the interferometer, the other one being considered as conveying the reference. A relationship holds between the chromatic distribution scaled in temporal frequency and optical delays that are functions of time. This suggests techniques of longitudinal measurements. Indeed one deals with Fourier transforms both in space and time, allowing to show some equivalence of the channelled spectrum with a Fourier hologram - the reconstruction of longitudinal information being performed at a transverse diffraction plane. This is illustrated in rugosimetry and in the assessment of calibrating tests.

b) Statistical properties of random surfaces by spectral analysis of scattered light.

By analogy with conventional speckle methods, we have proposed the concept of temporal speckle as the spectrum of the incident radiation spreads over a broad band. It can be stated that no statistical model is needed for surface rugosity and evaluation of the r.m.s. for Gaussian rugosities are presented as illustration.

The following text is very close to that of the original lecture and we did not feel the need for any proper introduction to it. A rough idea of the work, which has been given in the abstract, will be further recalled through the fundamentals in the first paragraph below.

1. Channelled spectra as temporal Fourier hologram :

Fig. 1 is a photograph of a channelled spectrum. The experimental device usually involves a spectroscope in cascade with two Young's slits (S_1, S_2) illuminated by white light point source S_0 (Fig. 2). A channelled spectrum is observed at the output of the spectroscope. Such a spectrum may be regarded as the result of the

Fig. 1. Classical channelled spectrum photographed at the output of the device sketched in Fig. 2

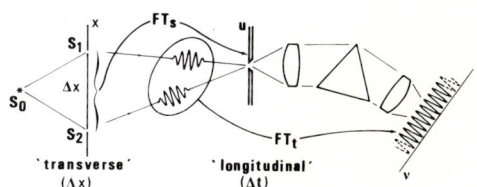

Fig. 2. As, for a given wavelength λ, the intensity is minimum in the corresponding particular pattern of interference fringes. Therefore a dark band is observed at the output of the spectroscope in the region of such a λ, or $\nu=c/\lambda$

spectral analysis of superimposed monochromatic interference patterns generated by the various white light spectral components. An alternative interpretation comes to consider the spectroscope as a temporal Fourier Transformer acting on two wave-groups of white light $f_1(t)$ and $f_2(t)$ incident on to the entrance slit of the spectroscope, and delayed one with respect to the other by Δt. Under such conditions, the distribution of energy $I(\nu)$ along a ν-axis (ν : temporal frequency) at the output of the spectroscope is expressed by the temporal Fourier Transform (FT_t) of the delayed wave-trains :

$$I(\nu) = |FT_t \{f_1(t) + f_2(t)\}|^2$$
$$= |FT_t \{f_1(t)\}|^2 \cdot \{1 + \cos2\pi\nu\Delta t\}$$
$$= B(\nu) \cdot \{1 + \cos2\pi\nu\Delta t\} \qquad (1)$$

with $B(\nu)$: power-spectrum of the white light source.

Thus the passage from the two slits plane to the entrance pupil of the spectroscope, together with what we have just discussed describe a two-step operation involving successive FT, in spatial and temporal domain respectively. It also substantiates a part of the title (Space-Time Optics) [2,3,4].

Fig. 3 depicts a similar situation when a two-beam interferometer illuminated with white light is adjusted for a path difference Δ. Any incident wave-train of white light $f_0(t)$ is splitted into two wave-groups $f_1(t)$, $f_2(t)$ delayed by $\tau = \Delta/c$ (c : velocity of light) ; the signal $f(t)$ emerging from the interferometer can be written as :

$$f(t) = f_1(t) + f_2(t)$$
$$= f_0(t) \otimes \{\delta(t - t_0 + \frac{\tau}{2}) + \delta(t - t_0 - \frac{\tau}{2})\},$$

t_0 being an arbitrary time corresponding for example to the propagation of light through the interferometer, and $\otimes$ standing for a convolution.

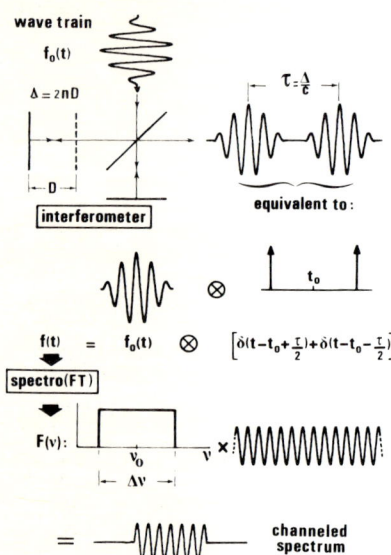

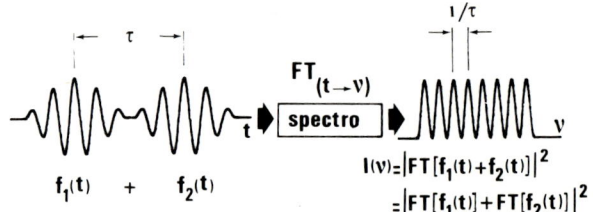

Fig. 4. Coherent superposition of two spectra.

← Fig. 3. Schematic diagram of image formation of a channelled spectrum illuminated by an impulse of white light.

The spectral analysis of $f(t)$, by means of a spectroscope, yields a channelled spectrum $I(\nu)$, that expresses a chromatic, or temporal filtering performed by the interferometer on the power-spectrum $B(\nu)$ of the incident light (in Fig. 3, for simplicity sake, the spectrum of the incident light is assumed to be uniform within a frequency band $\Delta\nu$ centered at $\nu = \nu_0$) :

$$I(\nu) = |FT_t\{f(t)\}|^2 = B(\nu) \cdot \{1 + \cos 2\pi\nu\tau\} \qquad (2)$$

Eq. 2 indicates that the fringe spacing along a ν-axis is inversely proportional to the delay τ. It is possible to go a little bit further. The channelled spectrum may be described [5] as the coherent superposition of the temporal spectra of the wave groups $f_1(t)$ and $f_2(t)$ - i.e. $TF\{f_1(t)\}$ and $TF\{f_2(t)\}$. The latter can be taken as a reference (Fig. 4) : this is the description of a Fourier hologram in the temporal domain. It is worth noting that the carrier frequency corresponding to the elementary fringe function $1 + \cos 2\pi\nu\tau$, is linked to the *longitudinal* interval Δ, or to the *time* delay τ between two signals - that is different in classical holography where the angular separation reference beam object beam is the parameter.

One may wonder how metrology can take advantage of temporal holograms. The answer lies in the very nature of such holograms which enable access to delay τ by diffraction in monochromatic light, whenever τ is not accessible through classical methods, such as monochromatic interferometry, which are in general limited to relative measurements. Experiments confer validity to the above statement.

2. Applications of temporal holography to measurements of optical delays :

i) profile of a ground surface :

Suppose one mirror of the Michelson interferometer concerned is replaced by a rough surface. The accidents of the surface introduce local delays between the wave-groups travelling through the interferometer. Fig. 5a shows the resulting channelled spectrum. The spacing of the fringes along a ν-axis codes the surface profile. The spectrogram behaves as a Fourier hologram : by diffraction in monochromatic light, an image of the surface profile is retrieved in the ±1 orders of diffraction. In Fig. 5b, the average roughness is about 1 μm.

ii) calibrating test :

Fig. 6a is an image of a channelled spectrum corresponding to a polished spherical surface set along one arm of the interferometer (radius of curvature : 30 cm).

Fig. 5a : One mirror of the two-beam interferometer being replaced by a rough surface, local path-differences are expressed by irregularities in the channelled spectrum.

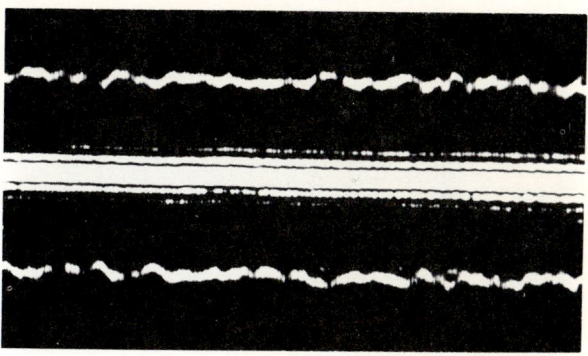

Fig. 5b : Reconstruction of the surface profile along coherent process.

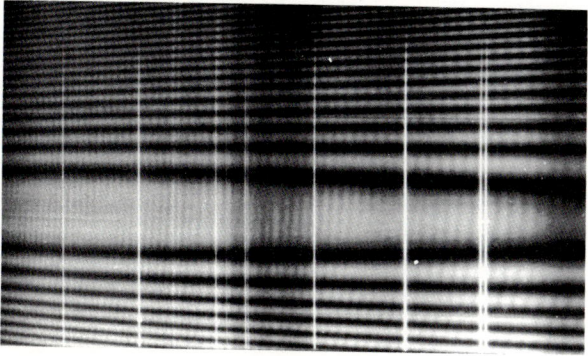

Fig. 6a : channelled spectrum obtained as a spherical surface is set in place of one mirror of the interferometer.

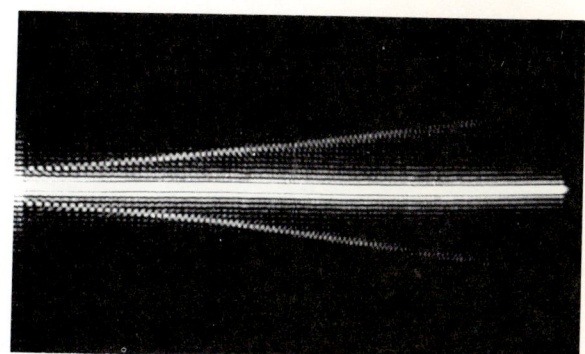

Fig. 6b : Reconstruction of the spherical profile of the surface.

Two images of the spherical profile of the surface are reconstructed by diffraction (Fig. 6b).

The method applies to various problems dealing with path-differences (6). The measures range from 1 μm to some cm, with an accuracy of $5 \cdot 10^{-3}$ or about.

3. <u>Speckle phenomena in temporal optics ; determination of statistical parameters of rough surfaces</u> :

(7) The transposition of conventional holography to temporal optics also applies to speckle phenomena (Fig. 7) (7). In the case of spatial speckle, the average size of the speckle grain generated by a diffuser at the focal plane of a lens is inversely proportional to the illuminated area Δx of the object. Let us transpose the experiment to the field of temporal optics. A random temporal signal corresponds to the diffuser ; the Fourier Transformer is now a spectroscope. As in the spatial domain, a speckled pattern is expected at the output of the spectrosocpe, its average size, along the ν-axis, being inversely proportional to the time interval Δt limiting the incident signal.

i) recording of a speckled spectrogram :

The above situation is encountered in Fig. 8(8) : a rough surface (S) is illuminated with white light, and a spectroscope set at the focal plane of a lens (L_c) analyzes the light reflected back by the surface. The light reflected in the z-direction is due to the delayed wavelets Wi, Wj,..., each arising from distinct microscopic elements of the surface. This sequence of wavelets, g(t), distributed at random is time-modulated by a determination function rect(t/Δt) corresponding to the envelope of the statistical distribution P(z) = rect(z/Δz) of surface accidents along the z-axis :

$$g(t) = s.(t) \cdot \text{rect}(t/\Delta t) \qquad (3)$$

with $\Delta t = 2\Delta z/c$ and s(t) standing for a stochastic process. The random signal g(t) being defined in a limited time interval Δt, its power-spectrum B'(ν) is randomly modulated :

$$B'(\nu) = |FT_t\{g(t)\}|^2$$

$$= |S(\nu) \otimes \text{sinc } \pi\nu\Delta t|^2 \qquad (4)$$

where $S(\nu) = FT_t\{s(t)\}$, and $\otimes$ denotes a convolution.

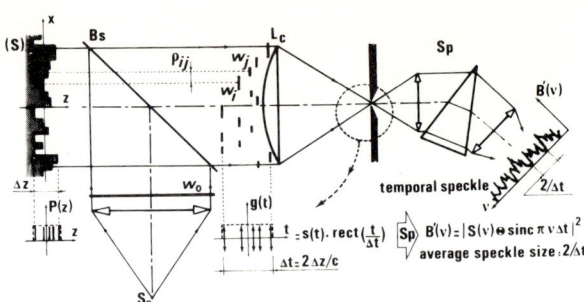

Fig. 8 : Recording of the temporal speckle due to a diffusing surface.

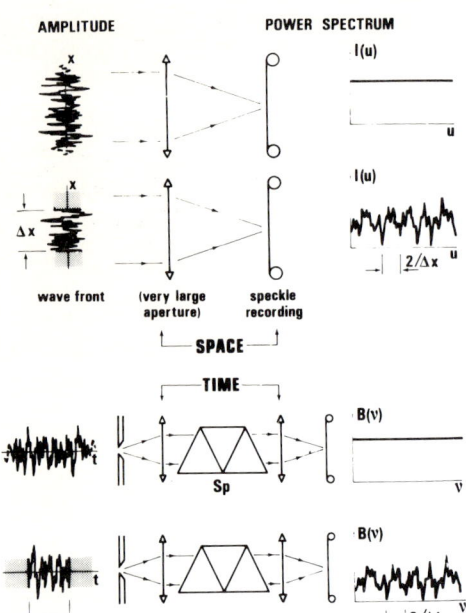

Fig. 7 : Comparison between spatial and temporal speckle : the effect of a limited extent (Δx or Δt) is the apparition of a graininess in the power-spectrum

Fig. 9 : Photograph of the temporal speckle in the ν-plane. Unfortunately the present black and white reproduction only gives a rough idea of the phenomenon

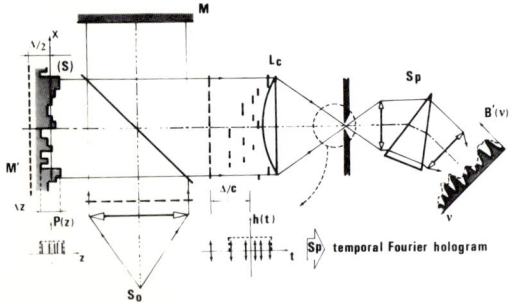

Fig. 10 : Addition of background to a temporal speckle. The result is actually a hologram.

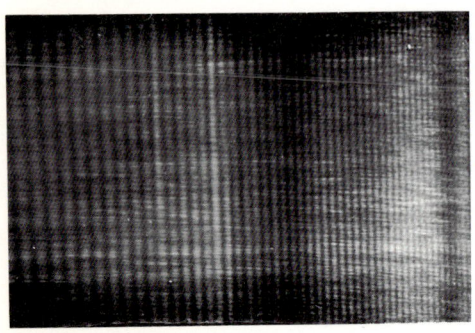

← Fig. 11 : Picture of a "speckle hologram"

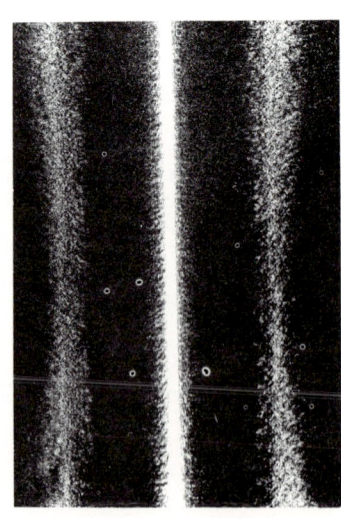

Fig. 12 : The ±1 orders of diffraction represents the probability density function of the surface.

The recording of that energy distribution is a speckled spectrogram (Fig. 9) ; the average size of the grain along the ν-axis is $2/\Delta t$ and then depends on surface parameters. Temporal speckle can also be used to measure surface roughness. An example is given now.

ii) application to the determination of statistical properties of a rough surface :
A hologram can be recorded by adding the temporal speckle to a spectrum of white light which behaves as a uniform background. In Fig. 10, the "background" is generated by a reference signal coming from mirror M The speckled pattern is now modulated by a set of fringes (Fig. 11a) : the result is a hologram of the random signal g(t) (one may notice that the speckle contrast is reduced, as in spatial optics). From such a temporal hologram taken as a diffracting pupil, two images of g(t) are reconstructed ($\pm$ 1 orders). They correspond to the histogram P(z) of the surface (S) along the z-axis (Fig. 11b).

The method seems to be of some interest whenever statistical parameters of surfaces have to be determined, without reference to any mathematical model, as usual in space-optics. A more exhaustive study is still in progress.

References

1. The channelled spectrum phenomenon has been known for a long time (Dutour 1773) ; see for instance H. Bouasse, Z. Carrière, Interférences (Delagrave Ed., Paris, 1923).
2. C. Froehly, A. Lacourt, and J.-Ch. Viénot, Nouv. Rev. Opt. 4, 1973
3. J.-Ch. Viénot, J.P. Goedgebuer, A. Lacourt, Applied Optics, 16,2, 1977
4. G. Bonnet, Nouv. Rev. Optique, 7, 4, 1976
5. J.P. Goedgebuer, A. Lacourt, J.-Ch. Viénot, Optics Commun. 16, 1, 1976
6. A. Lacourt, J.-Ch. Viénot, J.P. Goedgebuer, Jap. Jl. of Appl. Physics, 14, 1975, Suppl. 14-1
7. J.-Ch. Viénot, J.P. Goedgebuer, *Applications of Holography and Optical Data Processing*, Pergamon Press, New-York, 1977, pp. 95-104
8. J.P. Goedgebuer, J.-Ch. Viénot, Optics Commun., 19, 2, 1976

INTERFEROMETRIC MEASUREMENT OF HETEROGENEITIES IN SEMICONDUCTORS*

Waldemar Kowalik

Technical University of Wrocław, Institute of Physics
Wybrzeże Wyspiańskiego 27, 50-370 Wrocław, Poland

Abstract

Interference method of measuring distribution of refracting index continuous heterogeneities in semiconductors are described. Difference distributions of sample thickness and refracting index are calculated by scanning three interferograms and making use of computer analysis. Distributions of sample cuneiformity and refracting index gradient may also be found as well as the speed of variation of these quantities with respect to the variation of co-ordinates. Measuring results of silicon sample are given. Measuring accuracy 1×10^{-5} may be achieved for refracting index distribution in 1 mm-thick semiconductor sample.

Introduction

Good knowledge of features of semiconductors from which electronic elements are made is of a very great importance for working up and preparing technological production process. Selection of semiconductors before that process thus admitting to only these materials which show pre-determined features yield conntable economic advantages consisting in an increased share in the final product of elements with better and more repeatable parameters. The basic properties of a semiconductor are the knowledge of impurities, admixtures and material structure. Several complementary verification methods are used in order to determine these properties. One of the fundamental methods is refracting index measurement. Continuous distribution of refracting index differences in semiconductor plates may be determined very precisely by means of the described method. Linear changes are also detected in these continuous ones which are extremely hard or even impossible to detect by other methods. The measuring method and measuring accuracy are discussed in detail in a paper prepared for publication for Applied Optics.

Principle of measurement

The measurement is carried out by means of Twyman-Green's laser interferometre /Fig.1/. Three interferograms are made:
- for two-beam interference /Fig.2a/:
    1. interference between M1 and M2 mirrors - through the examined sample,
    2. interference between M1 and M2 mirrors - without sample,
    3. interference between sample surfaces,
- for multipass interference /Fig.2b/:
    1. interference between M1, M2 and M3 mirrors - through the sample,
    2. interference between M1, M2 and M3 mirrors - without sample,
    3. interference between sample surfaces.

Interferograms 1 and 3 are obtained at unmoved sample, while interferogram No 3 without M1, M2 and eventually, M3 mirrors are ordered to interference fringes, starting from wedge edge side. Each interferogram represents a flat interference field which maps information on optical pathy difference or, the so-called, interference order $m/x,y/$. Let us assume that subsequent interferograms represent suitable interference orders: $m_b/x,y/$, $m/x,y/$, $m_o/x,y/$. Because the analysis of two-variable function is timeconsuming to a high degree, let us assume that we are interested in getting information at some points only. Then the analysis of interferograms is made only along horizontal and vertical scanning lines, which form a scanning net. Measurement results are obtained at nodes of scanning net which should be sufficiently dense. Such a procedure gives a possibility of replacing two-variable functions by single-variable ones the number of which is equal to the number of scanning lines. Scanning of interferograms consists in ordering interference order, i.e. fringe number, to a suitable co-ordinate corresponding to the position of interference fringe centre on scanning line. Because such a relation is given only at some points, approximation should be made as to be better acquainted with it. The approximation may be carried out by polynomes and least square method. For a continuous function dependence /e.g. $m/x/$/ obtained in this way values of $m/x_j/$ function/ may be found as well as its first $\partial_x m/x_j/ = \partial m/x_j// \partial x$ and second

---

* This paper have been presented on the Congress by H. Kasprzak,
Technical University of Wrocław, Institute of Physics.

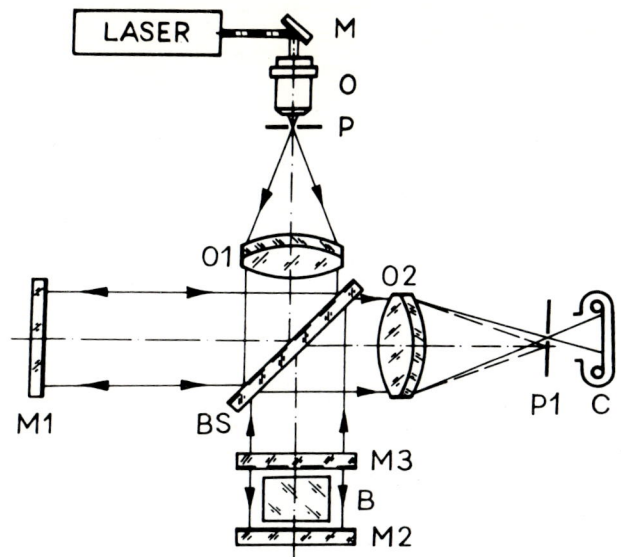

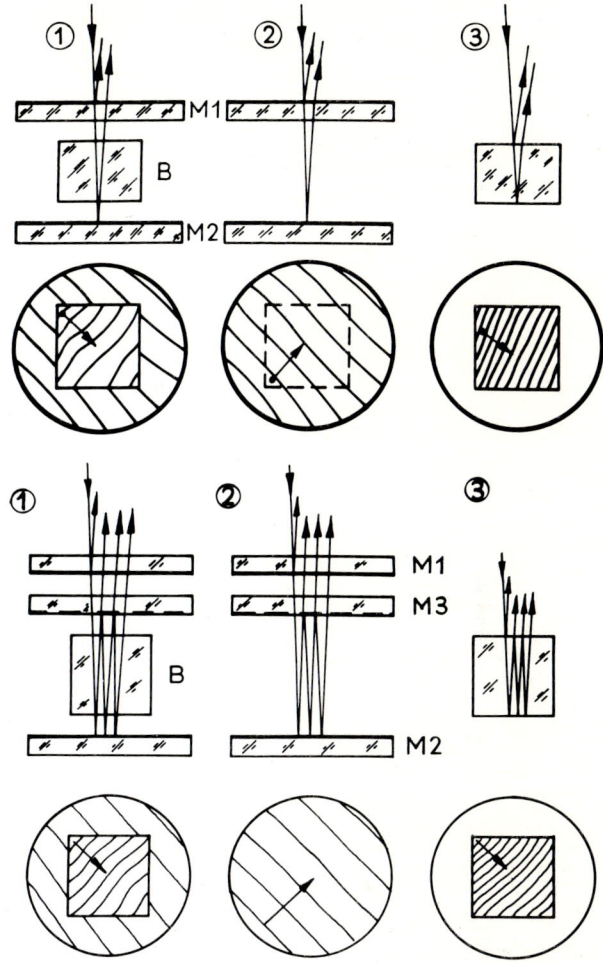

Fig. 1. Twyman-Green laser interferometer: M - directional mirror, OM - microscope objective, P - spatial filter diaphragme, O1 - collimator objective, O2 - telescope objective, BM - beam - splitter, M1, M2 - interferometer mirrors, M3 - partially transmission mirror for multipass interference, P1 - diaphragme for multipass interference order selection, C - photographic camera.

$\partial_x^2 m/x_j / = \partial^2 m/x_j //\partial x$ derivatives for all j^{th} points in which scanning net nodes are placed. Because our measurement is a relative one, a suitable point should be selected to which all the measurements will be referred /in our case the sample centre was this point/. The following relations follow from a detailed analysis of interferograms [1,2]:

Fig. 2. Three interferograms. Two-beam interference and multipass interference.

1. Geometrical difference of sample thickness is calculated from the knowledge of differences in interference order $\delta_x m$, $\delta_x m_b$, $\delta_x m_o$, where δ_x is the operator of interference order difference between the actual point and the reference one. This geometrical thickness difference is:

$$\delta_x L = \frac{\lambda}{2N} (\delta_x m - \delta_x m_b + \delta_x m_o) \qquad /1/$$

The next value is the "optical" thickness difference of sample, i.e. such a geometrical difference which results from optical testing at an assumption of sample homogeneity,

$$\delta_x L_o = \frac{\lambda}{2nN} \delta_x m_o \qquad /2/$$

and the third is refracting index difference

$$\delta_x n = \frac{n}{L} (\delta_x L_o - \delta_x L) \qquad /3/$$

where δn is the difference of averaged values of refracting index from the whole sample thickness. The same averaging is for values ∂n and $\partial^2 n$. In the above formulae λ denotes wave length of laser light used for interference, n - refracting index of sample material for λ wave length, L - thickness of the examined sample, whereas N - the so called multipass interference order.

Two-beam interference may be used for measurements /then the interferometre operates without a partially transmitting M3 mirror/ as well as multipass interference [3, 4] with the M3 mirror for making such an interference and P1 diaphragm for selecting interference order. Measuring accuracy may be increased by means of multipass interference for then the fringe spacing corresponds to the difference of optical paths equal to $\lambda/2N$ for two-beam interference N=1/. Because of this, for scanning lines parallel to y axis the above relations are true, and results for the same scanning net nodes should be indentical. Possible differences result from measuring errors/. It may be written dawn as:

$$\delta L = \frac{1}{2}(\delta_x L + \delta_y L) \qquad /1a/$$

$$\delta L_o = \frac{1}{2}(\delta_x L_o + \delta_y L_o) \qquad /2a/$$

$$\delta_n = \frac{1}{2}(\delta_x n + \delta_y n) \qquad /3a/$$

2. From the knowledge of first derivatives of interference order for x axis the following values are calculated:
- geometrical wedge

$$\varphi_x = \partial_x L = \frac{\lambda}{2N}(\partial_x m - \partial_x m_b + \partial_x m_o) \qquad /4/$$

- "optical" wedge

$$\varphi_{ox} = \partial_x L_o = \frac{\lambda}{2nN}\partial_x m_o \qquad /5/$$

- and refracting index gradient

$$\partial_x n = \frac{n}{L}(\varphi_{ox} - \varphi_x) \qquad /6/$$

similar dependencies will be valid also for scanning lines parallel to y axis. Because quantities describedby equations 4-6 concern a flat vector field, it may thus be written down that modules are:

$$\varphi = (\varphi_x^2 + \varphi_y^2)^{1/2} \qquad /4a/$$

$$\varphi_o = (\varphi_{ox}^2 + \varphi_{oy}^2)^{1/2} \qquad /5a/$$

$$\partial n = [(\partial_x n)^2 + (\partial_y n)^2]^{1/2} \qquad /6a/$$

whereas the direction of corresponding vectors /measured between positive axes of the sample from the horizontal to the vertical one/, will be;

$$\Theta_1 = \arctan \frac{\varphi_y}{\varphi_x} \qquad /4b/$$

$$\Theta_{1o} = \arctan \frac{\varphi_{oy}}{\varphi_{ox}} \qquad /5b/$$

$$\Theta_{1n} = \arctan \frac{\partial_y n}{\partial_x n} \qquad /6b/$$

respectively.

3. Knowing second derivatives of interference order $\partial_x^2 m$, $\partial_x^2 m_b$, $\partial_x^2 m_o$ one may calculate the speed of geometrical wedge change in respect of co-ordination change:

$$\partial_x \varphi = \partial_x^2 L = \frac{\lambda}{2N} (\partial_x^2 m - \partial_x^2 m_b + \partial_x^2 m_o) \qquad /7/$$

and the same for "optical" wedge

$$\partial_x \varphi_o = \partial_x^2 L_o = \frac{\lambda}{2nN} \partial_x^2 m_o \qquad /8/$$

and refracting index gradient

$$\partial_x^2 n = \frac{n}{L} (\partial_x \varphi_o - \partial_x \varphi) \qquad /9/$$

For "y"-scanning lines the formulae will be analogous. Corresponding resultant modules and directions of the above quantities can be derived on a base of formulas /7/, 8/, /9/ similar as /4a/, /5a/, /6a/, /4b/, /5b/, /6b/.
As it was previously stated, interferogram analysis process takes much time. A suitable shortening program has been drawn up enabling operators to do it automatically. It is sufficient to feed into a computer data on interference fringes located on scanning lines for three interferograms. A scanning device is being designed now which will make it possible to automatize completely the measuring process. As in all the interference measurements, every factor that would affect results should be eliminated. Such factors are: variable temperature gradients in interferometre arms /varying according to air movements/; temperature gradient in the sample under examination and interferometre vibrations. The used measurement technique has a decisive meaning for measuring accuracy /i.e., two-beam or multipass interferometry/. The next factors are accuracy of: interference fringe centre positioning, co-ordinate measurement as well asscanning nets coincidence on every interferogram. An additional improvement of measuring accuracy may be reached by appropriately selected approximation and by the use of equidensit technique [5]. Errors of quantities characterizing refracting index variations are inversely proportional to the thickness of the examined sample.

Results of measurements

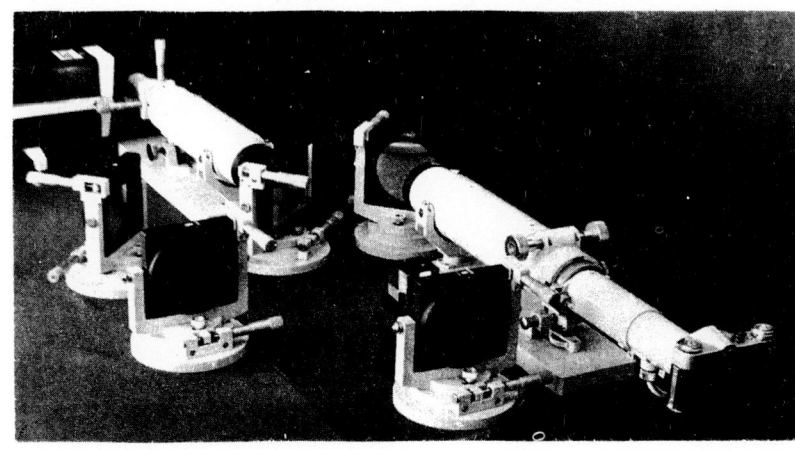

Fig. 3. Twyman-Green interferometer used for measurements.

Fundamental measurements for which the present method has been designed and which have been processed are tests of continuous heterogeneities of optical materials, i.e. optical glass and laser rods. Semiconductors may also be measured. For this purpose some changes have been made in the interferometre, viz.: objective lens replaced, telescope equipped with opto-electronic transducer and a 1152 nm wave length laser used. For the reason of unfavourable measuring conditions for this wave length /residual sensitivity of opto-electronic transducer and small or even zero semiconductor transmission/ only a silicon sample has been examined. The measurement in slightly more distant infrared is possible, e.g. for $\lambda = 3.39$ μm or $\lambda = 10.6$ μm [6,7] which wave lengths are well transferred by semiconductor samples. A special image detector, nevertheless, is needed here, being a part of mirror interferometre.

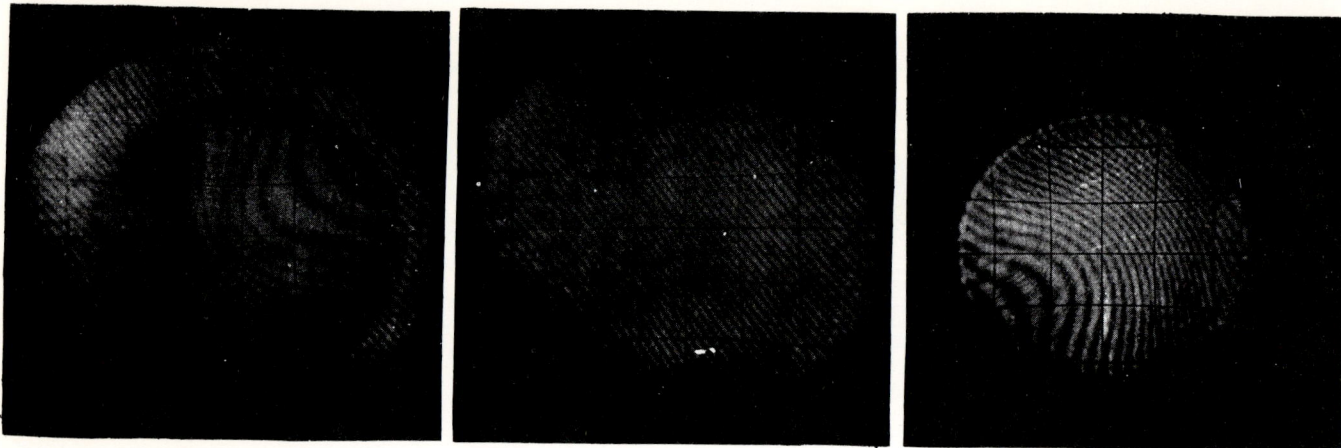

Fig. 4. Interferograms for silicon sample.

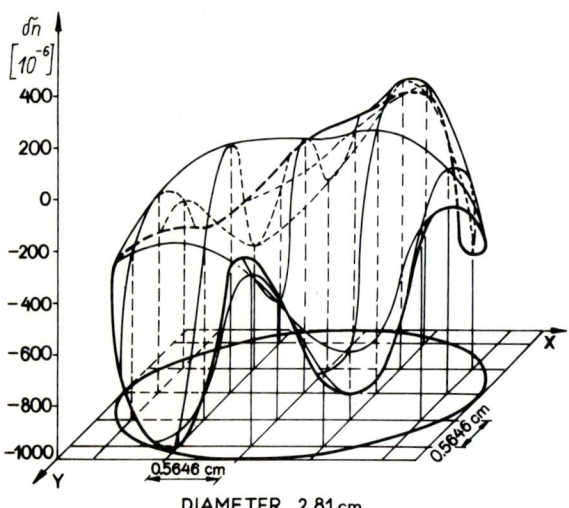

Fig. 5. Distribution of refracting index for silicon sample.

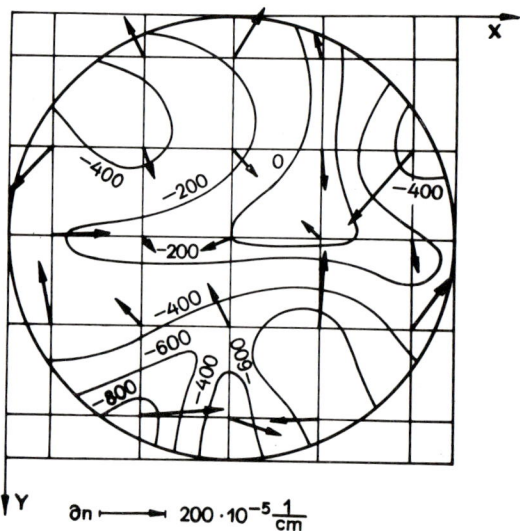

Fig. 6. Distribution of refracting index and its gradients for silicon sample.

The examined silicon sample was of 2.81 cm diametre and 0.355 cm thick. Two-beam interferometry was used when testing and it was proved that this method is quite sufficient regarding the great heterogeneity of the examined material. The obtained interferograms are presented in Fig. 4. A high image distortion may be observed because of the use of opto-electronic transducer. In order to get rid of this distortion during the measurement, a plate with straight and rectangular lines was placed before transducer photocathode; scanning was performed along these lines. Amendments were also introduced compensating variable value of enlargement obtained at transducer output. Thus the measurement results are free of opto-electronic transducer errors. Measurement results are given in Fig. 7, whereas refracting index heterogeneities are presented in Figs. 5 and 6. Measuring errors resulting from analysis of measurement accuracy are $\Delta /\delta L/ = 63$ nm, $\Delta /\delta L_0/ = 7$ nm, $\Delta /\delta n/ = 7 \times 10^{-5}$, $\Delta \varphi = 0,47$ μrad, $\Delta \varphi_0 = 0.135$ μrad, $\Delta /\partial n/ = 1.1 \times 10^{-5} \frac{1}{cm}$. As it was confirmed by more detailed examinations, actual measuring errors are almost three-fold smaller than calculated and given ones as a result of the approximation and result smoothing. Measuring accuracy may be even greater for better homogeneous material by the use of multipass interferometry and equidensit technique.

Acknowledgment

The author wishes to thank Henryk Kasprzak /of the Technical University of Wrocław, the Institute of Physics/ for presenting this paper on the Congress.

NODES OF SCAN GRID COORDINATES [cm].

	1	2	3	4	5
1		0.8404 / 0.2758	1.4050 / 0.2758	1.9697 / 0.2758	
2	0.2758 / 0.8404	0.8404 / 0.8404	1.4050 / 0.8404	1.9697 / 0.8404	2.5343 / 0.8404
3	0.2758 / 1.4050	0.8404 / 1.4050	1.4050 / 1.4050	1.9697 / 1.4050	2.5340 / 1.4050
4	0.2758 / 1.9697	0.8404 / 1.9697	1.4050 / 1.9697	1.9697 / 1.9697	2.5340 / 1.9697
5		0.8404 / 2.5343	1.4050 / 2.5343	1.9697 / 2.5343	

VARIATIONS OF REFRACTIVE INDEX [10^{-6}].

J=	1	2	3	4	5
I= 1		-385.485	-188.632	18.671	
I= 2	-337.236	-527.969	-270.387	112.846	-510.765
I= 3	-227.853	-28.079	-0.000	26.052	-123.688
I= 4	-333.460	-329.864	-538.221	-740.869	-252.891
I= 5		-952.906	-323.891	-715.037	

MODULUS [$10^{-6} \frac{1}{cm}$] AND DIRECTION OF REFRACTIVE INDEX GRADIENT.

J=	1	2	3	4	5
I= 1		1228.119 / 239°59`	1721.082 / 298°55`	830.236 / 257°10`	
I= 2	1945.467 / 133°05`	814.469 / 72°00`	1142.913 / 46°22`	1207.502 / 87°51`	3006.135 / 131°09`
I= 3	1660.026 / 358°29`	472.925 / 53°03`	527.908 / 158°38`	771.585 / 221°33`	862.867 / 83°03`
I= 4	1913.115 / 262°09`	975.675 / 223°24`	1140.593 / 239°41`	2388.227 / 273°03`	2139.830 / 306°06`
I= 5		2545.424 / 353°23`	1593.498 / 17°12`	1514.228 / 178°03	

SPEED OF MODULUS [$10^{-6} \frac{1}{cm^2}$] AND DIRECTION VARIATIONS OF REFRACTIVE INDEX GRADIENT.

J=	1	2	3	4	5
I= 1		1115.948 / 11°18`	2873.089 / 172°56`	7827.177 / 195°08`	
I= 2	2949.768 / 319°31`	4676.747 / 279°18`	6407.9831 / 257°36`	1461.223 / 251°07`	6140.665 / 197°49`
I= 3	2688.931 / 253°08`	7395.999 / 271°30`	9795.0381 / 277°51`	4934.311 / 280°09`	5507.483 / 317°24`
I= 4	3305.215 / 263°11`	6803.140 / 274°44`	8124.3901 / 281°31`	1421.667 / 282°52`	6797.489 / 297°15`
I= 5		3602.462 / 220°31`	6390.072 / 183°20`	9843.318 / 176°50`	

Fig. 7. Refractive index variations. Results of measurement for silicon sample.

References

1. Kowalik, W., Interference measurement of continuous heterogeneities in optical materials, **Submitted for publication in Appl. Opt.**
2. Kowalik, W., Doctor's dissertation - Paper No 229 of Technical University of Wrocław, Institute of Physics, 1974.
3. Langenbeck, P., Appl.Opt. 6, 1425 /1967/.
4. Langenbeck, P., Appl.Opt. 8, 545 /1969/.
5. Lau, E., Krug, W., Die Aquidensitometrie, Akademie Verlag, Berlin 1957.
6. Roberts, F.E., and Langenbeck, P.H., Appl.Opt. 8, 2311 /1969/.
7. Munnerlyn, C.R., Presented as the 1970 Annual Meeting of the J.Opt.Soc.Am., September 30, 1970.

APPLICATION OF INTERFERENTIAL CORRELATION OF SPECTRA TO THE DETECTION OF POLLUTANTS IN THE ATMOSPHERE

G. Fortunato, A. Maréchal, Melle Wolfer

E.N.S.E.T. Cachan, Institute of Optics Orsay, C.I.R.N. Lacq

Introduction

Modern spectrometry offers rich possibilities by convenient use either of interferometric devices (Fourier transform spectrometry) or of selective modulation (grid spectrometry). Progresses in resolution and luminosity have been spectacular in the last decade, and new domains of high resolution and low luminosity have been explored. In those cases of extreme performances, the instruments are generally highly sophisticated, delicate, and expensive. Nevertheless, Fourier transform spectroscopy is now developing for practical applications : we have examined the possibilities of interferometric devices to routine spectral analysis in chemistry, biology, pollution detection etc... and are now aware of the interesting characteristics of those mountings by the fact that they are luminous, flexible and very simple. They need no computer and are very suitable for low resolutions. We shall describe first the basic principle, and later focus on the various possibilities resulting from the direct access to the interferogram and the application of the mathematical properties of the Fourier transform : Fourier derivation, Fourier correlation with a reference spectrum, Fourier correlation of derivatives etc... All those possibilities result from techniques of matched filtering for spectral recognition, obviously similar to analog techniques for pattern recognition.

A) BASIC PRINCIPLE OF THE SIMS

The "Spectromètre Interférentiel à Modulation Sélective" (SIMS) combines the use of interferences and selective modulation. Its principle has first been pointed out by R. Prat (1), and G. Fortunato (2) has shown that it is the only way to obtain a high "étendue" and consequently a high optical signal.

The leading idea of the device is
- the production of very luminous interference fringes in a plane.
- the analysis of those fringes by a moving grid and the production of a photoelectric signal by synchronous detection.

1 - The Interferometric Device

In order to obtain luminous fringes with an extended source, it is necessary to manage in order that the position of the fringes should not depend upon the position of the emitting point on the source. Fortunato has shown that the only solution is the Prat mounting comprising Fig. 1.
- a doubling device producing two laterally shifted images of the source,
- a converging lens.

The fringes located in the focal plane of the lens, and produced by various coherent coupled images of the points of the source, do not move when the point source moves ; as a conséquence, the flux can be important by the fact that there is no fundamental limitation on the solid angle of the beams.

S.I.M.S.

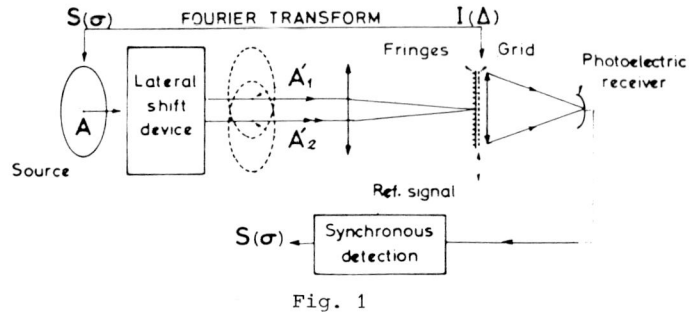

Fig. 1

2 - Selective Modulation

If now we put a periodic grid in the fringes and move it, the outgoing flux is modulated only for the wavelength for which the fringe separation is matched to the grid spacing : it is then possible to modulate selectively one wavelength (or small spectral region) and use synchronous detection in order to produce the spectral signal. The modulated flux will be proportional to the source luminance for the wavelength for which fringe separation, $i = \lambda/\alpha$ (where α is the angle between the two interfering rays), is matched to the grid period. Changing the wavelength can easily be performed by acting on the angle which depends upon the lateral shift produced by the interferometer : we use now a very simple polarizing interferometer invented by Nomarski Fig. 2. and based on the properties of two Wollaston prisms located between polariser and analyser. The adjustment of the distance between the two prisms allows the variation of the wavelength.

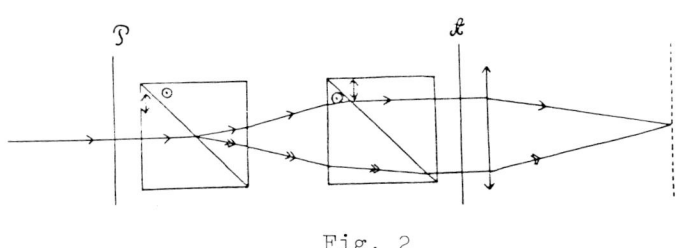

Fig. 2

3 - Luminosity

The high gain in "étendue", with respect to classical devices, leads to appreciable advantages. Nevertheless, we have to take into account the increase of noise due to the unmodulated flux, and a detailed discussion done by Fortunato has led to various encouraging conclusions.

4 - Resolution

The response of the apparatus to monochromatic light (laser) has shown to be in agreement with theoretical predictions : we perform the analog Fourier analysis of a sample of N fringes, and the resolution is equal to N and does not depend directly on the luminosity : if we increase the number of fringes, we increase the resolution at the same time, and in fact, luminosity and resolution do not depend upon the same parameters and are practically independent.

B) OPTICAL DATA PROCESSING ON THE SIMS

It turns out that the SIMS is a luminous spectrometer, but another feature can also be significant : we have in the focal plane of the lens the interferogram of the source, i.e. the Fourier transform of the spectrum. It is then possible to operate on this interferogram in order to obtain various signals representing linear operations on the spectrum. In other words, we have at our disposal The Fourier transform and we can take advantage of this situation.

1 - The Spectrum

If $S(\sigma)$ is the spectrum (as a function of the wave number σ) the interferogram is $I(\Delta)$, where Δ is the optical path difference and they are related by

$$i(\Delta) = I(\Delta) - I_0 = \int S(\sigma) \cos 2\pi\sigma\Delta \, d\sigma \quad (1)$$

where $i(\Delta)$ is the variable part of the interferogram and I_0 the average of I. If we only need $S(\sigma)$, we operate and analog Fourier analysis of $i(\Delta)$ by using a movable periodic grid.

The response function is normally a sinc function, but procedure of apodisation can be applied by using a convenient smoothing screen on the interferogram.

2 - Derivation

If, instead of wishing to obtain the spectrum $S(\sigma)$, we prefer to obtain directly one of its derivatives, we can use the general properties of the Fourier transform ; as an exxample, the F. T. of $S'(\sigma)$ is proportionnal to $\Delta I(\Delta)$; which means that , in order to obtain a signal representing S', we should multiply the interferogram by Δ, what is very simple : we put in the interferogram a "two triangles" mask, and replace the ordinary periodic grid by a grid made up of two zones : for $\Delta>0$ and $\Delta<0$, the black and white bars are interchanged in order to multiply the signal by a negative factor for $\Delta<0$. Fig 3 represents the result obtained on the transmission spectrum of NO_2 where the carves $S(\sigma)$ and $S'(\sigma)$ are represented.

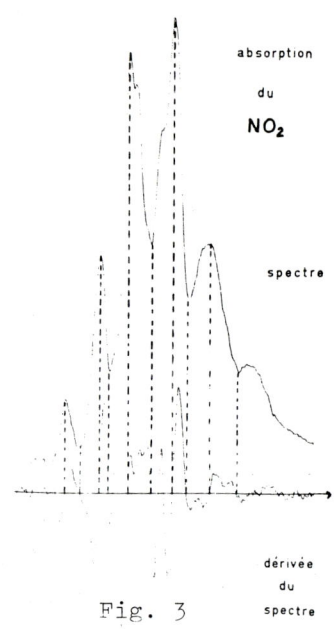

Fig. 3

We should notice that happily the Fourier derivation does not operate the derivation of the noise which is also the case of synchronous detection methods for obtaining the derivative).

3 - Correlation

In order to recognize a reference spectrum $(S(\sigma)$, is useful to correlate the observed spectrum $S(\sigma)$ with the reference S_R . This means that we should multiply yjr interferogram $i(\Delta)$ and $I_R(\Delta)$, which is possible if we put in the interferogram plane a screen representing $I_R(\sigma)$, i.e. the interferogram of the reference spectrum. If we move the interferogram, the signal obtained ia a linear combination of the signals produced by the various wavelengths and we act at the same time on the various elements composing the spectrum.

4 - Correlation of Derivatives

Spectral signals S or S_R are always positive. This means that even if S and S_R are totally different, they can be both non-zero in some spectral domain and the correlation will be positive, which is misleading. On the other hand, derivatives are positive or negative and it seems to be safer to correlate the derivatives rather than the spectra ; in order to perform this correlation, we have to multiply $\Delta i(\Delta)$ by $\Delta i_R(\Delta)$ which means that we have to use a Δ^2 filter on the interferogram i_R . It is noticeable that this operation tends to eliminate the central part of the interferogram, which brings no useful information, and use the parts of the interferogram that correspond to an appreciable value of Δ .

5 - The Case of Quasi Periodic Spectra

The Case of interesting molecules (SO_2, NO_2) can be quasi periodic Fig. 4

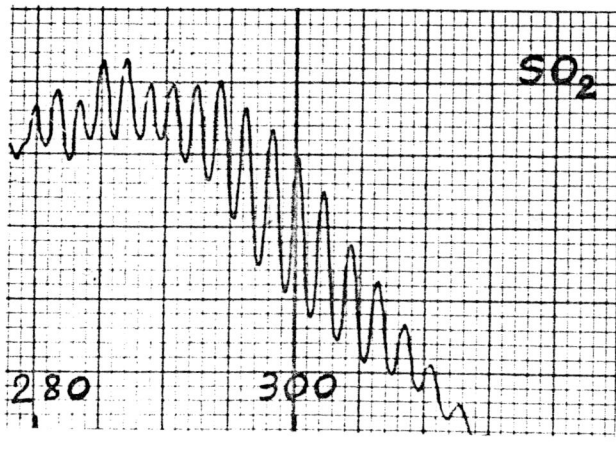

Fig. 4

and this has an immediate consequence on the interferogram Fig. 5., which has a noticeable contrast (fringes reappearing) for a given value of Δ. In those conditions, we can select by a convenient mask this value of Δ in the interferogram and base the correlation on 2 conditions :

- existence of a critical value of Δ where the fringes have an appreciable contrast,
- correlation of spectra.

It seems that this procedure is safer than the simple correlation of spectra through multiplication of interferograms.

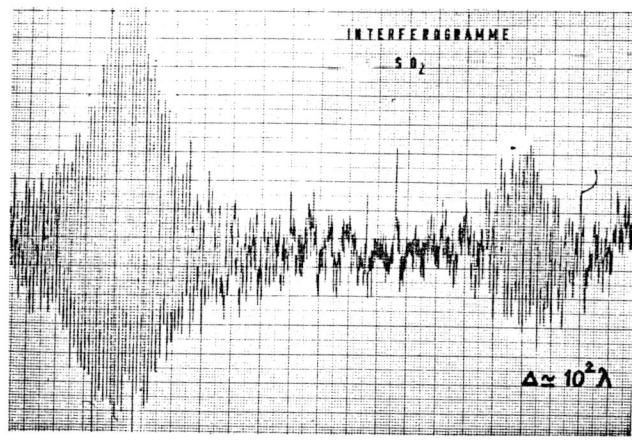

Fig. 5

6 - Practical device

This study conduces us to the realization of a first device in order to detect the pollutants in the atmosphere such as SO_2. This device comprises fig. 6

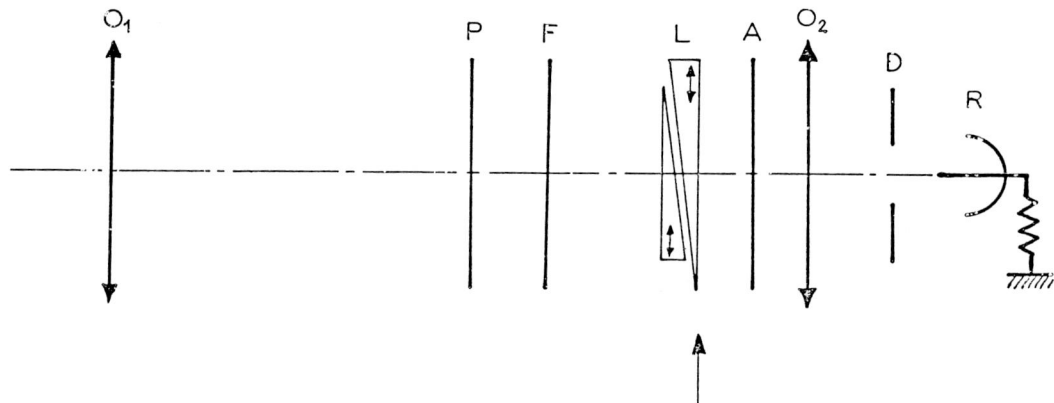

Fig. 6

- O_1 : silica lens
- P : u.v. polarizer
- F : broad band filter
- L : quartz plate which the thickness is ajustable in order to obtain interferences in vicinity of characteristic optical path difference of SO_2
- A : rotating analyzer
- O_2 : silica lens
- D : diaphragm isolating the center of the interference pattern adduced by the plate L
- R : photoelectric receiver

This device pertmits us to study the evolution, in a real case (chimney-stalk), of the concentration of SO_2. The sensibility is 30 p.p.m. for a middle concentration equal to 3.000 p.p.m.

CONCLUSION

Once more in the domain of optics, the Fourier transform relation between two characteristic quantities allows simple and efficient operations. In the present case, it should be obviously very useful for the realization of simple and sensitive devices used in automatic recognition of spectra : we hope to present, in the near future, more detailed results.

BIBLIOGRAPHY

Prat, R., J. Appl. Phys. Suppl. 1,448 (1965)
Optica Acta 13,2,73 1965)

Fortunato G. et Maréchal A., C.R. Acad. Sci. (France) 274 B(1972) 931 and B(1973) 527

Fortunato G. Thesis Orsay June 76, to be published in the Nouvelle Revue d'Optique.

MEASUREMENT OF SMALL ROTATIONS

Aline Huard and Christian Imbert
Institute of Theoretical and Applied Optics
Bat. 503 — Centre Universitaire d'Orsay
B.P. N° 43 91406 Orsay Cedex, France

Abstract

We describe two apparati: one serves to measure, the other to stabilize the rotations of light beams with a sensitivity which may surpass a hundredth of a second of arc.

Introduction

We present a pointer designed to measure small deflections of light beams (about ten seconds of arc). As a direct application of this apparatus we have also produced a directed beam stabilizer of the same sensitivity. This sensitivity may attain a hundredth of a second of arc.

Optical Principle

The principle used is the following: let there be a light ray of constant polarization arriving on a plane dioptric. This dioptric separates a medium of index n from the air (figure 1); the incident ray is propagated in the medium n.

The curves representing the variations of the Fresnel coefficients as a function of the incidence (figures 2a and 2b) show that if the ray arrives at a variable incidence i near the limit incidence i_ℓ, the intensities of the reflected or transmitted ray vary quite rapidly as a function of this incidence. Thus we can associate to any rotation Δi a variation of the transmission factor ΔT, that is, a variation of flux of the transmitted beam, which will be seen by a photodetector receiving this beam. The knowledge of the relation

$$\Delta \phi = f(\Delta i) \tag{1}$$

permits us to calibrate the system. But this relation is not linear in the case of geometrical optics. Nonetheless the presence of experimental constraints such as the divergence of the beam and the absorption (even if weak) of the medium n tend to linearize this relation over a limited range to the detriment of the sensitivity. For reasons to which we will return, we have represented (figure 3) the variations of flux transmitted by a plane dioptric, the incident beam being a non-polarized laser beam. Its characteristic divergence is three minutes of arc. In these conditions we see that the slope of the curve $\tau = f(i)$ is first attenuated, and secondly becomes constant with a good approximation in the vicinity of a given angle of incidence i_0. The calculation gives a defect of linearity of one percent over a range of twenty seconds of arc centered on i_ℓ. In these conditions we can legitimately write:

$$\Delta \tau = \tau_0 \Delta i \tag{2}$$

where τ_0 is the transmission factor of the dioptric for the average incidence i_0, $\Delta \tau$ and Δi the small increases of these quantities around this functioning point defined by τ_0, i_0.

As stated above, we chose a laser source for several reasons: its power is used over a very small geometrical extension and its radiation is highly monochromatic. These advantages are accompanied by inconveniences, however. In fact, this source can fluctuate in intensity, in direction, and accessorily in polarization. Thus we must determine a setup such that if the laser beam arrives at a system (S) which turns it, the resulting measurement of the experiment should depend only on rotations of (S).

Principle of the Measurement

In the setup sketched in figure 4, one of the two identical detectors gathers the transmitted beam and the other the attenuated reflected beam. Let A be the characteristic constant of the two photodetectors; the signal due to the transmitted beam is:

$$V_T = A \tau \phi \tag{3}$$

That due to the attenuated reflected beam is:

$$V_R = A N(1 - \tau) \phi \tag{4}$$

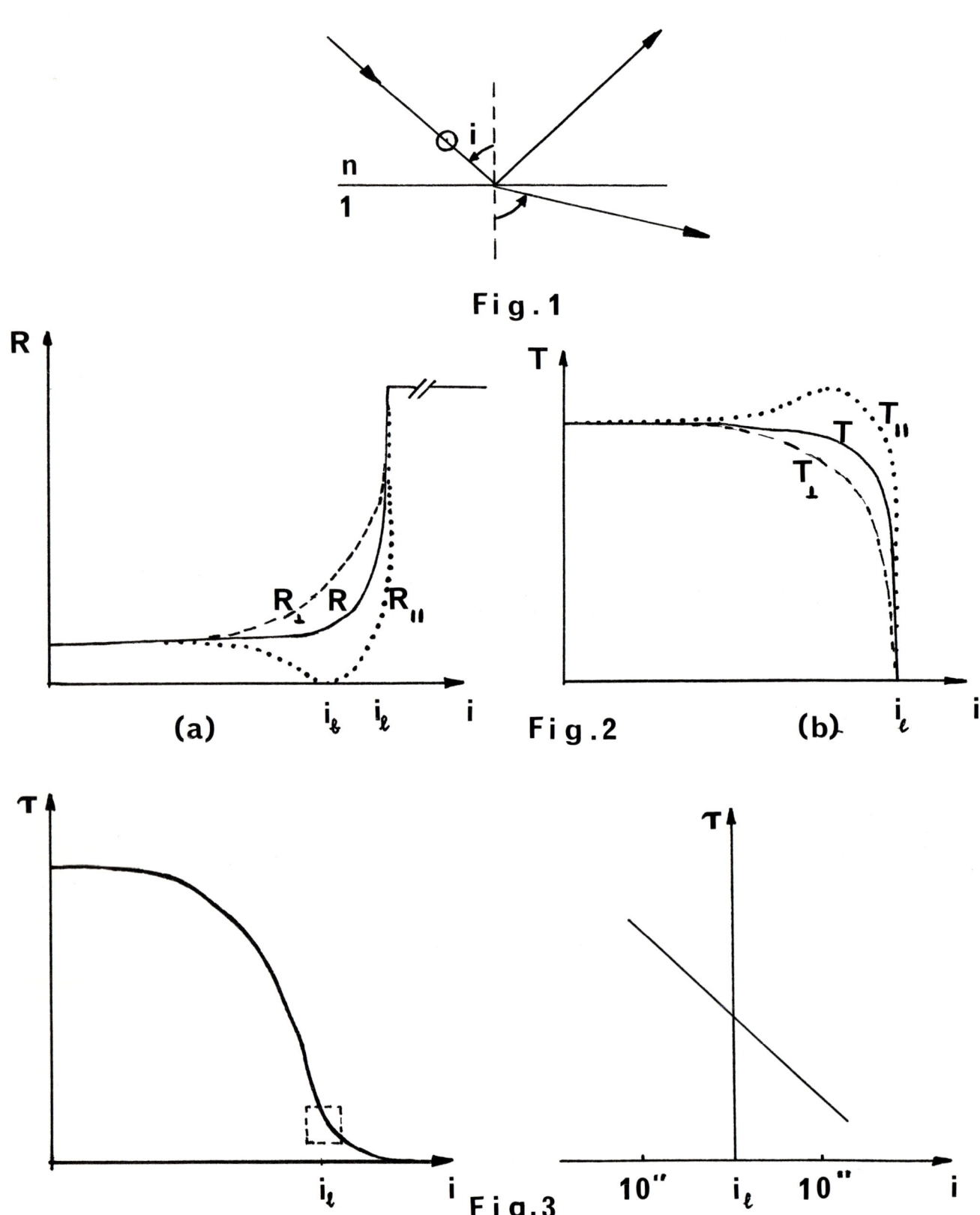

Fig.1

Fig.2

Fig.3

where:
- ϕ represents the flux of the beam falling on the dioptric
- τ the transmission factor of the dioptric for the incidence i of the beam
- N is the local transmission of a photometric wedge placed in the reflected beam.

For an incidence i_o which we can choose we have:

$$V_T - V_R = 0 \tag{5}$$

Let us suppose that the beam turns beginning from this position under the effect of the phenomenon being studied:

$$V_T - V_R = A(1 + N) \phi \Delta \tau \tag{6}$$

That is, from relation (2),

$$V_T - V_R = A(1 + N) \phi \tau_o \Delta i \tag{7}$$

$V_T - V_R$ then depends on the rotation to be measured through coefficients which may vary. In particular τ_o can vary because the source beam fluctuates in direction and polarization. We must thus polarize the beam, for example, rectilinearly and control it in direction. The setup presents a functioning point $\Delta V \equiv 0$ independent of the flux. We must then complete the comparison device by an electromechanical system of control. The assembly will then be able to function correctly. In fact, let there be a device deviating the beam by an angle α under the effect of a tension (V); $\alpha = h(V)$ is known, the device having been calibrated (figure 5).

We connect the tension $V_T - V_R$ to the controls of the electromechanical device through an integrator: as long as $V_T - V_R \neq 0$ the system turns the beam. It stops as soon as $V_T - V_R = 0$.

We thus see that it is possible to make:
- a directed beam stabilizer
- a pointer whose signal depends only on the rotations to be measured. The information should be taken under the control of the deviator system.

Optical Scheme of the Apparatus

Keeping in mind these remarks, we give the optical layout of the prototype of the pointer and the stabilizer (figure 6). We can essentially distinguish two functions:
- the $O_1 L O_2$ part which is the optical part of the operating mechanism;
- the continuation of the device, which corresponds to the comparison system.

The controlling device is at the same time a rotation demultiplier system. O_1 and O_2 are two objectives of the same focal distance f; the ensemble is afocal. A parallel-face plate L of optical thickness (ne) lies between O_1 and O_2. If the plate turns through $d\theta$, the beam emerging from the ensemble turns through $d\alpha$, with:

$$d\alpha = d\theta \frac{e}{f} (n - 1) \tag{8}$$

After this device comes the comparison system, very near the principle presented. Nonetheless, let us point out some details:
- It is the same photodetector which receives the transmitted beam and the reflected beam attenuated by means of an adjustable photometric wedge. (We can choose the functioning point of the apparatus.)
- A mechanical modulator lets through alternately the reflected beam, then the transmitted beam. The signal emerging from the photodetector is represented in figure 7.
- Let us remark finally that the beam reflected at the level of the dioptric P is relayed to the cell either by a plane mirror M_2 (as in the case of the pointer) or by a semi-reflecting surface (as in the case of the stabilizer). A part of the beam stabilized in direction is thus transmitted towards the exterior.

Treatment of the Signal

The treatment of the signal can be schematized in the following manner (figure 8): the incident beam, after traversing the modulator, arrives at the photodetector. The emerging signal is treated by a synchronous demodulation amplifier (SDA) whose frequency of modulation is equal to that of each of the beams. This is done by the intermediary of the external reference signal taken at the level of the modulator. The SDA gives at the exit a constant signal proportional to the difference of the levels $V_T - V_R$. This signal is transmitted through an integrator to a piezo-electric bi-plate controlling the rotation of the plate L. A measuring device (voltmeter, recorder, etc.) permits us in the case of the pointer to know V and thus α by the relation $\alpha = h(V)$.

The integrator is a conventional element of control devices: it permits us to stabilize the beam in direction with an error of practically zero.

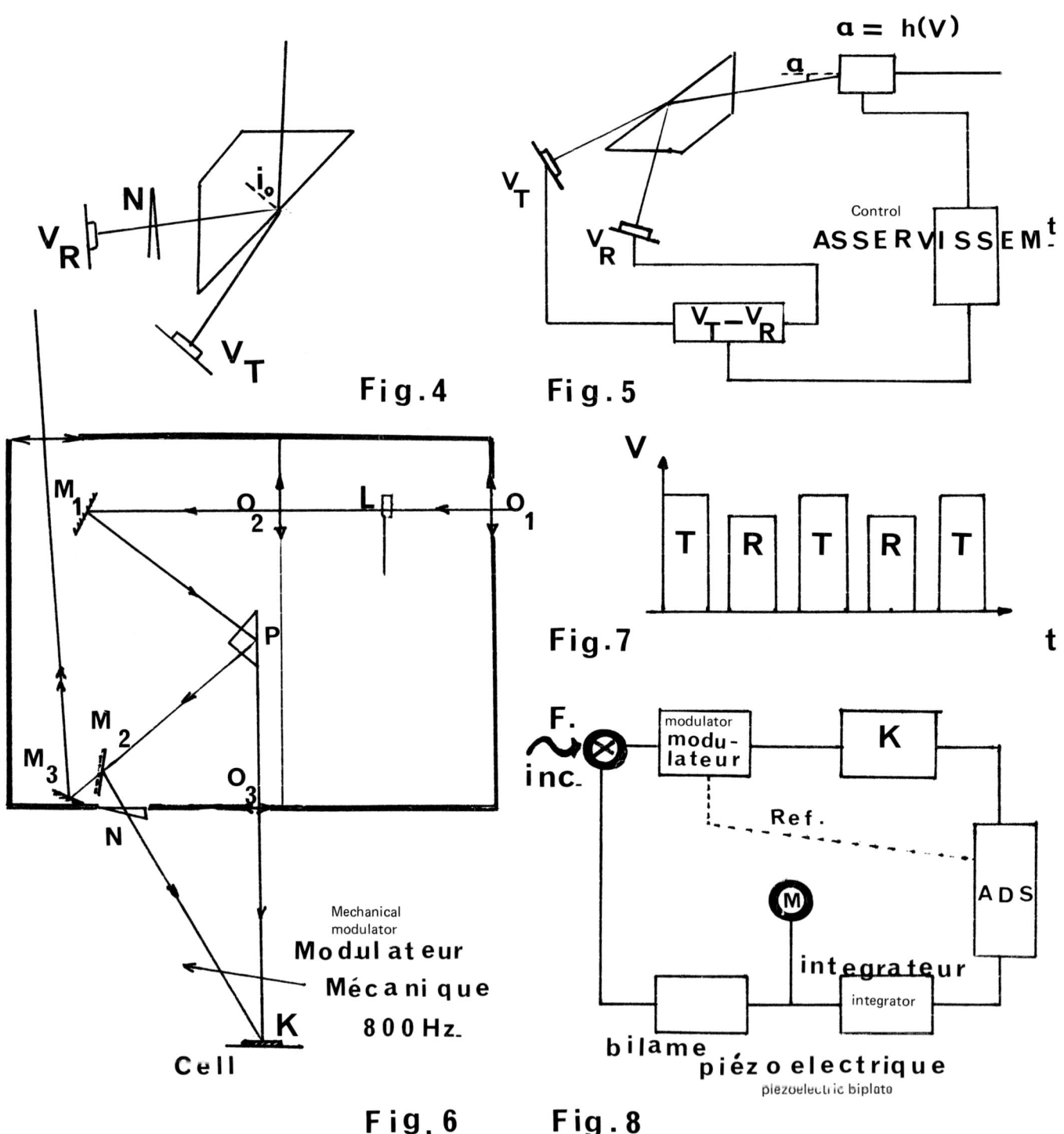

Fig. 4 Fig. 5

Fig. 6 Fig. 7 Fig. 8

Performance

We characterize the performance of the apparati by the sensitivity ε_1 and the bandwidth Δf. In a first approzimation they depend:
  - on the time constant t of the SDA
  - on the gain A of the system functioning as an open loop.

$$t = 0.3s \quad A = 140 \quad f = 1Hz \quad \varepsilon_i = 0.06"$$

$$A = 5000 \quad \Delta f = 8Hz \quad \varepsilon_i = 0.002"$$

The dynamic of these apparati is 100, as it is limited by the class of the SDA.

For the present the pointer can be used only at low frequency, that is, for frequencies between 0.5 and 10 Hz, since in static the piezo-electric ceramic presents a very marked phenomenon of hysterisis.

Conclusion

This equipment shows good performance as to sensitivity. This is true on condition that the environment be properly adapted. It is necessary to filter very severely the vibratory and thermic disturbances in order to hope to make measurements of around 0.01 second of arc.

1st EUROPEAN CONGRESS ON OPTICS APPLIED TO METROLOGY

Volume 136

SESSION 1.2

MEASUREMENT THROUGH OPTICS

Session Chairmen
Lohmann
Marechal

DETERMINATION OF THE INDEX PROFILE OF A DIELECTRIC PLATE BY OPTICAL METHODS

André Roger and Daniel Maystre

Laboratoire d'Optique Electromagnétique, Equipe de Recherche Associée au C.N.R.S. n° 597,
Faculté des Sciences et Techniques, Centre de Saint-Jérôme, 13397 Marseille Cedex 4, France

Abstract

We deal with an inverse scattering problem : knowing some properties of a diffracted wave, what can be said on the diffracting object ? This leads to strong mathematical difficulties. Using new methods, we show it is possible, for instance, to deduce the index profile of an inhomogeneous plate from its reflection coefficient.

1) Introduction

Making direct scattering is starting from the optical properties of an object and calculating the diffracted wave. This requires big computer programs, since generally, it cannot be done rigorously by using analytical formulas. Unfortunately, these numerical algorithms are not inversible and the inverse problems are mathematically much more difficult. So, as a first trial, we have studied a rather simple device, the inhomogeneous dielectric layer.

2) The physical device (Figure 1)

An inhomogeneous layer, whose permittivity ε depends only on z, is backed by a perfectly conducting plane. Consider a plane incident wave on this medium, with the angle of incidence θ and the amplitude 1. The electric field is supposed to be parallel to Oy. The reflected wave has the amplitude $B(\theta)$, with $|B(\theta)| = 1$, so that $B(\theta) = \exp[i\phi(\theta)]$. Our inverse problem is the following one : knowing $\phi(\theta)$, find $\varepsilon(z)$.

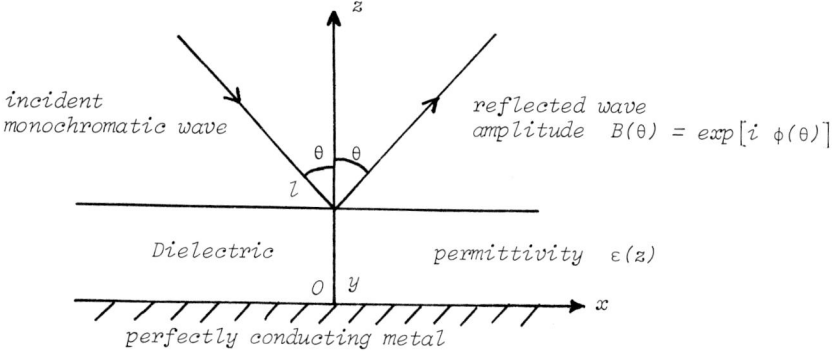

Figure 1. The physical device

3) Inverse scattering and parameter optimization

It is worth noting that inverse scattering is fundamentally different from the "parameter optimization" methods.

3.1. Parameter optimization.

When searching a function, in physics, one can often dispose of precise informations, so that the research is restricted inside a family of functions, depending on a small number of parameters. One fits these parameters in order to reproduce the results of the experiences as precisely as possible. This is the parameter optimization method.

3.2. Inverse scattering.

When preliminary informations are lacking, it is necessary to look for a function as a mathematical object belonging to an infinite dimensional space. This is not a superficial difference, because the infinite dimensional spaces are much more difficult to handle than the finite dimensional ones.

4) The formalism

4.1. Non linearity.

The basis equation, obtained from Maxwell's equations, is non linear. We have got over this difficulty by using a perturbation technique. This leads to make an iteration, where each step is a linear equation.

4.2. Fundamental instability.

At each step of the iteration, one must solve an integral Fredholm equation of the first kind, and with a regular kernel :

$$\int_0^1 \underbrace{K(\theta, z)}_{\text{known kernel}} [\underbrace{\varepsilon(z)}_{\text{unknown}} - \underbrace{\varepsilon_m}_{\text{known}}] \, dz = \underbrace{B(\theta)}_{\substack{\text{known (for instance} \\ \text{from measured data)}}} - \underbrace{B_m(\theta)}_{\substack{\text{analytically} \\ \text{known}}}$$

ε_m is a rough estimation of the average value of $\varepsilon(z)$.

This equation appears to be an ill posed problem at Hadamard's sense. This is the fundamental difficulty of the inverse problem.

5) Ill posed problems

5.1. The fundamental instability.

An ill posed problem is roughly a problem where the solution is not continuously dependent on the data. Using the diagrams of sets, figure 2 tries to show the meaning of this property.

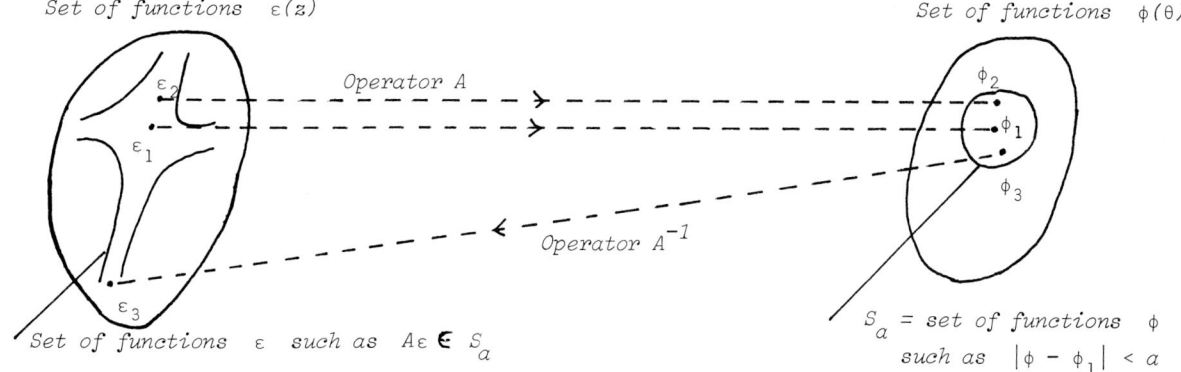

Fig. 2.   Ill posed problem at Hadamard's sense

The direct problem is starting from a function ε and computing the function ϕ. This problem is well conditionned, i.e. if you start from two adjoining functions ε_1 and ε_2, you will get two adjoining functions ϕ_1 and ϕ_2. Mathematically speaking, one can say that the direct operator A is continuous.

The inverse problem is starting from a function ϕ and computing the function ε. Unfortunately, it is possible to start from two adjoining functions (for instance ϕ_1 and ϕ_3 in figure 2) and to obtain two functions ε_1 and ε_3 quite different and very far from eachother.

One can say that the inverse operator A^{-1} is not continuous. The most important fact is that one cannot hope to get rid of this "instability" by taking functions ϕ_1 and ϕ_3 nearer and nearer, i.e. by increasing the experimental precision - this is shown by figure 3.

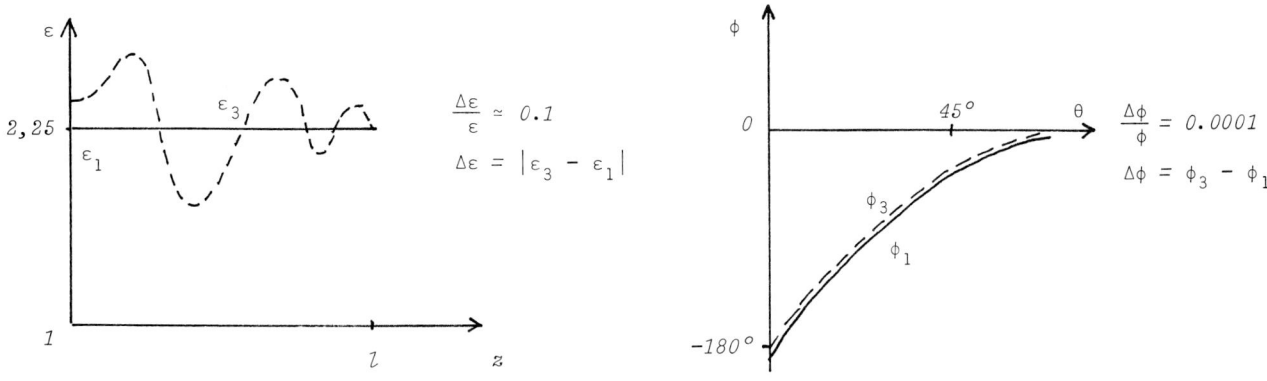

Fig. 3.   "instability" of the inverse problem

5.2. Tikhonov's method of regularization.

To get over this difficulty, we have used a method of "regularization", following the russian mathematician Tikhonov. Even lacking of preliminary informations, each physicist knows that the permittivity profile "must not be too big" or "does not vary too quickly" (if we exclude the case of discontinuous profiles). By using sophisticated mathematical methods, one can eliminate a priori many profiles and choose the smoothest one among all the profiles which reproduce the experimental data $\phi(\theta)$ with a given precision $\delta\phi$. This is done by rewriting the problem in a variational form and searching the profile as a function which minimizes a well-chosen functional [1] [2].

These methods have appeared to be numerically efficient and allow us to make theoretical reconstitutions of index profile. The index profile $\varepsilon(z)$ being given first, one compute the corresponding function $\phi(\theta)$. Then a random error is added to $\phi(\theta)$ and one tries to reconstitute $\varepsilon(z)$. (Figure 4).

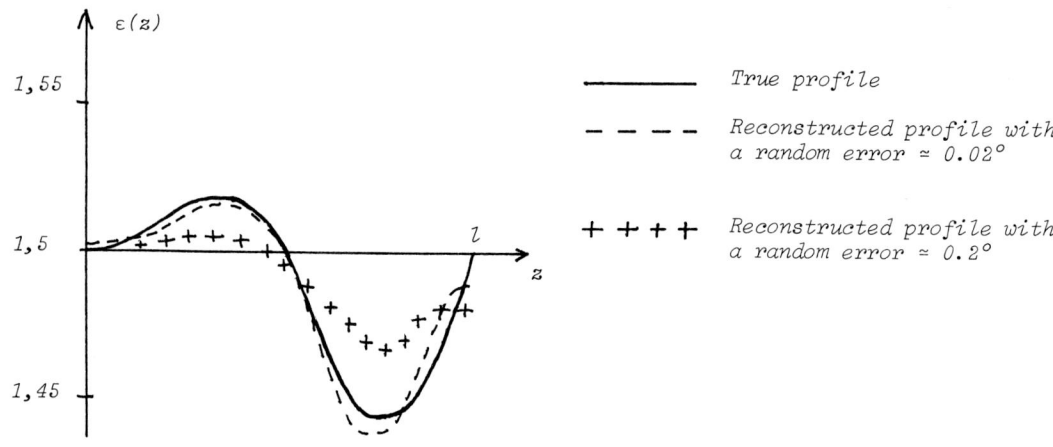

Fig. 4.   Reconstitution of an index profile

6) Conclusion

On a physically simple example, we have been able to carry out a rigorous analysis of the difficulties. Our work now will be to apply this method to other cases, especially to the gratings.

References

1. Tikhonov, A., Arsenine, V., <u>Méthodes de résolution de problèmes mal posés</u>, Editions de Moscou 1976.
2. Roger, A., Maystre, D., Cadilhac, M., <u>On a problem of inverse scattering in Optics : the dielectric inhomogeneous medium</u>, J. Optics (Paris), vol. 9, n° 1, pp. 27-35, 1978, to be published.

OPTICAL ANALYSER OF VIBRATIONS

Alain Constans
Technological Laboratory of Physics of Surfaces
National Conservatory of Arts and Trades
3, Boulevard Pasteur 75015 Paris

Abstract

An optical analyser of vibrations has been developed and produced. It permits the measurement of the amplitude and phase of the vibration of a mechanical part. It is characterized by a very small feeler area, on the order of several ^{2}, depending on the power of the lens used. The frequency domain extends from 20 to 100,000 hz. The amplitudes measured are also a function of the power of the lens and range from 10 m to 10. The apparatus offers 1000 measurement points. The principle of measurement is as follows: a displacement of the surface examined entails a defocusing which is expressed by the modification of the light flow received by a differential photodiode. If the surface vibrates, the light flow is modulated and the vibration is translated by a modulation of the photoelectric currents.

Working Principle

An objective O1 conjugates a source hole lit by an electroluminescent diode with the plane of the surface examined. By means of a semi-transparent plate, the image obtained by the reflection of the light flow on the surface examined is relayed to a second objective O2. The axis of O2 is slightly displaced with regard to that of O1. This second objective O2 forms the image of the source hole on a differential cell. When the examined surface is displaced longitudinally, the image of the source hole at the level of the cell is displaced with a highly amplified movement. But, by reason of the displacement of the optical axis of the objective O2, the image of the source hole is also displaced laterally, with a smaller amplitude, in the plane of the cell. The two parts of the differential cell each provide an electrical signal. Let A and B be these signals. The difference A-B is proportional to the displacement of the examined surface, but it is also a function of its reflecting power. The sum A+B is independent of the displacement of the surface, but is also a fuction of the reflecting power. On the other hand, the ratio A-B/A+B is proportional to the displacement of the surface and independent of the reflecting power. It is thus this ratio which is generated and displayed.

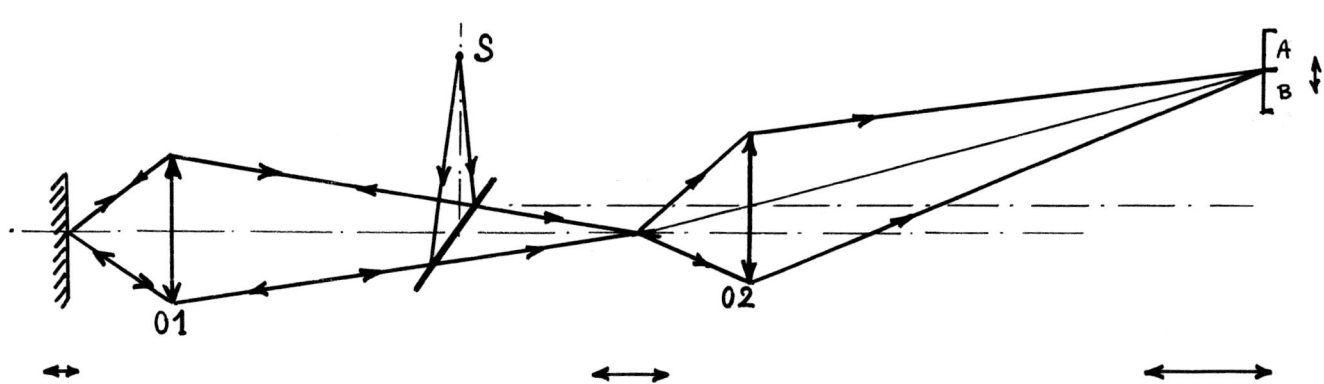

Fig. 1. The optical system is the same as that already produced in the laboratory for static feelers.

Electronic Measuring System

The block diagram of the electronic measuring system is represented in figure 2:

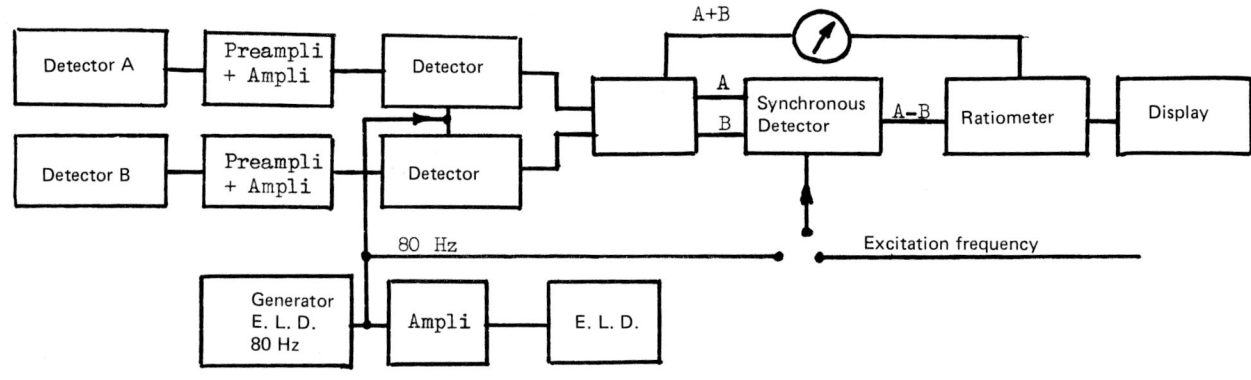

Fig. 2.

a) Dynamic Function of the Feeler

The electroluminescent diode (E.L.D.) is modulated at 80 Hz by a generator followed by an amplifier. The difference A - B is thus twice modulated, by the 80 Hz and by the frequency of the surface. The sum A+B is modulated only by the 80 Hz. The signals A and B are each amplified and detected at 80 Hz. Next the sum A+B is generated, then relayed to a ratiometer. At the same time the signals A and B arrive at a synchronous detector driven by the frequency of excitation of the vibrating surface. This detector emits the signal A-B, which is also sent to the ratiometer. When the ratio A-B/A+B is effected it is displayed on a digital display. The second synchronous detector also displays the phase difference between A-B and the excitation frequency of the vibrating surface. Thus, the apparatus measures the response in amplitude and phase of a mechanical part submitted to a frequency of excitation.

b) Static Function of the Feeler

Since the examined surface does not vibrate, the signals A and B are modulated only by the frequency of modulation of the E.L.D. at 80 Hz. In this case, the second synchronous detector must be driven by the 80 Hz, and produces the difference A-B, which is a continuous signal, just as A+B. Next, the sum and the difference are sent as before to the ratiometer and the ratio A-B/A+B is displayed.

Results Obtained

The performance of the apparatus depends on the power of the objective O1. The table below shows the range of measurement and the resolution obtained with objectives of magnification 8, 24, and 33. Magnification 33 is the maximum which can be used under normal conditions of measurement.

Objective	Range of Measurement	Resolution
8X	100 μm	0.1 μm
24X	10 μm	0.01 μm
33X	5 μm	10 nm

The range of frequencies extends from 20 Hz to 100,000 Hz with a short interval (several Hz) around 80 Hz within which it is impossible to measure, unless we slightly displace this frequency from 80 Hz.

The two following figures show two examples of measurement:

Figure 3: Response of a laser ceramic: the amplitude is shown as a function of the frequency on a logarithmic scale.

Figure 4: Amplitude and phase of a vibrating membrane as a function of the position of the point of examination. We can observe a maximum and a minimum of amplitude, as well as the shift of phase which accompanies the passage through the average amplitude.

It is also possible to measure the periodic vibration of a mechanical piece even if we do not have the excitation frequency of the piece as a reference. It is then necessary to associate the feeler to an averaging sampler.

Fig. 3. Response of a laser ceramic: the amplitude is shown as a function of the frequency. The three curves correspond to three different levels of excitation (20, 50 and 100 volts). Maximal amplitude 1.5   objective 33X

Fig. 4. Measurement of the amplitude and the phase of a vibrating membrane as a function of the displacement of the measurement point; the phenomenon of stationary waves appears clearly. Maximal amplitude 2.5   objective 33X

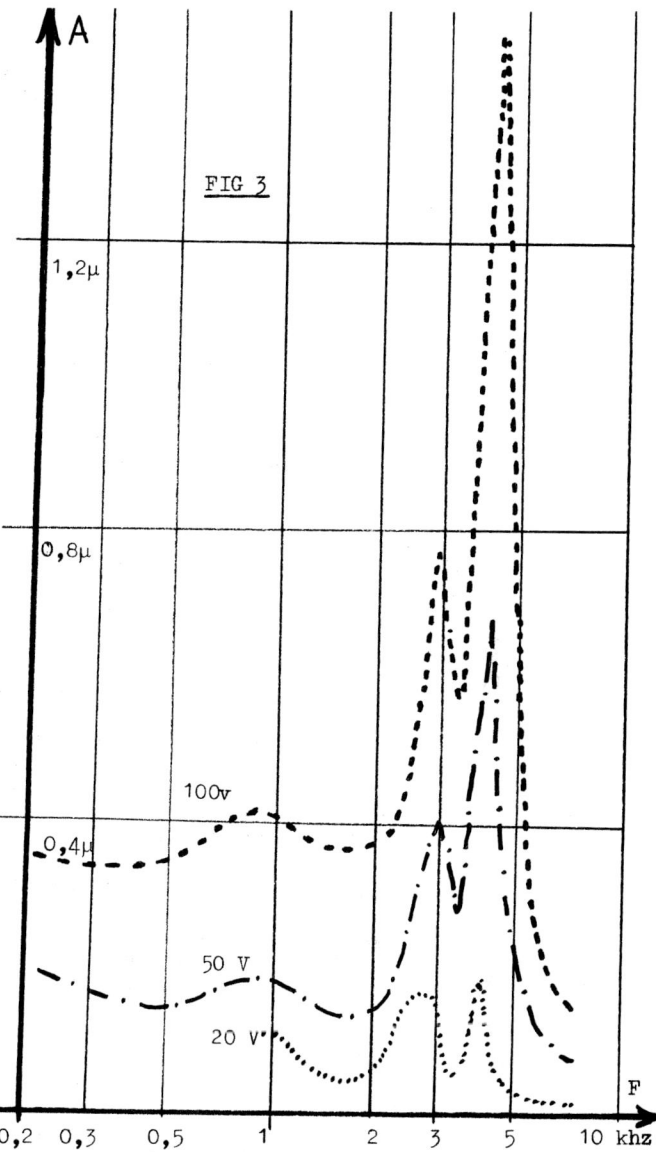

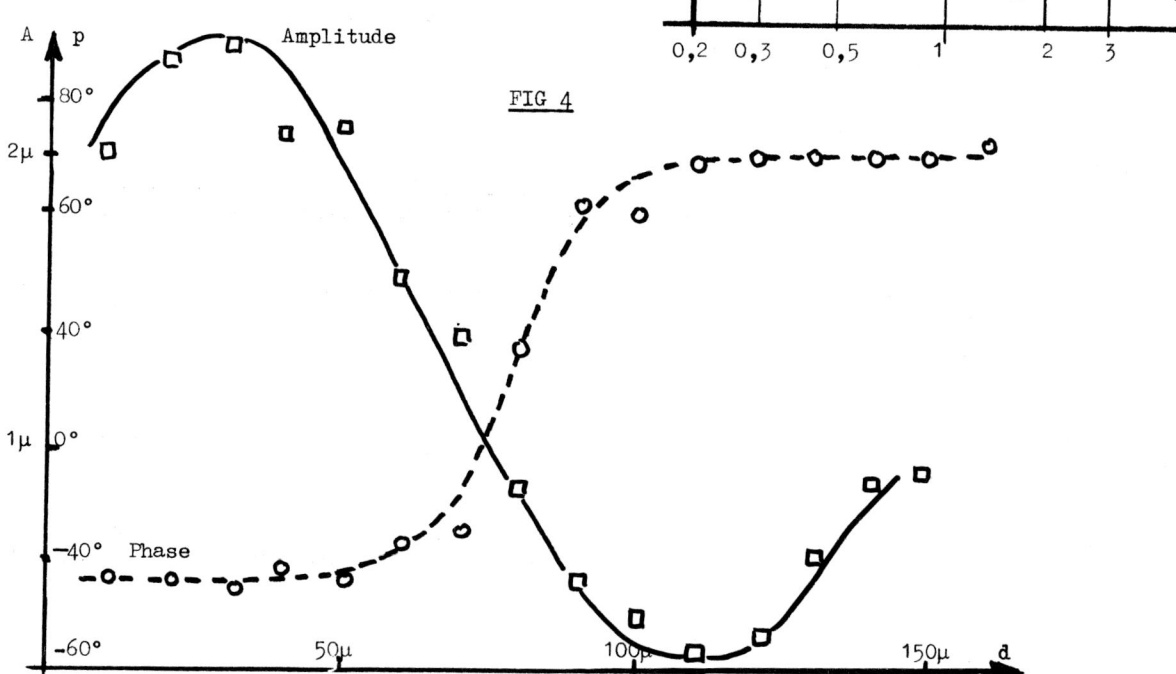

MODEX
OPTO-NUMERICAL MEASURER OF EXTERNAL DIMENSIONS

J. P. Elissalde, F. Pointeau, B. Turlier
MATRA Inc., Optical Division
93, Avenue Victor Hugo
92500 Rueil Malmaison, France

Abstract

The gathering of information by optical methods permits rapid and non-disturbing intervention at the level of the elements being tested. The measurements can be made dynamically (rapid sampling) and in difficult environments.

In general the data taken must be treated so as to permit exploitation in real time, and this in a flexible form so as to be adaptable to different needs.

The system proposed achieves an optimized synthesis of the present technological possibilities by using, as an optronic captor, point detector cameras coupled to a piloting and treatment unit using a micro-processor. The material is entirely static.

Besides its duties of dimensional testing (surveillance of production line), the system can be connected to the production tool, in order to constantly optimize the tolerances.

The simultaneous control of several production lines is possible.

The profitability of the system is assured by the gain in material, decrease in rejection rate and savings on laborious tasks.

1. Goal of the System

The purpose of the MODEX system is the gathering without mechanical contact and the treatment of dimensional parameters, the marking of parts which are mobile in space with regard to a given referential and, under certain conditions, the detection of defects of aspect or surface.

In most applications, the gathering of the data is made "in flight", on parts or sub-assemblies in movement (for example, in translation on a production line); the rate of operation may reach 500 per second.

2. Overall Characteristics of the System

It is conceived in modular fashion, so that it can be adapted gradually to the solution of more and more complex problems. The basic module is composed of a central unit (CU) of control/treatment which takes account of the information of one to four elementary captors, as is shown in diagram 1. The technical description of the constituent elements of the module is given below. Each module can carry out a certain number of missions defined in the following paragraph and it is important to note that several modules can be coupled to a central control system. The latter (a mini-calculator, for example) then assures the piloting and surveillance of the n modules which are joined to it. It is a priori limited only by the treatment capacity of the central organ. This central organ may very well be a computer already in service for other tasks; the connection is made according to usual standards (V24 transmission lines or current loops) since the outlets of the MODEX are intended for that.

3. Missions Fulfilled by the MODEX

A piloting system can fulfill principally a mission of testing and surveillance. In fact, the information captured can be in permanence, that is, in real time, compared to optimal values and a control system put in place by connection with the production tool (as in the case of repetitive parts produced in series). The module represented in the diagram gives:

(a) the display of the dimensional parameters proceeding from the captors $C_1,...C_4$. The rate of presentation of these values is adjustable and each individual value represents, in general, the average:

$$\frac{\Sigma N_i}{N}$$

of N elementary measurements, which improves the precision in the proportion $\sqrt{N}$ in all cases of random variation. The rapidity of the system (we can for each captor gather and treat several hundred units of information per second) permits us to do this work without problems.

(b) the inscription of the values taken on a printer, there also according to an adjustable mode and frequency. These values are gross and/or treated with an eye to statistical exploitation (calculation of standard deviation or variance).

(c) comparison of the values measured with the desirable values stored in memory in the CU.
This comparison is exploited in two ways:
(c.1) production of an alarm signal when the error surpasses a tolerance fixed with regard to the

order. The ordered level is adjustable in the CU. The alarm signal, audible or visual, appears either locally, in immediate proximity to the part being tested, or at a distance, on a centralized surveillance board for example.

(c.2) production of an error signal, in analogical or numerical form, which is used to automatically react by correcting an adjustment until the parameter under its control returns within its region of acceptibility.

(d) the possibility of entrance of diverse information proceeding from other modules or from any external organ. We can cite, for example, for the case of a line with automatic rejection of parts outside tolerance, the notation of the rate of waste by calculation of the difference between the number of pieces at the beginning and end of the line, etc.

(e) under certain conditions, those of contrast in particular, the possibility of detection of surface defects, above all in the case where they appear as heterogeneous zones on a continuous background.

(f) the possibility, as indicated above, of connection to a centralized organ of surveillance where the "control panel" of the factory will be located. This organ possesses its own informational tool (calculator and visualization console with alpha-numerical keyboard for dialogue). Here we reach the most sophisticated level of the MODEX system (see diagram 2).

At this stage, to increase reliability, we can consider diverse combinations such as the surveillance of the proper functioning of n-1 modules coupled by the nth, with hierarchization of the degrees of alarm in production incidents, etc.

4. Technical Description of MODEX

We will describe, along general lines, the two basic elements of the module: the captor and the central unit (CU).

(4.1) The Captor

This is a camera using as detector an appropriate arrangement (a function of the size and the geometry of the object to be studied) of rows of photodiodes with a spacing of 25 μm.

In most applications, an image of varying magnification of the object to be tested is formed in the plane of the photodiodes by an objective. A small connected lamp facing the camera provides a parallel beam of light in which the object is placed.

The lighting conditions differ in the case of detection of surface defects since we then use the light diffused by the zone to be tested.

Each photodiode delivers a pulse whose height is proportional to the quantity of light received. The diodes are coupled to a shift register. The scanning is effected at a typical frequency of 1 MHz. The pulses are compared to predetermined thresholds: the diode sufficiently illuminated gives level 1; that which has not received much light gives a signal below the threshold and remains at level zero.

We thus immediately quantify the information, which considerably increases the immunity to noise.

The electronics directly associated with the camera includes principally clock circuits, circuits for determination of exposure time, and for signature analysis. The latter is made by means of counters which record the number of diodes at state 1 and those at the zero state.

The choice of objective lens results from an optimization for each case bearing on the resolution required, the field to be covered, the tolerance admissible on the optical defects (distorsion).

(4.2) Central Unit

This unit includes the majority of the electronic circuits of servitude and control, as well as the treatment organ, which is a microprocessor for words of 8 bits. The mechanical presentation is that of a standard 19 inch rack (about 5U).

The memory capacity permits the coupling of four cameras to the CU (four Koctets suffice), keeping in mind the duties described in 3. Nonetheless, optional extensions are possible in the modularity optic indicated above.

5. Principal Typical Performances of the MODEX System

Measurable sizes: from some hundredths of a mm to several tens of cm (with the aid of one or more captors, depending on the case).

Typical resolution: 0.01 mm

Rate of gathering of information by each captor: up to 200 per second.

Calculation of the average over n values ($2 < n < 200$), n being an adjustable number. In effect, n depends on the variance, the latter being related to n by the classic relation

$$\sigma^2 = \frac{V(x)}{n}$$

where σ represents the standard deviation. σ conditions the confidence interval within which we can situate the precision of the estimation of the true value X of the parameter considered by its average value:

$$X = \frac{1}{n} \sum_{i=1}^{n} X_i$$

There exist, moreover, instrumental causes of error which are, principally:
-the distorsion of the camera objective
-the uncertainty in the adjustment of the enlargement
-the "blur" introduced in certain cases of measurement by the movement of the part being tested (as in conventional photography everything depends on the period of the movement compared to the exposure time). If the phenomenon is bothersome, we consider a local device to stabilize the part at right angles with the measurement zone.

Diverse corrections introduced in the calculation (which include a previous calibration on a perfectly known part), combined with appropriate adjustments, permit us in a wide range of measurement to bring the instrumental error below the resolution.

Industrial Environment

The environment is sometimes severe but should not risk disturbing the quality of the measurements. In extreme cases (excessive temperatures), the captors are protected. All the technological precautions (quantification at the level of the captor, transmission in differential mode by twisted pairs in a protected cable) are taken to avoid the interference of radiated parasites (presence of strong currents near the MODEX), in particular when several modules are coupled to one centralized control unit.

5. Metrological Applications

The flexibility of the MODEX system and its very great reliability (no sub-assemblies in movement) make it suited for multiple applications, among which we can cite:

-manufacture of tubes, wires, cables: testing of diameter, bending, length, ovalization of tubes. Continuous measurement of diameter of wires and cables.
-glass industry: continuous testing of tubes, bulbs, flasks, bottles (classification and forms).
-metallurgic industry: surveillance of lengths and thicknesses of sheet metal, etc.

In all cases, the profitability of the MODEX system will be assured by a reduction of rejection rates and testing time and a parallel improvement of the quality of the products.

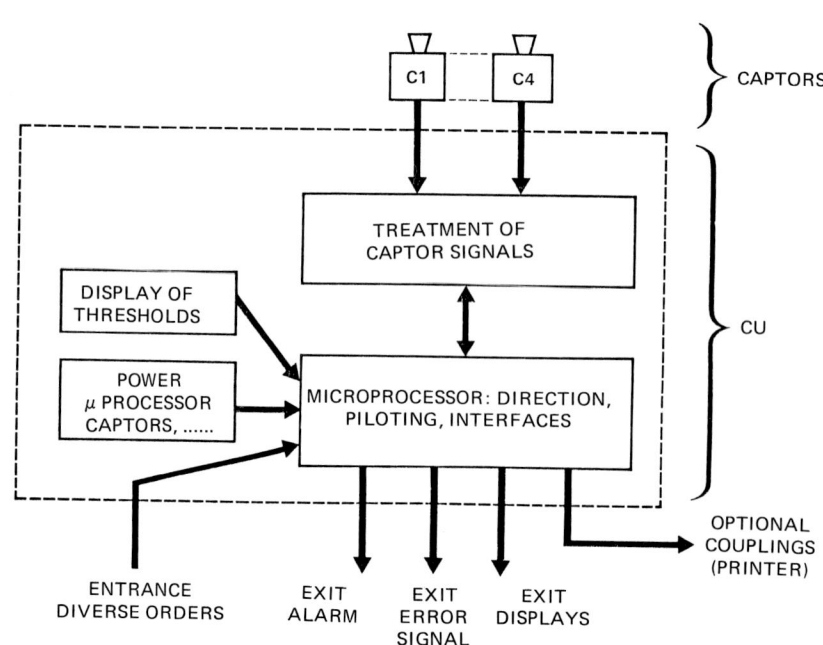

Diagram 1

MODEX, OPTO-NUMERICAL MEASURER OF EXTERNAL DIMENSIONS

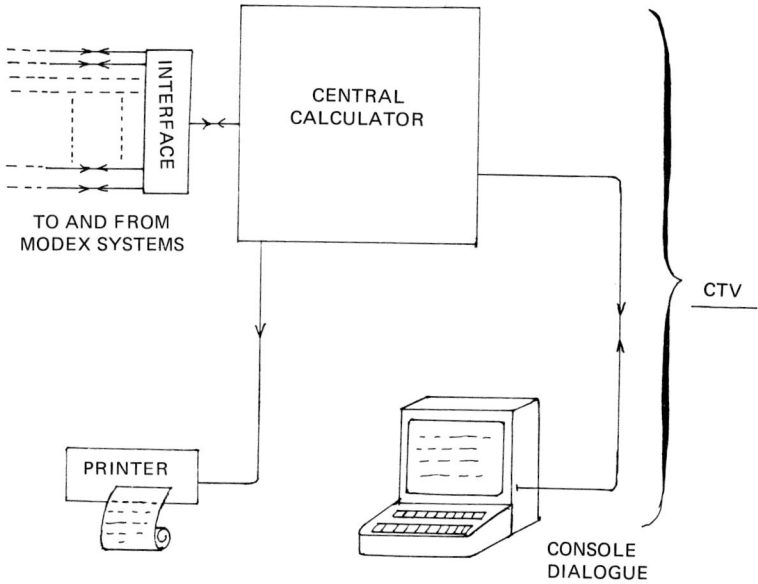

Diagram 2

Photo of the MODEX system

1st EUROPEAN CONGRESS ON OPTICS APPLIED TO METROLOGY

Volume 136

SESSION 2

INTERFEROMETRY

Session Chairmen
Denegre
Ratterink

COMPARISON OF WAVELENGTHS TO THE PRIMARY STANDARD AT THE FRENCH NATIONAL INSTITUTE OF METROLOGY

P. Bouchareine and A. Janest

Institut National de Metrologie, 292, rue Saint-Martin, 75141 Paris Cedex 03

Abstract

We describe some properties of the two beam interferometer that we have built at the "Institut National de Metrologie" for wavelength comparison. We describe particularly the phase shifts between visible wavelengths at zero path difference.

Introduction

The National Institute of Metrology of the National Conservatory of Arts and Trades is the primary laboratory of the National Bureau of Metrology for lengths. It is thus responsible at the national level for the comparison of measurements of length to the primary standard, that is, to the wavelength of the orange radiation of krypton 86. To this end we built a two-wave interferometer of the Michelson type, working in a secondary vacuum to eliminate any index correction. The separating plates are in Brewsterian incidence to avoid any parasitical interference capable of dephasing the fringes.

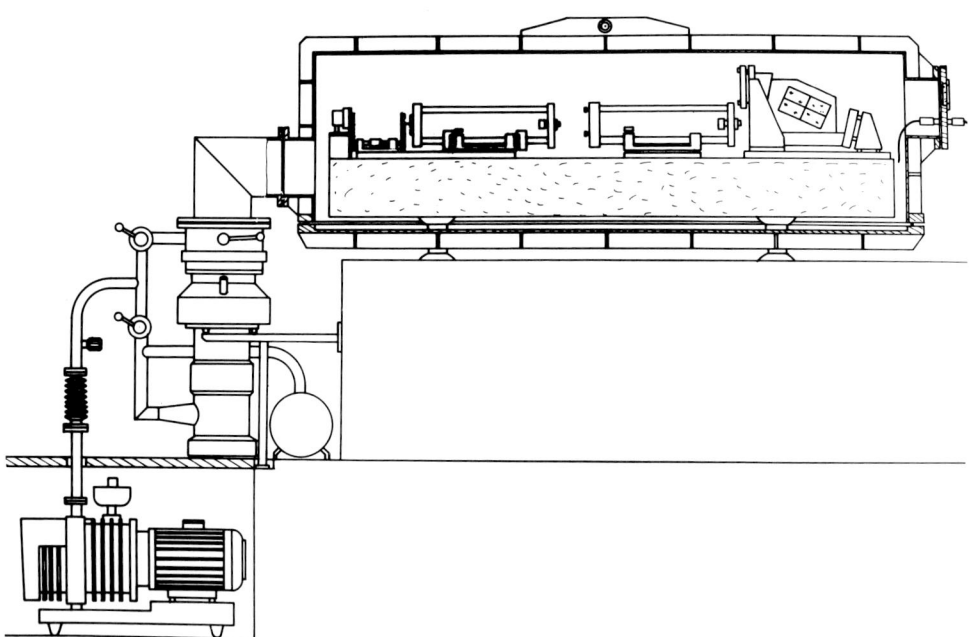

Fig. 1. General view of the interferometer in its airtight case. The light beams enter and leave by the ports at right. At left is the oil diffusion pump and its vacuum reservoir; the primary pump is under the floor. The interferometer rests on a diabase slab 0.2 x 0.5 x 2 meters, insulated by flexible shock-absorbing supports.

1. The Compensated Field Interferometer

This is a cat's-eye interferometer, which we have already described[1,2], derived from those which P. Connes and his colleagues have applied to spectrometry by Fourier transformation. We have installed a compensation of the field of interferences proposed in 1967 by M. Cuisenier and J. Pinard[3]. We thus benefit from fringes at infinity with large angular field (flat tint) at the path differences which we have chosen: 25 cm and 50 cm. The light flow received by the photomultiplicator is practically limited only by the monochromator of the Ebert-Fastie type with a focal length of 1m charged with isolating the krypton radiations. The signal/noise ratio given by simple 1P 28 photomultipliers allows us to work with a time constant of several hundredths of a second, and thus to multiply the measurements.

Another innovation of this interferometer is the absence of adjustment of the position of the supports of the optical parts. One property of the cat's-eyes is to send back an incident ray always parallel to itself. The localization of the fringes at infinity is thus assured provided that the two interfering rays are mixed by a semi-reflecting surface rigorously parallel to the separating plate. This parallelism was obtained by the use of brass studs screwed forcibly into a piece of cast aluminum which is thermically stabilized. Once the adjustment is definitevely made, the interferential flat tint is given by the superposition of the optical centers of the cat's-eyes, which we obtain by displacement of one of them on the marble slab which supports the assembly. This means of adjustment is certainly less convenient than a reduction device with elastic return, but the long-term stability of the interferometer is thus improved spectacularly: adjusted and put in vacuum at path difference 0 on August 16, 1977, our interferometer still offered a coefficient of visibility of the fringes superior to 90% on the 21st of October, despite the almost daily mechanical disturbances of the primary pumping. No access to the interferometer other than the translation of the mobile cat's-eye exists when the assembly is in vacuum.

2. Study of Phase Shifts at Zero Path Difference

The comparison of the wavelengths is made by the ratio of the numbers n and n_o of the fringes given for a single path difference by the radiation studied and the standard radiation. When the wavelengths of the radiations compared are very different, the diverse phase shifts undergone in the interferometer are expressed by a different displacement of the position of the mobile mirror giving the interferential maximum of the zero order fringe. This phase shift at the origin is customarily compensated for by making a difference between measurements made at the path differences Δ and $-\Delta$.

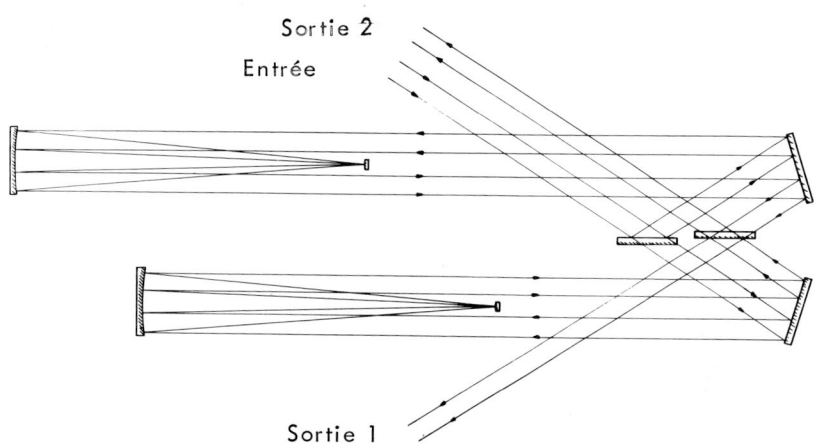

Fig. 2. Optical setup of the cat's-eye interferometer: the separating and mixing plates give this type of interferometer a better symmetry than that of the conventional Michelson interferometer. Moreover, the change in the radius of curvature of the secondary mirror of a cat's-eye permits us to compensate the interference field and replace the rings at infinity by a flat tint.

In our interferometer, whose optical layout is given in figure 2, the very good symmetry of the two arms brought about by the mixing plate, which is distinct from the separator plate, minimizes these phase shifts at zero path difference. One consequence of this quality is the very good symmetry of the fringes in white light (figure 3). This is the reason we chose to measure these small phase shifts (some hundredths of a fringe) and to use them as corrections on the measurements made at great path differences.

We evaluated this chromatic dispersion of the origin of the fringes by two distinct methods, of which only the first has given results so far:

1st: by measuring, one by one, the phases of the fringes given by diverse monochromatic radiations emitted by a krypton lamp;

2nd: by a complex Fourier transformation effected on the signal of the white light fringes of figure 3, where the fringes given by the totality of the visible radiations of the spectrum figure simultaneously.

3. Measurement of Phases of Monochromatic Radiation

This is the customary measurement of a fractional excess on a sinusoidal signal. The interferometer delivers simultaneously the signals given by the fringes of a reference beam (stabilized laser) and of the radiation studied. The reference signal controls the sampling of the signal of the radiation studied with a periodicity of 633 nm. Since we already know very well the ratio between the wavelength of the radiation studied and that of the laser, we know with a negligeable uncertainty to within a thousandth of a fringe the variation of the fractional excess from one sample to the next. We can thus calculate for each sample a theoretical value of the fractional excess compared to the first. The dispersion over the values thus calculated informs us about the quality of the sampling and the translation.

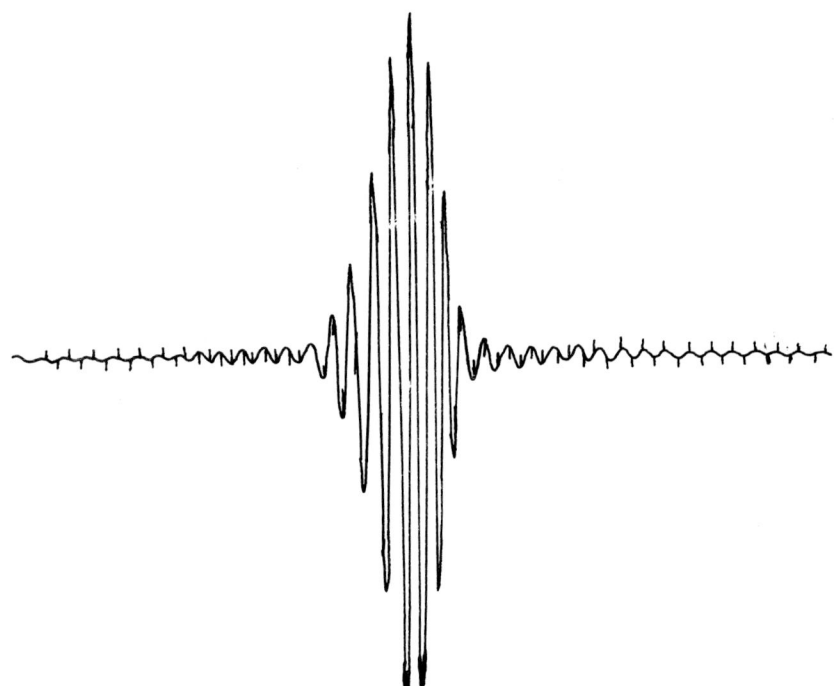

Fig. 3. White light fringes given by the interferometer at zero path difference. Their symmetry bears witness to the good achromatism of the apparatus; the pulses are given by the interference fringes of the reference laser and define the scale of the path differences.

Figure 4 represents the fluctuations of these excesses gathered in the course of a translation of the cat's-eye whose amplitude is about fifty fringes, at different stages of the focusing of the translation device. In a, the reduction lever arm moved by a differential micrometer screw exerts leverage by the intermediary of a fixed bearing pressing on a polished glass. The periodicity of the displacements (fifteen fringes) corresponds to a complete rotation of the adjusting screw and reveals the lateral forces related to the inevitable decentering of the stud and transmitted by the bearing. The defect in translation induces differences in path difference between the reference channel and the measurement channel on the order of 30 mm, or 5 hundredths of a fringe.

In b, we have added a bearing resting on two sapphires, which transmits practically no transversal forces. The translation is very much improved and the individual standard deviation over this series of samples is several thousandths of a fringe.

In c, two parallel modifications have been effected: on one hand, the elastic support of the cat's-eye, consisting of four spring bands, is hyperstatic and, despite the precautions taken upon installation of the springs, badly defines a position. It has been replaced by a device with three plates. On the other hand, the automatic acquisition of the data under the control of a microprocessor has been improved by separating two functions: the surveillance of the reference signal which detects the instants of sampling and the acquisition on the measurement signal of the values of the maximum and the minimum. To avoid the influence of the noise, which was expressed by an excess of the maximum and an underestimation of the minimum, we were led to take the average of four successive acquisitions. These operations carried out parallel to the sampling caused in the latter random delays which were manifested by a phase noise which is probably not negligeable. In its last version, the data acquisition program evaluates the extrema of the signal at the beginning or the end of each recording in the course of a translation of a supplementary fringe. The individual standard deviation over this last series is 13 ten-thousandths of a fringe. In the ensemble of these three diagrams, some values are not taken into account in the calculation of the standard deviation, since they correspond to neighboring points of the extrema of the sinusoid for which a given photometric uncertainty leads to a very great uncertainty about the phase. The selection criterion is that the absolute value of $\cos \varphi$ be inferior to 0.96.

We thus measure successively the fractional excesses of several types of radiation, among them the standard radiation. To correct a slow drift in the phase relation between the reference channel and the measurement channel (which may attain a hundredth of a fringe over the duration of a series of measurements, or one hour), it is necessary to make several measurements on the standard radiation. Once these excesses are measured, we determine by calculation the number of fringes it would be necessary to pass so that all the fringes return approximately in phase. This is a conventional calculation of coincidence which presents no ambiguity over several hundred fringes with about ten wavelengths and uncertainties inferior to several hundredths of a fringe. The result of this calculation should lead to a fractional excess of zero for all wavelengths. In reality we find a residue of several thousandths of a fringe characterizing a phase shift at the zero path difference.

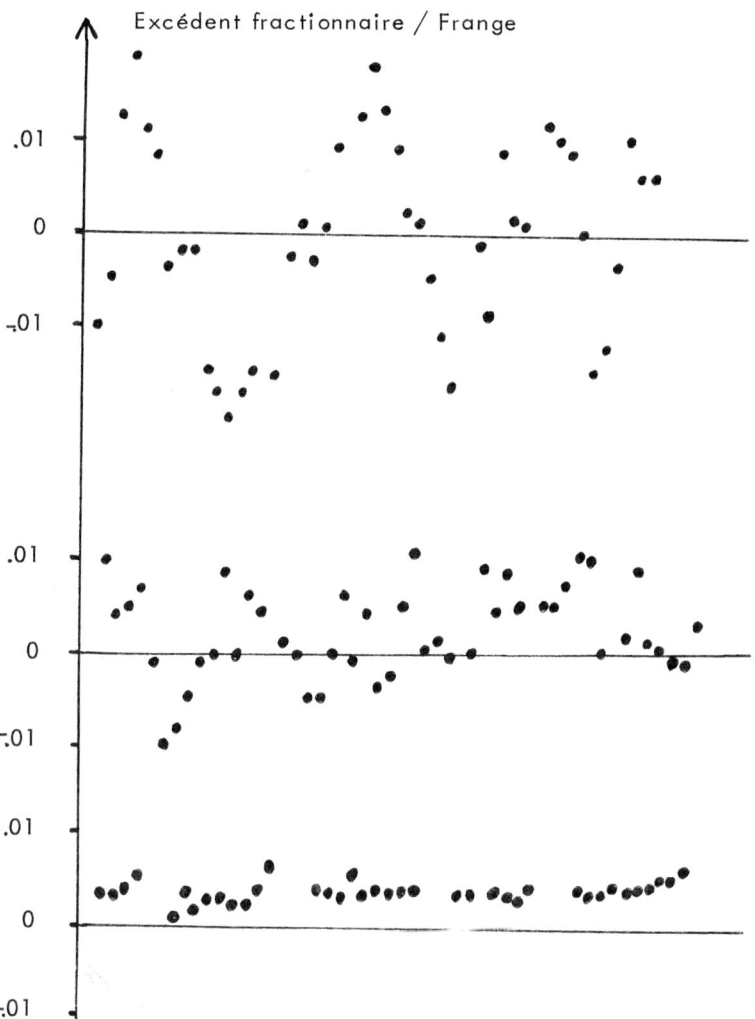

Fig. 4. Dispersion of calculated values of the fractional excess at the origin point in the course of a translation of fifty fringes. (a) direct contact between the adjusting screw and the mobile cat's-eye (b) decoupling by a bearing which eliminates the lateral forces (c) improvement of the elastic support of the mobile cat's-eye and of the automatic data acquisition.

Figure 5 presents the results obtained over 10 recordings. The standard radiation (605.7 mm) defines the zero path difference with an uncertainty of plus or minus 2 thousandths of a fringe, the standard deviation of the phase fluctuations observed between different recordings. We see that no significant phase shift is revealed by these measurements. Their dispersion is nonetheless unsatisfactory since each fractional excess is obtained by an average of 5 to 10 samples, whose dispersion is characterized by a standard deviation inferior to 2 thousandths of a fringe. The cause is very probably at the level of the operations effected between two different radiations: rotation of the grating of the pre-monochromator, change of gain of the photometric circuit, defect in reproducibility of the translation of the mobile cat's-eye.

4. Measurement of the Chromatic Dispersion From White-Light Fringes

The defect in reproducibility indicated at the end of the preceeding paragraph is eliminated if we effect the phase measurements on a simultaneous recording of the fringes given by the different radiations, that is, by the ensemble of a transformation in cosine and a transformation in sine. The ratio of the imaginary part (transform in sine) to the real part (transform in cosine) of the transform for a given value σ_o of the wave number is the tangent of the angle of phase φ to the origin of the monochromatic fringe σ_o:

$$\varphi = \text{Arctg} \left\{ \frac{\int_{\Delta_1}^{\Delta_2} \phi(\Delta) \cdot \sin(2\pi \sigma_0 \Delta) \, d\Delta}{\int_{\Delta_1}^{\Delta_2} \phi(\Delta) \cdot \cos(2\pi \sigma_0 \Delta) \, d\Delta} \right\}$$

The choice of the limits Δ_1 and Δ_2 implies an *a priori* choice of the original path difference. If φ is a linear function of the wave number σ, this origin has been badly chosen and it is advisable, to obtain the irreducible phase shift, to subtract from $\varphi(\sigma)$ the part porportional to σ. The choice of the coefficient of proportionality is equivalent to choosing as path difference reference the summit of the zero fringe for a particular radiation.

The calculation of the Fourier transforms in sine and cosine is made numerically on a sampling of the fringe signal given by the fringes of the reference laser. But, on one hand, the positive and negative pulses are not equidistant; on the other hand, Shannon's theorem shows that, to avoid confusion between the radiations situated on one side or the other of 632.99 nm, a narrower sampling is necessary. The latter, effected by interpolation between the pulses, suffers fluctuations in the speed of variation of the path difference. The samples thus interpolated introduce into the calculation a phase noise which we have not been able to reduce, so far, to less than several hundredths of a fringe, above all at the extremities of the visible spectrum, where the module of the Fourier transform has a small value. We are looking, on one hand, for a specific technique, and on the other, we study the problem of phase fluctuations in a Fourier transformation as a function of the signal/noise ratio in the original function and errors of positioning of the samples, a problem which to our knowledge has not been treated.

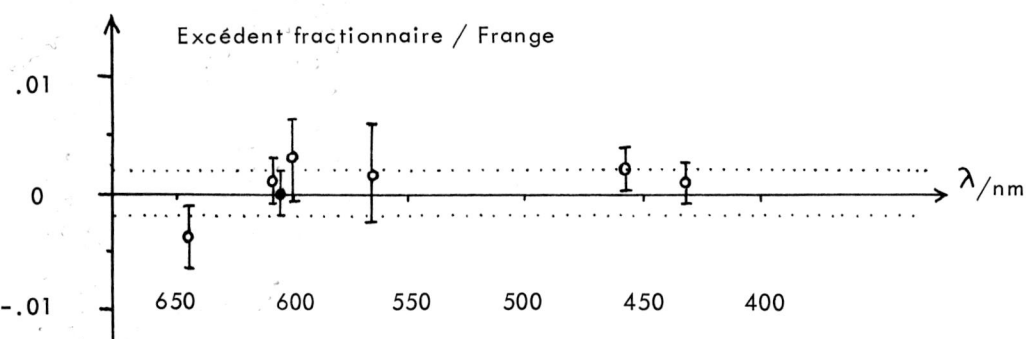

Fig. 5. Fractional excess of some monochromatic radiations emitted by the krypton lamp at the summit of the fringe, origin of the standard radiation (605.78 nm).

Conclusion

The provisional results which we present do not permit us to affirm it, but we are convinced that the information about the relative phase between the fringe systems of different wavelengths at the zero path difference is primordial and that it can be obtained with a significantly smaller uncertainty than at large path differences. This is essentially because the measurements can be made with intense sources whose wavelengths do not have to be defined with great precision. We even hope to make this measurement with an incandescent lamp, provided that a valid indication is given for the sampling of the path difference.

References

1. Bouchareine, P., A Field Compensated Interferometer for Wavelength Comparison, AMCO, vol. 5, Sanders and Wapstra, p. 417, 1976.
2. Bouchareine, P., L'interféromètre de l'Institut National de Métrologie pour la Comparaison des Longueurs d'onde Lumineuses, Bulletin of the Bureau National de Métrologie, n° 24, 1976
3. Cuisenier, M., Pinard, J., Spectomètre de Fourier à oeil-de-chat et à balayage rapide, J. Phys., vol. 28, C2, p. 97, 1967.

INTERFEROMETER FOR TESTING INFRARED MATERIALS AND OPTICAL SYSTEMS

Michel Lamare
Laboratory of Instrumental Optics
Institue of Theoretical and Applied Optics
91406 Orsay Cedex, France

Abstract

We describe an interferometer functioning in the region of 10 μm, developed for the examination of the homogeneity of IR materials and for the Twyman-Green testing of the wave surface of IR optical systems. This interferometer used a CO_2 laser as source and a diffraction grating as separator. A scanning camera associated with a point detector reconstructs the interferogram on the screen of an oscilloscope. We also present a second-generation instrument recently produced in collaboration with an industrial company as well as some typical applications of these interferometers, in evoking their possibilities of extension to the domain of common mechanics for the demonstration of the form of parts.

We also present an experiment of linear reconstruction of the phase in order to directly obtain the wave surface, effected with the infrared interferometer, whose path difference is put to use.

Introduction

The infrared radiation around 10 μm has been for a long time the object of particular interest, by reason of its potential applications. This is for two reasons: this radiation corresponds on the one hand to the maximum of emission of bodies at ambient temperature, and on the other hand is situated in a good atmospheric transmission window. Nevertheless, until the last few years technological reasons related to the detectors slowed the development of these applications. With the recent appearance of high-performance quantic detectors and pyro-electric detectors, punctual or in the form of a layer using a readout system of the "vidicon" tube type, the infrared domain at 10 μm has rekindled interest.

Still very recently, the best associated optical systems were constituted of reflecting elements. The perspectives for development of thermic imagery at 10 μm led us to envisage purely dioptric solutions able to avoid inconveniences inherent in the catadioptric combinations (central occultation, clumsiness, mediocre reliabllity of adjustments) Nonetheless the majority of the materials which can be used to produce such systems are not transparent in visible light, and it is important to examine their homogeneity in the spectral region of use before fabricating them. It is also important to check the optical quality of the system produced to verify its conformity with the calculation and to correct its eventual defects, and to determine a priori its influence in a complex system of visualization of infrared images.

The infrared interferometer at 10 μm studied and produced at the Institute of Optics responds to this double objective, and we propose to present here a panorama of the activities of the "Laboratory of Instrumental Optics" in this field. We will limit ourselves in this presentation to recalling briefly the instrumental productions by emphasizing their applications and we will invite the reader desiring more detailed information on particular points to refer to more specialized articles which have appeared previously.

Infra-red Interferometer at 10 μm

The particularity of this interferometer resides in the use of a diffraction grating serving at the same time as beam splitter and reference surface. This grating is adjusted in auto-collimation in the -1 order with regard to the incident beam; this wave serves as reference wave. The measurement wave of the interferometer is constituted by the wave reflected specularly in the 0 order. The measurement wave is recombined with the reference wave after reflection on an auxiliary plane mirror and a new specular reflection on the grating (figure 1). We remark that this configuration implies a large path difference leading to the use of a CO_2 laser stabilized by means of a feedback loop. An afocal assembly constituted of a germanium meniscus and an off-axis parabolic collimator permits us to dilate the laser beam and to obtain an incident wave of 150 mm in diameter, essentially plane.

The visualization of the interferometric field is effected by the intermediary of a semi-reflecting surface placed in the collimator which removes the reference and measurement waves to direct them towards an optico-mechanical system consisting of a mirror oscillating in two orthagonal directions associated to a pyro-electric point detector of the T.G.S. type. A spherical mirror assures the sample-detector conjugation. The interference field is reproduced point by point (128 points- 128 lines) on the screen of an oscilloscope with memory on which the position of the spot is piloted by the movement of the oscillating analysis mirror, while the electrical signal delivered by the pyro-electric detector controls, after treatment and formation, the luminance of this spot. The production time of an interferogram is 5 s [1].

From a practical point of view, in the case of a test for homogeneity the material cut in the form of a plane-parallel plate is inserted in the measurement arm in front of the auto-collimator plane mirror. To test the wave surface of a complete optical system by the Twyman-Green method, however, the system is associated to a spherical caliber whose center of curvature coincides with its image focus, which is disposed on the measurement arm of the interferometer. The instrumental precision is around $\lambda/4$ at 10.6 μm. Let us point out, moreover, that we take advantage of the polarizing properties of the diffraction gratings to act on the photometric equilibrium of the two arms of the interferometer. In fact, for a better photometric use of the setup, we use a grating for which only the 0 and -1 orders exist, and we can then show

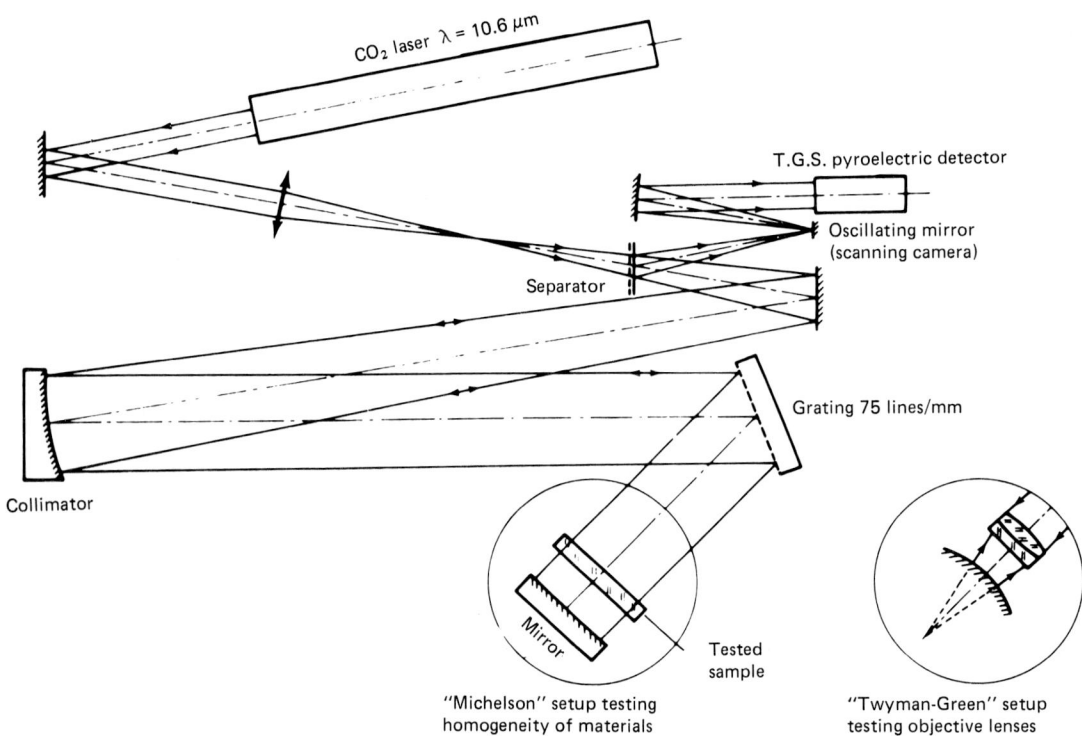

FIG-1 Diagram of infrared interferometer at 10 μm

that for such an echelette grating of a given angle of blaze, the energy efficiency in the zero order does not vary when this grating undergoes a rotation of 180° in its plane (2).

By assimilating, in a first approximation, the interferometer to a two-wave system, we show that the ratio of the energies of the measurement wave to the reference wave is a function of the efficiency in the 0 order. This can be modified, for a given grating, by acting on the polarization state of the incident light. This property permits us, with the grating we adopted (75 lines/mm blaze: 6°30'), to conserve a fringe contrast superior to 60% when the transmission of the sample examined varies from 10 to 100% (3). The rotation of the polarization plane is obtained by combining the rectilinear polarization of the laser with a half-wave plate in CdS. Let us note that this arrangement is particularly advantageous in the infrared domain, where we find high-index materials presenting great losses by Fresnel reflection.

<u>SORO Scanning Interferometer</u>

Recently the company "SORO Electro-Optics", in collaboration with the Institute of Optics, has produced an instrument directly derived from the infra-red interferometer described previously, and presenting specific advantages over the latter due to the scanning of the field of measurement. In the preceeding instrument, the sample examined is permanently lit by the laser beam greatly enlarged by an afocal device, and only the analysis of the interferometric field is effected point by point; there results a "waste of energy" which requires the use of a laser of average power (20-25W). In the case of the SORO interferometer, this disadvantage has been avoided by using a combination of two orthogonal oscillating mirrors, controlled in position, which simultaneously effect the scanning of the sample with the laser beam and the reconstruction of the interferogram. This arrangement offers a considerable photometric gain which permits us to test highly absorbent samples whose transmission is inferior to 5% while still using a laser of weak power (5W) (figure 2).

Moreover, the use of controlled oscillating mirrors permits us to easily vary the dimension of the field analysed, creating an "electronic zoom" effect while still conserving good conformity upon reproduction of the interferogram. Let us also add that the measuring capacity of the instrument has been carried from 20 to 75 fringes in 150 mm. The application of the instrument is thus facilitated; the preliminary adjustments to acquire the interferograms are not critical and we can test optical components whose wave surfaces present considerable unevenness (aspherical surfaces, for example).

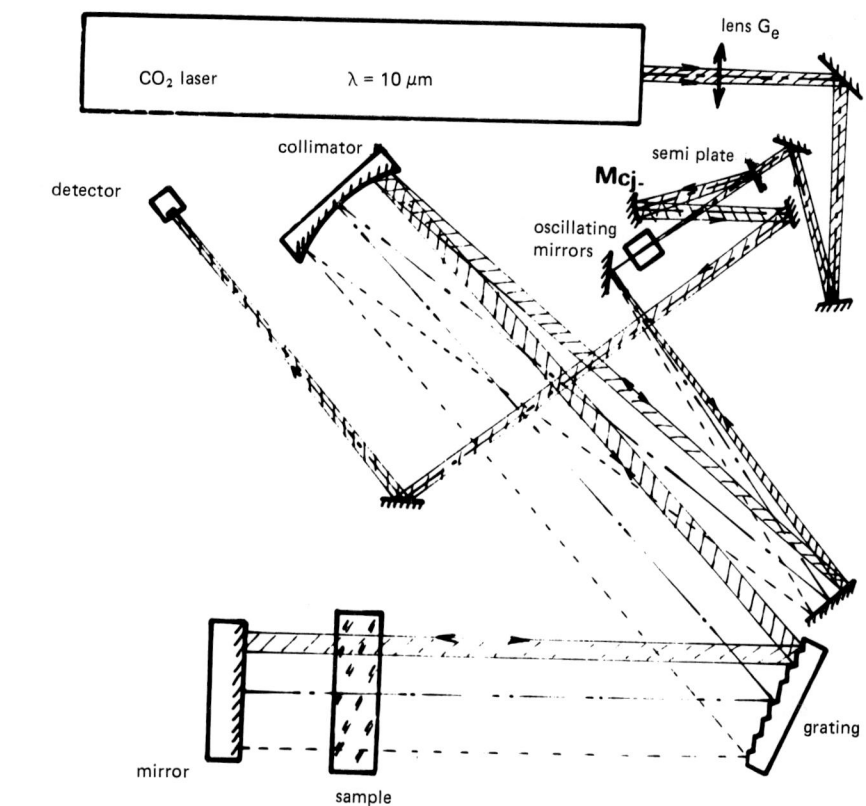

Infrared scanning interferometer

SORO – IOTA

FIG. 2

Some Examples of Applications

The field of application of these interferometers covers essentially the field of optics with testing of homogeneity of index of infrared materials and the examination of the wave surface of optical systems, but it can also be extended to the field of common precision mechanics, as certain examples will show.

Testing of Homogeneity

Figure 3 represents the interferogram of an untreated germanium disk 50 mm in diameter and 25 mm in thickness. The deformation of the fringes is derived from the inhomogeneities of the material and from the residual geometric deformations of the sample, whose faces have been tested in visible light by conventional techniques of interferometry. Let us note that, taking account of the high refractive index of germanium, the planeity of the surfaces must be examined with care. Let us also note that the sample is slightly prismatic (about 15') in order to eliminate the Fizeau fringes. Since it involves these diverse elements, the material presents a slow fluctuation of the refractive index on the order of 4×10^{-4} with a decrease from the center toward the edge. This last information on the direction of the deformation can be obtained by measured shifts of the autocollimator plane mirror of the measuring arm, or more simply by interposing on this arm a dephasing element: mylar band, hot wire, thin local deposit on the autocollimator mirror, etc.

Testing the homogeneity of an infrared material

Germanium
⌀ 50mm   e = 25mm
$dn \leq 4.10^{-4}$
$n_\searrow$ for $R_\nearrow$

FIG. 3

Testing the Wave Surface of an Optical System

Figure 4 shows the Twyman-Green interferograms of an aspherical lens in Irtran II (f = 75mm) open to f/1. The examination was effected on the (y' = 0) axis for a parallel incident beam and for a focus plane x'o corresponding to the least deformation of the wave surface. On both sides, the interferograms figure in intra and extra focal position at ± 200 μm from the x'o plane.

Twyman-Green Interferometry 10 μm

aspherical lens   _ IRTRAN II   f = 75 mm _ f/1 _ y' = 0

x'o − 200 μm   x'o   x'o + 200 μm

FIG. 4

Application to Mechanics

Conventional interferometric techniques in visible light occupy a particular place in the metrological arsenal, since their very great sensitivity reserves them for the most part for the testing of pieces of very great quality of form and surface state. In infra-red interferometry, the substitution of the infrared wavelength ($\lambda = 10 \mu m$) for the visible wavelength ($\lambda \simeq 0.5 \mu m$) is expressed by a desensitization of

the phenomena by a factor of 20 which permits us to extend interferometric methods to the domain of mechanical parts of common precision (several μm), while conserving the specific advantages of these methods:
-absence of material contact; an interesting possibility for soft or fragile materials whose use in mechanical construction of average precision is spreading; possibility of surveillance at a distance of inaccessible parts (situated in a hostile environment, for example).
-quasi-perfect reference surface; with the possibility of examing, besides essentially plane surfaces, spherical, even cylindrical surfaces, by the method derived from that of Twyman-Green.
-visualization of a surface by its contour lines with direct sampling by a physical scale with a spacing of 5 μm ($\lambda/2$).
-rapid and global acquisition of the measurement in a tested zone of large dimensions.
From a study undertaken at the Laboratory to specify the practical limits of this method, the following points emerge:
-the "macrogeometric capacity" of the instrument is considerable (superior to 350 μm over 150 mm for the scanning interferometer); we can examine parts presenting great unevenness and deviations of form.
-the precision, on the order of 1 to 2 μm, is compatible with present requirements of mechanics.
-the great coherence length of the laser permits us to simultaneously examine parts which are essentially parallel, but not coplanar.
-the microgeometry of the parts must be fairly fine; the upper limit of the rugosity Ra is situated for the metallic pieces between 1.6 and 3.2 μm depending on the surface characteristics (rugosity, ondulation, etc.) determined by their mode of manufacture. In particular, the behavior of anisotropic surfaces (obtained by plaiting, turning, planing, etc.) when their ondulation is of the same order of magnitude as the infra-red wavelength, is assimilable to that of diffraction gratings, whose efficiency varies with the orientation of the polarization of the incident beam. The practical limits adopted here correspond to a fringe contrast around 25 to 30%, which constitutes the detection threshold of our laboratory model; by modifying the instrumental parameters (more powerful laser, better performing detector, for example) we could tolerate slightly rougher surface states.
-the examination of spherical surfaces, and especially cylindrical ones, remains handicapped up to now by the fact that the infra-red optical systems which are associated to them must be of excellent quality, and with a large aperture, for the solid angle of exploration of the piece to be sufficiently large.

Some examples

Figure 5 illustrates the possibilities of interferometry at 10 μm for the testing of mechanical pieces. These roughly plane pieces have been placed on the measurement arm of the interferometer in place of the autocollimator plane mirror.

It will be noted, concerning the interferogram of the plaited square, that the upper zone and the lower zone correspond to roughly parallel planes about 90 mm apart.

It will also be noted, concerning the interferogram of the grooved plate, that the upper and lower parts each present a defect in parallelism with regard to the central part of about 40".

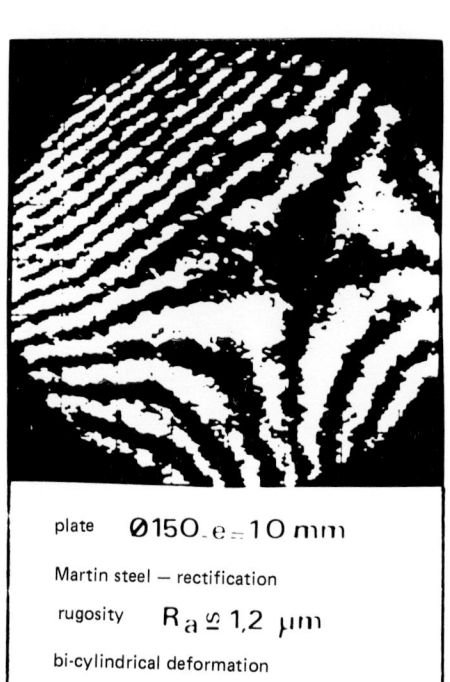 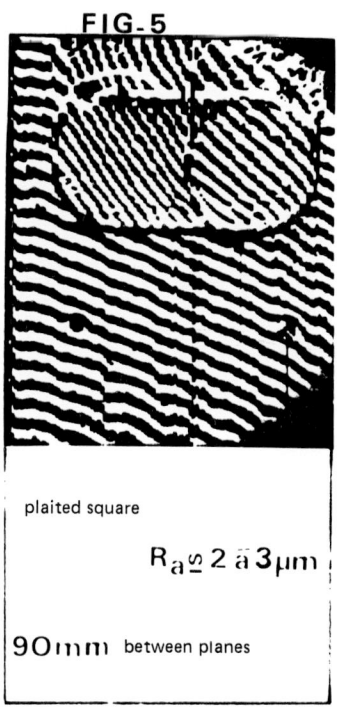 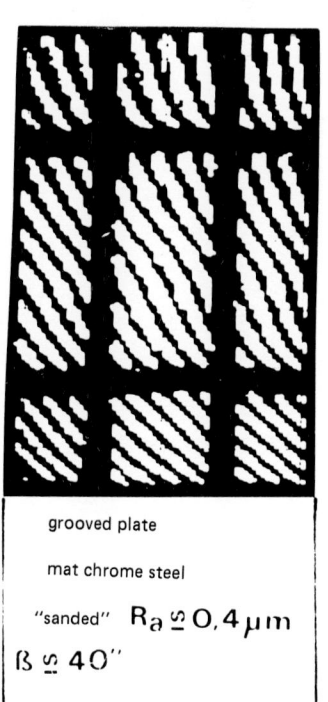

FIG-5

plate ⌀150 e=10 mm
Martin steel — rectification
rugosity $R_a \simeq 1.2$ μm
bi-cylindrical deformation

plaited square
$R_a \simeq 2$ à 3 μm
90mm between planes

grooved plate
mat chrome steel
"sanded" $R_a \simeq 0.4$ μm
$\beta \simeq 40''$

Infra-red Interferometer with Linear Reconstruction of Phase

The interpretation of interferograms, which reduces to the measurement of the algebraic variation of the order of interference, often proves to be very delicate in the case of irregular fringes. Interferometry with linear reconstruction of phase, which permits the direct recording of the profile of the wave surface, without ambiguity of sign, seems to be the most general solution of this difficulty.

Based on the infra-red interferometry developed in the laboratory, two methods have been tried[4].

Modulation of Path Difference

In this method we control the path difference (Δ) of an interferometer to remain constant, either at a fixed point of the field as a function of the time, or at every point of the field in the course of its analysis by a mechanical or electronic scanning device. If we consider a Michelson interferometer this control will compensate for all variations in Δ by displacement of one of the mirrors; this displacement represents at every instant the local variations in phase compared to a phase taken arbitrarily as origin.

The determination of the phase by control of the path difference presents the following advantages:
  -no reference detector is needed;
  -the phase variation is obtained algebraically and not by means of one of its trigonometric lines;
  -as in all "zero" methods, the measurement is not affected by fluctuations of intensity of the source.

In this method, in which the phase discrimination (necessary to know the direction of the variations) consists in the oscillation of one of the mirrors of the interferometer, we can show that when the path difference tends to stray from a designated value Δ, a signal is produced at the frequency of the oscillation of phase whose amplitude represents the phase shift; this signal is used to control the position of one of the mirrors of the interferometer by maintaining the path difference at the designated value; the measurement of the displacement of the mirror directly provides the variation in phase.

Figure 6 shows the application of this principle to the infra-red interferometer. A piezo-electric ceramic in one piece with the measurement mirror assures at the same time the modulation and the compensation of the path difference. An oscillator controls the measurement mirror and serves as reference to a synchronous detector receiving the measurement signal output by the detector, which analyses the interferometric field in one direction. An integrator connected to the output of the synchronous amplifier controls an H.V. power supply controlling the ceramic. This integrator voltage is recorded; it relates the variations of the phase to the phenomena of hysterisis and the non-linearities of the ceramics (around 20% in this case).

Nonetheless, while the ceramics constitute an interesting solution for the modulation of the path difference where the amplitude is limited to a few μm, they are not adapted to the problem of the compensation, which they limit to around 1.5 fringes (7.5 m).

We can nonetheless remedy these inconveniences by replacing the modulation of the path difference with that of the wavelength of the source, on condition that the interferometer present a sufficient decompensation.

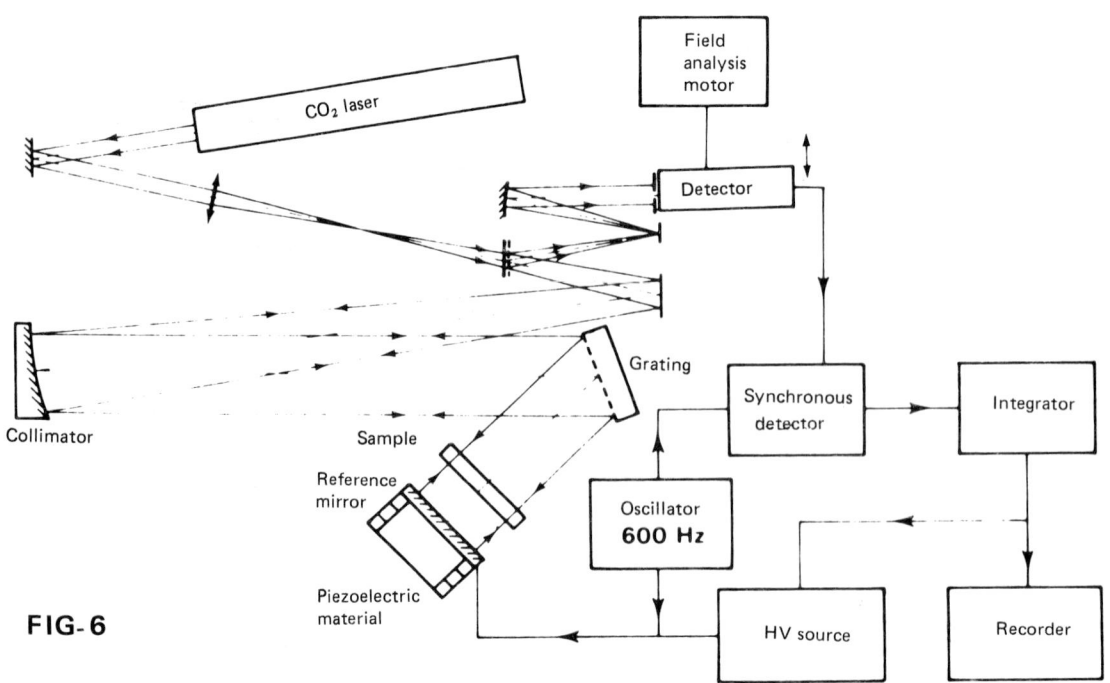

FIG-6

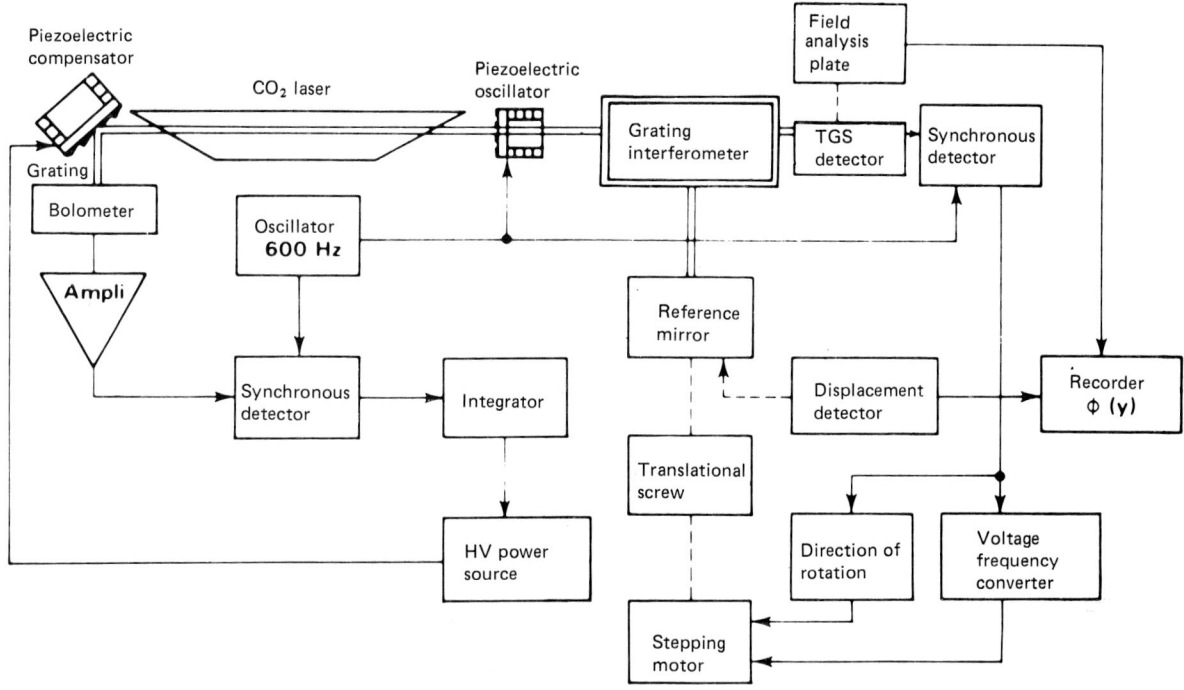

FIG.7 Block diagram: Laser frequency modulation interferometer

Modulation of Laser Frequency

The modulation of the frequency emitted by the CO_2 laser is obtained by periodically varying the length of the resonating cavity; the latter is formed, on one hand, by a diffraction grating (152 lines/mm), Littrow mounted, and on the other hand, by a germanium plate; these two elements are fixed on a piezo-electric element. The ceramic in one piece with the plate serves to slightly modify the frequency emitted by the laser, while the one fixed to the grating assures the compensation of the slow drifts of the cavity. Taking account of the compensation length of the interferometer ($\Delta \simeq 2m$), we see that the modulation of the frequency of the laser creates a modulation of phase corresponding to about 0.13 fringe ($\lambda/15$). The compensator device of the interferometer, which constantly maintains the value of the path difference at a designated value, is composed of a stepping motor driving a screw, controlled by the error signal from the synchronous detector; this screw operates the measurement mirror of the interferometer with an elementary displacement of 1 μm ($\lambda/10$).

The movement of this mirror is directly measured by an inductive feeler (resolution 0.1 μm with ± 1000 points) which delivers an analog signal whose variations represent the variations of path difference. This signal is sent to one of the axes of an XY recorder whose other axis is connected to a linear potentiometer reproducing the translation movement of the TGS detector analysing the interferometric field (figure 7).

Some Results

<u>Twyman-Green Setup</u> The optical system examined is an aspherical meniscus ($f \simeq 75$ mm - f/1) in IRTRAN II. The associated spherical caliber is mounted on the compensation plate. Figure 8 relates to an examination on the axis, in the vicinity of the best focus plane, with a lateral displacement of the sphere of reference (corner). The wave surface is examined along a diameter (48 mm) which is essentially perpendicular to the direction of the fringes. The wave is divergent at the center and convergent at the periphery. We note a deformation on the order of 9 μm resulting from a residual aberration.

<u>Michelson Setup, Application to Mechanics</u> Figure 9 corresponds to the examination of a roughly straightened steel plate (rugosity $R_a = 1.2$ μm); this plate, mounted on the compensation plate, serves directly as a "measurement mirror". The profile of the plate was examined along three inclinations, and we observe a concavity whose depth is on the order of 5 μm over 86 mm.

In conclusion, "phase" interferometry by control of the path difference presents interesting advantages, but important technological problems have yet to be solved before we attain an operational stage: the compensator device should have a precision to 1 μm; no element of important demensions, and in particular the piece to be examined, must be displaced; the recording time of a profile must be reduced (10 min. for

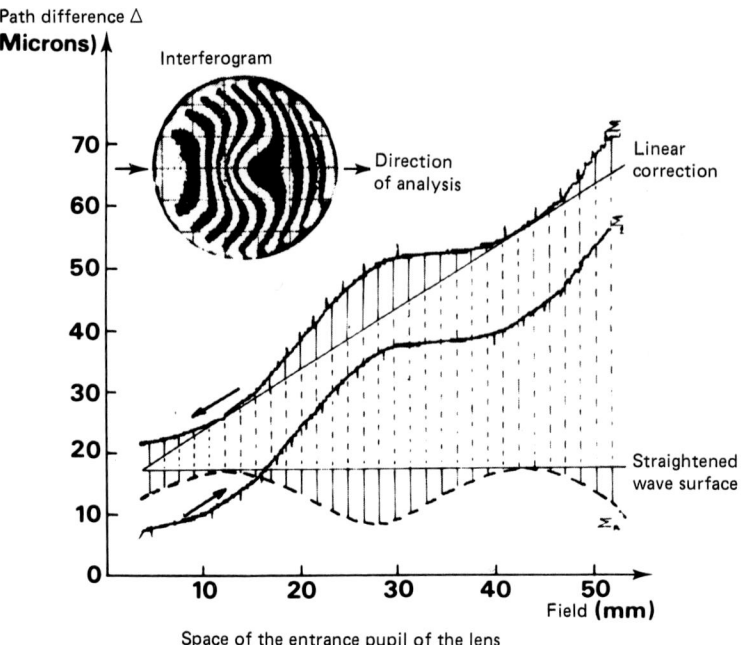

FIG. 8 — Twyman-Green setup. Aspherical lens
($IRTRAN - f \simeq 75$ mm), $excentrement \simeq 3'$;
focus plane point: x'_0.

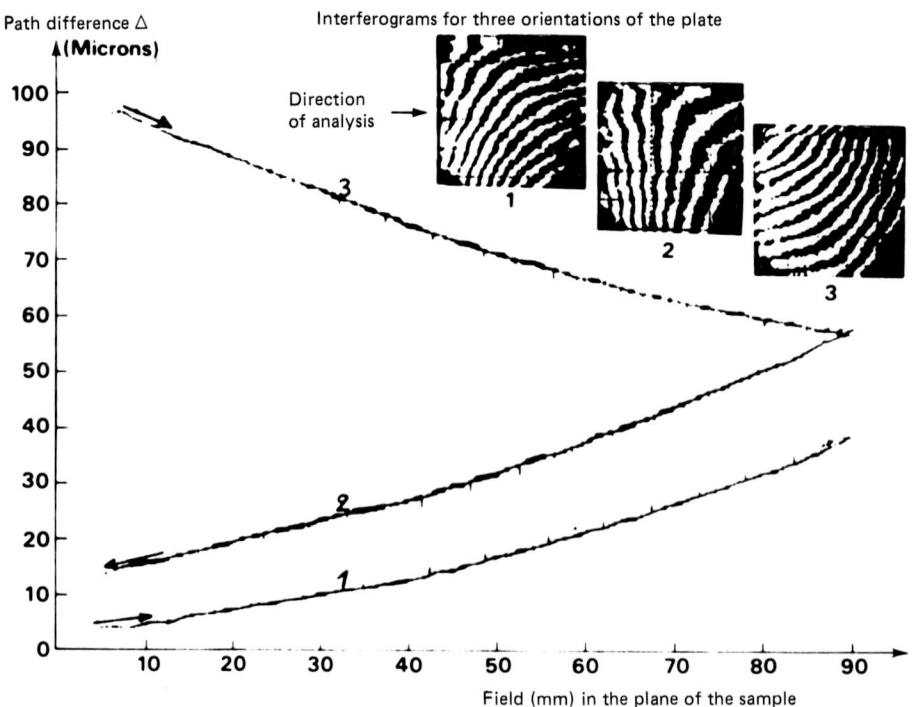

FIG. 9 Application to mechanics — profiles of a steel plate
(rough rectification $Ra \simeq 1.2$ µm) (Michelson setup).

150 mm presently).

Further, despite its specific advantages, this method will not completely replace the conventional global method of acquisition of the field of fringes (5 sec. for 150 mm), to which it is complementary.

* * *

We wish to thank the D.R.M.E. and the D.G.R.S.T. for their support in these studies.

References

1. Charlot, A., Corno, J., Lamare, M., Simon, J., Interféromètre infrarouge à laser, Mesures régulation automatismes, Jan., Feb. 1975.
2. Petit, R., t. 2, n° 2, pp. 115-120, N.R.O.A. 1971.
3. Charlot, A., Corno, J., Lamare, M., Simon, J., Une application des propriétés polarisantes des réseaux a l'interférométrie dans la région de 10 microns, t. 6, n° 2, pp. 97-101, N.R.O.A.
4. Corno, J., Lamare, M., Simon, J., Une expérience d'interférométrie a restitution linéaire de phase a 10 μm vol. 8, 1977, J. Optics.

DISPLACEMENT AND VIBRATION MEASUREMENT BY LASER INTERFEROMETRY

L. E. Drain, J. H. Speake and B. C. Moss
Materials Physics Division, AERE, Harwell
Didcot, Oxon, OX11 ORA, England

Abstract

A description is given of an interferometer capable of non-contact measurements of the displacement and velocity of solid surfaces. The interferometer may be used with any C.W. laser in the visible region and no treatment of the surface is usually necessary. To overcome the problem of directional ambiguity, the reference beam is frequency shifted electro-optically using a Kerr cell device. The output is a frequency modulated signal that may be processed to obtain instantaneous velocity by a ratemeter or more sophisticated frequency recording device. Displacement measurements may be made by counting cycles and continuously subtracting counts derived from the shift frequency. For instantaneous movement a precision of one count or a quarter of a wavelength is obtained but mean displacements may be registered with greater accuracy. In another mode of operation, the output of the interferometer is fed back to control the frequency shift to "phase-lock" the interferometer. This enables very small movements at high frequencies to be recorded in the presence of the much larger low frequency vibrations normally encountered in the laboratory. This technique has been successfully used in the calibration of ultrasonic transducers.

Introduction

Optical interferometry is a very useful technique in the study of the displacement and vibration of objects to which mechanical probes cannot be attached. The use of laser light sources makes the application of the technique very practical. The main advantage over other techniques is that no contact with the object is required and thus the dynamics of the motion are not affected. Generally no preparation of the surface is required. Measurements may be made remotely and very small displacements recorded. Calibration may be easily done by reference to the wavelength of the light.

Velocity is related to the frequency of the interference signal and displacements may be obtained by counting cycles. However for the instrument to be of practical value, some means of discriminating direction of motion is essential. In the instrument to be described, this is achieved by an electro-optic device that shifts the frequency of the reference beam by a known amount. A stationary object thus gives rise to an output signal at the shift frequency. A moving object gives a signal of higher or lower frequency depending on the direction of motion.

The Optical System

The arrangement of the interferometer system is shown in figure 1.

The Frequency Shifter

Polarized light from a laser (most C.W. types are suitable) is first passed through an electro-optic frequency shifting device of a type previously described[1]. As a high conversion efficiency is not required, the simplest design is adequate for this application. It consists of a suitably oriented quarter-wave plate to convert the light to circular polarization and a sequence of two electro-optic cells inclined at 45° to each other, driven by voltages at the required shift frequency but differing in phase by 90°.

Liquid Kerr electronic cells have proved reliable and have good extinction properties. The required combination of cells can be conveniently constructed in a single envelope. Only about 100 volts R.M.S. are required at the shift frequency but a D.C. bias of 7-10kV is necessary. This would not be needed with solid state Pockels cells.

Most of the light passing through the cell retains its original polarization and frequency but the electro-optic effect in the cell generates a component of the reverse circular polarization which is also shifted in frequency. This component is conveniently about 5% of the total. The circular polarizations are reconverted to plane polarizations by a second quarter-wave plate and separated by the polarizing beam splitter which forms part of the interferometer, the frequency shifted component being selected as the reference beam.

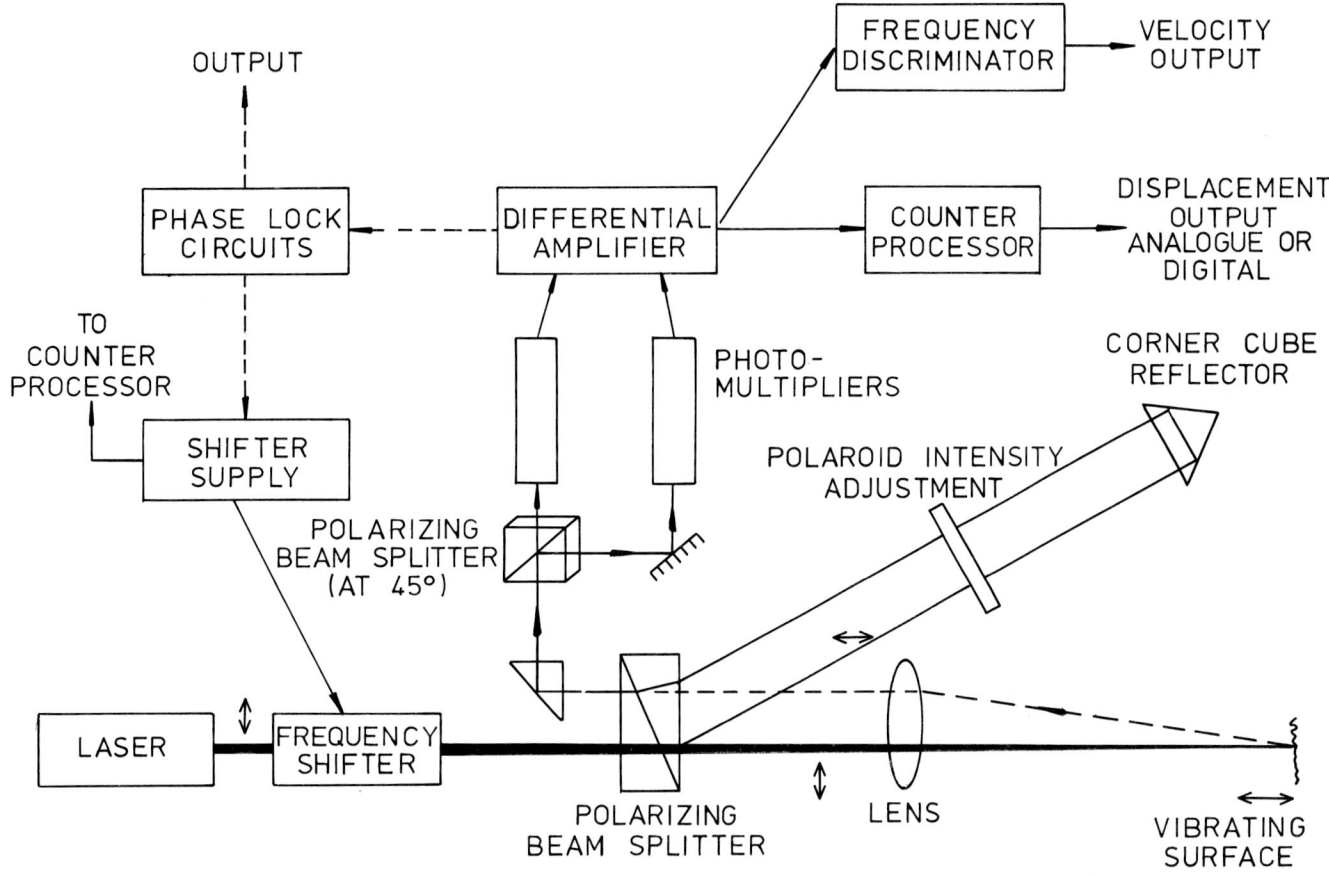

Fig. 1. The interferometer system

The Interferometer

The reference beam is returned to the beam splitter on a parallel but slightly displaced path by means of a corner cube whilst the main beam is focussed onto the object. Scattered light is returned and recombines with the reference beam in the beam splitter and is deflected to the photo-detectors. The intensity of the reference beam can be controlled by a rotatable polaroid plate in the reference beam path. An intensity 5 or 10 times that of the scattered light is generally found to give the best results.

It will be noted that the scattered light and the reference beam, when recombined, have orthogonal polarizations and thus will not interfere to give intensity fluctuations unless further polarizing components are used. This situation may be used to advantage by having two detectors in a balanced arrangement. After passing through an aperture limiting hole, the recombined beams are split by a polarizing beam splitter whose axes are at 45° to the polarization axes of the scattered light and reference beam components. The effect of this is to provide two interference channels whose phases differ by 180°. The output signal is obtained by subtracting the outputs of the two photomultipliers in a difference amplifier. This system is necessary for the phase-locked mode of operation of the interferometer and was originally introduced for this reason, but it has the general advantage that the effects of intensity fluctuations in the intensities of the beams are minimized and much cleaner output signals are obtained.

Coherence Considerations

The spatial coherence of the reference beam and scattered light is better if the illuminated spot on the object is as small as possible. Best signals are therefore obtained when the surface of the object is at or near the focus of the lens. In practice there is a range of a few centimetres where good signals can be obtained. Focussing lenses of focal length 100-300mm are commonly used giving, with a 1mm diameter laser beam, focal spot diameters of 60-200μm.

For long distance work, beam expansion is desirable to obtain the small focal spot required. Distance itself does not reduce the signal to noise ratio obtainable, so long as the illuminated spot can be made adequately small by telescopic optics, but of course atmospheric turbulence, absorption and scattering will reduce the signal quality.

If retroreflecting tape or paint can be applied to the surface, signals are greatly increased and the focussing condition is not so important. With corner cube reflectors, parallel outgoing and return beams are used. No lens is required except for beam expansion to reduce the effect of diffraction in long distance work.

The reference path length is adjustable so that the reference and scattered beam path lengths may be equalized to obtain good interference. However, the coherence lengths of laser commonly used are several centimetres so that adjustment is not critical. Alternatively, the difference between the optical paths may be made close to an integral multiple of twice the laser cavity length. This also ensures coherent interference with a multimode laser and enables the reference beam path to be reasonably short in all circumstances. If continuous interference over a long distance is required, in metrology for example, it is necessary for the laser to operate in a single (longitudinal) mode but to obtain this condition, some light power has to be sacrificed and the stabilization of the laser is an additional complication.

Signal Processing

Two basically different methods of operating the interferometer are currently used. The first is applicable to comparatively large, low-frequency vibration and uses a counting technique. The limit of resolution (in vibration studies) is one count, which can correspond to $\lambda/4$ or about 0.16µm. The second method is appropriate for high frequency (greater than about 10kHz) vibrations of small amplitude.

Counter Processing

In the first method of processing, a constant shift frequency is applied to the reference beam. The shift should be higher that the doppler shift corresponding to the maximum velocity of the surface. This ensures that the beat signal from the interferometer does not fall to zero at any time. A shift frequency of 5MHz is our standard and with a helium-neon laser source, this allows a maximum velocity of 1.5 metres/sec.

Velocity may be measured from the output of the photodetectors by frequency discrimination by a ratemeter or a frequency tracker as used in laser doppler flow measurements. The shift frequency corresponds to zero velocity.

Displacement may be measured by counting the cycles of the output of the photomultipliers and the cycles of the shift frequency and subtracting. The subtraction is done virtually continuously and the difference converted to analogue form for oscilloscope display, spectrum analysis or other type of processing required. As the counting is usually based on zero crossings of the signal, there are two counts per cycle and thus one count corresponds to a quarter of a wavelength displacement approximately 0.16µm for a wavelength of 0.633µm. This is the limit of resolution in the fast tracking of vibrations.

If average displacement are required, higher precision is possible. The presence of the shift enables interpolation to be made to fractions of a count. This is done by averaging the difference between the signal counts and shift counts over a large number of samples over an interval of a second or two. This gives a true time average provided the sampling frequency is not a submultiple of the shift frequency. The computation can be performed by a microprocessor which can also convert to millimetres or inches for a direct read-out. A resolution of 10 to 20 nanometres may be expected.

Optical Phase Lock Operation

For extremely small displacements at high frequencies as are present on the surface of an ultrasonic transducer, another mode of operation is used. This uses the interferometer close to its balanced condition where it is most sensitive. The difference output of the detectors is given by:

$$i_D = A \sin(\frac{4\pi x}{\lambda}) \qquad (1)$$

where x is the distance moved by the surface from a position giving equal output from the detectors. The parameter A is a function of the scattered light intensity. For values of x, small compared with the wavelength, the relation to the detector output may be taken as linear.

DISPLACEMENT AND VIBRATION MEASUREMENT BY LASER INTERFEROMETRY

To maintain the interferometer in the balance condition requires fine adjustment of the path length of the reference beam. This is a very delicate adjustment to do mechanically and unless anti-vibration mountings are used, the adjustment would be continuously upset by low frequency laboratory vibrations. This problem may be overcome by the use of the frequency shifter to control the effective path length of the reference beam automatically. This is achieved by feeding-back the output of the balanced detectors to control the frequency of shift. This is shown by the dashed lines in figure 1. One of the two possible balance points of the interferometer is stable in the sense that small displacements of the object (or other optical components) produce a change in shift frequency which changes the phase of the reference beam in the direction to bring the interferometer back in balance. Thus the interferometer system may be made to "lock" in this condition.

The frequency shift applied in the locked condition is very low, reproducing the laboratory vibrations and has both positive and negative values. The power amplifiers driving the Kerr cell must therefore be D.C. coupled. To obtain the required drive voltage (200V peak to peak), the circuits employed high voltage transistors but at these low frequencies the power requirement is small. The drive frequency is obtained by mixing the output of a variable capacity oscillator operating around 10.3MHz and controlled by the feedback signal, with fixed R.F. oscillators. The mixing system was designed to minimize intermodulation products in the required pass-band.

The feed-back is naturally more effective at low frequencies as the phase control is derived by integration of a frequency change. Thus with moderate gain in the feed-back circuit, low frequency vibrations are compensated but high frequencies are not much affected. In the low frequency range, the differential output of the detectors is proportional to velocity but at high frequencies, it is proportional to the displacement of the surface. The transition frequency should be above the range of laboratory vibrations but substantially below the range where quantitative displacement measurements are required.

The sensitivity does of course depend on the intensity of the scattered light reaching the detector. With a diffusing surface, there is a random variation as the illuminated spot moves across the surface. The sensitivity in terms of output per unit displacement of the surface thus varies and a calibration must be performed for each spot on the surface for which quantitative measurements are required. Fortunately this is not difficult. It is only necessary to cut the feed-back loop so that the interferometer runs out of lock. The peak amplitude of the sinuosiodal variations of the output then immediately gives the constant, A, in equation (1) from which the calibration constant can be calculated. The possibility exists of an automatic system for adjusting the gain of the amplifier to keep the sensitivity constant. This would be necessary if automatic scanning of the surface were envisaged. It is planned to develop this.

Calibration of Ultrasonic Transducers

An important application of the interferometer system has been to the investigation and calibration of a series of piezoelectric ultrasonic transducers. Typical ultrasonic displacement amplitudes are of the order of 10 nanometres and below and the interferometer was operated in the phase-lock mode to achieve this degree of sensitivity. This method of measuring the ultrasonic displacement of a surface has enabled an absolute calibration to be made.

A schematic diagram showing the experimental arrangement for exciting the probes and measuring the spectral response of the transducers is shown in figure 2. The transducers are driven by a sinusoidal voltage swept over a frequency range from D.C. to 1MHz. This frequency is obtained by mixing the output of the heterodyning oscillator of a 1L5 spectrum analyzer with an external oscillator. The frequency of this oscillator could be adjusted so that the frequency driving the transducer is always the same as that giving the maximum response of the spectrum analyzer. The output of the laser interferometer was recorded by the spectrum analyzer in this way. For this series of experiments, the feed-back gain of the interferometer phase-lock system was chosen so that the output was proportional to displacement for frequencies over 5kHz.

The diameter of the focal spot of the laser beam on the probe face is only about 200μm and so very localized areas of the face can be examined. This allows very good spatial resolution and the variation in transducer behaviour across its active face can be examined in detail. Figure 3 shows the displacement response at two points on the face of a transducer having a main resonance at 2.5MHz. There is a considerable difference demonstrating that the device is not behaving as a piston.

Displacement amplitudes as small as 0.01 nanometres have been measured by this technique using a 5mW helium-neon laser. Greater sensitivity could be obtained by the use of a higher power laser. The signal to noise ratio could also be improved by the use of phase-sensitive lock-in detection. By this means, the effective band width can be made very narrow and the phase of the displacement relative to the drive voltage can be registered.

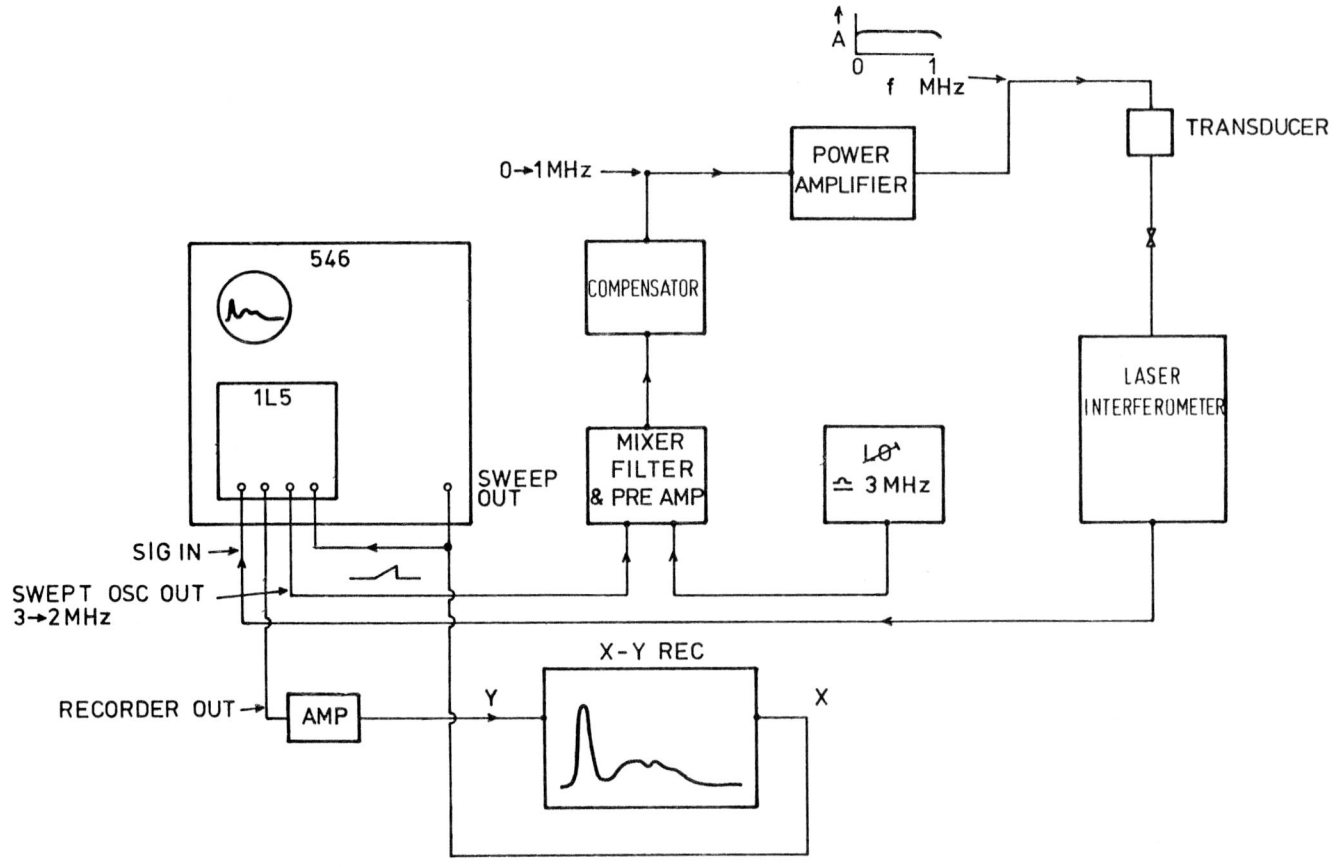

Fig. 2. Circuit arrangements for measuring transducer response characteristics

The pulse response of transducers has also been recorded. It is more difficult to obtain very good signal to noise ratio in pulse work as considerable bandwidth has to be allowed to avoid distorting the waveform, but considerable improvement may be obtained by signal averaging over a large number of identical pulses.

The interferometer is an invaluable tool in monitoring the transducer response, i.e. whether it is an analogue of the displacement, velocity or acceleration of the front face. This information may be obtained by measuring the phase difference between the measured displacement and the electrical drive. A programme of research is currently being undertaken to measure the absolute calibration of ultrasonic transducers in their working environment.

Conclusions

The Harwell laser interferometer is a novel optical instrument capable of making precise measurements on the position or movements of engineering components and structures of all sorts. It is able to measure vibration characteristics remotely with no need for reflectors on the structure of interest. Measurements can be made at distances of 60 metres to detect vibration amplitudes as small as nanometres. With a stabilised laser the same instrument could be used to monitor the position change of an object moving over a distance of about 60 metres. The range of frequencies which can conveniently be measured extend from D.C. to 20MHz. The large dynamic range of distance and frequency which the instrument can cover, together with the interest accuracy of 1 in 10^6 using a stabilised laser, make the instrument a very useful measuring tool for diverse application. Some of the users to which the interferometer have been put are: (i) vibration measurement in the nuclear industry; (ii) metrology; (iii) civil engineering applications such as monitoring building vibration; (iv) turbine blade monitoring; (v) loudspeaker studies; (vi) ultrasonic transducer studies. It is fully expected that the use of the device will continue to grow.

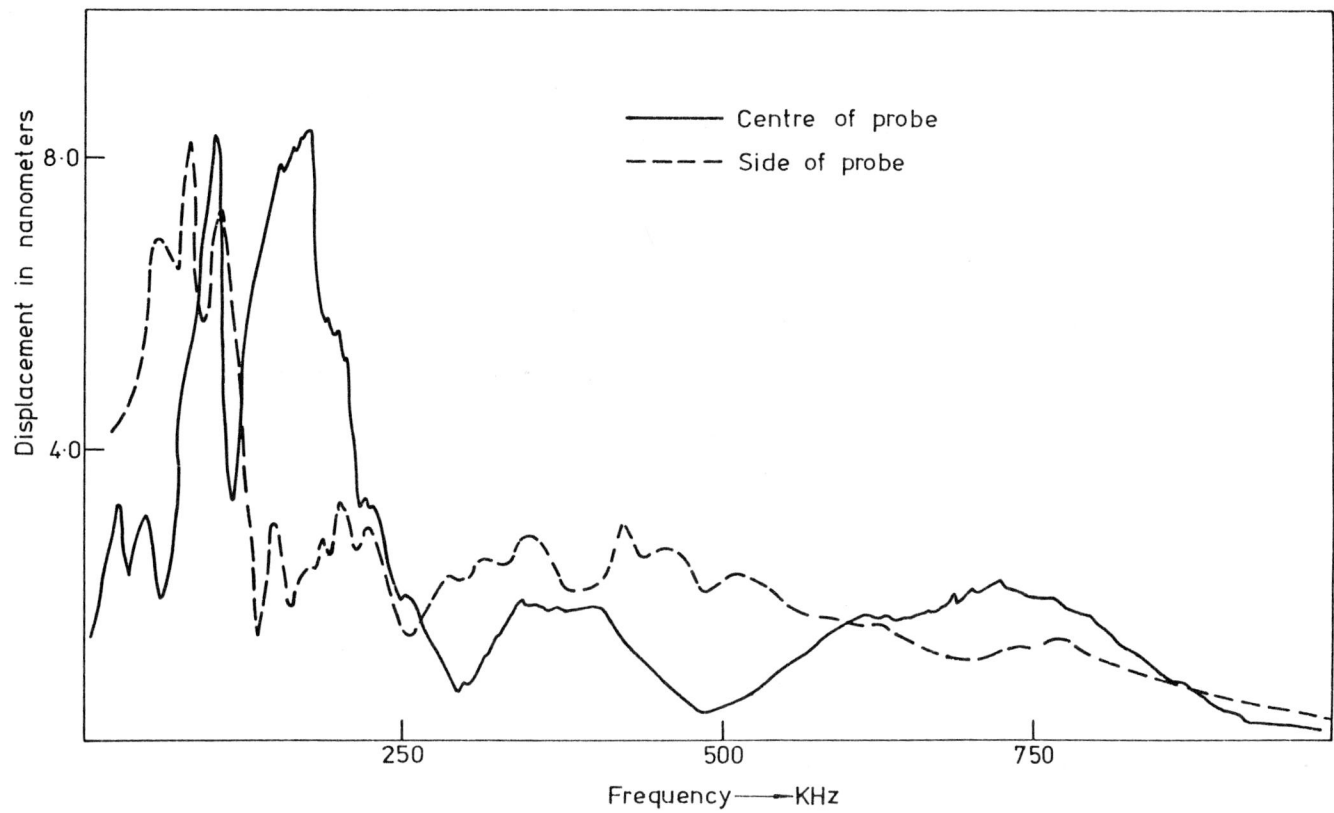

Fig. 3. Displacement of ultrasonic waves at two points on a transducer

References

1. Drain, L.E. and Moss, B.C., Opto-electronics, 4, pp. 429-439, 1972.

APPLICATIONS OF PHASE MODULATION INTERFEROMETRY TO THE CHARACTERIZATION OF MATERIALS AND TO DIMENSIONAL METROLOGY

G. Roblin

Institute of Optics
BP 43 - 91406 Orsay Cedex, France

Abstract

The characterization of materials requires the knowledge of the refractive index and its variations under the effect of a given parameter. Whether we are dealing with transmission by a transparent substance or reflection on a metal surface, the measurement of the phase difference introduced between the entrance and exit waves permits us to discover this characteristic of the material.

In dimensional metrology, a phase difference at transmission is as much a function of the index as of the thickness, and the metrology of length standards requires the knowledge of the loss of phase at reflection. Thus the lengths considered as possible optical paths, or even the rotations, can also be expressed as a phase difference.

These phase differences, demonstrated by normal or differential interferometry, can be measured with a gain in time and sensitivity if this interferometry is done by phase modulation. The very small uncertainties permitted by this process make it particularly interesting to apply to measurements of small phase differences expressing small variations in index or thickness. Its possibilities will be illustrated by examples of characterization of photosensitive materials, determination of loss of phase at reflection, longitudinal localization of objects, dilatometry and torsiometry.

This process also permits us to obtain simply the rate of the variations of complex amplitude of a phase object as a function of space or time.

Introduction

The measurement of phase differences as magnitudes derived from phenomena of all kinds is probably the most important field of optical metrology. In what follows we will present a retrospective of diverse results obtained in this field, in the characterization of materials as well as in dimensional metrology. These two aspects of the problem, for that matter, are not always independent. In fact, if we consider for example the case of a dielectric material, the phase variations which appear at transmission are due to the variations in the optical path followed; that is, to variations of thickness as well as variations of index. It is then necessary to effect two measurements to determine these two unknowns, either by transmission, in immersing the object successively in two media of different rfractive indices, or one by transmission and the other by reflection, notably to characterize by these two magnitudes a partially deposited layer on a material. Thus it is possible to characterize any physical phenomenon whose effect is to vary one or the other of these two magnitudes, or even both. In another field, that of metallic materials, the phase difference which they introduce at reflection, either in normal incidence with regard to a dielectric or other metal, or in oblique incidence between two perpendicularly polarized waves, permits their characterization or the determination of their optical constants and their variations under the effect of any phenomenon.

After giving these few examples we will recall the methods we have experimented with permitting us to demonstrate the variations of phase and to measure them.

Methods

The general problem at the experimental level thus requires the previous exposition of the variations of phase due to some phenomenon. These variations, linked to those of path difference at transmission or reflection, are detectable only by interferometry (figure 1). The wave transmitted or reflected by the object is compared to a reference wave, the phase variations φ being transformed into variations of lighting $\sin^2 \varphi/2$ in an interferogram. The metrology effected is thus in fact a photometry, that is, a comparison of signals. The measurement can be made by compensation, the lighting at the point considered

$$S = \sin^2 \frac{\varphi - \varphi_K}{2} \qquad (1)$$

being reduced to a given value, zero for example, the resulting phase

$$\phi = \varphi - \varphi_K \qquad (2)$$

being then also zero, this choice corresponding to the stationarity of the signal as a function of the compensation phase:

$$\frac{dS}{d\varphi_K} = -\frac{1}{2} \sin(\varphi - \varphi_K) = 0 \qquad (3)$$

APPLICATIONS OF PHASE MODULATION INTERFEROMETRY TO THE CHARACTERIZATION OF MATERIALS AND TO DIMENSIONAL METROLOGY

This simple, classic method, as it has been laid out here, presents all the inconveniences of zero methods in continuous current, and it is preferable to use methods employing a modulated signal, whose easier treatment permits a gain in time and sensitivity. This choice is equivalent in effect to a large number of measurements per unit of time with a notable increase in the signal/noise ratio [1].

To obtain a signal of this nature, that is, to modulate the lighting of the interferogram, the best procedure consists in modulating the phase to be measured, by adding to it a phase which is periodic in time. This can be obtained (figure 2) by the alternate displacement of one of the two mirrors of a Michelson interferometer, for example. If this added phase is of the form

$$\varphi_m = \varphi_0 \sin \omega t \tag{4}$$

the exit signal of the photoreceptor

$$S = \sin^2 \frac{\varphi - \varphi_K + \varphi_m}{2} \tag{5}$$

is modulated in time, and by synchronous detection at the frequency of modulation, we obtain a continuous signal

$$J_1(\varphi_0) \sin(\varphi - \varphi_K) \tag{6}$$

which, for the measurement, it suffices to cancel by the action of the compensator.

In reality, we did not use this type of normal interferometer, but rather lateral division, polarization interferometers of the Wollaston or Jamin type. In the first model (figure 3), the two de-phased and perpendicularly polarized waves provide an elliptical wave retransformed by a quarter-wave plate Q into a rectilinear wave, whose orientation, equivalent to the phase, is modulated by a vibrating half-wave plate D. In the second type (figure 4), which can moreover be associated to the preceeding one, where it advantageously replaces the half-wave, the interferometer is capable at the same time of producing the division of the wave and modulating the phase by modulation of the inclination of one of the two elements. Since this second device authorizes a larger phase modulation amplitude, it is more favorably employed, the optimal modulation amplitude being a quarter wave [2].

Applications

Such methods have been applied to the measurement of phase differences in microscopy, notably in order to characterize materials by their refractive index and their thickness, or the variations of these quantities under the influence of some phenomenon, for example the action of light on a photosensitive material. This type of material layered on a support constitutes a phase object whose phase varies under the action of light. Measurements are made on this type of object with the aid of a lateral division interferometer W (figure 5) associated with a phase modulator J. The exit signal of the synchronous detector can be used to control the rotation of the compensation polarizer. The latter is integral with an angle coder Cd. The rotation of the analyser, equal to half the phase, can be recorded in E as a function of the position of the object A.

With this instrument we have been able to detect, on different photosensitive materials, variations of index due to the illumination on the order of several units of the 3rd decimal with a precision of several units of the 5th decimal for layers around ten microns thick. These variations of index, which are situated at around 5 units of the 3rd decimal, are most often accompanied by variations of thickness on the order of a tenth of a micron, plus or minus a few angstroms.

This type of instrument can also be used for reflecting materials presenting a loss of phase at reflection, which is an important phenomenon, notably for the measurement of end gauges. We should note, however, that lighting by reflection requires the use of a vertical light source most often provided with a semi-transparent plate whose factors of transmission in amplitude are different depending on whether the wave is polarized perpendicularly or parallel to its incidence plane (3). Keeping in mind the relative orientation of the various polarizing elements of the instrument, this has the effect of provoking in the measurement a relative error which may reach considerable proportions, above all for the very small phase differences. Figure 6 illustrates this effect. We see here the measurement of the difference between the phase shifts at reflection of chromium and aluminum deposited on a glass slide. To make this measurement, the instrument was provided with a separator composed of a layer of titanium oxide deposited on borosilicate. The entrance polarizer was oriented parallel or perpendicularly to its incidence plane. The two results obtained differ, by more or by less, from the measurement made with the instrument provided with a mirror covering only its half-pupil, which eliminates this parasitical effect.

These methods of phase modulation interferometry have also been applied to interferential dilatometry of magnetic samples submitted to directed fields in order to determine the anomalies of dilatation and their anisotropy, and to accede to the thermic variation of the magnetostriction. This study was made or the magnetism laboratory at Grenoble.

We used an additive polarization interferometer permitting us to compare standards of end gauges proposed by K. Dorenwendt and G. Nomarski [4]. This instrument [5], besides its property of additivity of the phase differences, possesses a certain number of interesting degrees of immunity. It functions in partially coherent light, a Wollaston prism being conjugated to the one that plays the role of interferometer. The extremities of the gauge blocks and the sample are conjugated. The conjugation of the extremities of the gauge blocks is conserved whatever the position of the ensemble of the two blocks. The conservation of the optical path is not affected by the displacement of only one of the two blocks parallel to the optical axis. The direction of the normal to the wave surface at the exit of the instrument is not affected by a movement of the ensemble of the two blocks. The small lateral displacement of the interfering waves has only a negligeable effect on the variations of index along the path of the light. For the measurement, the two types of modulator previously described were used. The residual noise due solely to the elements of the electro-optical system corresponds to a tenth (half-wave plate) or a hundredth (Jamin modulator) of an angstrom, which permits us to attain dilatation coefficients on the order of 10^{-8}.

Other problems in the field of dimensional metrology can find a solution in this use of phase modulation interferometry linked to an appropriate treatment of signals, notably in mechanics. Thus the study of a longitudinal optical feeler [6] permitting us to locate the position of a surface leads to the conception of a profile recorder or a very sensitive (several angstroms) automatic focusing device. In the same way these methods can be applied to the detection of very small rotations (inferior to a second of arc) of a target around a sighting axis, and thus to torsiometry [7].

To conclude, we will cite one more interesting possibility of this phase modulation interferometry. Up to now we have described only applications to measurement. But in fact some experimenters wish only to be informed qualitatively about the form of the wave surface transmitted or reflected by the object. This phase object can be expressed only by an interferogram whose simple global observation does not permit us to distinguish equal changes of phase, but of contrary signs. It is possible to do this only by proceeding at each point to the compensation, which amounts to effecting the measurement before obtaining the information about the sign of the phase, which increases the access time for this information. In absence of all compensation ($\varphi_k = 0$), we have seen (formula 6) that we obtain by synchronous detection a continuous signal proportional to the sine of the phase, thus of the same sign as the latter. Consequently (figure 7), by associating a method of scanning the interferogram, by flying spot for example [8], to a phase modulation interferometer, we can display this information for the whole of the object. This display can be made either in the form of curves or in the form of images, in the presence of modulation, which gives the information about the sign of the phase, as well as in the absence of modulation, which gives a very different photometric analysis of the interferogram. If we obtain information about the sign of the phase, there eventually remains an ambiguity as to its quadrant. To remove this last indetermination, it suffices to exit the signal on a second path de-phased by 90° (figure 8), which amounts to extracting the two components, real and imaginary, of the complex amplitude, thus to completely reconstruct it.

References

1. Roblin, G., "Electronically generated limitation of the sensitivity of optical measurements using time modulation", Optics and Laser Techn., vol. 9, pp. 161-167, 1977.
2. Allain, M., Roblin, G., "Mise en oeuvre et optimisation d'un modulateur de retard optique", Optics Comm., vol. 11, pp. 196-200, 1974.
3. Roblin, G., "La mesure des déphasages en microscopie par interférométrie à modulation de phase", J. Optics (Paris), vol. 8, pp. 309-318, 1977.
4. Dorenwendt, K, Etude d'un interféromètre additif à polarisation, applications à la comparaison des étalons de longueur à bouts, D.U., Paris, 1968.
5. Roblin, G., Souche, Y., "Interféromètre à polarisation destiné à la dilatométrie différentielle", Nouv. Rev. Optique, vol. 5, pp. 287-293, 1974.
6. Nomarski, G., Roblin, G., "Localisation longitudinale des surfaces par interférométrie à modulation de phase", Nouv. Rev. Optique, vol. 2, pp. 105-113, 1971.
7. Nomarski, G., Roblin, G., Géléoc, P., "Sur une méthode précise de détermination photoélectrique d'une direction de polarisation", C.R. Acad. Sci. (Paris), vol. 269 B, pp. 290-293, 1969.
8. Allain, M., Prevost, M., Roblin, G., "Procédé expérimental de restitution de la partie imaginaire de l'amplitude complexe d'un objet de phase", Optics Comm., vol. 15, pp. 384-388, 1975.

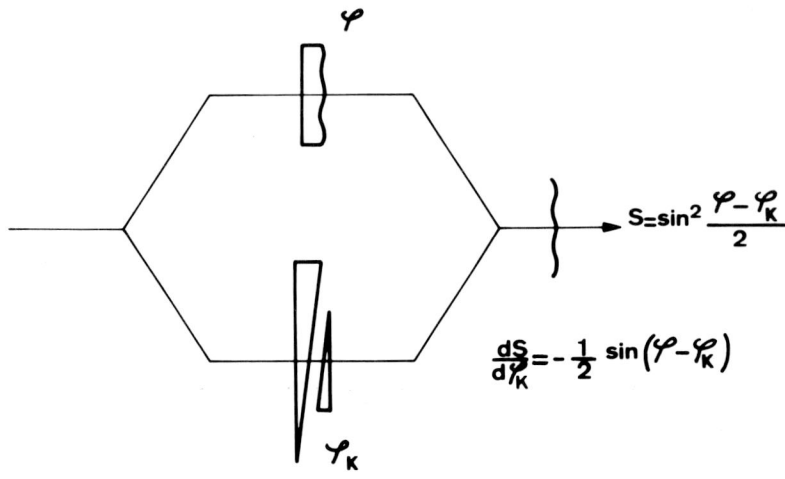

Fig. 1. Principle of interferometry

APPLICATIONS OF PHASE MODULATION INTERFEROMETRY TO THE CHARACTERIZATION OF MATERIALS AND TO DIMENSIONAL METROLOGY

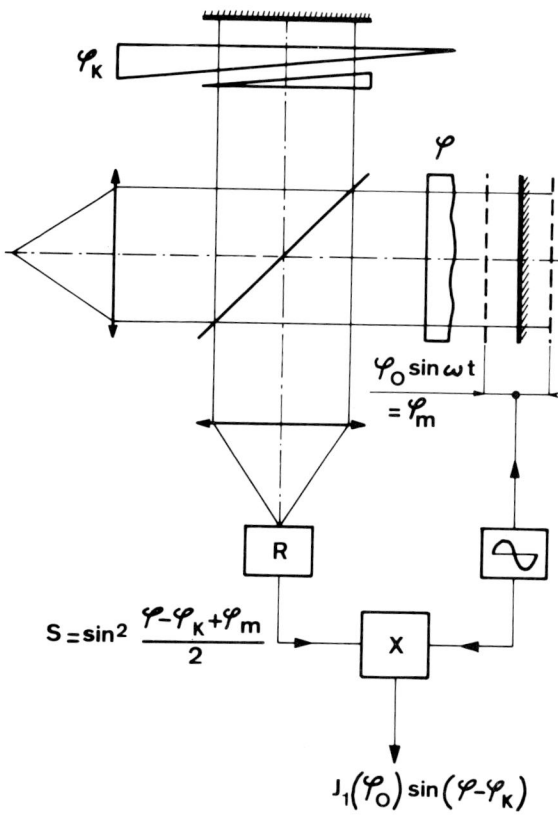

Fig. 2. Principle of phase modulation interferometry

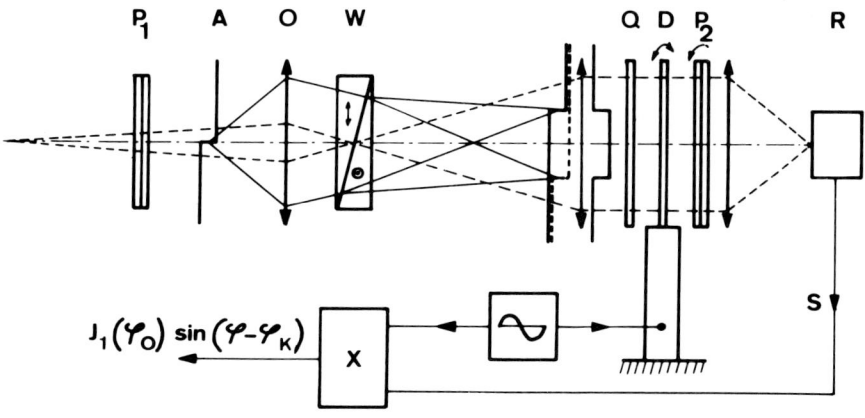

Fig. 3. Wollaston interferometer

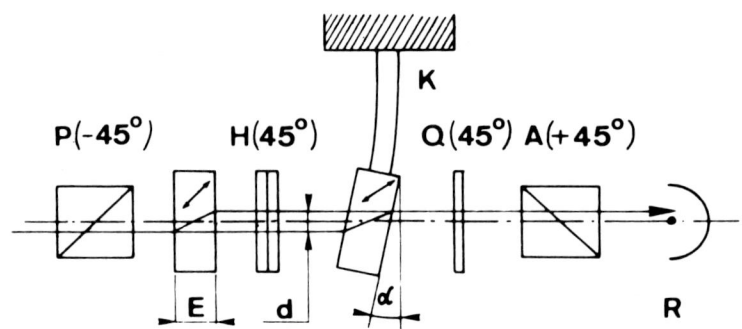

Fig. 4. Jamin interferometer

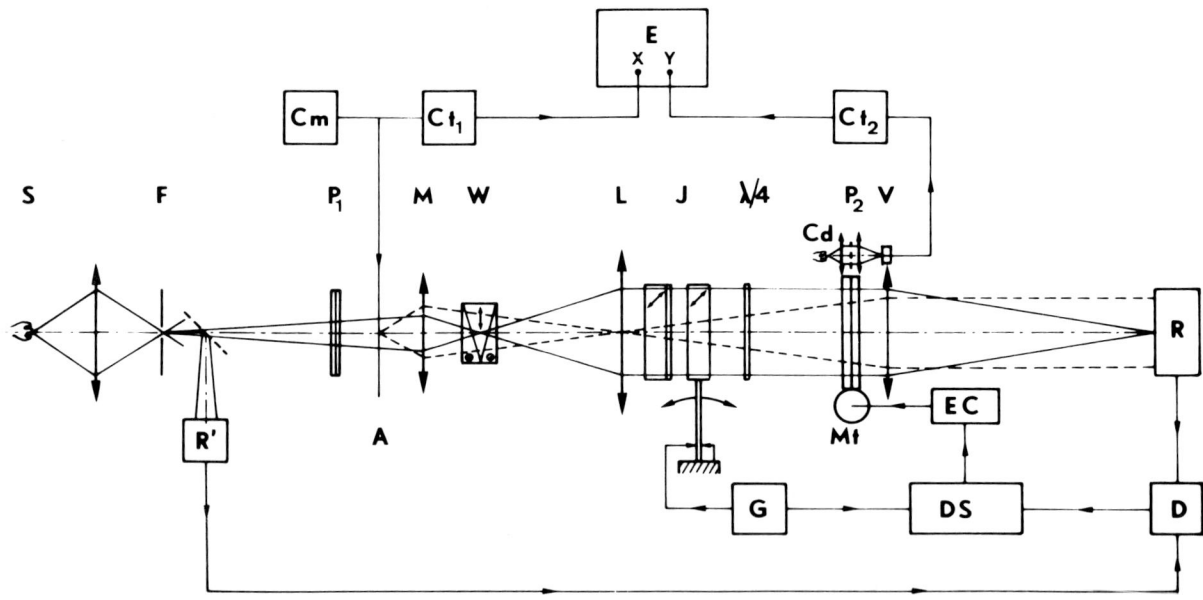

Fig. 5. Bench for measuring phase differences

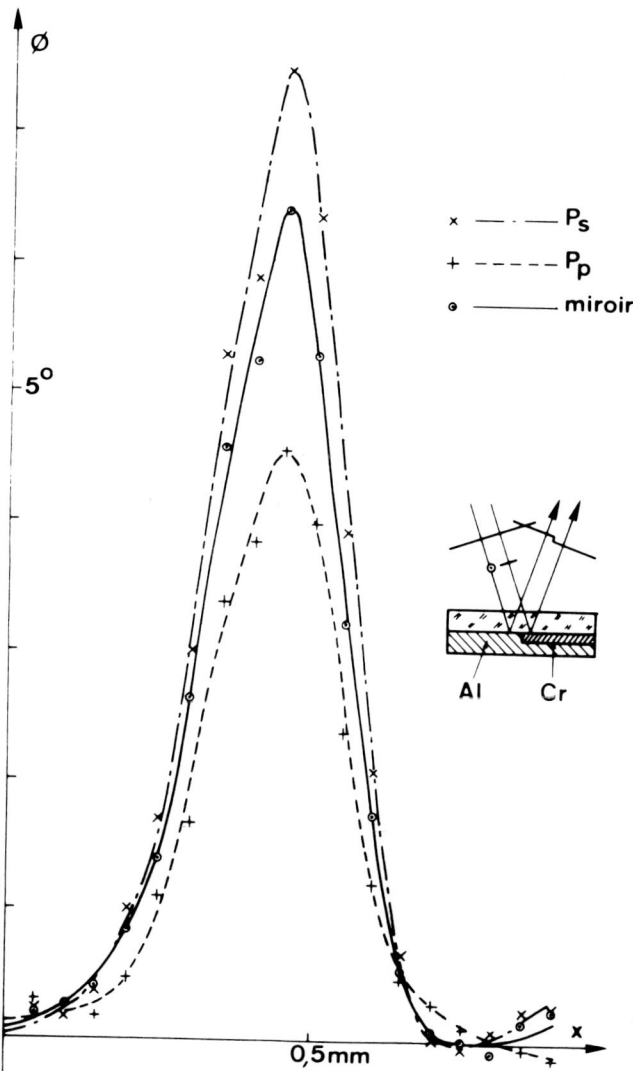

Fig. 6. Error in measurement due to the semi-transparent plate of the light source

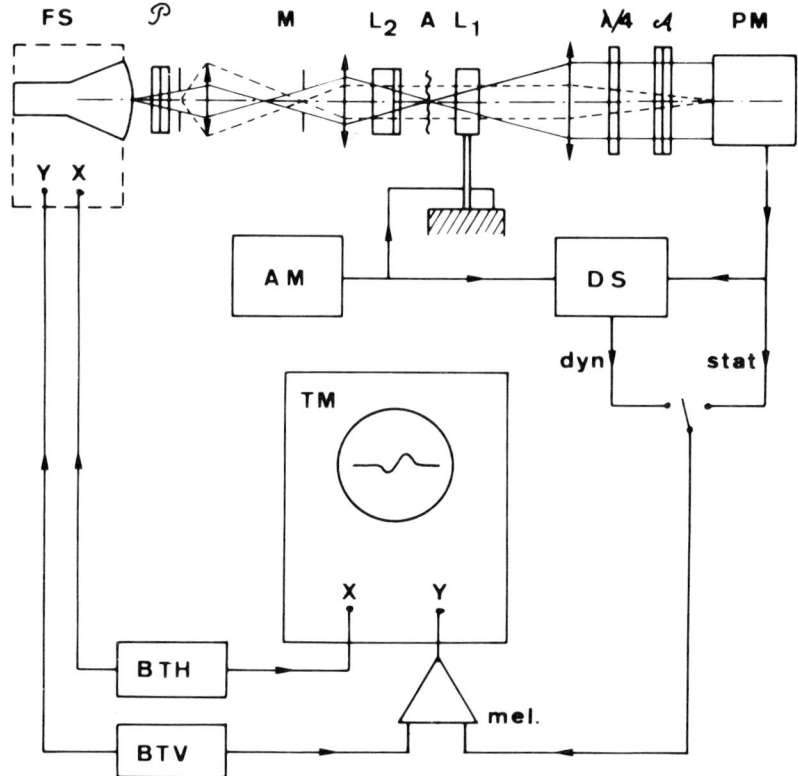

Fig. 7. Diagram of the scanner interferometer

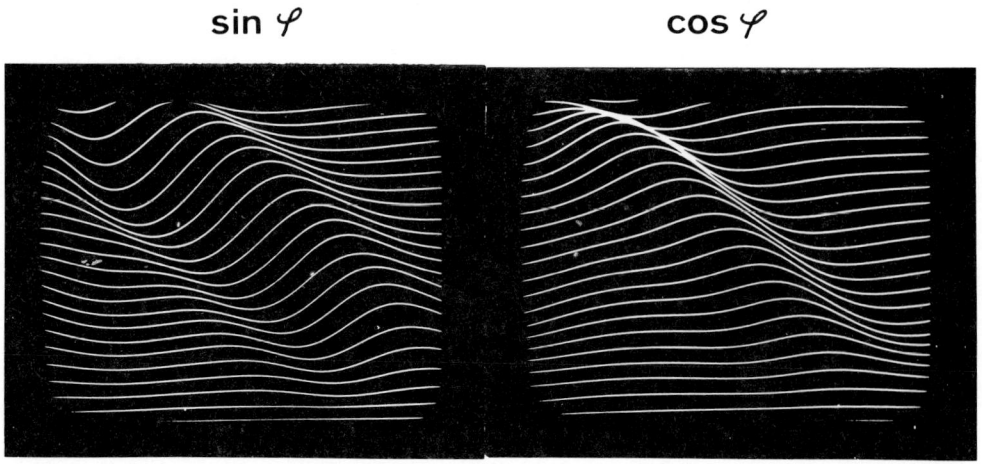

Fig. 8. Reconstruction of the complex amplitude

IMAGE SPECTROGRAMS OF THREE-DIMENSIONAL OBJECTS: METROLOGICAL APPLICATIONS

Alain Lacourt
Laboratory of General Physics and Optics
University of Franche-Comte 25030 Besancon Cedex, France

Abstract

We propose an imagery device presenting a double particularity: (1) one of the axes of the image space is an axis of the temporal frequencies $O\nu$; that is, the image is inscribed in a projection of the spectrum of the white light emanating from the object; (2) the image is anamorphosable at will.

The light radiated by the achromatic object is dispersed by an auxiliary grating, then projected on the entrance slit of a spectroscope. The spectroscope, in again dispersing the chromatic components removed by the slit, reproduces a unique image of the object.

By playing with the position of the auxiliary grating, for example, we arbitrarily modify the geometrical dispersion of the device; it results that the size of the image in one of its dimensions (that corresponding to the $O\nu$ axis) is anamorphosable at will.

Besides the possibility of anamorphosis, the realization of image spectrograms opens interesting possibilities in metrology (absolute measurements of thickness, in particular).

Introduction

The imagery device described here presents three particularities: (1) one of the axes of the image space is an axis of temporal frequencies of the light [1,2]; the image is inscribed in a spectrum of chromatic decomposition of white light; (2) the image is anamorphosable and deformable at will; (3) since the image appears in the form of a juxtaposition of quasi-monochromatic luminous slices, we can produce interferences which are useful in making, for example, measurements of thickness.

Principle of the Production of an Image Spectrogram [3,4]

A sketch of the device is shown in figure 1. The object radiates white light, spatially coherent or incoherent. The objective O_1 conjugates the object plane with the plane π materialized by the entrance slit F of a spectroscope. F is parallel to the lines of the grating R.

Under the dispersive effect of the grating R, we observe in the plane π a continuum of monochromatic images of the object, displaced from each other.

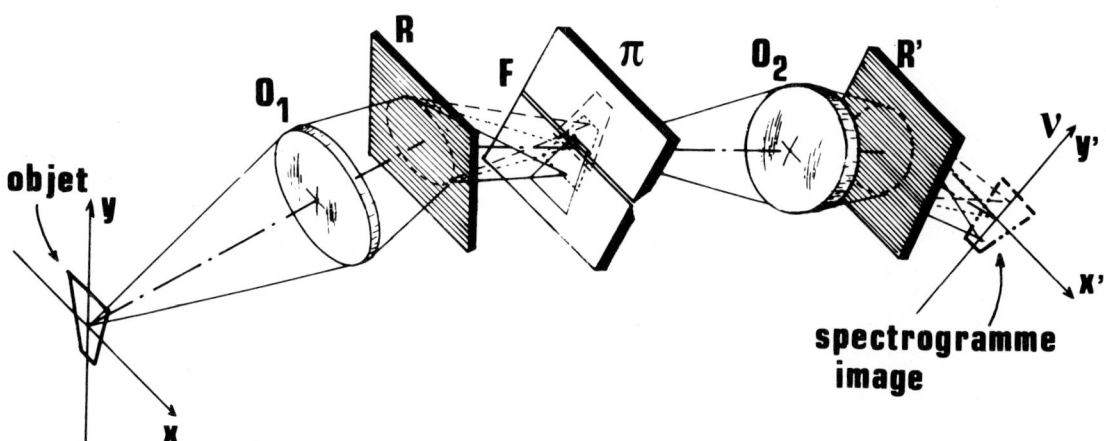

Fig. 1. Recording of an image spectrogram. O_1 and O_2: objectives; R and R': gratings; F: slit

We have shown only three images in the figure, although they are of course infinite in number in practice. The entrance slit of the spectroscope removes from each monochromatic image a slice corresponding, for each wavelength, to a different slice of the object. The spectroscope, in re-dispersing the light which passes through the slit, juxtaposes in its exit plane the different slices removed and consequently provides an image spectrogram of the object (fig. 2a).

Let, in fact, $B(\nu) \, O(x,y)$ be the spectral density of the object light, where $O(x,y)$ is a function of spatial distribution, supposed independent of ν. In the case where the spectroscope also includes a grating R', we have shown that the lighting in its exit plane is expressed by:

$$E(x',y') \propto \frac{1}{x'^2} \, B\left(\frac{cD'N'}{x'}\right) \cdot O\left[x' \, \frac{DN}{D'N'} - x'_0, \, y'\right]$$

where: c is the velocity of light,
N and N' are the spatial frequencies of the gratings R and R',
D represents the distance from the plane π to the grating R,
D' the distance from the plane of the spectrogram to R',
x'_0 is the abscissa of the entrance slit of the spectroscope.

$B\left(\frac{cD'N'}{x'}\right)$ corresponds to the extension of the spectrum of the source. It may be convenient to filter $B(\nu)$ so that

$$\frac{1}{x'^2} \cdot B\left(\frac{cD'N'}{x'}\right) = C^{te} \, .$$

We see that $E(x',y')$ does reproduce an image of $O(x,y)$.

Principle of Anamorphosis (4)

The anamorphosizing property of the device is derived immediately from the expression of the lighting $E(x',y')$. The coefficient DN/D'N' constitutes in effect a scale factor which depends at the same time on the spacing and the position of the gratings intervening in the device. In particular, we easily modify D by a longitudinal translation of the grating R, from which results the effect of anamorphosis. Figure 2 shows a non-anamorphosized image spectrogram of the LOBE trademark in (a); then its contracted (b) and dilated (c) images. Reasoning analogous to that developed above is applied when we substitute any disperser for the gratings R and R'. In particular, we used direct vision prisms which permit "in-line" setups. We can then continuously observe the "film" of the changes of state of the final image spectrogram when we displace one disperser or the other. The effect of the "zoom" anamorphosizer can be very large. In the example of figure 2b, the anamorphosis coefficient is inferior to 1/4; it is above 16 in image 2(c). These values in no way constitute impassable limits.

Deformation of Images by Correlation in Real Time

Up to now, the entrance pupil of the spectroscope has been a rectilinear slit, parallel to the lines of the grating or to the edges of the prism. Now we can show that in general the distribution of light in the exit plane of the spectroscope results from the product of the convolution or correlation* of the object distribution (here $O(x,y)$) and the distribution of transmittance of the entrance pupil of the spectroscope. This property will be discussed in detail in a future publication.

In the present case, we have limited ourselves, by way of example, to modifying the orientation of the entrance slit of the spectroscope by a rotation in its plane. The property of correlation mentioned above is then expressed by a deformation of the image spectrogram of the object. Thus, in figure 3a, the letters of the word LOBE are straightened up from their original angle (figure 2a). A new rotation of the slit leads again to a reversal of this inclination (figure 3b). More generally, if we give the entrance slit a twisted form, we observe a "flag" effect (fig. 3c). Displacing this slit in its plane we see the image spectrogram "wave" in real time.

---

*Depending on the choice of different geometrical parameters of the setup

(a)

(b)

(c)

Fig. 2. Image spectrograms   (a) non-anamorphosized; (b) contracted; (c) dilated.

IMAGE SPECTROGRAMS OF THREE-DIMENSIONAL OBJECTS: METROLOGICAL APPLICATIONS

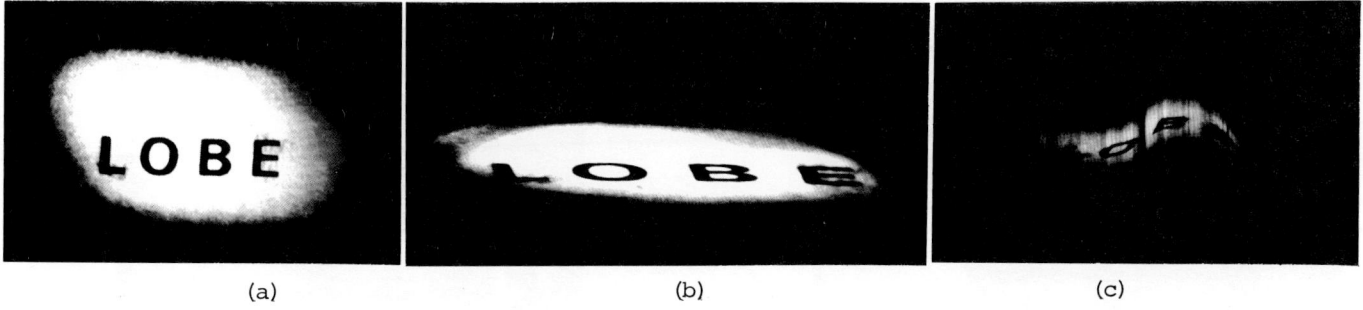

(a) (b) (c)

Fig. 3. Deformed image spectrograms of the LOBE trademark shown in figure 2a.
(a) Straightening of the letters
(b) Reversal of their inclination
(c) "Flag" effect

Application: Measurement of Optical Thicknesses (4)

The setup sketched in figure 4 leads to the production of image spectrograms displaying grooves, which permit measurements of optical thicknesses.

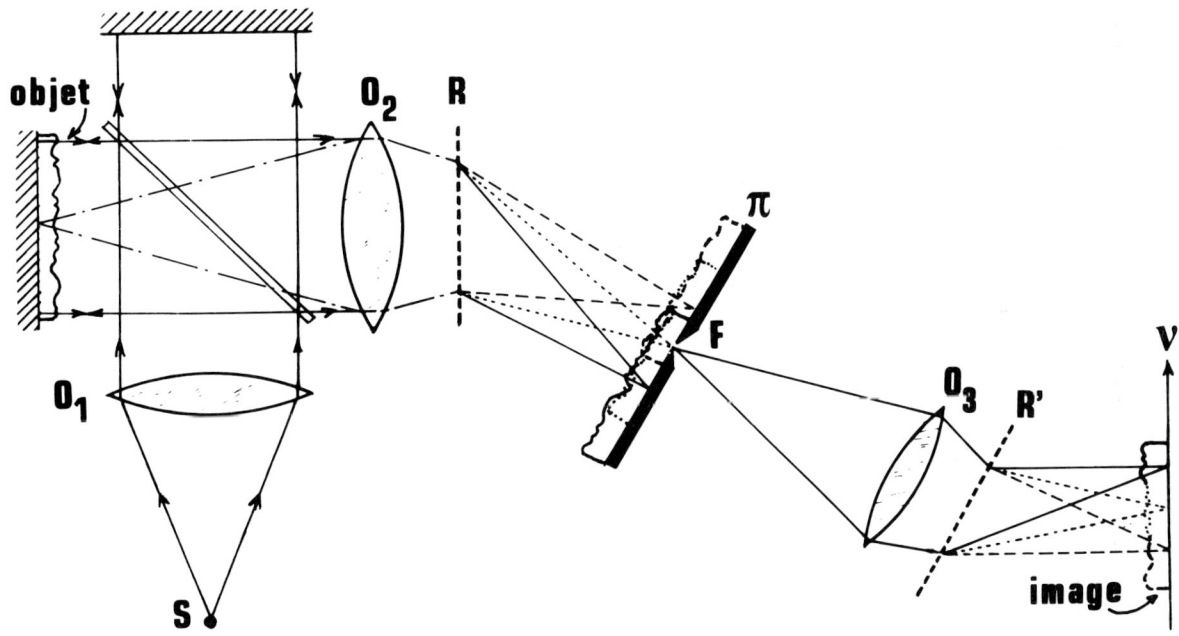

Fig. 4. Measurement of optical thicknesses by means of an image spectrogram.
O_1, O_2, O_3, objectives; R and R', gratings; F, slit.

In this case, the object is a transparent sample of which we form an image spectrogram using the method described above. Furthermore, it is placed near the mirror of a Michelson interferometer, previously adjusted to the optical contact. This time, the system is lit by a plane wave of polychromatic light. In the plane π the waves having traversed the two arms of the interferometer interfere wavelength by wavelength. No fringes are seen at this level since the systems relative to the different wavelengths mutually blur each other. But the spectroscope, following the process described above, removes a slice from each of the monochromatic fringe systems and provides a readable representation of the phenomenon.

Figure 5a, made by J.P. Goedgebuer, represents the image spectrogram of a piece of 16 mm motion picture film. Calculation shows that for a given local thickness, the interfringe in the image of the region considered is a strictly periodic function with regard to an axis graduated in number of waves.

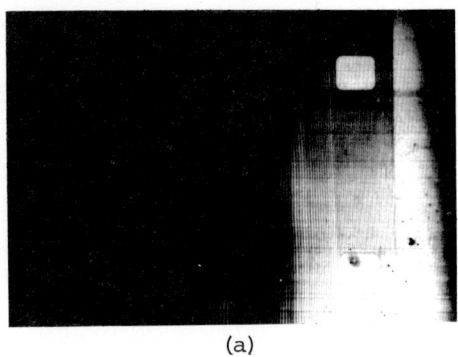 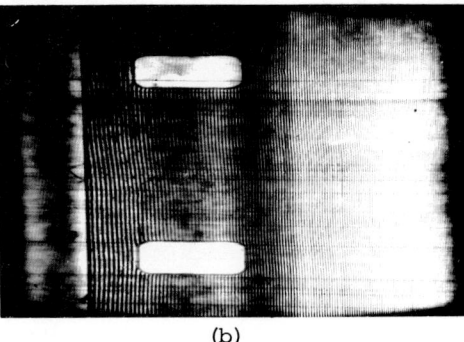

Fig. 5. Image spectrograms of transparent samples (16 mm movie film). The grooves provide an evaluation of the local thickness (measurements of the interfringe) and a global representation of the variations of thickness. (a) non-anamorphosized image; (b) anamorphosized image.

The value of the period is proportional to the thickness: the measurement of the interfringe thus provides an absolute evaluation of the thickness, point by point. Furthermore, the global observation of the fringe system gives an indication of the variations of thickness by specifying their sign: the reductions of thickness are expressed by a deviation of the fringes towards blue (σ increasing) and inversely the increased thicknesses deviate the grooves towards red (σ decreasing).

For a finer analysis, the spatial resolution of the measurement can be increased by an anamorphosis, which dilates the image without modifying the interfringe (figure 5b).

References

1. Froehly, C., Lacourt, A., Vienot, J.C.: Nouv. Rev. Optique, 1973, 4, pp. 183-196.
2. Vienot, J.C., Goedgebuer, J.P., Lacourt, A.: Applied Optics, 1977, 16, 2 pp. 454-461.
3. Armitage, J.D., Lohmann, A., Paris, D.P., Jap. Jal. of Appl. Phys., 1965, 4, suppl. 1, pp. 273-275.
4. Lacourt, A., Goedgebuer, J.P., Optica Acta, 1977, 24, 8, pp. 827-835.

INTERFEROMETRICAL SETUP FOR THE STUDY OF THERMIC TURBULENCE IN A PLANE AIRSTREAM

Alain Gagnaire and Albert Tailland
Laboratory of Physics — Ecole Centrale de Lyon
69130 Ecully, France

Abstract

The difficulties of investigating thermal turbulence using a classical hot wire anemometer in flows with very high mean velocity and temperature are well known. We are thus concerned in this paper with the possibility of studying such a turbulence by means of a laser beam propagating through a heated free plane jet of air. The beam undergoes in the flow random perturbations which tend to deflect it around the path that it would have taken if it had propagated in a vacuum. A formula is derived which relates the random deflection to the random temperature fluctuations. The experimental set-up is described. The values of the mean temperature and velocity have been chosen so that comparisons between the results obtained by this optical method and thoses obtained by the classical hot wire are made possible. A close agreement between both sets of results proves the validity of the model.

Introduction

A very convenient velocity and temperature measurement technique would be one in which no probe is introduced into the flow. The laser Döppler velocimeter presents such an advantage and is now a well-developed method which has attained a very high degree of accuracy. But it is not suitable for temperature measurements. The aim of the present investigation is, with a view to measuring temperature fluctuations, to relate these fluctuations to the perturbations that are experienced by a laser beam propagating through a turbulent flow. The basic idea consists in the following. The inhomogeneous distribution of temperature in the flow causes inhomogeneities in the spatial and temporal distribution of the refractive index, according to Dale-Gladstone's law. The parameters of an electromagnetic wave propagating through such a medium i.e. its amplitude, frequency, phase and direction will therefore generally undergo random fluctuations. In the present case, owing to the peculiar characteristics of a free plane jet and on the basis of Komogorov's theory of locally isotropic turbulence, only the deflection of the beam may be taken into account. This gives a possibility of deriving a relation between random deflection and random temperature fluctuations.

Experimental set-up

The optical system and its location with respect to the turbulent free jet are shown schematically in Fig. 1a. A light beam from a 1mW He-Ne laser is expanded in diameter by a first afocal system of lenses, after passing through the flowing fluid. It then falls onto two pinholes P_1 and P_2 close together (3,8mm) in a screen. By means of a second nearly afocal system, the interference pattern is formed over a plane where the aperture of a photomultiplier is placed.

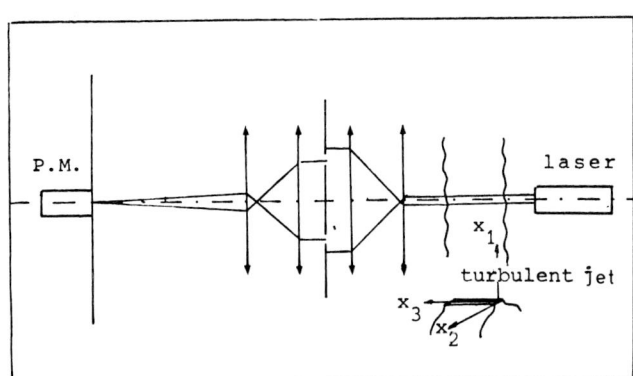

Fig. 1a. Schematic experimental arrangement

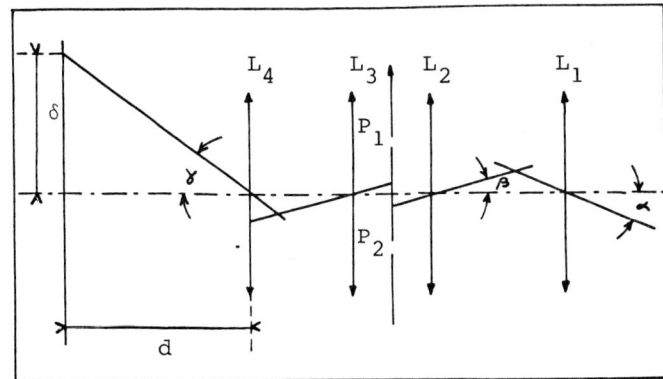

Fig. 1b. Optical system. Deflection of the beam and displacement of the interference pattern in the plane of the P.M. aperture

When no perturbations are present in the medium, the interference pattern is a rest. A deflection α (assumed to be small) of the beam is to be seen after it has passed along a path lenght ℓ through the flowing turbulent medium (Fig. 1b). This deflection causes a displacement δ of the whole interference pattern so that :

$$\alpha = K \delta \quad (1)$$

where K is a constant depending only on the geometrical characteristics of the optical system. The aperture of the P.M. is small enough (32μ) for the measurements of the light intensity to be considered as being performed on nearly a point.

The variation of the light intensity in the plane of the P.M. aperture is then proportional to the displacement δ and to the voltage output v of the P.M. A simple relation between δ and v can be obtained if δ is supposed to be smaller than a half fringe spacing and if the P.M. aperture is placed at the point where the light intensity I is equal to Imax/2 (Fig. 2). In the immediate neighbourhood of this point the light intensity may be considered as varying nearly linearly with δ. Taking into account (1), the relation:

$$v = K' \alpha \quad (2)$$

can be derived.

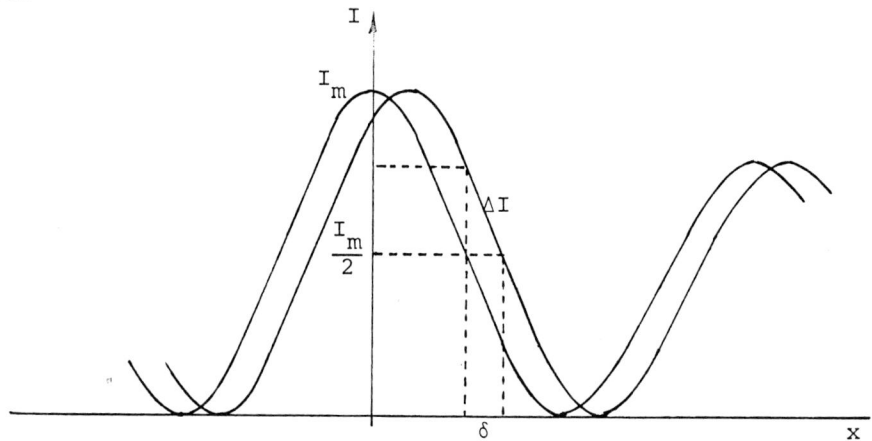

Fig. 2. Diagram illustrating the variation of light intensity on the aperture of the P.M. as function of the displacement δ of the interference pattern

The relation (2) was obtained on the assumption that the deflection α is located in the plane defined by P_1, P_2 and the optical axis of the system. Generally the deflection occurs randomly around the direction of this axis so that (1) must be replaced by :

$$\alpha = K' \sqrt{v_1^2 + v_2^2} \quad (3)$$

where v_1 and v_2 are the variations of the output of the P.M. corresponding to fringes running parallel respectively to the axes x_1 and x_2.

Theory

At any point $\vec{x}$ in the turbulent flow, the instantaneous value of the refractive index may be decomposed into a mean value $\bar{n}$ and a fluctuation $\mu(\vec{x})$ around the mean value so that :

$$n(\vec{x}) = \bar{n} + \mu(\vec{x}) \quad (4)$$

In air $\bar{n}$ may be taken as being equal to unity.

The fluctuations $\mu(\vec{x})$ are related to the temperature fluctuations according to Dale-Gladstone's law :

$$\mu(\vec{x}) = \frac{aP_o}{T^2} \theta(\vec{x}) \tag{5}$$

where, a is Gladstone's constant, P_o the static pressure in the flow, T the temperature and $\theta(\vec{x})$ the temperature fluctuation at the point $\vec{x}$. In addition, because of the randomness of the medium, spatial correlation function of the refractive index between any two points $\vec{x}$ and $\vec{x} + \vec{r}$ may be introduced :

$$R_\mu(\vec{r}) = <\mu(\vec{x}) \mu(\vec{x}+\vec{r})> = \left(\frac{aP_o}{T^2}\right)^2 R_\theta(\vec{r}) \tag{6}$$

where $R_\theta(\vec{r}) = <\theta(\vec{x}) \theta(\vec{x}+\vec{r})>$ is the temperature correlation function between the same two points.

It may also be shown that the smallness of the beam diameter with respect to the value of the integral scale of the turbulence allows us to neglect all phase variations which are not due to the direction fluctuations of the beam in the plane of the pinholes P_1 and P_2. [1]

The direction fluctuations have been evalueted from the ray equation. Taking into account for the geometrical approximation to be suitable that the wavelenght λ is small compared with the scale of the smallest inhomogeneities of the refractive index η ($\lambda \ll \eta$, where η denotes Kolmogorov's inner scale of the turbulence) [2],[3], the ray equation is written as :

$$\frac{d\vec{ns}}{ds} = \vec{\nabla}\mu \tag{7}$$

where $\vec{s}$ is the unit vector tangent to the ray trajectory and s the arc lenght of the curve.

Integrating along a small path ℓ we obtain :

$$n(\vec{s} - \vec{s}') = \int_o^\ell \vec{\nabla}\mu \, ds \tag{8}$$

that leads to :

$$<\alpha^2> = -2L <\mu^2> \int_o^\infty \vec{\nabla}^2 R_\mu(0,0,r_3) \, dr \tag{9}$$

Taking (3) and (4) into account, gives :

$$<\alpha^2> = 4 \left(\frac{aP_o}{T^2}\right)^2 L \, D_\theta <\theta^2> \tag{10}$$

where

$$D_\theta = \int_o^\infty \frac{1}{r} \frac{\partial}{\partial r} R_\theta(0,0,r_3) \, dr \tag{11}$$

L is the path lenght of the beam through the jet.

We finally derive the relation between the r.m.s. of the temperature fluctuation $\sqrt{<\theta^2>}$, the r.m.s. of the output voltage from the P.M. and the various characteristics of the whole set-up :

$$\sqrt{<\theta^2>} = \frac{1}{2} \frac{T^2}{aP_o} L^{-1/2} D_\theta^{-1/2} K' \sqrt{<v_1^2> + <v_2^2>} \tag{12}$$

Experimental results

Figures 3 and 4 show the distributions obtained by a hot-wire anemometer, of the mean velocity $\overline{U}$ and temperature $\overline{T}$ and of the r.m.s. of the temperature fluctuations in the plane of symmetry perpendicular to the exit slit of the jet and in a plane parallel to the slit at a distance x_1 = 20 cm. The symmetrical shapes of the various curves show that the jet is satisfactorily two-dimensional. The comparisons with the temperature fluctuation distributions obtained by the laser method are presented in figures 6 and 7. A very good agreement is to be seen except in a region near the slit for values of x_1 up to 10 cm.

This discrepancy is due to a more strongly marked anisotropy of the flow in this region where the turbulence is not fully developped. The laser method which is sensitive to the temperature gradient accounts clearly for this anisotropy (Fig. 5).

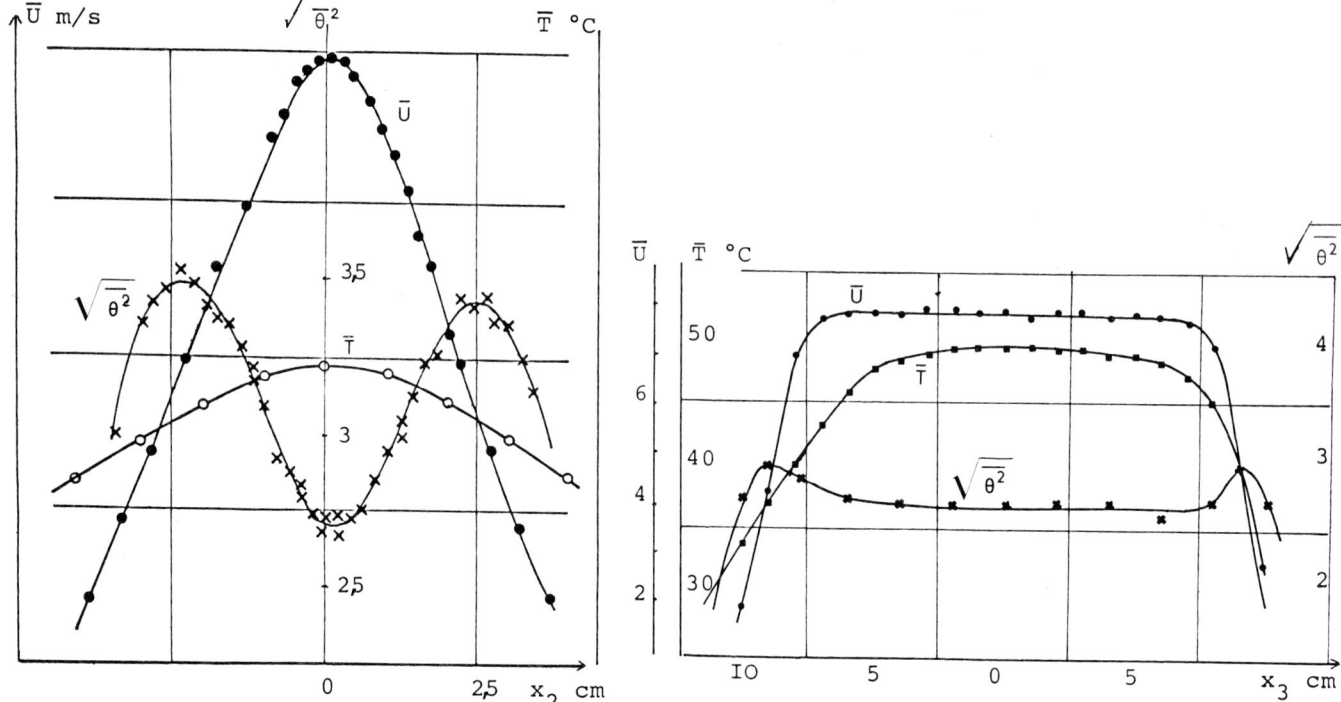

Fig. 3. Velocity and temperatures profiles as function of x_2 in the cross section x_1 = 200mm obtained by the hot-wire method

Fig. 4. Velocity and temperatures profiles as function of x_3 parallel to the slit in the cross section x_1 = 200 (hot-wire)

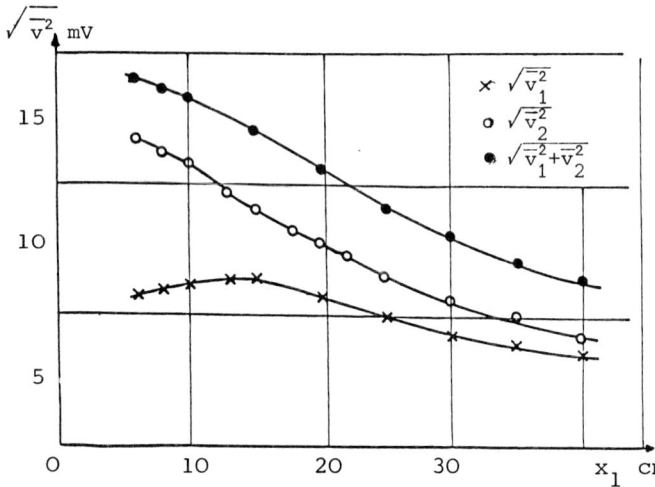

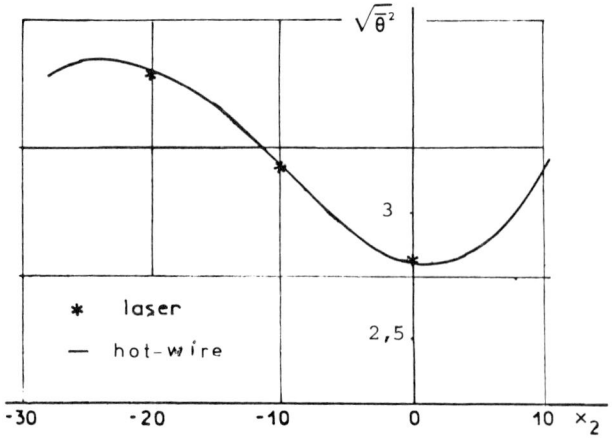

Fig. 5. P.M. output as function of x_1
v_1 : fringes parallel to x_2 axis
v_2 : fringes parallel to x_3 axis

Fig. 6. Comparison of the tranverse distribution of $\sqrt{\langle\theta^2\rangle}$ obtained by the two method

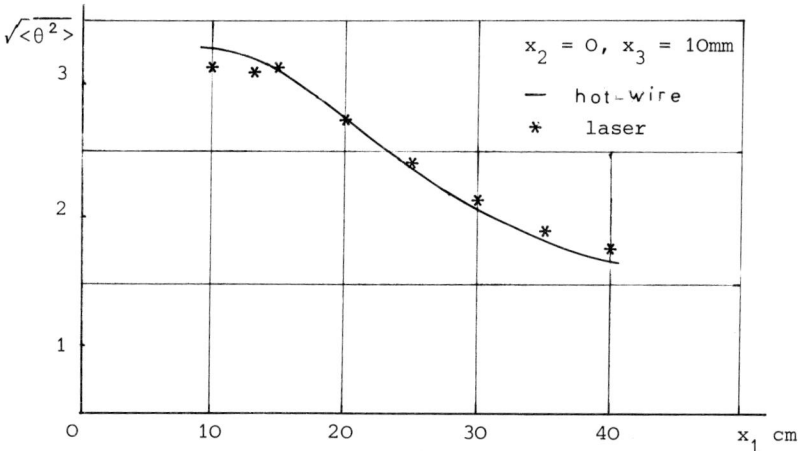

Fig. 7. Comparison of the $\sqrt{<\theta^2>}$ distribution versus x_1 obtained by the two methods.

For the calculation of $\sqrt{<\theta^2>}$, the values of

$$D_\theta = \frac{1}{2} \int_0^\infty \vec{\nabla}^2 R_\theta(0,0,r_3) \, dr$$

where R_θ is the temperature correlation coefficient along the x_3 axis, must be calculated. This may be done by using Kolmogorov's hypothesis of the local isotropy of the turbulence, i.e. for the very small eddies, that leads to :

$$D_\theta = -2 \, C_\theta^2 \, l_\theta^{-1/3}$$

where C_θ is a constant and l_θ the microscale of the turbulence. This microscale may be evalueted with a sufficient accuracy from the temperature fluctuation spectrum obtained by the laser method under the preceding conditions.

In the case of high $\overline{T}$ one may also assume the possibility of evaluating the values of D_θ from the curve $D_\theta(x_3)$ plotted, with a hot-wire anemometer, in a range of lower temperatures.

It must be furthermore noticed that, as most optical methods, the laser method used here yields values that are integrated along the whole path lenght. The good agreement between the results of both methods is due to the fact that, in the particular case of a plane jet, the temperature fluctuations are the same at every point along x_3, except at the two borders of the stream, the influence of which may moreover be neglected.

Conclusion

The validity of the model proposed in the present study, for the purpose of elaborating a method of investigation of the thermal turbulence for the cases where the other classical techniques are failing is thus satisfactorily proved.

The preceding discussion shows however that this model must be first extended to very higher temperatures and then eventually modified to be applied to geometrically more complicate flows (round jets, ...). In addition it will be finally necessary to aim at very local measurements.

References

1. Gagnaire, A., _Etude de la propagation d'un faisceau laser dans un milieu turbulent engendré en laboratoire_. Thèse Université Claude-Bernard - Lyon (France)
2. Chernov, L.A., _Wave propagation in a random medium_, Mc Graw-Hill 1960
3. Tatarsky, V.I., _Wave propagation in a turbulent medium_, Mc Graw-Hill 1961

1st EUROPEAN CONGRESS ON OPTICS APPLIED TO METROLOGY

Volume 136

SESSION 3

PHOTOGRAMMETRY AND HOLOGRAMMETRY

Session Chairmen
Stroke
Paraskevas

HOLOGRAMMETRIC PLOTTER
(APPARATUS FOR THREE-DIMENSIONAL MEASUREMENT ON AN IMAGE RECONSTRUCTED BY HOLOGRAPHIC PROCESS)

Gerard Bocquemo, Dominique Laroche, Bernard Turlier

MATRA Inc., Optical Division, 93 Avenue Victor Hugo
92500 Rueil-Malmaison, France

Abstract

Our holographic technique permits us to produce the equivalent of a photogrammetric stereomodel. The exploitation is made in two ways:
   -the direct method, in which we visually put a luminous point in contact with the three-dimensional holographic virtual image of the object;
   -the video method, in which we work with a camera on the real holographic image of the object.
   These methods permit the measurement of distance by successive sightings on the object with a precision better than 0.1 mm.
   The apparatus which we are developing presently works by successive sightings. It may evolve towards an automatic sequence of sightings, advantageously using repetitive details already existing on the object or curves drawn optically.

Introduction

At present, when one wants to make measurements of an object in three dimensions, two techniques are commonly used:
   -If we can have the object itself at our disposal throughout the duration of the measurement, we can use a mechanical feeler which we move over the object. The measurements are made by recording the coordinates given by the feeler during its movement. These measurements thus require a physical contact with the object which may be destructive.
   -In the case where no contact with the object can be tolerated and where we wish not to immobilize the object itself, we can use the techniques of short-distance photogrammetry. Nonetheless, this method also presents inconveniences:
   --certain forms of objects are not easily photographed from two different angles of view;
   --it is difficult to produce stereoscopic plates for mobile objects with large translation velocities in normal ambient light;
   --phase phenomena cannot be photographed.
   Since the MATRA company is oriented, in a large proportion of its activities, toward the production of materials for photogrammetric interpretation, and moreover carries on a great deal of activity in research on applications of holography, we naturally became interested in the possibilities which holography could offer in the field of three-dimensional measurement.
   In effect, holography is a technique permitting the recording and reconstruction, in simple fashion, of a three-dimensional object. If certain precautions are taken, a stigmatic image with magnification 1 is reconstructed and we thus dispose, by means of a very simple setup, of the equivalent of a "stereomodel" given by the conventional processes of photogrammetry.
   The exploitation of this image can be made using two distinct techniques which have been studied.

Hologrammetric Plotter Working on the Virtual Image

This technique is very similar to that used in photogrammetry. It consists in bringing a spatial marker into contact with the point of the holographic image whose spatial coordinates we wish to know (figure 1).

1. Working Principle

The hologram is taken in the conventional way with a continuous or pulsed laser. The virtual image is reconstructed in three dimensions with magnification 1 from a single photographic plate (or film).
   The observer looks at the hologram with the naked eye or with a binocular optical system. The enlargement of the optical system increases the impression of relief and permits us to better evaluate the "contact" between the point aimed at and the spatial marker.
   The luminous spatial marker is brought into contact with the target point of the image by means of a coordinatograph with 3 axes.

2. Limits of the Study and Description of the Laboratory Model

2.1 Limits of the Study
The coordinatograph used permits a maximum displacement of :

OX axis: 230 mm
OY axis: 300 mm
OZ axis: 230 mm

The study of holographic volume has thus been limited to 200 x 200 x 200 mm.
The holograms were made with an He-Ne 15 mW laser; the plates used were of the 9 x 12 cm format.

2.2 Description of the Model The model was separated into 4 sub-assemblies:
-an optical assembly to reconstruct the hologram, including:
  --the laser
  --the beam splitter
  --the support for the hologram,
as well as the adjustments of position relative to these elements;
-the luminous spatial marker(consisting of the end of an optical fiber of 250 μm) linked to the three-axis coordinatograph, which is controlled by electric motors connected to coders;
-the system to control the movements, linked to the coordinatograph:
  --the X and Y movements are grouped on a single manual sensor or "ball";
  --the Z movement is generated by a cylinder type system. The X,Y and Z movements are displayed on digital counters.
-the optical observation system, consisting of an 8X binocular system, equipped with interchangeable shades 20 mm in diameter, for aiming at finite distance. A mechanical system permits the convergence of the optical axes on the point studied in order to obtain stereoscopic fusion.

3. Evaluation of the Virtual Image Plotter

3.1 Control of the Movement on the Coordinatograph The gap between the real mechanical displacements measured with feelers to the micron and the displacements recorded by the coders and displayed is :

        X axis: 0.03 mm
        Y axis: 0.03 mm
        Z axis: 0.02 mm

This limits, in certain cases, the intrinsic precision of our system.

3.2 Estimation of the Intrinsic Precision of the Sightings The hologram is reconstructed in exactly the same conditions as at recording and we aim at a determined point. The measurement is made by two methods:
1st method: successive sightings on a point with all adjustments destroyed between measurements.
2nd method: aiming in sequence at a series of points on an object.

Axis	Global dispersion over 10 sightings
X	± 0.03 mm
Y	± 0.03 mm
Z	± 0.2 mm

The dispersion represents the maximum deviation in fact recorded, and not an average which would have given better results.
  The measurements of the distance between two points which we made on object types (echelette, curve, cylinder, etc.) made for these trials and measured carefully, gave us a precision corresponding to two times the global dispersion of the sightings on a single point, or a composition of the following precisions:

        in X: ± 0.06 mm
        in Y: ± 0.06 mm
        in Z: ± 0.4 mm

3.3 Effect of the Precision of Repositioning of the Plate on the Sightings In order to avoid distorsion of the image, the precisions of repositioning of the hologram with regard to the recording are:

$\Delta \theta$: ± 70 minutes of arc (rotation of the TY axis)

$\Delta \phi$: ± 30 minutes of arc (rotation of the TX axis)

Translation in X or Y: ± 2mm

(The distorsion thus obtained remains within the limit of the precisions of measurement)
  A rotation in TZ has no reason to be in the setup intended for a precise repositioning in this rotation. The effect would be a movement, then disappearance of the image.
  A translation along TZ must be limited to ± 10 mm if we do not want to modify the enlargement between the object holographed and the image reconstructed. The change in enlargement is quickly accompanied by a considerable degradation of the quality of the image.

3.4 Effect of a Change of Wavelength We reconstructed, at the He-Ne wavelength (6328 Å), an object which we had entrusted to the Franco-German Institute of Saint Louis to make holograms with a krypton laser (λ = 6471 Å and 6761 Å). While the apparent quality of the reconstructed holographic image is good, there is a deformation of the object (distorsion type) as well as the enlargement ratio foreseen by the simplified theory. This makes the change of wavelength delicate for precision sightings.

4. Conclusion

The study of the model of the hologrammetric plotter working on the holographic virtual image of an object gave us results close to the objectives fixed at the outset. However, we perceived a certain number of limitations:

-the direct visual stereoscopic sighting creates a subjective aspect related to the operator, who must be a specialist trained in stereoscopic exercises;

-the precision of the sighting is related to the stereo base of operation, fixed by the optical system and limited by the "aperture" of the plate. This is even more bothersome since the image is virtual.

-the use of an optical system of observation with considerable magnification and thus a small relative aperture increases the speckle phenomenon, which becomes very bothersome for precise sightings.

This led us to study a hologrammetric plotter for the real image.

Hologrammetric Plotter Working on the Real Image

1. Principle (figure 2)

To obtain a stigmatic real image of an object, it is necessary to have a parallel reference beam and to reconstruct with a beam identical to it.

The interest of a real image formed by the rays diffracted by the hologram is that we can physically approach it with the optical receptor without being bothered by the diaphragm and the stop constituted by the holographic plate in the case of a virtual image.

The detection system which we had at our disposal was a silicium photocathode tube, integrated in a closed circuit TV camera (it is to be remarked that the adaptation in wavelength was not optimized for the green ray of the argon laser). The receptor is mounted on the 3-axis coordinatograph. The operator observes a TV image, enlarged in the detector-visualization screen ratio; he focuses on the latter.

2. Description of the Laboratory Model

The laboratory model includes two distinct sub-assemblies:
-a bench for recording and reconstructing holograms
-a 3-axis coordinatograph, the same as that employed in virtual image hologrammetry.

2.1 Recording and Reconstruction Bench. The recording laser was an argon 2W laser equipped with a Fabry-Perot and used in its band at 5145 Å.

The diameter of the objective employed to constitute the collimated reference beam allows the 9 X 12cm plate to be covered.

The angle θp between the parallel beam and the normal to the plate is 30°. The object is placed approximately on the normal to the plate. We remarked that the virtual image made real by rotating the plate 180° around the vertical axis Y'OY is of better quality than the true real image. We thus worked on this image.

The reconstructed image is either observed directly or conjugated on the silicium tube by means of an objective of aperture superior to that of the hologram.

The precision of sighting obtained on a detail of the holographic image is limited by the phenomenon of speckle or laser granularity. This led us:

-to try to attenuate this phenomenon without losing information about the observed image.
-to develop a system of stereoscopic sightings introducing an improvement in sighting but also a limitation in the observable volume.

2.2 Attempts to Reduce Speckle After fixing the magnification of the objective to conjugate the image on the tube, we approached the problem of speckle reduction. The speckle is bothersome, since it "drowns" the details of the holographic image.

We tried to reduce the speckle which was spotting our image by introducing, upon reconstruction of the hologram, different systems permitting us to reduce the coherence of the reconstruction beam. The result obtained was a weakening of the image, a loss of contrast and a loss of resolution.

The trials we ran creating a loss of coherence at recording seriously affected the holographic image. We then considered an electronic filtering of the video signal, treated before its reconstruction on a TV monitor.

The frequency analysis showed us that the spatial frequency of the speckle grains does not at all have the aspect of a spectrum of rays. We were then led to produce a stereoscopic sight.

2.3 Method of Sighting by Coincidence of Images The sighting method consists in observing the interesting detail of the object from two different angles by means of two separate optical systems. The treatment of each of the images provided by the two optical systems permits us to make a sighting at the level of the TV reconstruction monitor, by fusion of the two images or by obtaining a determined distance which is realized between the two images of the detail observed on the TV monitor.

The practical realization of this system was the object of an industrial patent taken by MATRA.

The use of this stereoscopic system permits an improvement of the precision of the sightings in OZ in a more limited volume than by the conventional sighting.

3. Evaluation of the Precision of the Sightings

The procedure used for the evaluation of the precision of the sightings with this new setup is identical to that of the setup for the virtual image. The precision obtained with a single optical system by a statistical series of sightings at a detail is:

$$OX: \pm 0.03 \text{ mm}$$
$$OY: \pm 0.03 \text{ mm}$$
$$OZ: \pm 0.08 \text{ mm}$$

The precision obtained with the stereoscopic sighting system is :

$$OX: \pm 0.03 \text{ mm}$$
$$OY: \pm 0.03 \text{ mm}$$
$$OZ: \pm 0.05 \text{ mm}$$

The exploitable volume with a single sighting system was limited by our coordinatograph, that is: 230 x 230 x 300mm.

The exploitable volume with the stereoscopic camera used can reach 100 x 100 x 100 mm for a plate 90 x 120 mm, the object being at about 200 mm from the plate. This limitation is of a geometrical nature and is due solely to the size of the plate.

The exploitable volume can extend to a dimension of 300 x 300 x 300 mm for a plate 203 x 254 mm (dimension of a commonly sold plate). The precision obtained in the measurement of distance is directly a function of the precision obtained for a sighting.

4. Influence of the Precision of Repositioning of the Holographic Plate

The reconstruction beam being parallel, the precision of repositioning in translation of the holographic plate is unimportant, as long as the latter remains within the beam. In rotation around OX and OY, the precisions of repositioning are on the order of 20 minutes of arc, to remain within the precision of the errors in sighting.

5. Creation of Details on the Object to be Recorded

Objects not presenting surface details do not easily lend themselves to a holographic sighting (likewise, they are poorly suited to photogrammetric sightings). We thus created fringes on the surface of our object and recorded them on our hologram.

This permits us to attain sighting precisions at the surface of the object equivalent to those existing for real details of the latter. A rapid theoretical preliminary study also permits us to consider a semi-automatic tracing of the fringes drawing a profile on the object, for electronic treatment of the video signal.

Conclusions

The experiments we performed show the possibilities offered by holography to be able to make precise measurements in space by means of a fairly simple apparatus.

We were also able to see the interest which lies in working on the real image, on one hand to improve the precision of the sightings, on the other because employing a video system of observation can permit us to effect a semi-automatic tracing of repetitive details existing or optically created on the object.

Devices to measure on the holographic image should be considered as complimentary to photgrammetric instruments for the study of objects which cannot be photographed from two different angles (for example, hollow objects, inaccessible objects, objects moving at great speed, phase objects). With plates which are selective in wavelength at recording, the hologram can be taken in an industrial environment with a pulsed laser.

Our real image apparatus is presently being evaluated by the Metrological Service of the Central Armament Establishment (E.C.A.).

Acknowledgements

We must first of all thank the DRME, which contributed to the financing of the study of these two hologrammetric plotters and permitted us to publish the results. We must also thank MM Abbe, Royer and Smigielski of the I.S.L., who helped us with their advice.

Bibliography

1. Agnard, J.P., Boivin, A. and Brandenberger, A.J., Obtention de pointés stéréoscopiques de précision sur l'image virtuelle: "Hologrammetrie", Proceedings of the International Symposium of Besançon, July 1970.
2. Laroche, D., Turlier, B., DRME report Marché n° 74/065.
3. Bocquemo, G., Laroche, D. and Turlier, B., DRME report Marché n° 75/372.

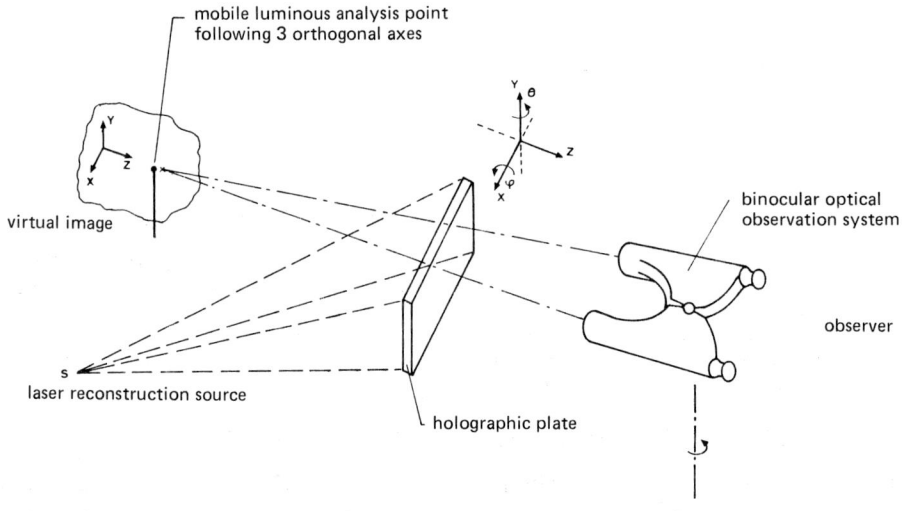

Figure 1

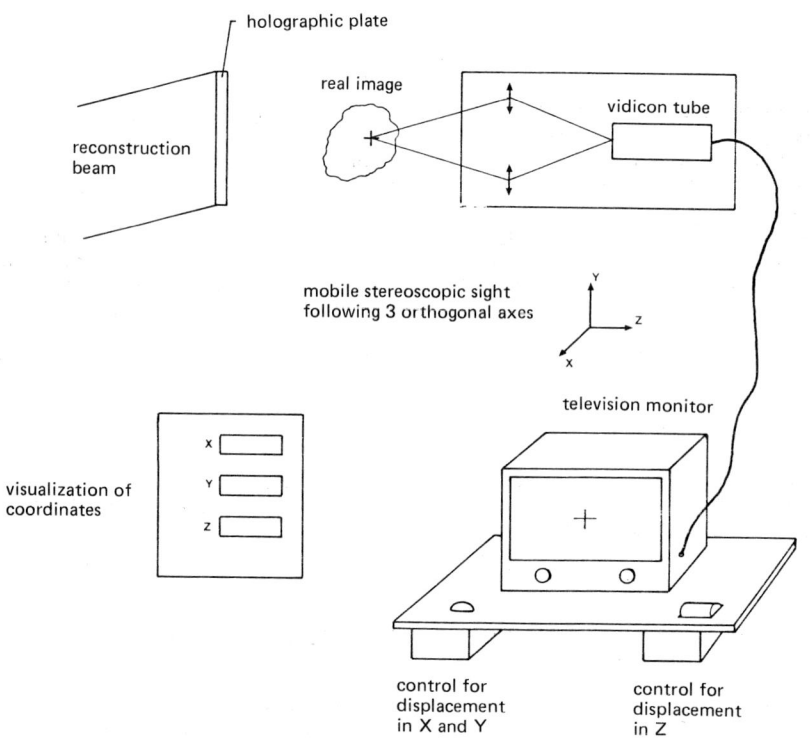

Figure 2

1st EUROPEAN CONGRESS ON OPTICS APPLIED TO METROLOGY

Volume 136

SESSION 4

PHOTOELASTICIMETRY

Session Chairmen
Monfils
Meyrueis

PHOTOELASTICIMETRY AND HOLOGRAPHIC INTERFEROMETRY: APPLICATIONS TO THE STUDY OF STRESSES AND DEFORMATIONS

J. Monneret, P. Rastogi, M. Spajer

Laboratoire de Physique Générale et Optique, associé au C.N.R.S. n° 214
"Holographie et Traitement Optique des Signaux"
Université de Franche-Comté, 25030 Besancon Cedex, France

Abstract

Various methods for the measurement of hydrostatic stress i.e. the isotropic component of the stress tensor, are presented. All these methods lead to a whole-field visualization.

In two dimensional problems, the hydrostatic stress is obtained by the measurement of the relative variation of thickness, caused by loading the model ; by using the technique of double exposure holographic interferometry. The birefringence effects are eliminated by making the object beam to pass through the model, and in-between the two passes, the beam is subjected to a 90° rotation of its plane of polarization. In the case of three-dimensional models, the method is based on the observation of the variations in the absolute index of refraction. The frozen-stress technique requires the use of those photoelastic materials which remain compressible at the "freezing" temperature ; but unfortunately, they are unavailable at present. Therefore, a non-destructive optical-slicing method is proposed and it is hoped that it shall give the required information concerning the hydrostatic stress. This method consists of an interferometric comparison of the two superimposed speckle patterns, each scattered from two neighbouring optically isolated light sheets.

Further, the measurement of the in-plane strain can be carried out using either holographic or speckle pattern interferometry ; the precision is found to be the same in both cases.

During the last few years our interest has mainly centered around the separation and determination of the stress in both two and three-dimensional photoelastic models. While carrying out these studies, we have preferred the technique of full-field visualization to that of point by point analysis.

Generally speaking, separation of stresses is equivalent to the complete determination of the corresponding tensor, i.e. not only the deviatoric component but also the isotropic component which is nothing but the hydrostatic stress. The major portion of this paper will be devoted to the determination of this last quantity.

In two dimensional cases, the determination of the hydrostatic stress is relatively simple to the extent where it is directly related to the relative variation in the thickness of the model during loading.[1-2] We use the technique of double exposure holographic interferometry in which the first exposure is made with the model in the unstressed state and the second exposure after the model has been stressed. The only difficulty which need be resolved consists of eliminating the birefringence effects which necessarily accompany the relative variation in thickness. To meet this end, we have made the object beam to pass twice through the model and in-between these two passes, the beam is subjected to a 90° rotation of its plane of polarization by means of a Faraday Cell. Thus, the vibrations which had traversed the model along the fast axis, while going out, return along the slow axis and this results in the elimination of the birefringence.[3-7]

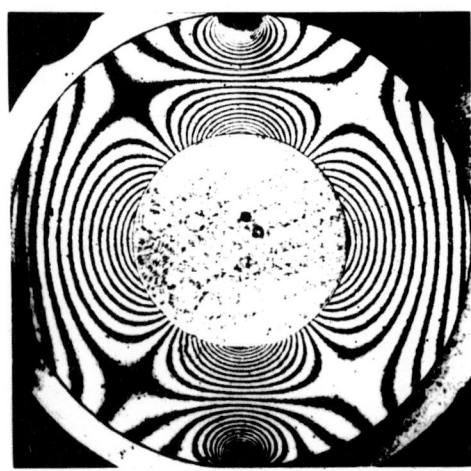

Fig. 1 a. Fringes resulting from the superposition of the effects of birefringence and the relative variations of thickness

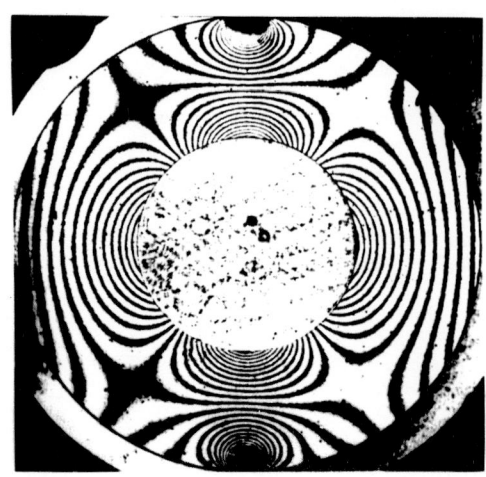

Fig. 1 b. Isopachic fringes, as observed after the elimination of the birefringence effects.

Fig. 1 shows the results obtained. On the left is the interferogram of a circular ring under diametrical compression as observed in conventional holographic interferometry. The pattern shows the superposition of the birefringence effects (isochromatics) and the effects of the variation in thickness (isopachics). On the right is the interferogram of the same ring but with the double passage of the beam through the model and the use of a Faraday Cell. Here, as we see, the only fringes that remain are those of isopachics which are the locus of the points of equal hydrostatic stress.

In order to be able to simultaneously benefit from the advantages of whole field visualization and the precision ellipsometric measurements, a second holographic method for obtaining the isopachics fringes has been developed. It consists of transposing the local variation of thickness into the variation of polarization of the object beam. To do this, we proceed as before but using two different orthogonally polarized reference beams for each exposure (fig. 2). The object is illuminated by the circularly polarized light.[8] At the time of the first exposure, the object is in the unstressed state and the reference beam is polari-

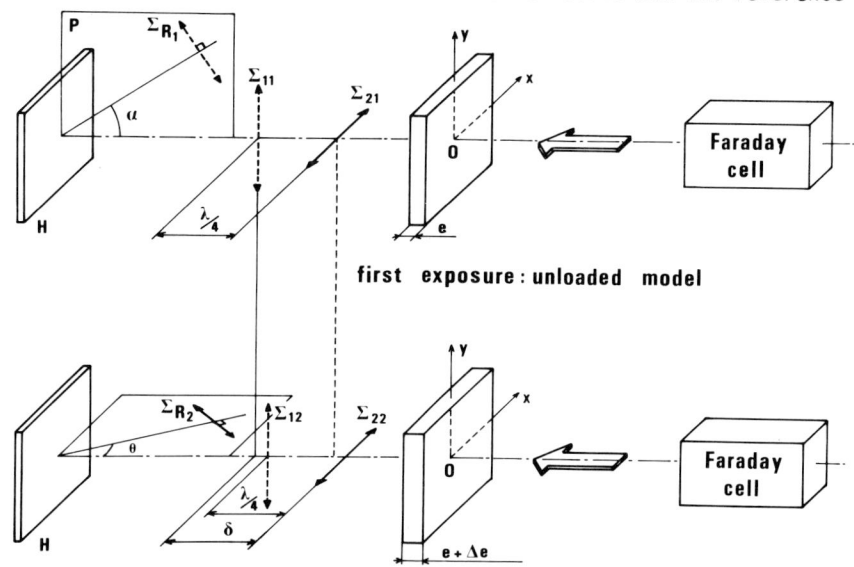

Fig. 2 The principle of the method which allows to transform the local variations of thickness into the variations of polarization of the object beam (the first passage through the model has not been shown).

zed in the plane P. Hence only the similarly polarized component of light Σ_{11} transmitted by the object shall produce an interference pattern in the hologram plane. For the second exposure the object is deformed and the reference beam is now polarized in the plane orthogonal to the plane P. As before, only that component of the transmitted light Σ_{22} shall be recorded holographically which is polarized in the same plane as the reference beam. During the reconstruction, the two orthogonal components Σ_{11} and Σ_{22} combine together to form an elliptic vibration, the ellipticity of which gives the variation in the thickness of the model. We may, therefore, observe the isopachics as before, while looking at the image reconstructed through a polarizer. In order to achieve more precision, we may study certain localized regions by making measurements of the ellipticity of the reconstructed beam, using an ellipsometer (9). The relative simplicity of the above method must in no way divert our attention from the difficulties faced during its experimental realization. The diffraction efficiency of the hologram should rigorously be the same for the two reference beams and the phase difference between the two reference beams should be maintained constant during the recording, the development and the reconstruction processes respectively. Practically speaking, a precision of the order of 1/10 of the fringe has been attained by using a holographic plate holder in immersion and the interferometric part of the system being kept enclosed in order to avoid the effects of atmospheric turbulence, temperature variations, etc...

Figure 3 illustrates the results so obtained. The figure on the left shows the isopachic fringes, obtained while observing the reconstructed beam by means of a polarizer ; and for the sake of comparison, the figure on the right is presented showing the same fringes but as observed by the first method.

In the case of three dimensional bodies, the study of hydrostatic stress is evidently much more difficult and complex in nature. Here, one is rather obliged to work on "slices" of the material, which need to be sufficiently thin so that along the light path the variations of stress should be negligible. We suppose that this slice of material has been isolated by mechanically cutting the three-dimensional model in which the deformations and the birefringence, characterising the loading to be studied, have beforehand been "frozen" in remanence by thermal processing (10).

Here the question of measuring the relative variations in thickness does not arise without having first subjected the material to long thermal cycles followed by its relaxation. As this method is quite laborious and difficult to realize in practice, we propose that the required information relating to the hydrostatic stress can be extracted from the measurements of the absolute indices of refraction, as the hydrostatic stress is proportional to χ, where $\chi = n_1 + n_2 + n_3 - n_0$

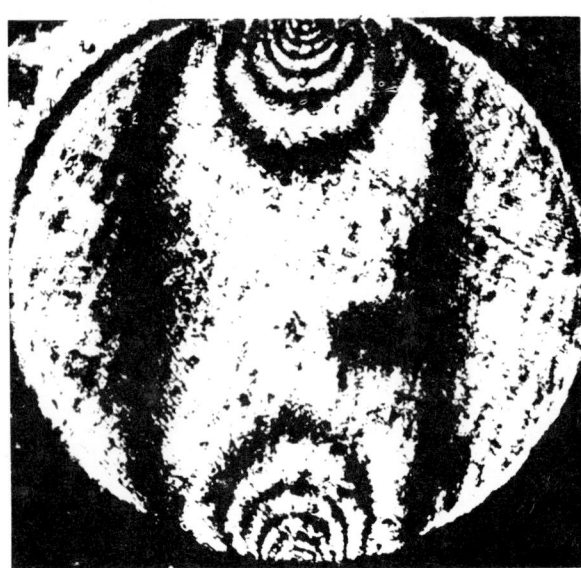

Fig. 3 On the left are shown the isopachic fringes obtained while observing the form of the reconsctructed beam by means of a polarizer ; the hologram being recorded following the principle described in Fig. 2. On the right are shown the same fringes but as obtained by the first method.

For illustrating the said method, we shall apply it to a circular disc under diametrical compression where the variations of the thickness have already been levelled, by milling, after "freezing".

We still insist with the double exposure holographic interferometry and the experimental set-up of Figure 2, except that here the model is immersed in a liquid of the same refractive index (n_o) as the photoelastic material in the unstressed state. This permits not only in the elimination of the diffusion from rough surfaces but more important, it also allows us to compare the "slice" of the frozen-stress model with its "equivalent liquid". During the first exposure the slice is immersed. The passage through the slice corresponds to optical path length $n_1 e$ (beam passing in on direction) + $n_2 e$ (beam passing in the inverse direction), where e is the thickness of the model at the point under consideration. Next the slice is taken out from the immersion tank and the second exposure is made. The double passage through the "equivalent liquid" of the slice corresponds to an optical path length $2 n_o e$. In the reconstructed image, the fringes are observed which represent the locus of points where $n_1 + n_2 - 2n_o$ is a constant. As it can be easily seen, this quantity is proportional to the hydrostatic stress $\frac{\sigma_1 + \sigma_2}{2}$. Figure 4 presents the results so obtained with a circular disc and a ring respectively, each under diametrical compression.

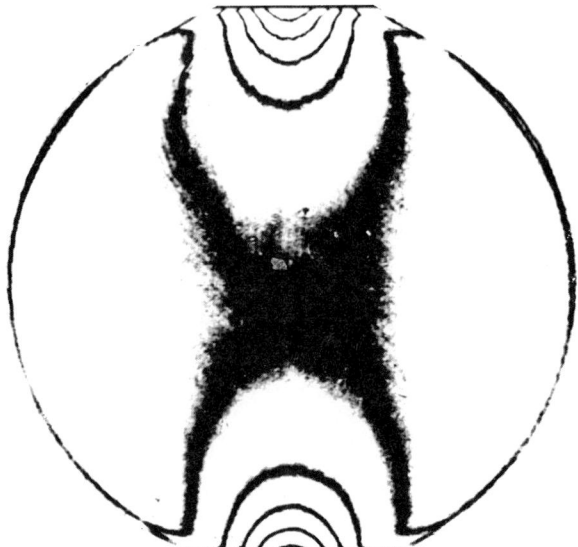

Fig. 4 Isopachic fringes obtained with a "frozen" circular disc and a circular ring, each under diametrical compression. Index matching is achieved with the help of a prism, of the same material as the model, kept inside the inner-circle of the ring.

At this point, it is necessary to stress that the method we have just described can only be applied to those materials which remain compressible at the freezing temperature, such as those used by A.J. Durelli (11). We should also point out that the usual materials which are incompressible at the freezing temperature cannot be used for the determination of the hydrostatic stress. The results we have obtained with the resin photoelastic PLM 4 B (8)go to confirm the results obtained by R.J. Sanford and V.J. Parks (12) with the resin Hysol 4290.

The last results published by the same authors (13), relating to the compressible materials studied by A.J. Durelli (11), bring certain questions to mind : From the point of view of photoelasticimitry,.these materials behave as incompressible, which bring forward an element of incompatibility between the laws of Maxwell and that of Neuman. At present, taking up the said work cannot be considered as one of the constituents of the material in question is no more commercialized. Thus, presently, the separation of the stress components,in three dimensional models,are dependent on the commercialization of the compressible materials in the domain of photoelasticimetry as much as in Mechanics.

Methods to avoid the frozen-stress technique. In order to achieve the said goal, two directions have been followed. The first consists of studying the stress by measuring the displacements and the strains undergone by the object under study. The second, a more direct approach, depends on the measurement of the variations in the absolute index of refraction caused by the deformation which, in turn, shall give us the required information about the hydrostatic stress.

Measurement of displacement and strain. As far as the measurement of the displacements and the strains is concerned, we shall limit ourselves to presenting only a few results pertaining to the in-plane displacement measurements by holographic or speckle interferometric techniques. In both cases, we use the technique of double illumination as proposed by J.A. Leendertz (14). The correlation between the two scattered speckles gives rise to the fringes of equal displacement which follow the vector $\vec{k}_1 - \vec{k}_2$ lying in the tangent plane of the scattering surface (fig. 5).

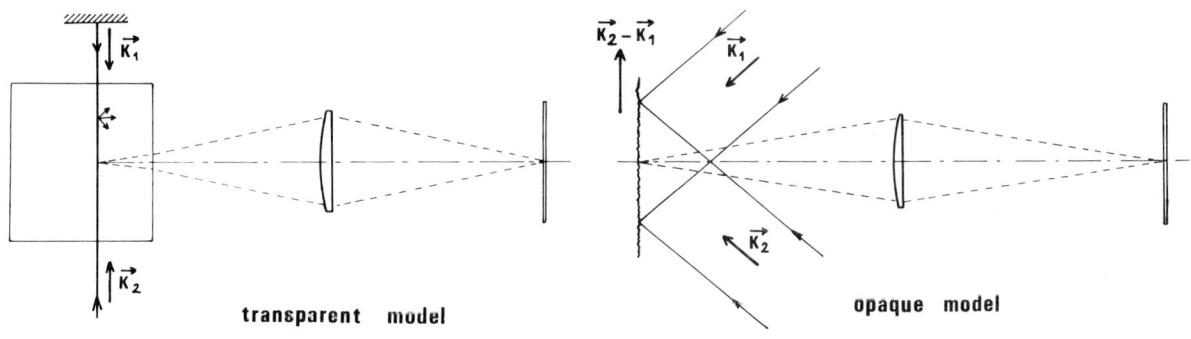

Fig. 5 Principle used in the measurement of the in-plane displacement

In the case where the measurements on an arbitrary interior plane of a transparent model are of interest (15), the double illumination is obtained by the autocollimation of a thin sheet of light on the plane of the mirror.

Whether they are obtained by holography or speckle interferometry techniques, these correlation fringes present a granular structure, as shown in figure 6. This aspect results from the limited aperture optical filtering systems which are necessary to increase the contrast of the fringes.

Figure 7 clearly shows the difference between the ordinary holographic fringes and the correlation fringes (zones of zero contrast), which give information about the longitudinal and the in-plane displacements respectively. As we can see on the photograph b, the quality of correlation fringes is dependent on the density of initial fringes. To increase this density, we add a system of auxiliary fringes (16). The high density initial fringes, in our case, are generated by translating the holographic plate, in its own plane, between the two exposures. These fringes are naturally localized on the object, whatever may be the amplitude of the translation. The resultant increase of contrast and the elimination of the auxiliary fringes between the photographs (c) and (d) has been obtained by optical filtering.

In order to deduce strain from these results, it is necessary to obtain the first derivative of displacement. This is performed by forming moiré between the fringes of displacement. But in order that the moiré should be of good quality, the primary fringes i.e. the fringes of in-plane displacement, should be of high fringe density distributed all over the surface. As one cannot increase the deformation beyond a certain limit, one is obliged, here too, to add the auxiliary fringes, but which in this case are the correlation fringes. The quality of moiré fringes shall be governed by the maximum frequency obtainable.

Fig. 6 Correlation Fringes representing the horizontal component of the in-plane displacement for a disc under diametrical compression and for a rubber membrane under strain.

Fig. 7 Interferograms of a rubber membrane under strain ; a) the object is illuminated by one beam only ; b) the object is illuminated by two beams symmetrically situated about the normal to the model , c) the same as b except that, in addition, the photographic plate has been translated in its own plane between the two exposures, in order to improve the quality of moiré ; d) increase in the contrast of moiré fringes by optical filtering.

In speckle interferometry this frequency limit is imposed by the average size of the speckle in the filtered image. In holography, the frequency limit is imposed by the fringe spacing in the auxiliary pattern. But as we shall further see, this limit reduces to the same as in speckle interferometry in the case of Image-Plane Holography. The reconstruction, carried out in white light, reduces the speckles, at least, its contrast. In figure 8, we have photographed the auxiliary fringes, the spacing of which is equal to the size of the speckle. On the left, where the reconstruction has been done in coherent light, the correlation fringes are almost not observable. On the other hand, the fringes appear to have an acceptable contrast under white light reconstruction, as shown on the right.

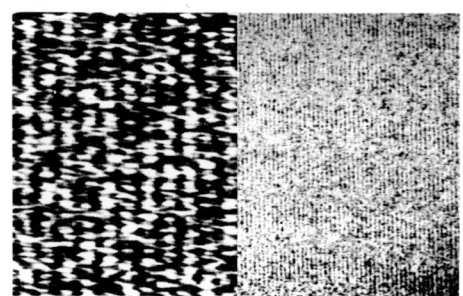
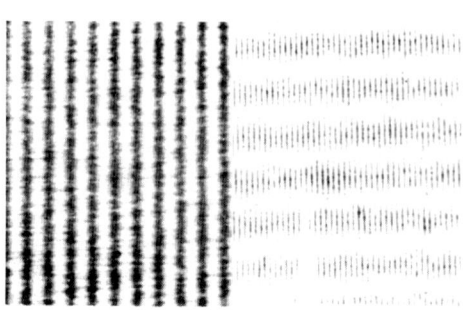

Fig. 8 Auxiliary fringes and the corresponding correlation fringes obtained on the reconstruction of the image-plane hologram. On the left is shown the reconstructed image in coherent light and on the

right is the same image but in white light reconstruction. The spacing of the auxiliary fringes is equal to the size of the speckle.

Thus we may conclude that in holography as in speckle, it is the size of the speckle in the filtered image which sets the frequency limit on the correlation fringes. We show that the minimum size of the speckle should be about four times the maximum local displacement. We have seen that the performance of the two techniques are the same vis a vis the measurement of displacements and strain. But as in holography the recording conditions are much more restrictive and delicate, in our opinion, it should be reserved for the study of only those problems where the light flux scattered or transmitted by the model, is weak.

<u>Techniques for the measurement of hydrostatic stress</u>. In order to be able to study the hydrostatic stress on any arbitrary interior plane of a three-dimensional photoelastic model, we shall make use of the optical isolation method [17][18] (fig. 9). The method which we propose does not rest on comparing the different states of polarization of the scattered speckles but is rather akin to the method used in the measurement of the in-plane displacement by double illumination, a technique described elsewhere in this paper. Same as before, we make a double exposure, the first exposure relating to the model in the unstressed state and the second after the model has been stressed. Here, each of the two speckles, scattered by the two light sheets, (fig. 9) play the same role as the speckles produced by each illuminated beam in the measurement of the in-plane displacement. Thus, in region where the coherent addition of the speckles has taken place, speckle correlation fringes will appear relating to the variations of phase between the two combined speckles. In other words, the fringes are related to the variations of absolute indices of refraction caused by the stress, introduced, between the two exposures. Once knowing these variations, we can easily deduce the values of hydrostatic stress inside the slice lying between the two light sheets.

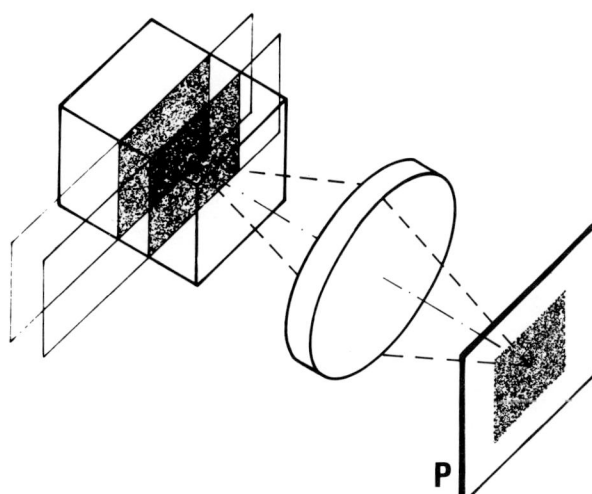

Fig. 9 Schematic of the "Optical-Slicing" technique, where the slices have been obtained by "cutting" the three dimensional photoelastic model by two light sheets. The speckles, scattered in the direction orthogonal to the plane of the illuminated beam, are polarized with respect to each other. The speckles are superimposed in the plane P. As a result of the effects of the birefringence inside the "Slice", the superimposed speckles shall interfere coherently in the regions where they are in the same state of polarization and incoherently in the regions where they are in an orthogonal state of polarization. The technique of double exposure permits the visualization of the variations of the absolute index of refraction, caused by the loading, in the regions of coherent superimposition.

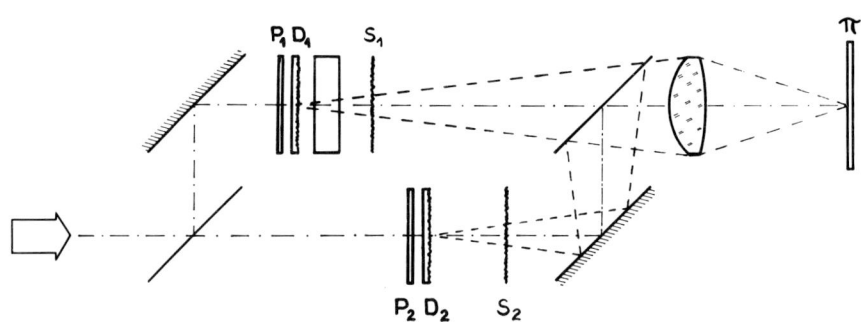

Fig. 10 Simulation of the system described in fig. 9, which shall allow the use of a low power light source ; P_1, P_2 are two parallel polarizers and D_1, D_2 are two diffusers. After traversing the model, the resulting two speckle patterns S_1 and S_2 are superimposed in the π plane.

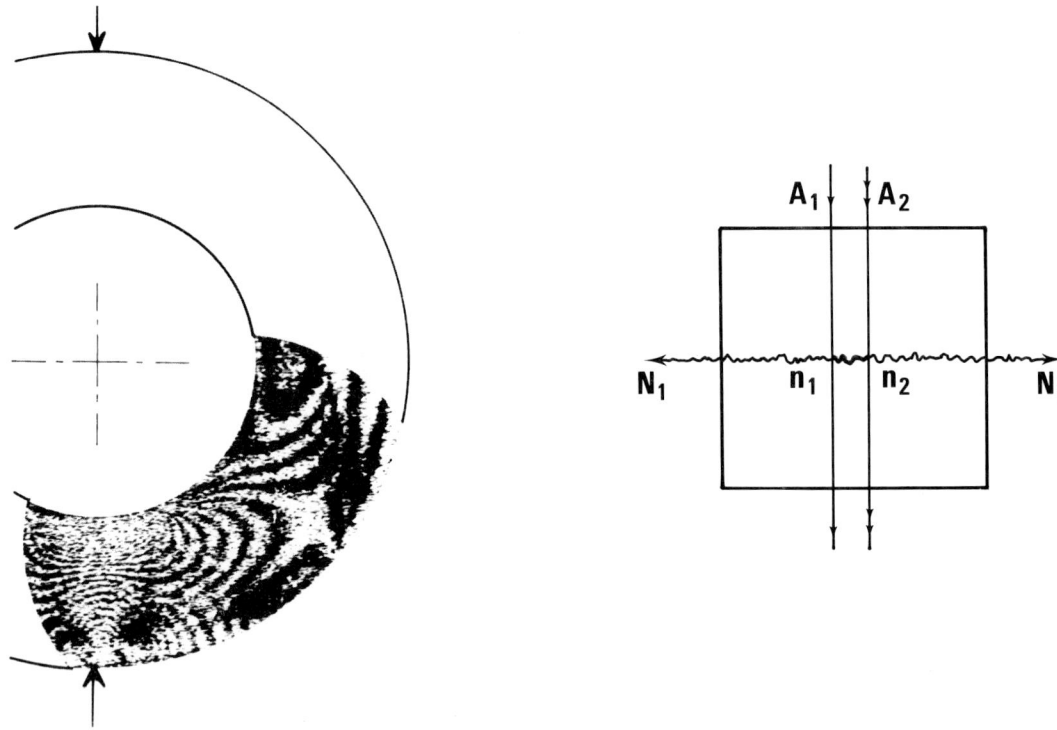

Fig. 11 Correlation fringes are observed in the region "visible" to the Mach-Zenhder interferometer described in figure 10. Between the two exposures, the photographic plate is given a small translation which allows us to apply spatial filtering technique

Fig. 12 Principle depicting the elimination of the effects of path difference along the incident light beams A_1M_1 and A_2M_2

As, at present the experiments are in progress, and so we cannot but present only the first results obtained by simulation technique, the set-up for which is shown in figure 10. In order to have enough light, we have simulated the two diffusing light sheets by introducing two diffusers, one in each arm of the interferometer. The resulting two polarized speckle patterns S_1 and S_2 are superimposed in the π-plane. Between the two exposures, the model is loaded and the photographic plate given a small translation in its own plane. This allows us to carry out spatial filtering, resulting in the contrast enhancement of the correlation fringes. The fringes thus obtained are presented in figure 11. We shall be able to determine the influence of displacement on the diffusing centers, lying within the two light sheets, only after our present experiments with the three dimensional models have been completed. Finally, we shall be able to eliminate the indices variations, between the light sheets A_1M_1 and A_2M_2, by comparing the interferograms obtained by means of diffused light in the directions N_1 and N_2 respectively (fig. 12)

References

1. Nisida, M., and Saito, H., "A new interferometric method of two-dimensional stress analysis" Proc. SESA, Vol. 21, pp. 336-376, 1964.
2. Fourney, M.E., "Application of holography to photoelasticity", Exp. Mechanics, Vol. 8, pp. 37-38, 1968.
3. O'Regan, R. and Dudderar, T.D., "A new holographic interferometer for stress analysis", Exp. Mechanics, Vol. 11, pp. 241-247, 1971.
4. De Bazelaire, E., "Photoelasticimétrie par holographie et polarimétrie", Thèse de Docteur-Ingénieur, Paris, 1972.
5. De Bazelaire, E., "Contribution à l'étude de la polarisation et de la cohérence temporelle en interférométrie. Représentation des phénomènes, applications", thèse de Doctorat d'Etat, Paris, 1974.
6. Chatelain, B., "Photoélasticimétrie holographique : Etude expérimentale de deux méthodes d'observation simultanée ou indépendante des réseaux d'isochromes et d'isopaches relatifs à un modèle unique soumis à une seule sollicitation", Thèse, Besançon, 1972.
7. Chatelain, B., "Holographic photo-elasticity : independent observation of the isochromatic fringes for a single model subjected to only one process", Optics and Laser Technology, Vol. 5, pp. 201-204, 1973.
8. Monneret, J., and Spajer, M., "Photoélasticimétrie holographique : séparation des contraintes dans des matériaux photoélastiques ; possibilités d'application aux modèles à déformations figées", Optica-Acta, Vol. 24, pp. 843-857, 1977.
9. Robert, A., and Guillemet, E., 1964, "New scattered light method in three-dimensional photoelasticity", British Journal of Applied Physics, Vol. 15, pp. 567-578, 1964.

10. Oppel, G., "Photoelastic investigation of three-dimensional stress and strain conditions", Forschung Gebiete Ing. 7, 1936, pp. 240-248 ; NACATM 824, 1937.
11. Durelli, A.J., "Complete experimental solution of three-dimensional elastic problems", Journal of Strain Analysis, Vol. 10, p. 42, 1975.
12. Sanford, R.J., and Parks, V.J., "On the limitations of interferometric methods in three-dimensional photoelasticity", Experimental Mechanics, Vol. 13, p. 464, 1973.
13. Parks, V.J. and Sanford, R.J. "On the role of material and optical properties in complete photoelastic analysis", SESA Spring meeting, Silver Spring, Maryland, 1976.
14. Leendertz, J.A., "Interferometric displacement measurement on scattering surfaces utilizing speckle effect", Journal of Physics E : Scientific Instruments, Vol. 3, pp. 214-218, 1970.
15. Barker, D.B., and Fourney, M.E., "Displacement Measurements in the interior of 3-D bodies using scattered-light speckle patterns", Experimental Mechanics, Vol. 16, pp. 209-214, 1976.
16. Sciamarella, C.A., and Gilbert, J.A., "A Holographic Moiré technique to obtain separate patterns for components of displacement", Experimental Mechanics, Vol. 16, pp. 215-220, 1976.
17. Desailly, R., "Visualization of isoclinics and isochromatics in a birefringent slice optically singled out in a three dimensional model", Optics Communication, pp. 61-64, 1976.
18. Desailly, R., and Lagarde, A., "Application des propriétés des champs de granularité à la phtoélasticimétrie tridimensionnelle", C.R. Acad. Sc. Paris, t 284 (10 jan. 1977) Série B, pp. 13-16.

1st EUROPEAN CONGRESS ON OPTICS APPLIED TO METROLOGY

Volume 136

SESSION 5.1

HOLOGRAPHIC INTERFEROMETRY

**Session Chairmen
Vienot
Dandliker**

DIMENSIONAL METROLOGY OF LENGTH STANDARDS BY HOLOGRAPHIC INTERFEROMETRY WITH PHASE HETERODYNAGE

M. Grosmann, P. Meyrueis
Laboratoire de Spectroscopie et d'Optique du Crops Solide
Associé au C.N.R.S. n° 232, Université Louis Pasteur
5, rue de l'Université, 67000 Strasbourg, France

Introduction

Our goal was to use the technical accomplishments of the L.S.O.C.S., notably in spectroscopy of modulation in the applications undertaken in the applied optics department of this laboratory. Our efforts aimed at elaborating a system permitting us to optically compare with precision different objects. Since the problems becomes more complicated with the complexity of the form of the object, above all in the domain of curved shapes, the study of a comparator of gauge blocks was the first stage of the program. We hope that many of those who are presently using an older system of measurement will benefit from this one, which will be commercialized shortly.

It is impossible to compare two ordinary diffusing objects by holographic interferometry. The microscopic differences of structure prevent any interference. Holographic interferometry is possible with diffusing objects only to study deformations or displacements. There is then a point to point correspondence between the two objects before and after they have changed. Measurements of dimensions changes by interference fringes are possible.

In the problem of grametric comparison of different objects by holographic interferometry it is necessary to use objects with polished surfaces, which limits the applications, since roughly polished objects are the most common in mechanical industry. But we can, by an optical artifice, transform a roughly polished surface into a reflecting surface by lighting it obliquely (with an angle of incidence of θ, if the rugosities are inferior to $\lambda/4\cos\theta$), as shown by Ennos. Most of the rays are reflected and interferences are produced. They will have the same facility of analysis as in classical interferometry. The interfringe becomes $\lambda/2 \cos$ instead of $\lambda/2$. This diminution of sensitivity can be largely compensated for by very sensitive devices to analyse the signal.

The principle of this method combines the advantages of classical interferometry (ease of analysis of the fringes, since they are not localized and completely comparable to those obtained with classical interferometry) and the advantages of holographic interferometry (flexibility, ease and economy of use, and above all the richness of possibilities of comparison of forms). Complete automation of the measurement is possible thanks to modern electronics.

The ends of the gauge blocks being rather well polished, they need only a small angle of incidence in order to behave as good mirrors.

A wavefront reflected by an end of a gauge block is recorded in the form of a hologram by the addition of a reference beam on a sensitive material. In this way the lighting of the hologram by the reference beam reconstructs the reference object which appears superimposed with the measured object.

Theory

The two wavefronts R1 and R2 which interfere produce fringes at a point in the plane of observation, lighting which varies according to the expression:

$$E = E_o (1 + \cos \frac{2\pi\delta}{\lambda}) \tag{1}$$

Our goal is to obtain a precise measurement of the path difference δ by means of an appropriate analysis of the signal. That is, we hope to obtain a practical and precise procedure for interpolation between two fringes.

With this in mind we introduce a phase modulation of frequency ω on one of the wavefronts R1 or R2. We then obtain an expression for the lighting E of the form:

$$E = E_o \left| 1 + \cos(\omega t + \frac{2\pi\delta}{\lambda}) \right|$$

We see that the lighting E will be modulated temporally with a phase $\psi_o = \frac{2\pi\delta}{\lambda}$. The optical signal can then be converted easily into an electrical signal. The measurement of S becomes the measurement of the phase of a sinusoidal electrical signal.

DIMENSIONAL METROLOGY OF LENGTH STANDARDS BY HOLOGRAPHIC INTERFEROMETRY WITH PHASE HETERODYNAGE

Measuring Device

Phase modulation of an optical signal

All interferometric devices, classical or holographic, include a reference wave and a measurement wave. If the interfering wavefronts are perfectly plane and parallel, an interference phenomenon is obtained in which the distribution of intensity is the same throughout the zone of interference. If the wavefront undergoes a global modification of phase which is related linearly with the time, the intensity in the zone of interference will vary sinusoidally with the time. If the wavefronts remain plane and parallel the phase of these periodic variations will be the same at a given time throughout the interference phenomenon. If one of the wavefronts includes a local disturbance of phase constant with the time (an advance, for example), the phase distribution of the phenomenon will no longer be uniform. The translation of one of the wavefronts with regard to the other will cause the zone of disturbance to show a phase variation with regard to the rest of the phenomenon. We thus see that a geometrical phase disturbance independent of time, stable over a wavefront undergoing a global variation of phase and interfering with a reference wavefront, entails a local modification of the phase of the interference phenomenon (advance or delay, according to whether the disturbance is ahead of or behind the rest of the interfering wavefront).

Let us place two detectors in the plane of the fringes, one being arbitrarily chosen as reference, the other being located in the zone of disturbance. The displacement of the fringes in this zone will be converted into electric current. The local disturbances of the wavefront to be studied, an advance for example, will be expressed by an advance in phase of the electric signal of the captors. It is easy to measure electronically this advance or delay of phase of the electric signal. The measurement is made independently of the static distribution of intensity of the interference phenomenon.

From this we can deduce the geometrical characteristics of the disturbance of the measurement wavefront with a precision superior to 1/100 of the wavelength. We can replace the global and linear phase variation of the studied wavefront with a cyclical variation of the phase of the reference wavefront.

A phase disturbance on the wavefront to be studied, such as the one previously considered, will be expressed (with a global cyclic variation of the reference wavefront) as a phase modification of the modulation in phase of the reference wavefront, analysed by the captor system previously described. This process of interpolation of the fringes is the one we have used.

Optical Phase Modulation by Mobile Grating

The modulation is fixed. Useful methods are birefringent crystals, Kerr cells in a particular setup, Pockels cells, a Piezzo electric cell system or ultrasonic diffraction cells. These electro-optical and acousto-optical methods require relatively expensive drivers and their adjustment and reliability are delicate. Let us consider a linear grating being displaced in a direction perpendicular to the spatial modulation in transmission. We suppose that the grating and the light beam are infinite. This beam is diffracted in two orders. For this system, the distribution of amplitude in the Frauhenhauffer region can be deduced from the Fourier Transparence of the transmission function $T(G1,t)$:

$$T(G1,t) = 1 + \cos \left| (2\pi/a)(G1 - vt) \right|$$

The resulting amplitude is :

$$(S,t) = \exp iz\, V_o t \cdot \delta(S) + \frac{1}{2} \exp \left| -i(\frac{2\pi}{a} vt) \delta(S + (\frac{2\pi}{a})) \right| + \frac{1}{2} \exp \left| (i\, \frac{2\pi}{a} vt) \cdot \delta(S - \frac{2\pi}{a}) \right|$$

where V_o is the optical frequency of the incident ray.

Thus the frequency of the light in the +1 and -1 orders has been increased and diminished respectively by v/a. The 0 order is unchanged. With a lens and a mask we can select the desired frequency.

A radial grating is different from the preceeding grating in that the speed and the spacing of the grooves vary with the distance from the axis. At distance r from the latter the frequency in the +1 order will be :

$$v(r) = V_o + (v/a)$$

But here we have $v = rw$ and $a = Kr$ where W is the angular velocity and K a positive constant of the grating.

The parasitical effects arise from errors in tracing of the grating, from variations in angular velocity, and from vibrations caused by poor quality bearings or the fact that the grating is not perfectly flat. All these defects increase the extent in frequency of the modulated signal. The system is, on the other hand, insensitive to errors of alignment.

The equation for the diffraction by a grating with small angles of incidence is $(\theta - \theta_o) = m t/a$, where θ is the angle of diffraction, θ_o the angle of incidence and m the order of diffraction.

Also the net deviation $(\theta - \theta_o)$ and the modulation in frequency are independent of the angle of the grating with the incident ray.

The radial grating must be mounted on precision bearings and powered by a synchronized motor by means of a belt. The grating had a diameter of 8 cm and 500 lines. The modulation was tried from 100 to 500 kHz.

In our method, the image recorded on the hologram has a modulation in phase, while the object is lit with light unmodified in phase.

During the recording the grating is immobile and the two beams have the same frequency. It is not until reconstruction that the disk is made to rotate. The optical phase difference is converted into the phase of the frequency of modulation. The two photodetectors are placed at two points P and R of the object, R being the reference. The relative phase of the modulation is determined by $\psi = \Delta\phi(P) - \Delta\phi(R)$ where $\Delta\phi P$ is the optical phase difference of the two interfering waves at point P. ψ can easily be determined by electronic measurements of phase. (See sketch 1a and 1b).

Acousto-optical Modulator

An ultra-sonic wave applied to a certain type of crystal produces periodic variations in the refraction index of the crystal. For an incident light beam this crystal behaves as a diffraction grating.

A sound wave is applied to the crystal by means of a transductor which transforms a HF signal into a sonic signal.

When the amplitude of the HF wave is modulated the efficiency of the grating is modulated in the same way as a mobile grating. The same occurs with the intensity of the diffracted beam.

The spacing of the grating changes with the frequency of the HF wave, as does the direction of the diffracted beam. (See sketch 2).

Use of Modulators in the System

(See sketch 3)

The best frequency of modulation ω is situated around 100 MHz. To obtain the frequency ω we use two modulators M1 and M2.

At recording the modulators function at the same frequency, which gives 0 as the resulting modulation and permits us to have the same angles as at reconstruction. At reconstruction, M1 gives 70 MHz and M2 gives 70.1 MHz ; the resulting ω is 0.1 MHz.

The phase difference between the two interfering wavefronts is converted in phase of the modulation ω as we have already seen. Thus the length measured between two positions will be deduced from the phase difference ψ of the modulation ω between the two positions.

The quality of the measurement depends to a great extent on the phasemeter, on the signal/noise ratio of the detectors, and on the mechanical stability of the assembly.

The multiples of 360° give the number of fringes. The stability and the sensitivity of the phasemeter should be superior to 0.1°. The signal/noise ratio Sb of the phasemeter should be superior to 20 dB to avoid the multiple zero crossings. The noise on the signal from the captors introduces an error caused by the variations in the zero crossing.

Besides the error of phase due to the amplitude noise, additional phase fluctuations can be produced in the signal. They are caused principally by optical instabilities in the paths of the beams or by instabilities in the positioning of the detector. These two effects are averaged over the integration time of the phasemeter. The global precision of the measurements is measured experimentally. It is around 10^{-3} fringes. The phasemeter counts the multiples of 360°, thus of the fringes, and interpolates on an angle of phase of 0.1°.

The detectors are photomultiplicators, in order to avoid phase noise that can occur with amplifiers working with photocells.

Experiment

The laser we used was a Spectra Physics 170 Argon with a wavelength of 4880 Å. The plates were 649 F Kodak. A hologram density of 0.5 and a ratio of the reference beam to the object beam of 2 were used.

Shrinkage of the photographic plates could in principle influence the results. To insure that such effects do not occur, we reposition the hologram with the reference gauge block still in position in order to compensate the gelatin deformations, or we use thermoplastics.

DIMENSIONAL METROLOGY OF LENGTH STANDARDS BY HOLOGRAPHIC INTERFEROMETRY WITH PHASE HETERODYNAGE

It is well known that the sensitivity vector of the fringe phenomena in holography is at a maximum for the direction of the bisector of the angle between the lighting direction and the observation. With our system, we observe in the direction parallel to the reflected light, so that the sensitivity vector is normal to the ends of the gauge blocks. The system works as a classical Michelson interferometer ; it is used according to the same principles but it is easier to operate.

The fringe spacing is $\lambda/2 \cos \theta$, θ being the angle of the object beam with the end of the gauge block, in our case 30°. The gauge blocks are parallel to the hologram to facilitate checking. Once the positioning is done, we count the fringes during the translation. Otherwise we can measure the difference of spacing of the fringes between the reference gauge block and the gauge block to be measured. The difference in length is $f(\lambda/2\cos\theta)$, f being the ratio of the difference between fringe spacing. At present, the accuracy is of $\lambda/10$. (Fig. 2,3,4).

Simpler and More Economical Fringe Counting System with Lower Accuracy

For measurements of lower accuracy our system can be used in the same way as an economical Michelson interferometer. The length of an object, a gauge block for example, can be measured with an accuracy of one fringe. A digital display gives the exact number of fringes scanned by the detector, even if (because of irregularities) the direction of the fringe movement is reversed.

The signal is produced by four diodes. Two of them are side by side in the fringe pattern ; the two others are outside of this zone and are used as references. The signal is then processed by two amplifiers. The output of these amplifiers is then fed into a logical system governing a two-way counter according to the direction of the scanning of the fringes. Then the signal goes to a display system (Fig. 5).

The detectors are diodes. An amplifier produces the voltage compatible with the TTL logic. A photodiode detects the light from the surroundings and gives a reference signal which is fed into the inverser input of the amplifier. Another diode gives the measured signal applied to the non-inverser input of the amplifier (Fig. 6).

We get + 5V when the diode is in a bright fringe and 0 V when it is in a dark one. In this case the measuring diode gives the same signal as the reference diode DR and the amplifier gives a zero output. In this way, we can work in binary logic.

To detect the direction of the scanning, it is necessary to have two fringe detection systems. The two measuring diodes are positioned side by side in a dark or bright fringe. During the scanning a succession of logical states appears for the amplifier. A and B are the diodes at the logical state 1 ; both are lighted. $\bar{A}$ and $\bar{B}$, the diodes at the logical state 0, are both dark. The scanning of a complete fringe is given by the following sequence : (A,B) , (A,$\bar{B}$) , ($\bar{A}$,B) , ($\bar{A}$,$\bar{B}$). We see that the problem of the direction of scanning is a problem of sequential logic. We will have to take into account all the successive logical states that do not fit with the scanning of a complete fringe.

Since the scanning of a fringe corresponds to four successive states of the couple AB, we have to have four memories for one direction of the scanning and four others for the other direction.

Limitations

The test was done with an 8 cm gauge block. We changed the repositioning system for each size of gauge block. But it is possible to double the holographic setup for each end. The positioning system can be simplified. A slight error in the positioning, as with the other system, adds parasitical fringes and should be avoided. There may be a problem with vibrations that can modify the visibility of the fringes, their recording, or their analysis with a temporal modulation. This problem can be solved by having a rigid base made of one piece for all the components and a small damping system, or by using a piezoelectric compensator.

The positioning system for the gauge block consists of a ball system adapted to holography. The translating system is driven by a piezoelectric motor.

The hologram is held by vacuum on a plate holder with full rotational and translational movement adjustable to $1/10$ of a μ.

The holograms are processed normally, and then either repositioned as described before or processed in situ. In the latter case, there are some practical problems.

Prospects and New Developments

With this experiment, we demonstrate that holography can be used to measure dimensions with high accuracy. A prototype of a commercial instrument can now be developed. It will be completely automated, giving one measurement in less than one minute, using photopolymer or thermoplastic plates and temporal modulation. The system can reach an accuracy of $\lambda/100$, the theoretical limit being $\lambda/1000$.

Instead of using a reference gauge block, an artificial wavefront can be produced by computer. However, the geometrical data of the system must be known and stabilized. In this way, a standard holographic length can be manipulated without difficulties and can be used, for instance, to rapidly check the accuracy in length and flatness of ordinary standards. Furthermore, the standard hologram can be reproduced at low cost.

Conclusion

Although holography has been used very efficiently for the study of diffuse reflecting surfaces, quantitative use of the data was very difficult. Using specular reflection, good results have been obtained in comparing different surfaces. We work with gauge blocks most easily, but our method can be applied to roughly polished objects as well, if the surface profile is not too complicated. It is expected that our results will initiate developments of a new generation of metrological instruments.

Bibliography

1. P. Smigielski, C. Vienot, H. Royer - Holographie Optique, edit. Dunod (1971)
2. M. Françon - Holographie, edit. Masson (1969)
3. P. Meyrueis - Thesis, University of Strasbourg (1974)
4. A. Dandliker, Eliasson, Mottier - The Engineering Uses of Coherent Optics, Cambridge U.P. (1976)
5. Crane - New Developments in Interferometry V. Interference Phase Measurement Applied Optics, vol. $\underline{8}$, 538-544 (1969)
6. J.C. Perrin - L'holographie en métrologie. Bureau National de Métrologie, edit. Chiron (1975)
7. Archbold, Ennos - The application of holography to the comparison of cylinder Bore, J. Sci. Instr., $\underline{44}$, 489 (1967)
8. Dandliker, Ineichen - Quantitative Measurement Through Holographic Interferometry. International Conference on Holography and Optical Data Processing. Jerusalem, Israel (1976)
9. Schwab - Interferenztreifenzäaler mit Analogausgang. I.S.L. Notiz (1976)

DIMENSIONAL METROLOGY OF LENGTH STANDARDS BY HOLOGRAPHIC INTERFEROMETRY WITH PHASE HETERODYNAGE

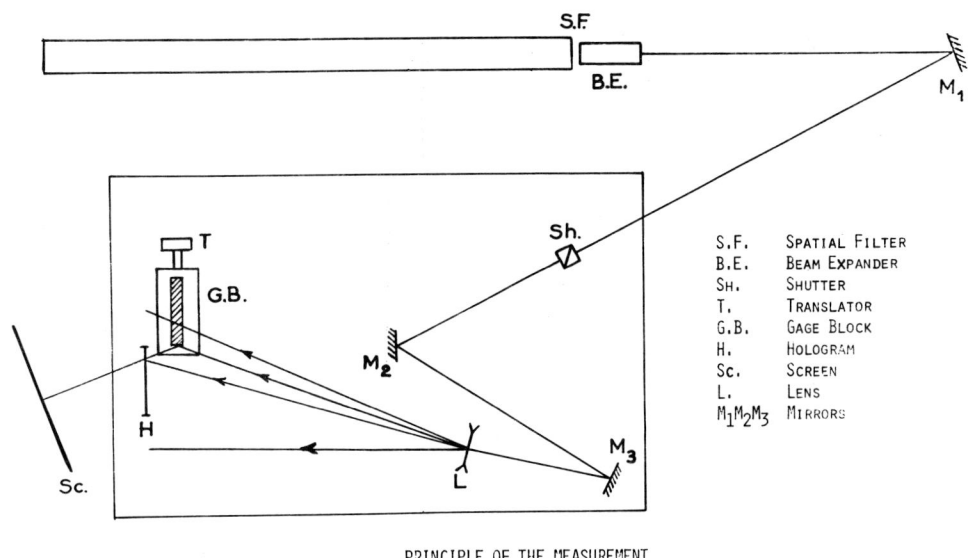

S.F.	Spatial Filter
B.E.	Beam Expander
Sh.	Shutter
T.	Translator
G.B.	Gage Block
H.	Hologram
Sc.	Screen
L.	Lens
$M_1 M_2 M_3$	Mirrors

PRINCIPLE OF THE MEASUREMENT

Fig. 1 a.

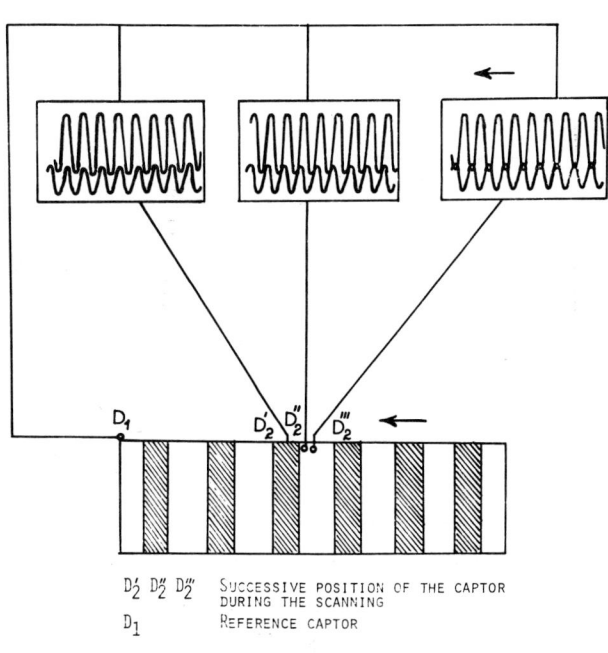

$D_2' \; D_2'' \; D_2'''$ Successive position of the captor during the scanning
D_1 Reference captor

FRINGES ANALYSIS WITH PHASE MODULATION

Fig. 1b

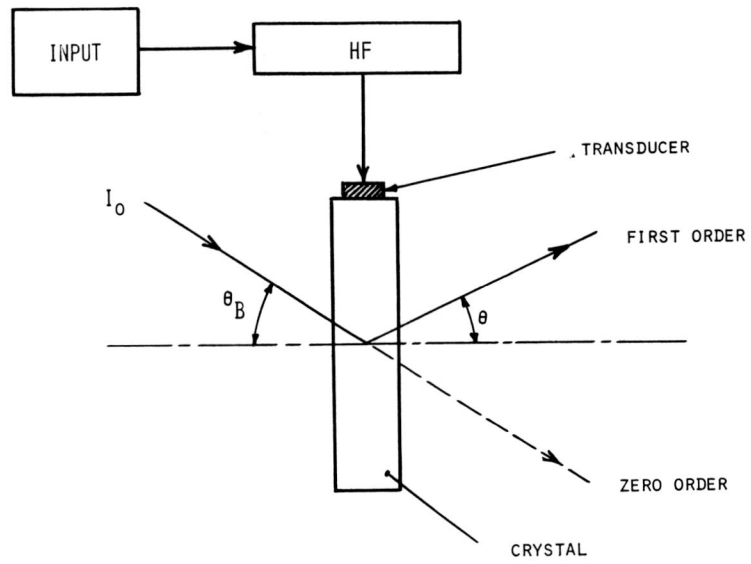

FIGURE 2

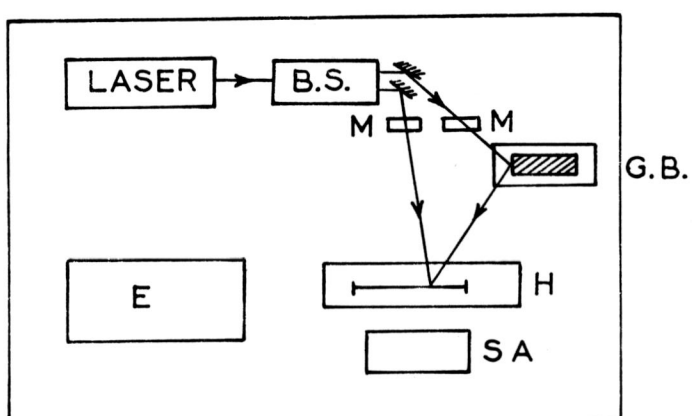

B.S. BEAM SPLITTER   S.A. FRINGE COUNTING SYSTEM
G.B. GAGE BLOCK   M. MODULATOR
H. HOLOGRAM -   E. POWER SUPPLY AND COMPUTER

USE OF A BEAM SPLITTER IN THE COMPARATOR
FIG. 3

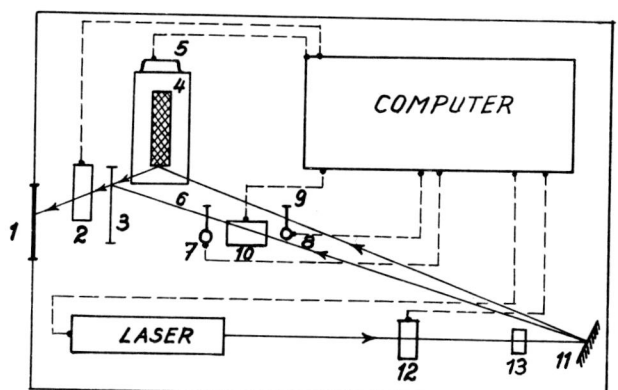

COMPARATOR WITH VARIABLE DENSITY AND WITHOUT BEAM SPLITTER
FIG. 4

1. DISPLAY SCREEN
2. FRINGE COUNTING SYSTEM
3. HOLOGRAM
4. GAGE BLOCK
5. POSITIONING AND TRANSLATOR SYSTEM
6. DENSITY FOR INTENSITY CONTROL REFERENCE BEAM
7. DENSITY AUTOMATIC CONTROL
8. DENSITY AUTOMATIC CONTROL
9. DENSITY FOR INTENSITY CONTROL OBJECT BEAM
10. MODULATOR
11. MIRROR
12. SHUTTER
13. SPATIAL FILTER AND BEAM EXPANDER

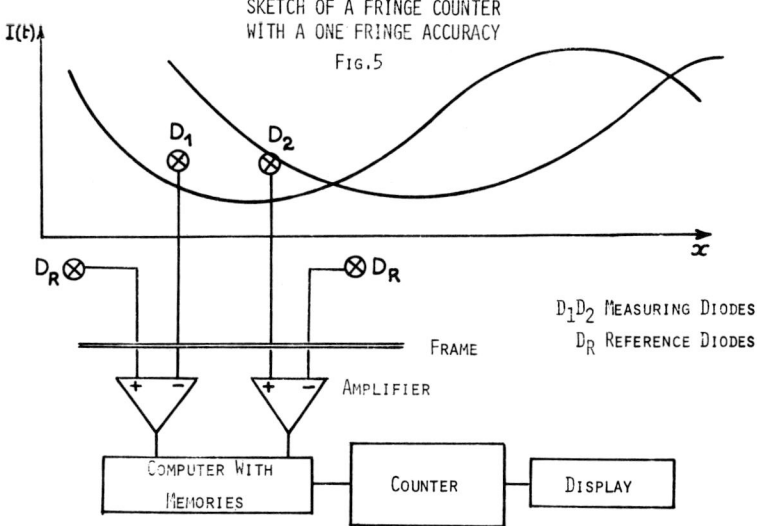

SKETCH OF A FRINGE COUNTER WITH A ONE FRINGE ACCURACY
FIG. 5

$D_1 D_2$ MEASURING DIODES
D_R REFERENCE DIODES

THIS SYSTEM IS CHEAPER AND EASIER TO MAINTAIN THAN THE TEMPORAL MODULATION ONE.

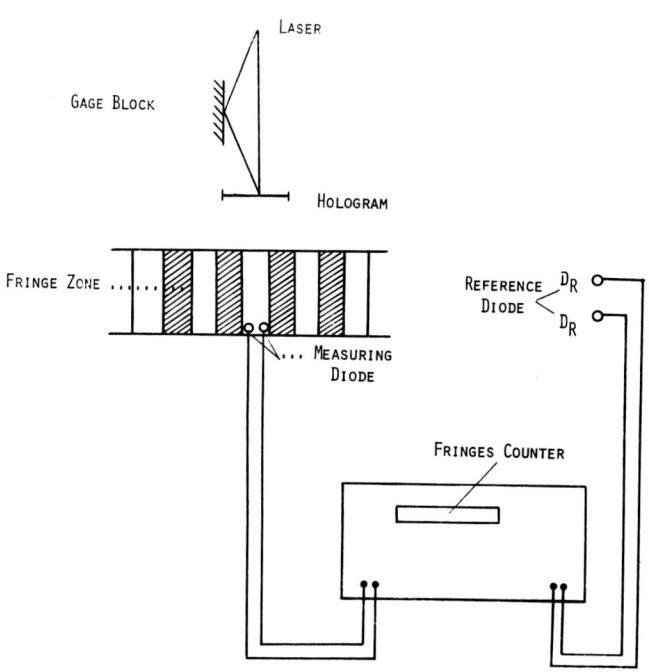

ECONOMICAL FRINGE COUNTING SYSTEM

FIG. 6

DOUBLE EXPOSURE HOLOGRAPHIC INTERFEROMETRY:
APPLICATION TO NONDESTRUCTIVE TESTING AND TO BREAKING POINT MECHANICS

Nicole Jolly, Jacques Poirier
Commissariat à l'Energie Atomique
Paris, France

Abstract

Holographic interferometry applied to non destructive testing is a powerful method, usable with any structure (here, pressure spherical reservoirs) provided that it can be acted upon, mechanically or thermically, so as to present two different dimension conditions corresponding to the two superimposed holograms.

By revealing the shape of the plastic zone, in the case of a breaking strength test piece acted on by a tractive effort, the method allows to validate a numerical simulation, the results of which are then introduced into a program for calculating the Rice integral, used as a criterion of the breaking point for elastoplastic materials subjected to a plane stress.

Introduction

Two applications in mechanics of the double exposure holographic interferometry method are presented. The first concerns the non destructive testing of spherical pressure reservoirs, the second relates to breaking point mechanics.

The non destructive testing of spherical reservoirs

The method was tried out on single and double layer reservoirs. The optical device seen in figure 1 is the conventional assembly used for double exposure holographic interferometry experimentation. At the output of the Helium-Neon laser (power 15 mW) a semi reflecting blade divides into two the coherent light beam. The two beams obtained : object beam and reference beam cause interference on the holographic plate. A first hologram of the object subjected to stress is recorded. The stress condition of the object having varied slightly under the effect of loading, a second hologram is recorded on the same holographic plate. On restitution, the two superimposed holograms give rise to a system of interference fringes which appears on the object. This network is characteristic of the dimensional evolution of the object between two exposures.

The non destructive testing of a single layer reservoir

The reservoir is formed from two thin steel hemispheres assembled in the equatorial plane by electron beam welding. It is acted on by hydraulic pressure by means of a system directly adapted on the holography bench (figure 1). A first hologram is recorded for an internal pressure P. Load-shedding ΔP of some 10^4 Nm^{-2} is carried out. A second hologram corresponds to the pressure $P - \Delta P$. The interference of the two holograms produces, on restitution, a network of fringes superimposed on the object. Figures 2, 3, 4 and 5 show the systems of fringes obtained for increasing load-sheddings : 1, 3, 8 and 13 10^4 N m^{-2}. The disturbed network reveals different zones on the surface of the reservoir. There can be distinguished the welded zone in the equatorial plane, on both sides, the zone thermically affected by welding then, moving away from the welding bead towards the two poles, the unaffected zone. The disturbances of the network are connected with the variation of the mechanical characteristics of the material in the welded zone or in the zones affected thermically by the welding which causes differences in the mechanical behaviour, so different movements of the surface of the reservoir acted upon under pressure.

Non destructive testing of a double layer reservoir

The reservoir comprises an internal layer formed from two light alloy hemispheres welded in the equatorial plane by electronic beam and an outer layer similar to the single layer spherical reservoir previously described. In manufacture, there is provided a clearance between the two layers of the reservoir at rest. When the reservoir is subjected to an internal hydraulic pressure, contact points appear between the two layers. This is what can be seen in figures 6, 7 and 8 which show the networks of fringes obtained by interference of two holograms produced with constant load-shedding of 2 10^4 N m^{-2} from three different pressures. The disturbances of the network which correspond to the contact zones between the two layers develop when the initial pressure increases.

With this method, the behaviour of each part can be observed individually and compared, for example, with that provided by the research department. It is a powerful non destructive testing method, adapted to any structure provided that it is capable of being acted upon, mechanically or thermically, so as to present two different stress conditions corresponding to the two holograms which will be made to interfere.

Application to breaking point mechanics

The holographic interferometry method is applied to a breaking strength test piece acted on by a tractive effort. The experimental results are compared with those of a numerical simulation carried out conjointly.

The breaking strength test piece

The notched breaking strength test piece, whose dimensions are noted in figure 9, was pre-cracked by means of a fatigue machine. The pre-cracking conditions are the following :

The load applied is equal to : $10\,000 \pm 5\,000$ N to initiate the crack, to $10\,000 \pm 4\,000$ N to propagate it over 6 mm for 150 000 cycles at 20 Hz.

The test piece is formed from Z2 CND 17 13 steel, a high consolidation material. In the course of a tensile test carried out on a cylindrical test piece taken from the same ingot as the breaking strength test piece, the rationalized tension curve of figure 10 was plotted.

The system of putting under load

For acting on the breaking strength test piece, a tensioning machine was designed and adapted on the holography bench (figure 11). The optical device used is the same as in the preceding application. To satisfy the demands of the test the load must be uniformly spread out over the edges of the test piece and impose thereon solely a movement of translation in the plane perpendicular to the direction of observation. The delicate adjustment was considered satisfactory when we obtained by double exposure two identical fringe systems on the mobile jaw and on the slightly stressed test piece ($2 \cdot 10^6$ N m^{-2}) : rectilinear fringes parallel to the axis of the crack, characteristics of the expected movement of translation.

Revelation of the plastic zone by holographic interferometry

For a certain value of the tensile stress, the plastification begins at the bottom of a crack and is propagated, causing locally a variation of thickness of the test piece which is not inconsiderable with respect to that reigning in the rest of the test piece, acted upon in the elastic region. It is this variation of thickness, connected to the development of the plastic zone which is revealed by holographic interferometry.

The loading corresponding to the appearance of the plastic zone at the bottom of the crack is empirically determined by the real time method. A hologram recorded for a tensile load T is developed and replaced meticulously. Load T increases progressively. The test piece observed through the first hologram is superimposed on a fringe network which develops with the load. The appearance at the bottom of the crack of a disturbance of the fringe system enables the tensile load T_p which causes the beginning of plastification to be found.

We proceed now by double exposure. The tensile stress T_p is applied to the test piece, a first hologram is recorded. The stress is increased by an increment ΔT and a second hologram is superimposed on the first on the same holographic plate. The interference of the two holograms gives, on restitution, the network of figure 12 (tensile stress : $4.2 \cdot 10^6$ N m^{-2}) in the test piece. Only the half test piece connected with the mobile jaw is presented. The disturbance at the bottom of the crack defines the plastic zone. The fringe systems presented in figures 13, 14 and 15 are obtained for increasing charges T, the increments ΔT remain the same. The tensile stresses in the test piece are then 7.2 9.2 and $10 \cdot 10^6$ N m^{-2}. The representational disturbance of the plastic zone extends when the loading increases.

Numerical Simulation

Conjointly with these experiments a numerical simulation was prepared. It calls on the finished element method (code PAM.NEPD) associated with a law of elastoplastic behaviour. From the geometry of the structure a mesh-work is defined. Figure 16 shows the mesh-work of the half test piece, refined in the zone of the crack, where the plasticity first develops during loading. The law of elastoplastic behaviour of the material used takes account of isotropic cold-drawing. It is taken into account point by point in accordance with the rationalized curve given in figure 10. For reasons of symmetry the calculation is carried out on a half test piece. The nodes located in the extension of the crack are compelled to move along the axis thereof, the nodes of the crack properly speaking are entirely free. This numerical simulation gives the movements at each node of the mesh-work, the distortion and stress tensors as well as the equivalent stress for each element. These results lead to the definition of the shape of the plastic zone for a given loading. Figures 17 and 18 show the shapes obtained for tensile stresses of 4.2 and $7.2 \cdot 10^6$ N m^{-2} in the test piece. They are compared with those obtained experimentally. In figure 19 the calculated shape is framed by two experimental plots. Since the fringe network is not very dense for 10^6 N m^{-2} the accuracy of the plot is not as good as in the case of figure 20.

The quite satisfactory agreement between the calculated shapes and those obtained experimentally allow us to validate the numerical simulation achieved.

Application to the calculation of the Rice integral

The results of the numerical simulation are introduced into a program for calculating the Rice integral J, used as criterion of the breaking point for elastoplastic materials subjected to a plane stress. When the value of this integral, which represents the potential energy variation P of the structure when the crack is propagated by a length dl ($J = -\frac{dP}{dl}$), exceeds a critical value J_c, an experimentally determined physical magnitude, we finish up at breaking point.

Applied to the breaking strength test piece, the holographic interferometry experimental method, by validating for a given material the numerical simulation presented, allows the risk of breaking to be determined by comparing the calculated value of J with the critical value J_c, a physical characteristic of the material.

Fig. 1. A device used for the non destructive testing of spherical reservoirs.

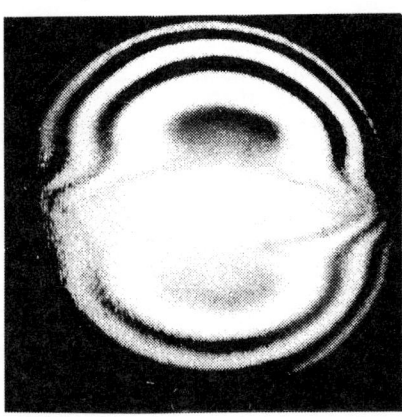

Fig. 2. A single layer reservoir - Superimposition of two holograms corresponding to a pressure variation of 10^4 N m^{-2}.

Fig. 3. A single layer reservoir - Superimposition of two holograms corresponding to a pressure variation of $3\ 10^4$ N m^{-2}.

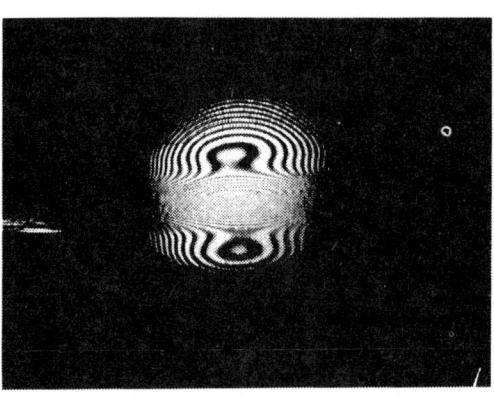

Fig. 4. A single layer reservoir - Superimposition of two holograms corresponding to a pressure variation of $8\ 10^4$ N m^{-2}.

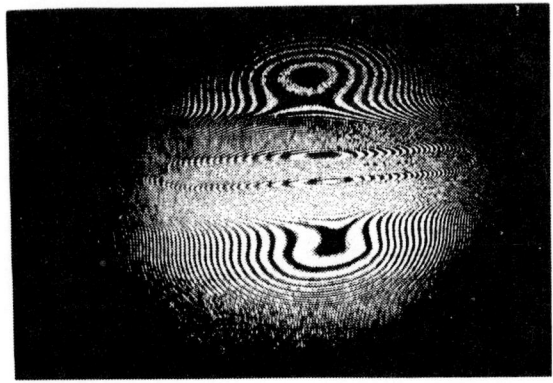

Fig. 5. A single layer reservoir - Superimposition of two holograms corresponding to a pressure variation of $13\ 10^4$ N m^{-2}.

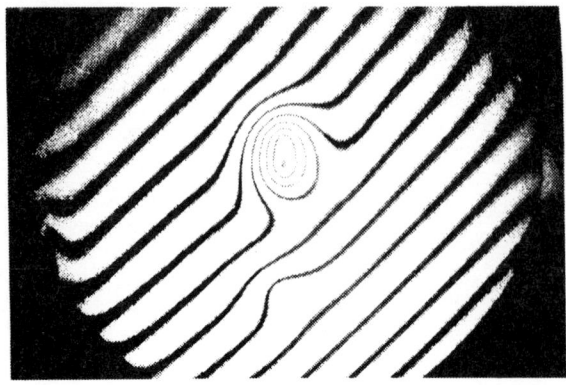

Fig. 6. A double layer reservoir - Superimposition of two holograms corresponding to a pressure variation of ΔP. The first hologram is recorded for a pressure $P = P_1$.

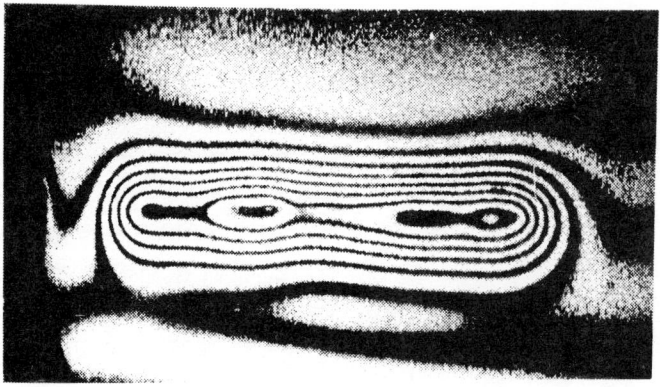

Fig. 7. A double layer reservoir - Superimposition of two holograms corresponding to a pressure variation ΔP. The first hologram is recorded for a pressure $P = P_2$ such that $P_2 > P_1$.

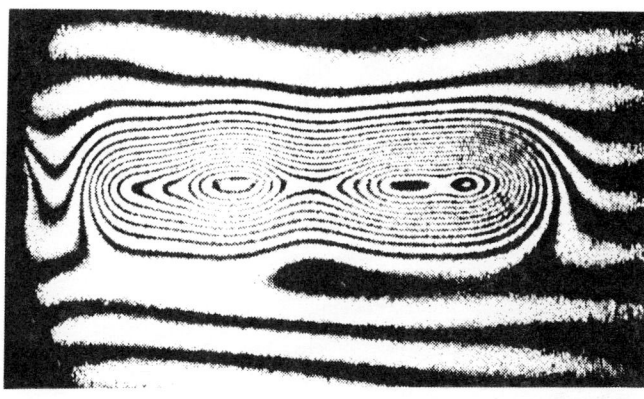

Fig. 8. A double layer reservoir - Superimposition of two holograms corresponding to a pressure variation of ΔP. The first hologram is recorded for a pressure $P = P_3$ such that $P_3 > P_2$.

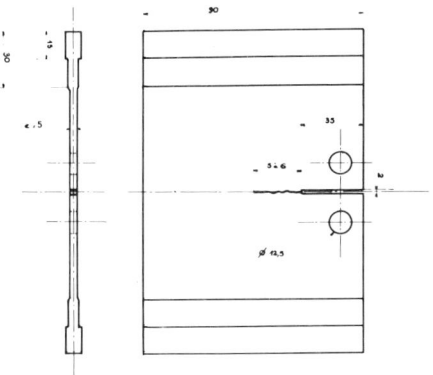

Fig. 9. Pre-cracked notched breaking strength test piece.

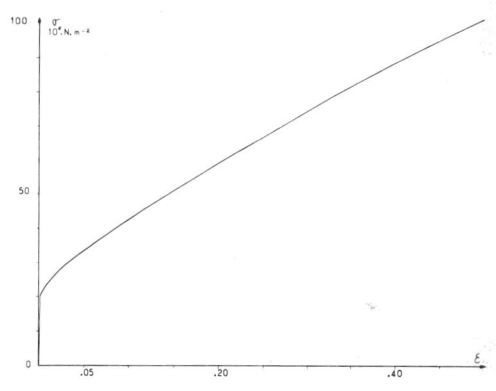

Fig. 10. Rationalized tensile curve for Z 2 CND 17 13 steel.

Fig. 11. A device used for making visible the plastic zone at the bottom of the crack.

Fig. 12. Interference fringe network obtained for a stress in the test piece of $4.2 \; 10^6 \; N \; m^{-2}$.

Fig. 13. Interference fringe network obtained for a stress in the test piece of $7.2\ 10^6\ N\ m^{-2}$.

Fig. 14. Interference fringe network obtained for a stress in the test piece of $9.2\ 10^6\ N\ m^{-2}$.

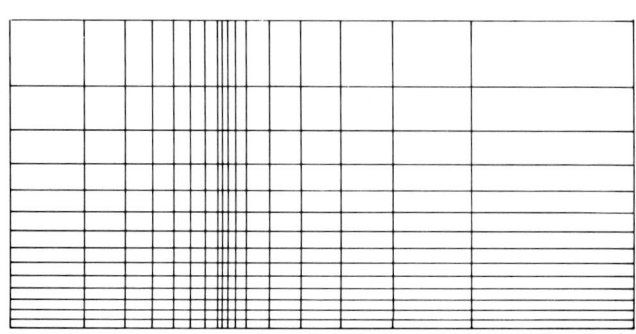

Fig. 15. Interference fringe network obtained for a stress in the test piece of $10\ 10^6\ N\ m^{-2}$.

Fig. 16. Mesh-work of the half test piece.

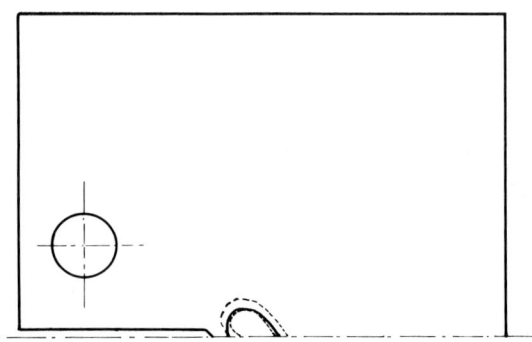

Fig. 17. Comparison of the plastic zone shapes obtained experimentally (___) and by numerical simulation (_ _ _ _) for a stress in the test piece of $4.2\ 10^6\ N\ m^{-2}$.

Fig. 18. Comparison of the plastic zone shapes obtained experimentally (___) and by numerical simulation (_ _ _ _) for a stress in the test piece of $7.2\ 10^6\ N\ m^{-2}$.

HOLOGRAPHIC INTERFEROMETRY APPLIED TO MINIMAL WEAR MEASUREMENT

J. T. Atkinson and M. J. Lalor
Department of Mechanical, Marine and Production Engineering,
Liverpool Polytechnic, Liverpool, U.K.

Abstract

The techniques of holographic contouring, holographic image subtraction and similar techniques are reviewed and discussed as methods for measuring minimal wear in engineering situations. These techniques enable the change in shape, size and surface microdeformation of wear test specimens to be measured, producing results which cannot be obtained using conventional methods.

Introduction

Wear properties of materials are usually measured in tests using pin and disc, and other similar machines; the operation of these tests involves the removal of the material under examination from a sample of known dimensions by rubbing the sample against a revolving disc, under known conditions[1]. The amount of wear is then related to the sliding velocity (u) and the applied pressure (p); the volume wear rate (V_r) is then calculated from:

$$p.u. = K V_r \tag{1}$$

where K is a constant of dimensions $Kg\, m^{-3}\, s^{-2}$. Other methods for examining wear (for example, radioactive tracing) are also available.

Standard tests then furnish the design engineer with the wear data of the available materials. It is then his task to minimize wear by controlling the so-called 'p.u.' curve. Unfortunately this simple procedure is complicated by the fact that wear rates vary unpredictably with design parameters and the wear process is little understood in most cases.

There is a need therefore for techniques which can measure wear rates in circumstances other than the standard test outlined above, i.e. in the real engineering situation. This paper is concerned with the application of holographic interferometry to the measurement of wear, in particular minimal wear. Before considering the optics it is necessary to look at the specification of wear more fully.

Consider the surface shown in Fig. 1A; S_1 is the original component surface and S_2 is the surface after the wear has occurred. These two surfaces are characterised by a surface height function:

$$S_1 = f_1(x,y)$$
$$S_2 = f_2(x,y)$$

For the purpose of this paper, two degrees of wear are defined as:-
(i) Complete refinishing
(ii) Incomplete refinishing.

<u>Complete refinishing</u> is said to have occurred when the distance between the centrelines

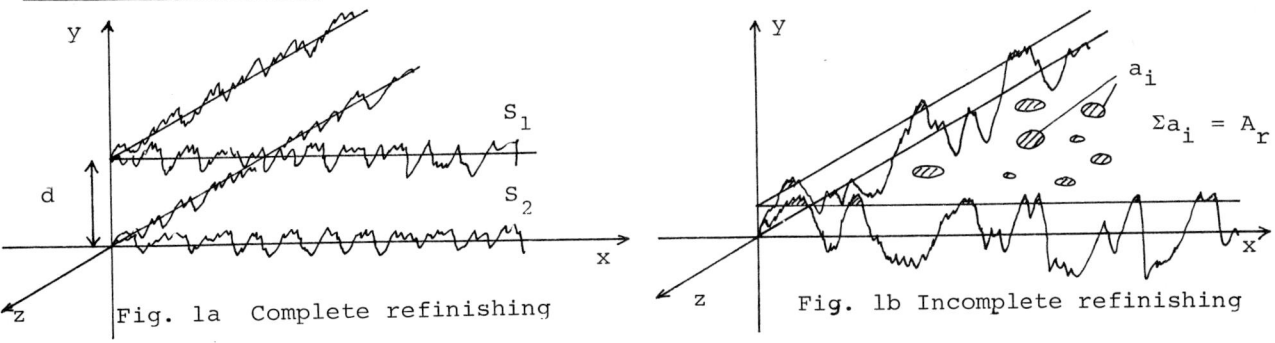

Fig. 1a Complete refinishing   Fig. 1b Incomplete refinishing

of the two surfaces (d) is greater than R_z (the peak to valley height). The total wear in this case can be considered as the volume between S_1 and S_2.

Incomplete Refinishing is said to have occurred when the wear process has removed those parts of S_1 which lie above a certain height from the centreline of the surface; as shown in Fig. 1B. An essential feature of this degree of wear is that some portion of the surface remains exactly the same throughout the experiment. This type of wear will also be referred to as microstructure deformation. The wear in this case can be characterised by the total area of microstructure deformation A_r.

These two degrees of wear require two different measuring techniques:
 (i) Holographic contouring
(ii) Holographic matched filtering and/or holographic image subtraction.

The remainder of this paper is mainly concerned with the evaluation of the above techniques when applied to the problem of wear measurement in bioengineering materials.

Holographic contouring

Of the three main methods for producing contour maps using holography, the dual index method is the cheapest(2). A typical contouring arrangement is shown in Fig. 2; the object to be contoured is sequentially immersed in liquids of refractive index n_1 and n_2 during two exposures of the hologram. Upon reconstruction of this hologram, fringes which trace loci of equal distance from the plane of the window can be seen localised on the object surface. The height difference between any two adjacent fringes will be given by(2):

$$\Delta h = \frac{\lambda}{2|n_1 - n_2|} \qquad (2)$$

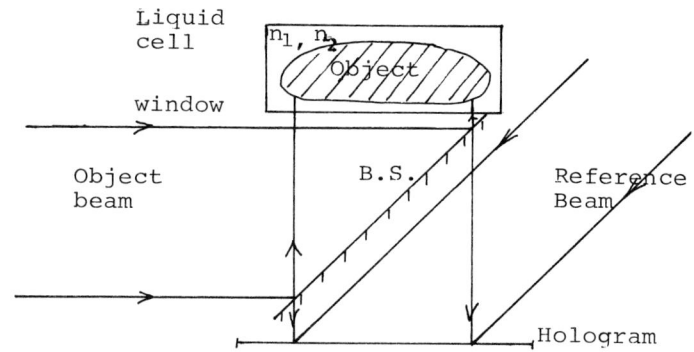

Fig. 2 Dual Index Holographic Contouring

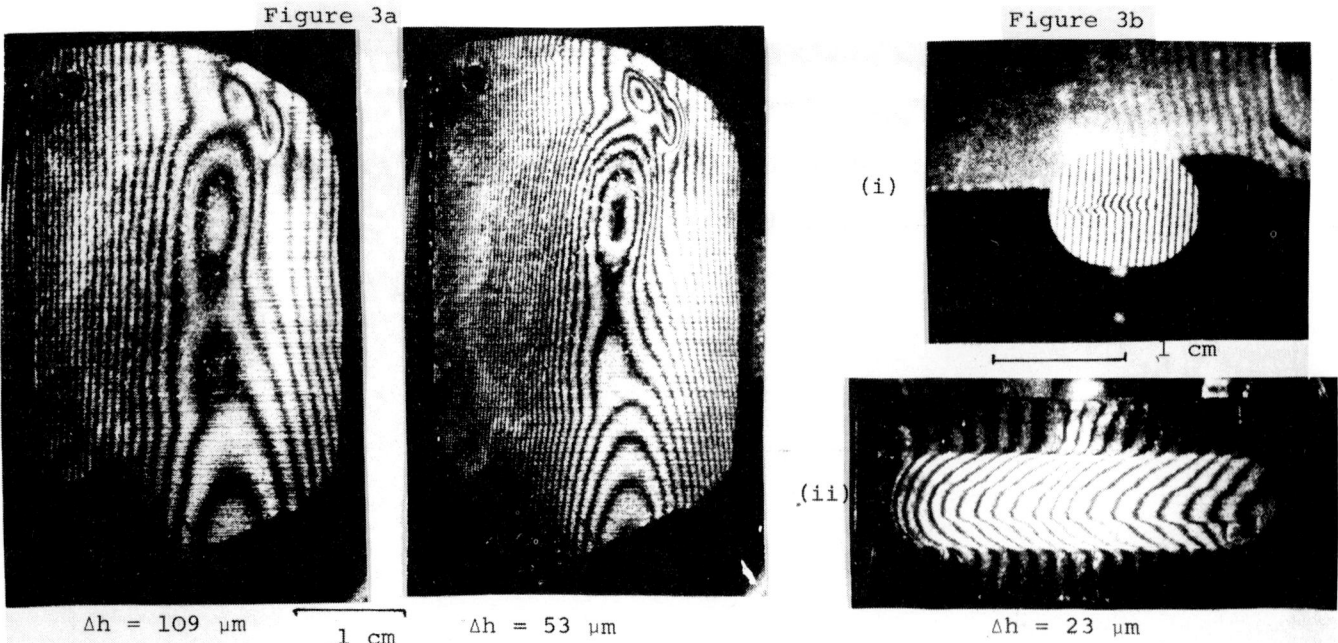

Figure 3a $\Delta h = 109\ \mu m$ $\Delta h = 53\ \mu m$
Typical tibial component of knee prosthesis

Figure 3b $\Delta h = 23\ \mu m$
Dental amalgam samples

The contouring depth, Δh, is then infinitely variable; the lower limit is dependent upon the surface roughness of the object.

Dual index contouring can be used to evaluate wear by contouring the test piece at intervals during the experiment and determining the change in volume using these maps. For more complicated shapes, a computer may be required to estimate this change in volume.

Fig. 3 shows some examples of maps made using the dual index contouring method. Fig. 3a shows two maps of a typical tibial component of a knee prosthesis. The contouring depths are 109 µm and 53 µm as indicated. The component is made of H.D.P. and was vacuum flash coated with aluminium before contouring to ensure good visibility of the fringes. No maps of the unworn component are available, but a number of conclusions can be drawn from examining the maps shown. The worn area can be identified as an ellipse 4 cm by 2.5 cm near the middle of the component, the major axis of the ellipse is not parallel to the posterior anterior direction (vertical) indicating slightly incorrect insertion. The depth of wear in the centre of the ellipse can be estimated as $\sim$ 150 µm by joining two extremes of a contour line which passes through the worn area with a straight line and counting the number of fringes between this line and the actual contour line.

Fig. 3b shows two samples of dental amalgam, worn in a simulation rig. The contouring depth of these maps is 23 µm. A rough calculation of the volume of material removed from sample (i) gives a value of $.5 \times 10^{-10}$ m^3. The surface of sample (ii) is slightly concave and the depth of wear can be roughly estimated as half a fringe i.e. 10 µm.

Holographic contouring may also be used to measure surface roughness as well as shape during wear tests [2,3,4]. Examination of Fig. 3a shows that where rubbing has decreased the R_a value of surface roughness the visibility of the contour fringes has increased. Tsuruta[2] and Ribbens[3] have formulated this dependence of fringe visibility upon surface roughness, viz.

$$\nu \simeq 1 - 8\pi^2 \sigma^2/\lambda_f \qquad (3)$$

where $\lambda_f = \lambda/|\Delta n| = 2\Delta h$ and σ is the r.m.s. surface roughness.

The visibility of contour fringes can be measured by scanning the real image of the object. As these fringes localise on the object surface, the surface roughness in equation (3) can be related to a defined area on that surface.

Holographic Matched Filtering

Marom[5] has used holographic matched filtering to detect metal fatigue. An arrangement similar to that used by Marom has been investigated. The arrangement is shown in Fig. 4. A hologram of the test piece is made in the usual way and repositioned kinematically. If only the "object wave" is allowed to reilluminate the hologram, then the reference wave will be reconstructed. The object is a diffuse reflector, therefore, for reconstruction to occur the hologram and object must be repositioned to within $\sim \lambda/10$. The intensity of the reconstructed beam (I_R) will be proportional to the intensity of the illuminating beam and the correlation between the construction and reconstruction object beams[5]. Marom expressed the correlation between the construction and reconstruction beams in terms of the cross correlation function of their light distribution, i.e.

$$M_{OK} = 1 - C_{OK}(O,O)$$

where M_{OK} represents the measure of correlation

and $C_{O,K}(\xi,\eta) = \iint_A f_O^*(x,y) f_K(x+\xi, y+\eta) \, dx \, dy.$

where $f_O(x,y)$ and $f_K(x,y)$ represent the light scattered from the surface(s) under examination.

An experiment was performed using a 1 cm x .5 cm ground aluminium sample as the object. Prior to reconstruction ten 1 mm x 5 mm strips of paper were attached to the surface of the object, covering it completely; these strips were removed during the course of the experiment, and were used to simulate various degrees of microstructure deformation. The intensity of the reconstructed reference beam was found to vary with area of microstructure deformation (A_r) as shown in Fig. 4. The results are shown for one experiment only.

It can be seen that, empirically, I_R decreases as A_r increases.

Ghost Image Filter

It is well known that a suitably recorded speckle pattern can act as a hologram[6]. Consider Fig. 5a. An object was illuminated as shown with a collimated beam, and the speckle pattern produced at the plate recorded photographically, the plate being returned to its original position kinematically. If a mask M is now placed in the illuminating beam so that only a small area of the object surface (called the reference area) is reilluminated, then a "ghost" image of the remaining area of the object will be seen when viewed through the plate. The quality of this image is strongly dependent upon the precise relocation of the plate and/or the object. The intensity of the image will depend upon the size of the reference area. The object is diffuse, and any change in the microstructure of the reference area should manifest itself as a decrease in the intensity of the ghost image, in much the same way as the reconstructed reference beam intensity drops with increased microstructure deformation in Marom's method. Fringes can be generated on the ghost image in the same manner as in ordinary holographic interferometry. Fig. 5b shows two ghost image reconstructions of the same object. The reference area is the bright central part of the image. Fringes were generated by moving the plate holder by 50 μm between two otherwise identical exposures of the plate. The upper and lower photographs were taken with

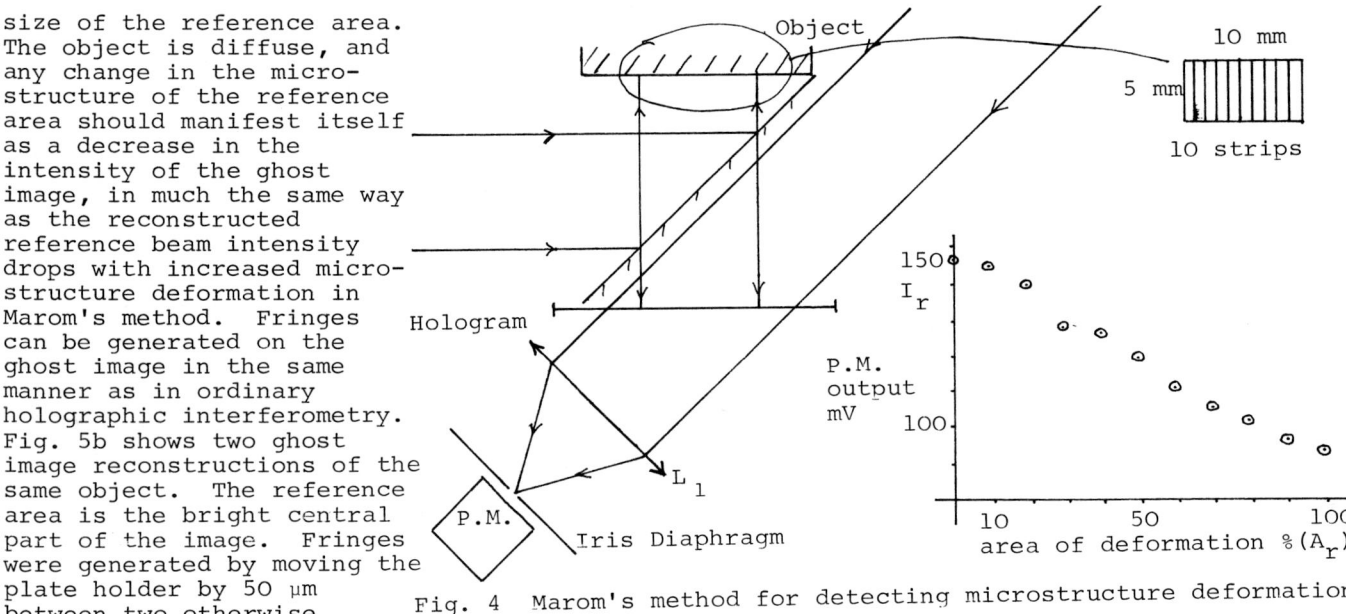

Fig. 4 Marom's method for detecting microstructure deformation

zero and 98% microstructure deformation of the reference area surface respectively. It can be seen that the quality and intensity of the image formed with zero microstructure deformation is greater than the image formed with 98%. But, however, the visibility of the fringes appears to be better in the latter case. No explanation of this phenomenon is offered here, but it is reasonable to suppose that measurement of these fringe visibilities, as well as the intensity of the images may help to assess the area of microstructure deformation of the reference area, after the plate has been taken.

As methods for measuring the area of real contact in wear tests holographic matched filtering seems to be inadequate in that the actual area of plastic deformation cannot be located. In the majority of cases this is desirable, and the next section deals with methods for doing just that.

Holographic Image Subtraction

<u>Introduction</u>

A correctly repositioned real time hologram will act as an image subtracting device[6]. For various reasons (e.g. emulsion shrinkage) this simple arrangement is not practical. Other methods for the subtraction of images in holography can be considered as direct or indirect.

<u>Direct Methods</u>. If, between two otherwise identical exposures of a double exposure hologram, the relative phase between the object and reference beams is changed by $\pi/2$,

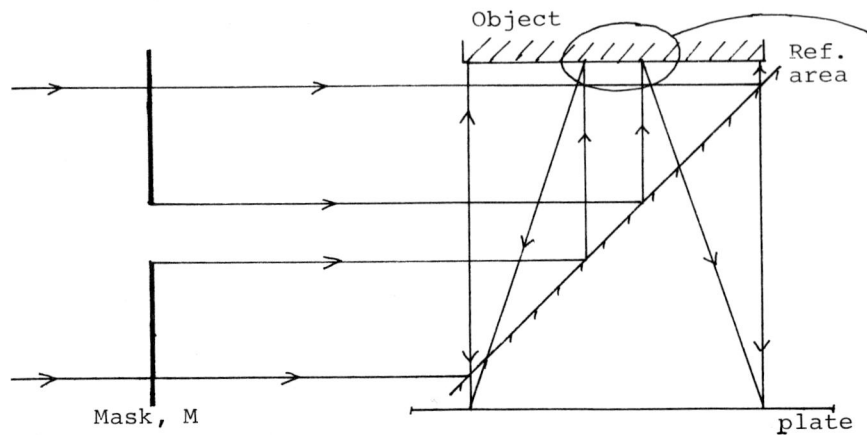

Fig. 5a Ghost image matched filter

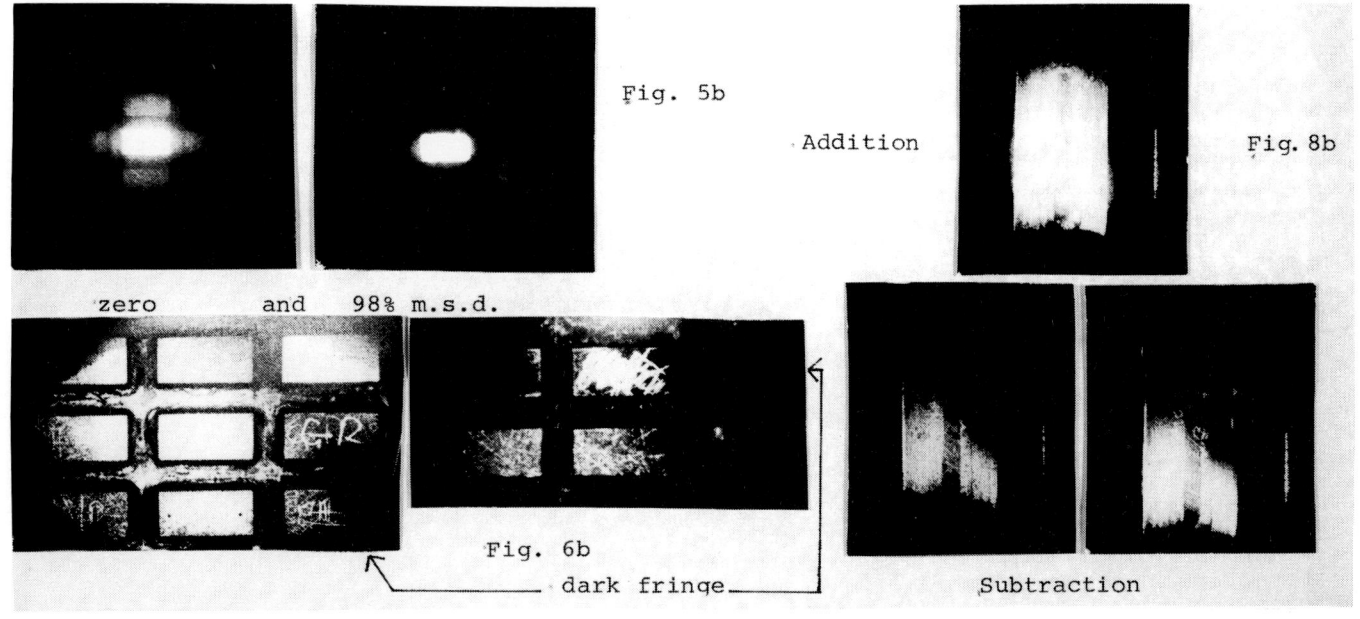

Fig. 5b zero and 98% m.s.d.

Fig. 6b dark fringe

Fig. 8b Addition

Subtraction

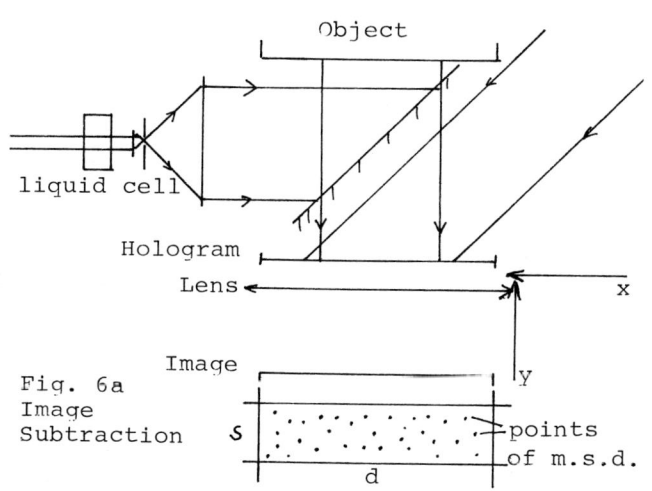

Fig. 6a Image Subtraction

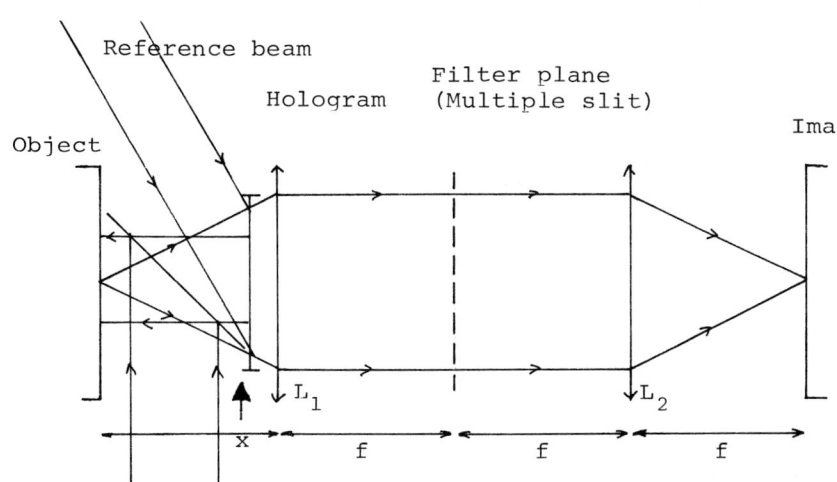

Fig. 7 Matsuda's Method

Fig. 8a

then the ensuing reconstruction will correspond to the subtraction of the two identical images; now consider that some deformation of the surface microstructure has occurred between the exposures (in most cases of interest this deformation consists of small areas of plastically deformed asperities[7]).

Consider Fig. 6a. The object is a plane but rough metal specimen located kinematically in the object position. If two double exposure holograms are made, one to serve as a reference for the intensity measurement (hologram A); the other (hologram B) is made with a $\pi/2$ phase shift between the two object beams; also the object is made to undergo some surface microstructure deformation between exposures such as that shown in Fig. 1b.

In principle, the area of plastic deformation can now be estimated by measuring the image intensity of the two reconstructions.

The ratio of these image intensities will be given by

$$\frac{I_{\pi/2}}{I_O} = \frac{\Sigma a_i}{x\,y} \frac{I}{I} = \frac{A_r}{A} \qquad (4)$$

where I = the intensity/unit area of the image surface for zero phase difference between the images,

$I_{O,\pi/2}$ = the reconstructed intensity of the image measured over the area of the photodetector, when the phase difference between the images is zero or $\pi/2$ respectively,

Σa_i = a_r = the area of plastic deformation,
A = $s\,d$ = the photodetector area.

The purpose of the techniques described in the remainder of the paper is to perform the measurement described above. The enormous problem in performing these experiments is achieving the $\pi/2$ phase shift over the entire object surface in a predictable and accurate fashion.

Fig. 6 shows some examples of subtracted images, using the set up shown, achieving the $\pi/2$ phase shift by changing the index of the liquid in the cell between exposures. The microstructure deformation in Fig. 6b consists of scratch marks on the surface of the object. These marks can be seen clearly against the background of a dark fringe.

Indirect Image Subtraction

(i) It is well known that the visibility of fringes formed in holographic interferometry decreases when microstructure deformation of the object occurs between exposures[4,8,9]. The authors have shown that when a set of Youngs fringes are formed on the object surface, then, in principle, the area of plastic deformation can be found from the ratio of the visibilities of the fringe patterns corresponding to the undeformed (ν_A) and deformed (ν_B) surface state[4] viz.

$$\frac{\nu_B}{\nu_A} = 1 - \frac{a_r}{S} \qquad (5)$$

Equation (5) can be seen to be directly equivalent to equation (4).

(ii) Matsuda et al have also devised a method for detecting surface microdeformation using holographic interferometry[8,9]. This method involves spatially filtering the image of the test piece, as shown in Fig. 7. The first exposure of the hologram is made with the undeformed test piece as the object and the hologram at x_0, the second exposure is then made after the surface microdeformation has occurred and the hologram has been moved to x_1. Upon reconstruction, the fringe pattern formed by the displacement Δx (= $x_0 - x_1$) of the hologram localises in the focal plane of lens L_1. Now by placing a multiple slit in this plane the images in the back focal plane of L_2 can, by suitable adjustment of the filter be made to correspond to the subtraction and addition of the deformed and undeformed images of the object. Problems encountered by the authors in using this method are that the localisation of the fringe pattern varies unpredictably when actual wear tests are performed, which means that individual spatial filters must be manufactured for every wear test. It is hoped to produce a simpler and faster method.

(iii) More recently, Dändliker et al[10] have suggested the use of two reference beam holographic interferometry as a method for subtracting holographic images. One possible experimental arrangement is shown in Fig. 8. The two reference beams are used sequentially to record the hologram of the object in its undeformed and deformed states. During reconstruction, if the hologram is relocated exactly, movement of the hologram in the X direction should result in the periodic addition and subtraction of the two images. A typical result is also shown. At first sight this result may seem poor, the microdeformation (the letters H.I.S. $\lambda/2$ were lightly inscribed on the surface) does not stand out as well as in Fig. 6. However, this method allows for post experiment manipulation of the fringe pattern and by slowly adjusting the X coordinate of the hologram the microstructure deformation can be seen quite easily as the dark fringe moves across the surface. Subtraction of the images reconstructed by beams 1 and 2 will occur when the hologram is moved by an amount which is a multiple of Δx, where

$$\Delta X = \frac{\lambda}{(\sin \alpha)\,\Delta \alpha}$$

if $\Delta \alpha$ is small.

HOLOGRAPHIC INTERFEROMETRY APPLIED TO MINIMAL WEAR MEASUREMENT

Discussion and conclusions

1. Dual index holographic contouring is a simple and reliable method for producing contour maps. These maps can be used for measuring wear rates. Restrictions on the depths of wear measured depend on the shape and size of the object, and in most cases can be overcome by suitable adjustment of the experimental optical arrangement, to ensure that the surface slope is small enough to allow for easy resolution of the contour fringes; the contour depth must be at least as small as the depth of wear to be measured. There is a relationship between the visibility of contour fringes and the surface roughness of the object, so that changes in the surface finish of the object can be detected if not measured.

2. Microstructure deformation of rough wear test specimens can be studied using various techniques of holographic matched filtering and image subtraction. These techniques are:
    (i) Holographic matched filtering
        (a) Marom's method[5]
        (b) Ghost image method.
    (ii) Holographic image subtraction
        (a) Direct methods
            1. Frozen fringe holographic interferometry
            2. Two reference beam holographic interferometry[10].
        (b) Indirect methods
            1. Measurement of visibility of Youngs fringe[4]
            2. Spatial filtering of fringes of equal inclination[9]
            3. Two reference beam holographic interferometry[10].

(i) Holographic matched filtering. The main drawback in using these methods for study of microstructure deformation as a wear parameter is that the actual areas of deformation cannot be located using these methods. Test piece relocation may also be a problem.

(ii) Holographic image subtraction. In theory at least these techniques provide a very attractive method for measuring the plastic deformation of surface asperities. Two reference beam holographic interferometry has been included as both a direct and indirect image subtraction technique as, in principle, all the other techniques can be performed using the two reference beam arrangement. It is felt by the authors that the development of this method may prove extremely useful in the science of tribology. As mentioned previously the problem of producing a $\pi/2$ phase shift between all the undeformed parts of a large (> 1 cm^2) wear specimen is great, furthermore to localise this single dark fringe on the specimen surface the problem is greater still (fringe localisation has been a problem with all the image subtraction methods). An obvious development of the method is the introduction of a feedback loop between the position of the hologram (and/or the reference beam direction) and the photodetector output to minimise and maximise the detector output. Should this prove successful it may be possible to study the shape and size distribution of the deformed areas, as well as the total deformed area. These studies could, perhaps, yield information about the generation of abrasive wear particles and other wear processes.

References

1. Lipson, C., *Wear considerations in design*, Prentice Hall 1967.
2. Tsuruta et al. *Jap. J. Appl. Phys.*, 6, 661. 1967.
3. Ribbens, W.B., *Appl. Opt.*, 13, 1085. 1974.
4. Atkinson, J.T. and Lalor, M.J., *Applications of holography and optical data processing*. Eds. E. Marom and A.A. Friesem. 1976.
5. Marom, E., *Bendix Tech. Journal*, Summer 1969, p. 39.
6. Collier, Buckhard and Lin. *Optical Holography*, Academic Press 1971.
7. Jones, A.M. et al. *Wear*, 31, 89-107. 1975.
8. Matsuda, K. et al. *Opt. Commun.*, 6, 2, 111-114. 1972.
9. Matsuda, K. et al. *Japan J. Appl. Phys.*, 14, Suppl. 14-1. 1975.
10. Dändliker, R., Marom, E., Mottier, F.M., *J. Opt. Soc. Am.*, 66, 1, 23. 1976.

APPLICATIONS OF HOLOGRAPHY TO THE STUDY OF STRUCTURES AND MATERIALS

Gaël Cadoret
Centre Experimental de Recherches et d'Etudes du Bâtiment et
des Travaux Publics
12, rue Brancion — 75737 Paris Cedex 15

Abstract

This paper presents two ways of using holographic interferometry, firstly for identification of materials such as metals as regards fatigue and failure behaviour and secondly for complete determination of the state of stress in transparent plane models.

1 - After outlining the principal parameters of the linear elastic theory of failure, we demonstrate the use of real time holographic interferometry in the quantitative determination of yield phenomena at fatigue crack tips.

2 - Under certain conditions elastic behaviour study of complex structures can be confined to analysis on two dimensional models. When the state of stress changes, each point in the model simultaneously undergoes a variation of index and thickness. The first factor can be linked to the difference in stresses whereas the second is a function of the sum of stresses. The optical effect from a double exposure holographic recording will not show a simple superposition of interference fringes of different origin but rather a complex intermodulation of different curves.

On the other hand, using a specific set-up and a non-reciprocal rotatory power with a FARADAY cell, it is possible to obtain simultaneously and separately the interference fringes representing the sum and the difference of stresses.

DETERMINATION OF THE SIZE AND SHAPE OF THE PLASTIC ZONE AT THE TIP OF A FATIGUE CRACK USING HOLOGRAPHIC INTERFEROMETRY

Introduction

Structural analysis based on current mechanical properties (strenght - elactic limit) fail to take into consideration the fracture toughness of materials, hence it fails to provide safety design from brittle fracture. This brittleness can be caused by sharp notches which bring about considerable local increase in stresses.

Amongst the factor whih contribute to the increase in stresses we must include "metallurgical" defects such as inclusions and cracks due to welding and "mechanical" defects such as drawing (threading) processing scratches, sharp angles, etc...) The service life of a casting or an assembly under fatigue stress is determined by the propagation (even the onset) of these cracks up to critical dimensions where sudden failure occurs. The spread of fatigue cracks is linked to the plastic deformation process in the aera surrouding the tip of the crack.

Linear elastic fracture mechanics

With the fracture mechanics theory it is possible to introduce characteristic parameters of local distribution of stresses and strains near a crack.

Failure modes

For a plane crack under a systeme of forces, the crack propagation can be represented by superposing three simple fracture modes :

Mode I (opening mode). The surfaces of the crack move perpendicular to each other.

Mode II (edge-sliding mode). The crack surfaces move in the same plane and in a direction perpendicular to the crack front.

Mode III (tearing mode). The crack surfaces move in the same plane and in a direction parallel to the crack front.

Dangerous fractures usually follow Mode I which is characterized by low plastic deformation and a flat failure surface. We shall study this type of failure more closely later.

Stress and strain fields

The distribution of stresses and strains near a crack tip can be obtained from results of the theory of elasticity, by assuming that the material is elastic, homogeneous and isotropic, and that the crack tip radius is zero. In the case of a crack stressed in compliance with Mode I, we obtain, in polar coordinates and disregarding upper order terms in r :

$$\sigma_x = \frac{K_I}{\sqrt{2\pi r}} \cdot \cos \frac{\theta}{2} \left[1 - \sin \frac{\theta}{2} \cdot \sin \frac{3\theta}{2} \right] \tag{1}$$

$$\sigma_y = \frac{K_I}{\sqrt{2\pi r}} \cdot \cos \frac{\theta}{2} \left[1 + \sin \frac{\theta}{2} \cdot \sin \frac{3\theta}{2} \right] \tag{2}$$

$$\tau_{xy} = \frac{K_I}{\sqrt{2\pi r}} \sin \frac{\theta}{2} \cdot \cos \frac{\theta}{2} \cdot \cos \frac{3\theta}{2} \tag{3}$$

and :

$$u = \frac{K_I}{8\gamma} \sqrt{\frac{2r}{\pi}} \left[(2k - 1) \cos \frac{\theta}{2} - \cos \frac{3\theta}{2} \right] \tag{4}$$

$$v = \frac{K_I}{8\gamma} \sqrt{\frac{2r}{\pi}} \left[(2k + 1) \sin \frac{\theta}{2} - \sin \frac{3\theta}{2} \right] \tag{5}$$

$$w = 0 \text{ and } k = 3 - 4\nu \text{ in plane deformation state} \tag{6}$$

$$w = -\frac{\nu}{E} \int (\sigma_x + \sigma_y) \, dz \quad \text{and} \quad k = \frac{3 - \nu}{1 + \nu} \text{ in plane stress state,} \tag{7}$$

with γ shear modulus, ν Poisson ratio

E young's modulus

$$\gamma = \frac{E}{2(1+\nu)}$$

The intensity factor of stress K_I is a constant independent of r and θ but linked to the shape of the test piece, the length of the crack and the system of loading. K_I characterises the singular stress field near the crack front.

K_I can be simply linked to the variation of potential energy caused by the extension of the crack. So, there is a critical value K_{IC} directly linked to the material which contributes to the crack formation energy. We will accept that failure occurs in the plane where normal stress is maximum when K_I reaches this critical value K_{IC} which quantitively characterises the resistance of a material to sudden propagation of a crack in Mode I. Using the mathematical theory of elasticity it is possible to make an analytical determination of the stress intensity factor.

Extent of the plastic zone

The shape and the size of the plastic aera can be evaluated by applying the TRESCA and Von MISES criteria and by taking the theoretical elastic state as reference. Figure 1 shows the limit of the plastic area obtained by use of these two criteria. Figure 2 presents the variation of the plastic area in a test of finite thickness cut by a crack.

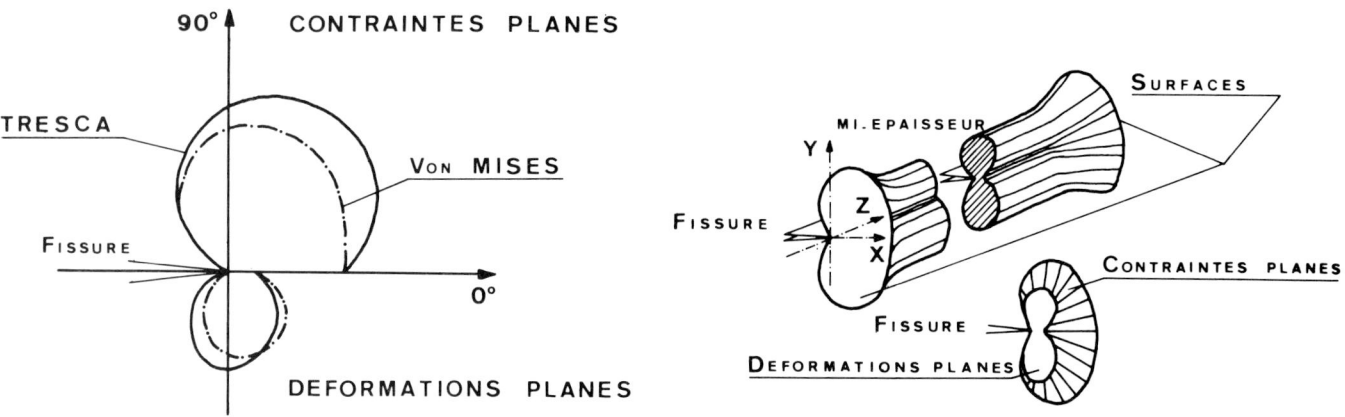

FIGURE 1

FIGURE 2

Generally the size of the plastic area is fairly proportional to $\left[\frac{K_I}{\sigma y}\right]^2$ where σy is the elastic limit of the material.

The theoretical solutions based on fracture mechanics are all the more realistic as the plastic area is small.

IRWIN assumes that the stress field outside the plastic area is the same as that determined in the elastic analysis with a "shift" of a quantity ry (radius of the plastic area) ry is then expressed :

$$ry = \frac{1}{\alpha \pi} \left[\frac{K_I}{\sigma y}\right]^\beta \tag{8}$$

with $\alpha = 2$; $\beta = 2$ in plane stresses
and $\alpha = 6$; $\beta = 2$ in plane deformations

Holographic interferometry offers the possibility of measuring the plastic aera, thereby also the means of finding an optimal value for parameters α and β

Description of the test piece

Our measurements were concentrated on compact test pieces of the ASTM-CT type, made of hyper-tempered austenic stainless steel type Z6 CND 17-12 whose mechanical characteristics are as follows.

$R_m = 560$ N/mm^2, $R_e = 230$ N/mm^2; A = 68%, Z = 74%

These test pieces were cut out of an ASTM-CT sample 20mm thick once it had been cracked under a $R_s = 0,1$.load ratio. The final amplitude of the stress intensity factor K is 15 MPa m with a Kmax of value 16,5 MPa m.

Figure 3 shows the failure surface of a precracked test piece, broken after thermal treatment. The 3mm thick test piece used (figure 4) was taken from the surface of a similar ASTM-CT 20 test piece. We found that the crack length, originally 3mm, was reduced to 0,25mm by the heat treatment.

FIG 3

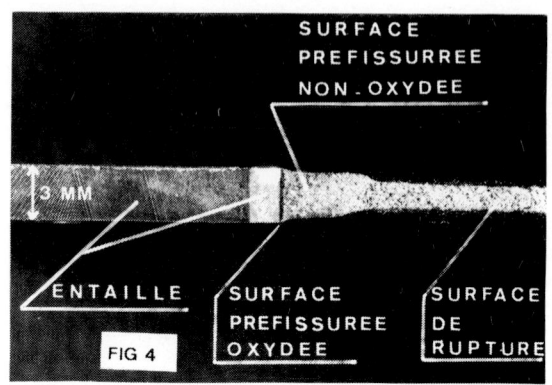

FIG 4

Experimental set-up used

Formation and interpretation of interference figures

With the technique of reconstructing a wave front by holography, it is possible to produce interference between two non contemporary waves. The measurement of the displacement to whitch each point of an object is subjected is obtained from the analysis of interference fringes produced by the addition of wave fronts coming from the object before and after stressing.

(4) shows that the spatial low frequency interference fringes (in practice the interfringe must be at least five times greater than the apparent diameter of diffraction elements) are linked by the expression.

$$\frac{2\pi}{\lambda} n_o \Delta = (2n + 1) \pi \qquad (9)$$

Where $n_o \Delta$ corresponds to the optical delay introduce into each point in the model, under load. The value of Δ can be expressed in terms of composents (U_i) of the displacement vector, that is in a contracted from

$$\Delta j = S_{ji} U_i \qquad (10)$$

The S_{ji} coefficients vary solely according to experimental conditions such as lighting and observation angles of the object.

With the experimental device used (figures 5 and 6) it is possible to obtain interference fringes which are linked only to the variations in thickness of the object examined.

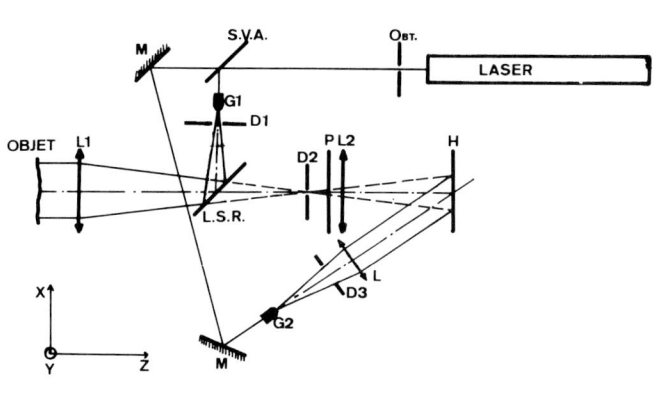

FIGURE 5

FIG 6

In the case of the test piece put to tensile stress, the measurement required, that is the variation of thickness under plane stress, is lower than the displacement in the loading plane. This situation led us to chose testing in real time where for each value of applied load, the waves diffracted by the object and those reconstructed by the hologram can easily be superposed. This is done by placing the holographic plate support on a micropositioning device which can be shifted in directions X and Y. Note that the reference beam must be plane (5) so as to keep the same lighting angle for all points on the hologram.
Figure 7 represents test piece F 114 observed through the hologram, whilst the load applied (about 22 daN) is identical to that applied during the recording. The absence of interference fringes indicates the precise repositioning of the hologram figures 8 and 9 visualize the interferograms obtained during the loading and for return to zero.

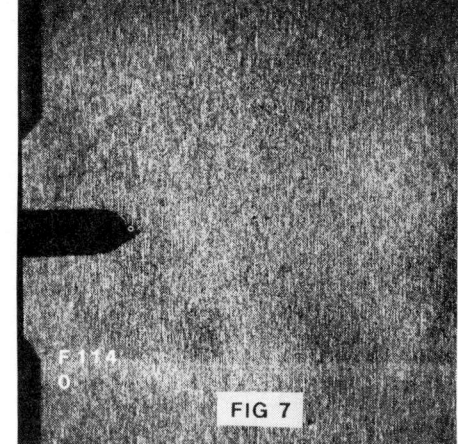

FIG 7

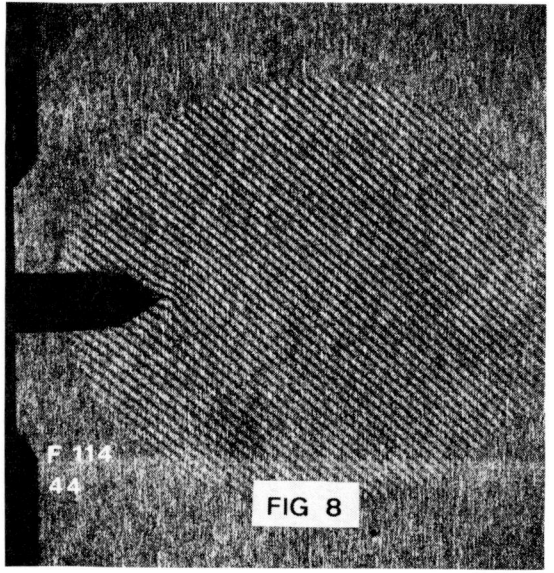

FIG 8

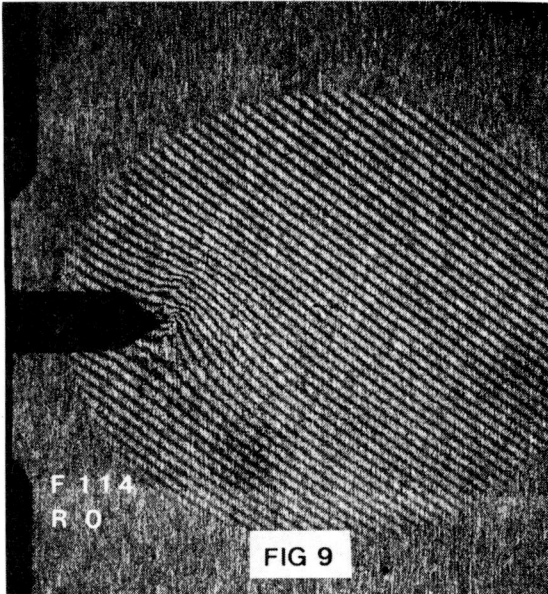

FIG 9

Measurements

Extent of the plastic area

The extent of the plastic area (6, 7, 8) can be obtained simply by delimiting the location of points where the interference fringes become non linear. Indeed a system of rectilinear fringes corresponds to the overall field of elastic deformations, the development of a plastic area disturbs this distribution by modifying the shape of the fringes, this modification supplies us with the information needed. Figure 10 indicates the importance of yielding in the different cases of loading.

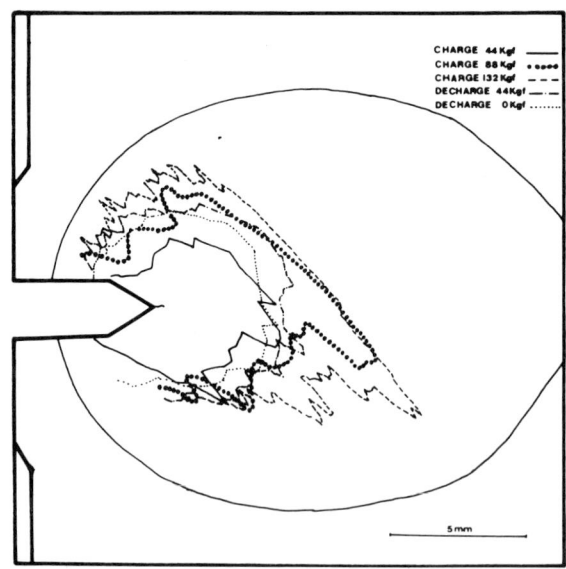

FIGURE 10

Shape of the plastic area

Once we knows that the displacement represented by an interfringe has a value of $0,3\mu m$ (cf equation 9) processing along parallel lines in the fringe direction is immediate. Figure 11 sums up calculation for a load of 88 kgf the different sections are indicated on figure 12.

If the thickness of the test piece is known we can deduce the mean value of relative cross deformation.

APPLICATIONS OF HOLOGRAPHY TO THE STUDY OF STRUCTURES AND MATERIALS

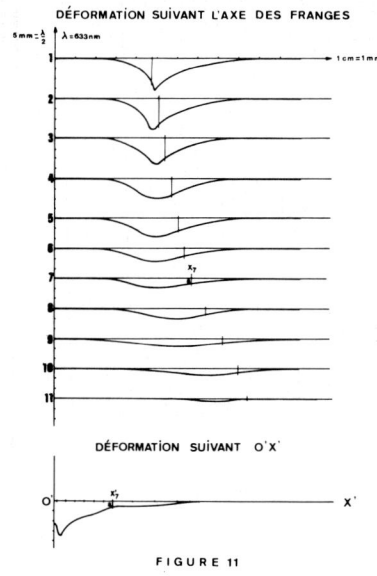

FIGURE 11

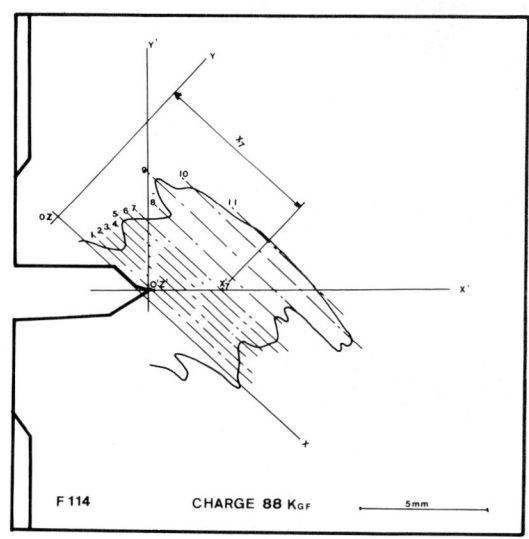

FIGURE 12

Nevertheless the lines of identical thickness variation can be obtained much more quickly and more accurately using an interferential setting. Figure 13 presents such a device. This lights up negatives (corresponding to figures 8 and 9) by two plane waves equally inclined on the optical axis, with this double diffracting device we may superpose differents conjugated orders so it is possible to obtain level lines with dual sensitivity (figure 14) if we use orders + 1 and - 1.

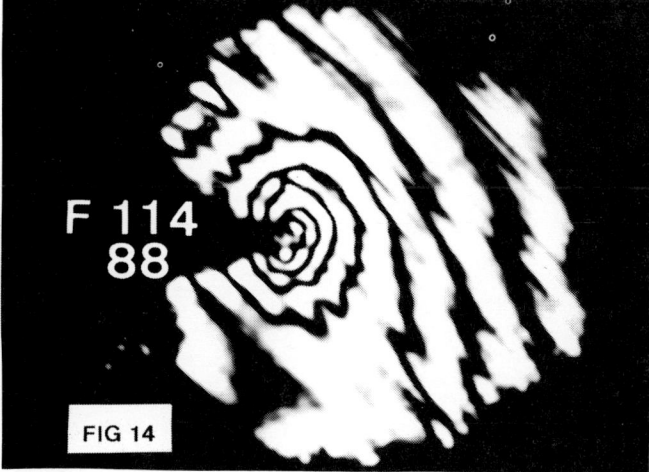

Conclusions

We have demonstrated that the use of holographic interferometry offers the possibility of making a quantitative study of the yield phenomene at a crack tip. The particular advantage of this method lies in the fact that it will not disturb the phenomenon measured and can be used on any kind of material, since surface state has little importance. Moreover the measurements (of high sensitivity 0,1 μm) are carried out simultaneously with load effects, when loading and also unloading.

STUDY OF TRANSPARENT MODELS BY PHOTOHOLOELACTICIMETRY

Introduction

The first septs in photoelasticimetry by holographic interferometry (photoholoelasticimetry) were based on work by LOHMAN (9) who demonstrated that holography offered a mean for recording states of polarisation ROGERS (10) in 1966 studied a piece of glass under stress and FOURNEY (11) a little later showed that the holographic method gave interference fringes linked to isoclinic, isochromatic and isopachic fringes.

Plotting the isochromatics

There is only one exposure. The model is stressed and at each point the incident vibration breaks up along the birefringent axes. Leaving the model there are two out-of phase wave fronts of orthogonal polarisation. There wave fronts interfere with the corresponding reference wave components. At reconstruction the two wave fronts recorded separately are reconstructed simultaneously, the amplitude then has a value of

$$Ar = K1 e^{i\varphi_1} + K1 e^{i\varphi_2} \tag{11}$$

hence the intensity

$$Ir = 4 K1^2 \cos^2 \frac{\varphi_1 - \varphi_2}{2} \tag{12}$$

that is in function of the stresses

$$Ir = 4 K1^2 \cos^2 \left[\frac{\pi e}{\lambda} (C1 - C2)(\sigma_{11} - \sigma_{22}) \right] \tag{13}$$

Plotting the isochromatics and the isopachics

We then undertake double-exposure, the model is not stressed, therefore not birefringent during one of these exposures. The amplitude at reconstruction is expressed:

$$Ar = K1 e^{i\varphi_0} + K2 e^{i\varphi_1} + K2 e^{i\varphi_2} \tag{14}$$

where $K1$ and $K2$ are constants dependent on the recording conditions.

With equal exposure times in each case, we obtain

$$Ir = K1^2 \left[1 + 2\cos\left(\frac{\varphi_1 + \varphi_2 - 2\varphi_0}{2}\right) \cdot \cos\left(\frac{\varphi_1 - \varphi_2}{2}\right) + \cos^2\left(\frac{\varphi_1 - \varphi_2}{2}\right) \right] \tag{15}$$

that is in function of the stresses

$$Ir = K1^2 \left[1 + \cos\left(\frac{\pi e}{\lambda}(C1 - C2)(\sigma_{11} - \sigma_{22})\right) \cdot \left[2\cos\left(\frac{\pi e}{\lambda}(C'_1 + C'_2)(\sigma_{11} + \sigma_{22})\right) \right.\right.$$
$$\left.\left. + \cos\left(\frac{\pi e}{\lambda}(C1 - C2)(\sigma_{11} - \sigma_{22})\right) \right] \right] \tag{16}$$

Interpretation of the interference fringes

Examining equation (16) shows that the interpretation of different fringes observed is far from obvious. It is not indeed a simple superposition of interference fringes of different origin (isochromatics and isopachics) but effectively a complex intermodulation.

Separation of isochromatics and isopachics

The difficulties linked with interpreting interference fringes obtained with the previous settings led us to investigate means of reading isochromatic and isopachic patterns separately.

The NICOLAS (12,13) method is based on polyester resin (stratyl A 16) property of having a photoelastic constant which changes symbol with temperature. The model is set under load and acquires a certain birefringence which can be cancelled at ambient temperature by applying a similar load effect. By doubling the value of this, the model develops birefringence opposite to that existing when it was set (frozen). This method is rather intricate to use and fails to offer sufficient accuracy.

HOLLOWAY, RANSON, TAYLOR (14) proposed a set-up (figure 15) where the surface of the model is made semi reflecting. Holography is then used to analyse displacements on the model surface. For complex loading conditions, it seens that the overall displacements, beyond the load effect, may mask the information corresponding to model thickness variations.

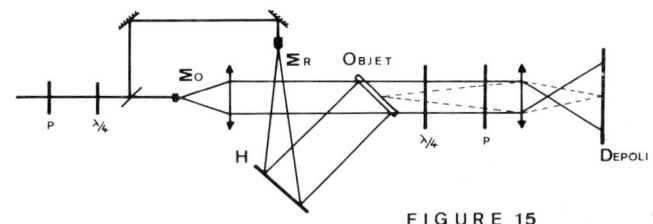

FIGURE 15

Another more general method, consists of making the light cross the model twice, the polarisation states or light forms undergo a 90° rotation between the two crossings. In this way, the resulting birefringence is zero and the isochromatics are cut out. Moreover the isopachics have dual sensitivity. The rotatory movement can be produced by a FARADAY rotator (figure 16) as proposed in 1968 by CHAU (15). It is also possible to use an quartz cristal (figure 17). From the historical point of view these ideas are not new, since SINCLAIR (16) in 1939 proposed an interferometric set-up using two models and a rotatory power (figure 18) and BUBB in 1940 (17) used a FARADAY rotator.

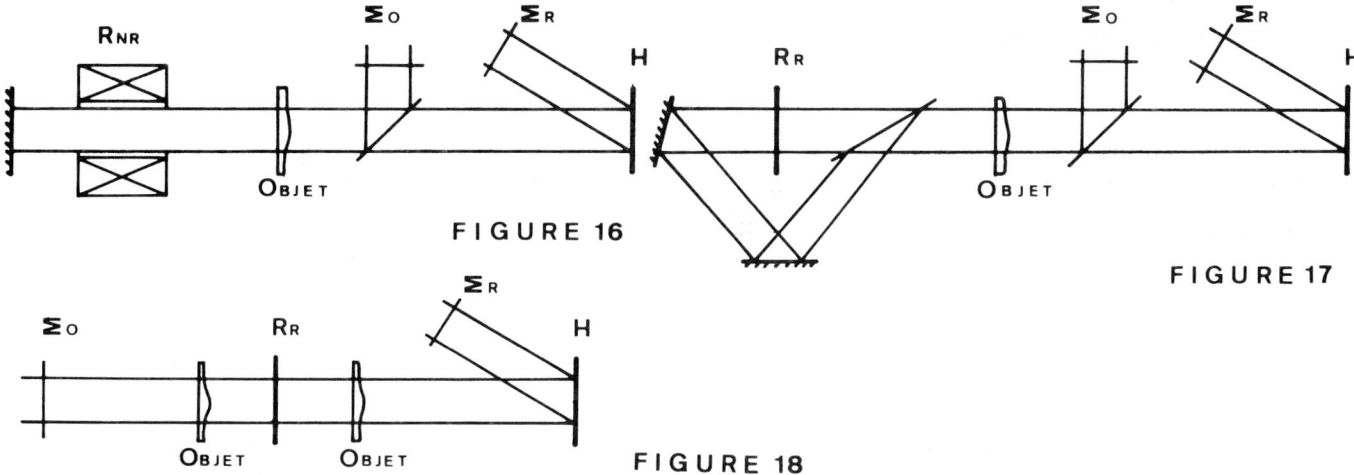

FIGURE 16

FIGURE 17

FIGURE 18

Some experimental results obtained with the aforementioned techniques were published more recently (18, 19). Taking into account experience acquired elswhere (20) we set up a holographic device using a FARADAY rotator.

Building a FARADAY cell

Making up a FARADAY cell implies bringing together a magnetic field which is sufficiently intense and transversely homogeneous and a medium with a hight VERDET constant. The apparatus built for this is show on figures 19 and 20. Figure 21 shows a radial section through the magnetic field at 85mm from the centre. At this distance the magnetic field intensity variation between the axis and at 17,5mm from it is below 0,3%.

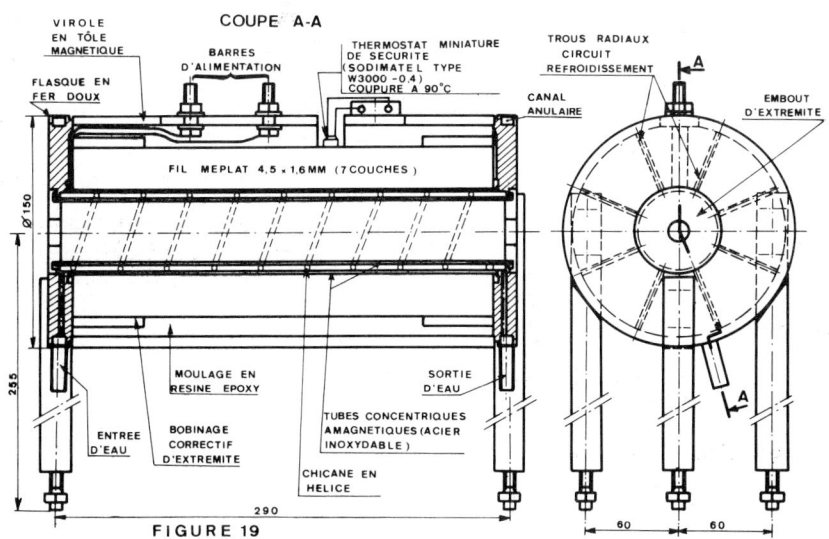

FIGURE 19

FIGURE 20

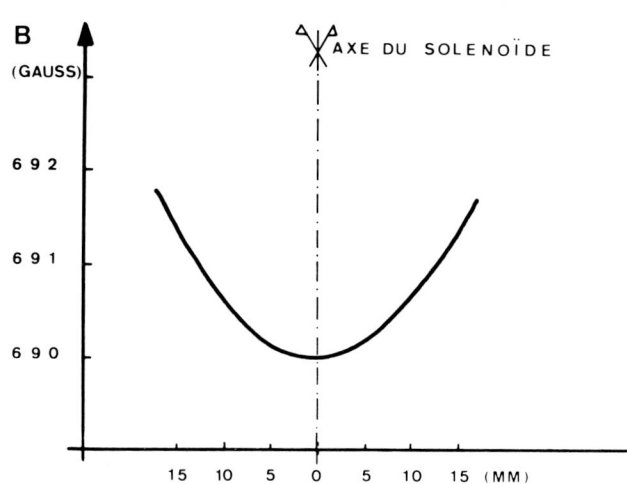

FIGURE 21

For the glass rod used in the cell we chose glass type SF 57, whose VERDET constant is 0,071 angle minute/Gauss.cm for a wave length λ = 633nm. The residual birefringence of this glass was reduced by annealing to a value below 1nm/cm so we are able to maintain constant at 0,3% on the rod diameter (35mm), the product of the magnetic field by the lengthcrossed (170mm)

Experimental device used

The device (figures 22 and 23) built on a large size granit slab (5m x 1,30m x 0,5m) was designed so that the object beam only undergoes reflexions under normal incidence, so the influence of the beam splitter dichroïsm and of the mirror M is negligible. Moreover the lens are only slightly open and placed so that the object is paired with itself thereby limiting the effects of diffraction on the model edge. The investigation field between lens L1 and L2 has a diameter of 200 mm.

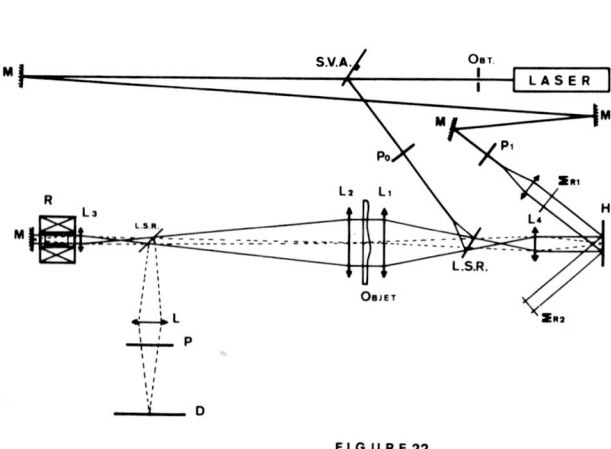

FIGURE 22

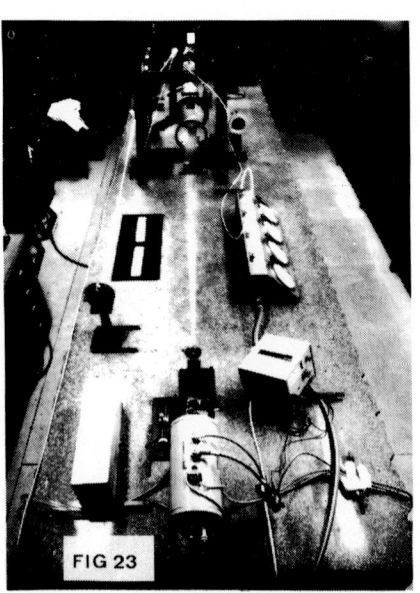

FIG 23

The recording is undertaken by double-exposure. For the first exposure the object is at rest and the rotatory power inactive (or the object is stressed and the rotatory power in action). In the second exposure the rotatory power is active and the object is under load effect (or is loaded differently).

At reconstruction, the amplitude wave is :

$$Ar = K1 e^{i\varphi_0} + K2 e^{i(\varphi_1 + \varphi_2)} + K2 e^{i(\varphi_2 + \varphi_1)} \tag{17}$$

by adopting equal exposure times, the reconstructed intensity is expressed :

$$Ir = 2Kl^2 \left[1 + \cos(\varphi_1 + \varphi_2 - 2\varphi_0) \right] \tag{18}$$

that is, in function of the stresses

$$Ir = 4Kl^2 \cos^2 \left[\frac{\pi e}{\lambda} (C'_1 + C'_2)(\sigma_{11} + \sigma_{22}) \right] \tag{19}$$

Where C_1 and C_2 are coefficients dependent on the mechanical characteristics (E, ν) of the material used as well as its piezo optical coefficients.

However, it is not necessary to know these different magnitudes separately, calibration on a structure of know analytical behavior will give accurate sensitivity of the model's structural material.

Applications

Investigation of contact in a gear cog

This work carried out on a model scale 7 is illustrated by figures 24 to 29 which show the interference fringes corresponding to isochromatics and isopachics for different cases of loading.

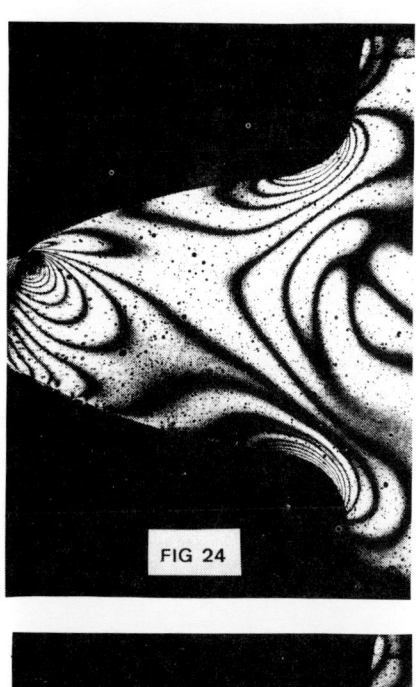

FIG 24

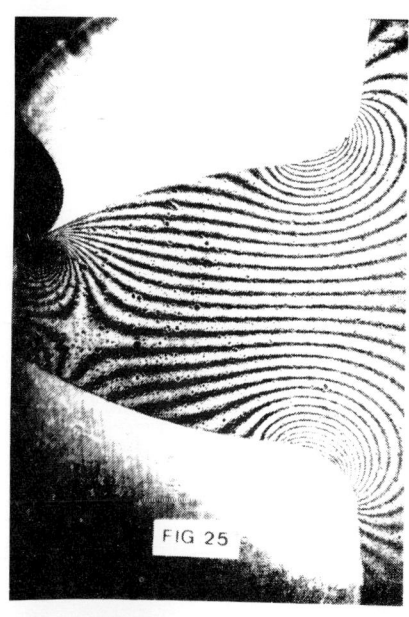

FIG 25

FIG 26

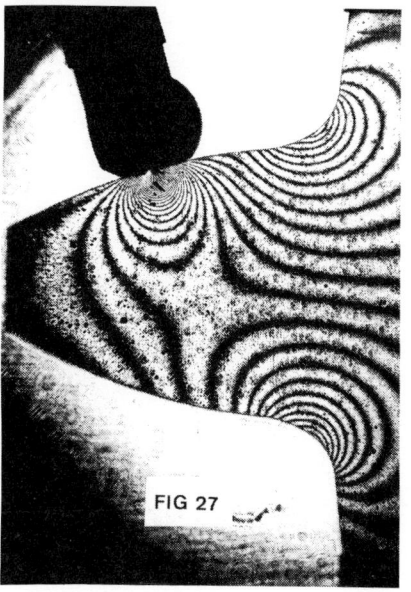

FIG 27

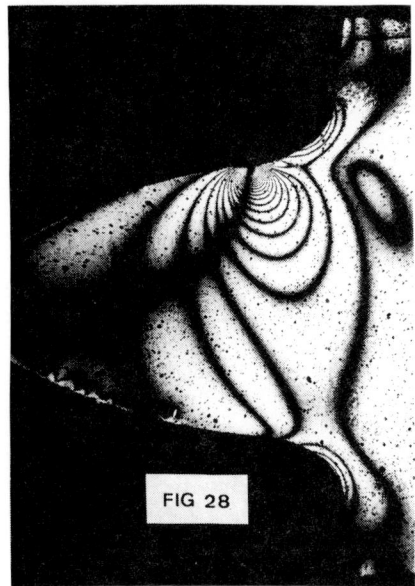

FIG 28

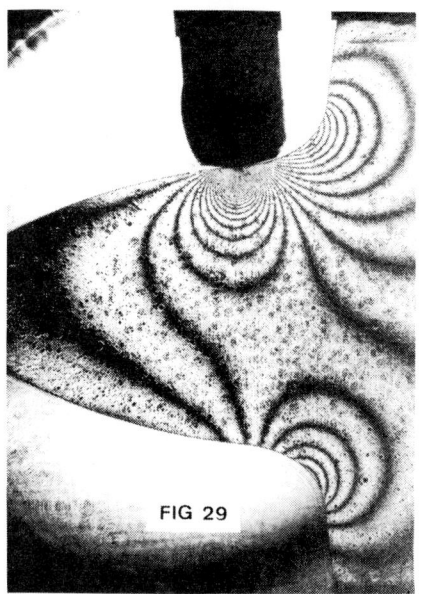

FIG 29

Study of stress distribution in the model of a bridge frame.

A 1/50 scale model (figure 30) is used for this test representing a plane section of a bridge frame, orthogonal to the longitudinal axis of the bridge.

FIGURE 30

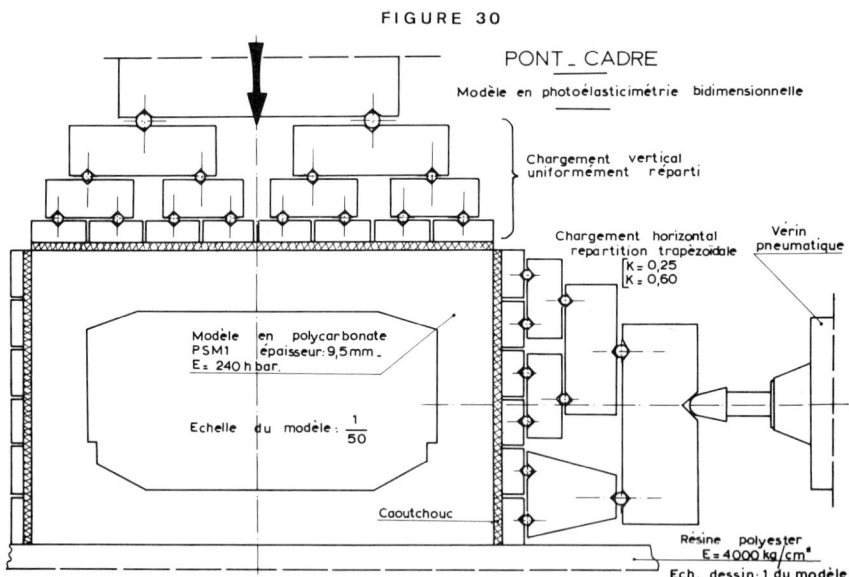

The loads are placed :

- on the upper beam extrados level, a uniform distribution of pressures calculated from the overlapping height.

- along the piers, a trapezoidal distribution of pressures from the overlapping height and a value of coefficient K of earth mass thrust.

- on the lower beam intrados the distribution of pressures is obtained using a resin plate, whose modulus of elasticity is determined to appropriately represent the foundation soil and the layer of compacted gravel.

Figures 31 to 34 show the isochromatic and isopachic patterns obtained for two different load effects.

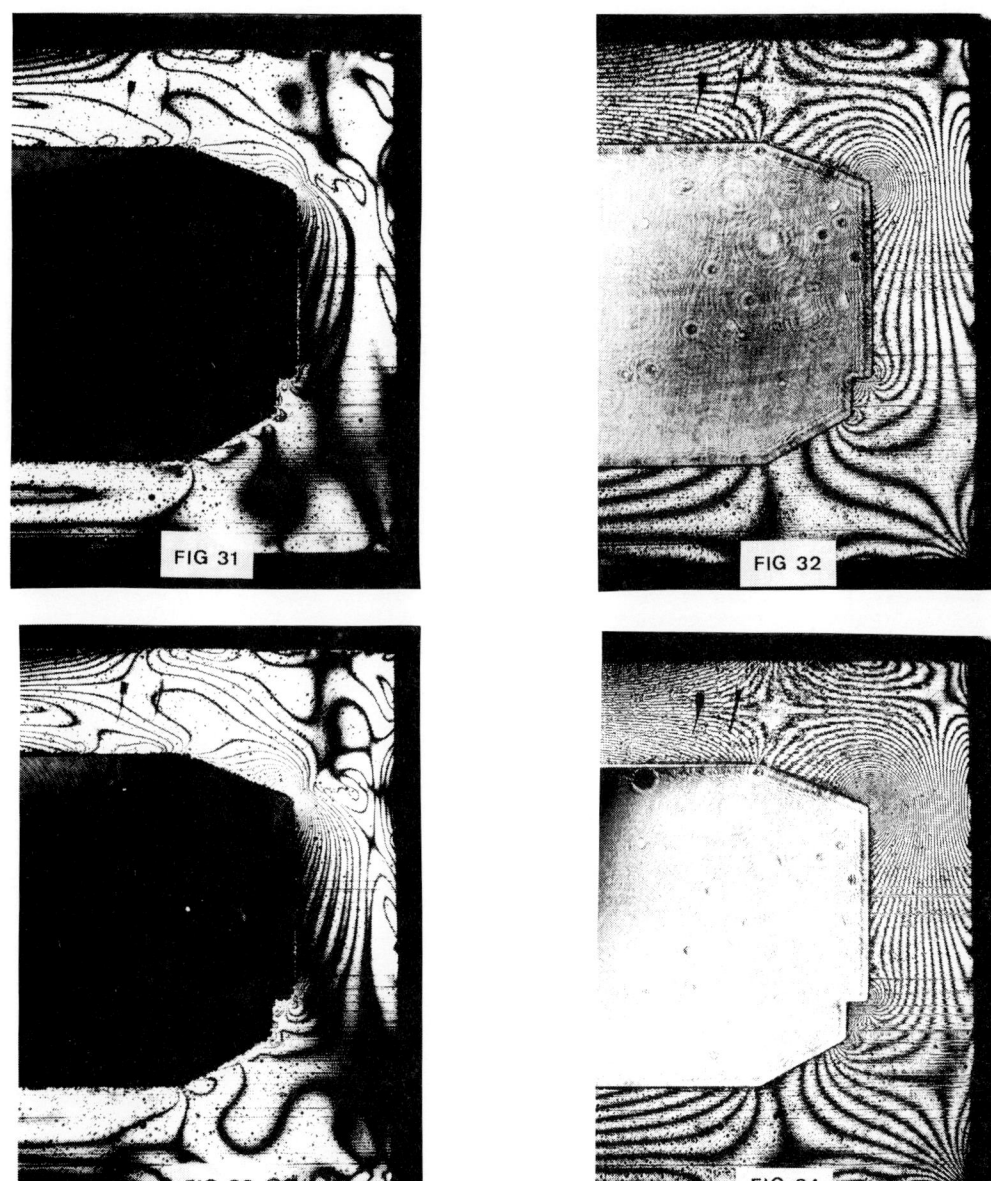

Calibration of the models structural materials

This calibration can be undertaken quite simply from analysis of isochromatic and isopachic patterns obtained for exemple on a disc under diametral compression. The analytical solution is well know for this structure and attributing a stress value to different lines observed raises no difficulties. Also note that on the free edges of the model, with the main stress know, it is not essential to number the isopachics, since there resetting can be carried out with the isochromatics. Figures 35 and 36 present the isochromatics and the isopachics for a disc under diametral compression.

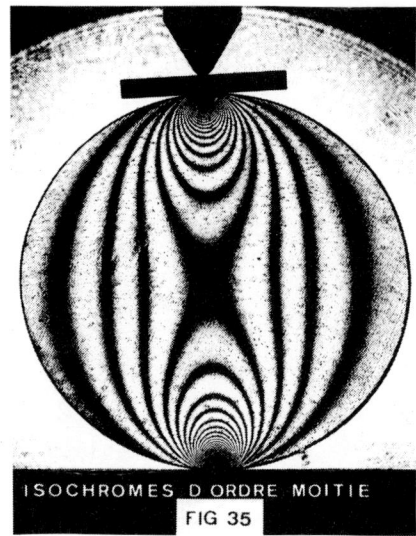

ISOCHROMES D ORDRE MOITIE
FIG 35

ISOPACHES
FIG 36

Conclusions

Holographic interferometry is at present operational in the quantitative study of stresses within transparent plane models.

This experimental method offers the possibility of studying complex structures at lower cost and within very competitive time limits. This is why it is in direct competition with mathematical desing methods prevalent up to present.

HOLOGRAPHICAL DISDROMETRY

V. M. Zakharov, L. N. Rasumov, I. N. Sisakyan
F.I.A.N. Moscow, USSR

Up till the present time characteristics of turbid medium were determined using indirect methods, such as optical fringe and photography methods as well as the following photoelectrical methods : shadowing, screening and scattering at small angles. Such contact methods as gravitational, electrical and inertial are significantly less perfect. Using the first two methods a parameter correlating with particle size and not size itself is found.

Lately researchers have been interested in the reliability of information obtained remotely. There has always been felt a need of an instrument the utilization of which would not be limited to a small size range. Thus, for example, the application range of an inertia method / the most widespread one / is limited to a 20-70 km range. Smaller particles are drawn away by a flow of the medium in which they are dispersed; larger particles are split up at colliding with the surface of a trap. Significant advance in the investigation of disperse system characteristics was made by the development of a new meteorological trend - holographical laser disdrometry. A distinguishing feature of the given trend consists in large sounding volumes / of an order of several cm^3 / at high spatial and temporal resolution. Pulsed holographical disdrometry is particularly universal in this respect. This method can be applied to measurements with frequencies of several tens of Hz /2/ at a "momentary" fixation / of several nsec. / of the medium state a volume under study.

Let us estimate the possibilities of using the holography method for the determination of the minimum size of disperse system particles and their permissible concentration. The minimum size of the detected particles is determined by the resolution of a hologram. Linear dimensions of the elements of the resolution in the object space $(\Delta x)_{ob}$ and in the image space $(\Delta x)_{im}$ on the basis of Meier's ratios /3/ are connected by the equation :

$$(\Delta x)_{ob} = \frac{1}{\mu} \cdot \frac{R_2}{R_{im}} \cdot (\Delta x)_{im} \qquad /1/$$

where R_2 and R_{im} are the distances of the object and image from the hologram. On the other hand, it can be demonstrated that in the image space the linear size of the resolution element satisfies the non-equality:

$$(\Delta x)_{im} \geq \delta \cdot \mu \cdot \frac{R_{im}}{f} \qquad /2/$$

where δ is the medium size of the photomaterial grain;
f is the equivalent hologram focus defined by the equation : $\frac{1}{f} = \left/ \frac{1}{R_1} - \frac{1}{R_2} \right/$
R_1 is the reference source distance from the hologram.
Combining the conditions /1/ and /2/ we obtain the following :

$$(\Delta x)_{ob} \geq \delta \cdot \frac{R_2}{f} \qquad /3/$$

The hologram resolution N is characterized by an inverse quantity :

$$N \leq \frac{f}{\delta \cdot R_2} \qquad /4/$$

and measured by lines/mm

Thus, the closer R_1 and R_2 are to each other, the hologram to the object, and the higher the recording medium resolution, the better the hologram resolution.

The upper limit of Eq./4/ is the theoretical limit of the hologram resolution. However, this limit is reached only when signal-to-noise ration exceeds a certain threshold value. Literature contains practically no information about the realization of a high hologram resolution exceeding the analogous quantity for recording material.

In pulsed holography there is a factor limiting the resolution of the above method, i.e. the motion of disperse system particles. It can be demonstrated that a hologram resolution cannot be less than a particle travel double projection into the recording plane. If during the exposure the travel exceeds $(\Delta x)_{ob}$, the deterioration of the hologram resolution is quite evident. To avoid this the fulfilment of the following non-equality is necessary :

$$\tau \cdot \nu \leqslant \frac{\delta R_2}{2f} = \frac{R_2}{4fN} \qquad /5/$$

where ν is the particle travel speed; τ is the exposure time /generally coincides with the length of a laser pulse/. Thus, a high resolution emulsion "notices" the motion of disperse system particles quicker.

Let us estimate counted particle concentration. In Papers /4, 5/ it was ascertained that the magnitude of the system transmission must be not worse than 0.8. As far as :

$$T = exp \{-2\sigma N l\} \qquad /6/$$

where σ is the area of a particle cross-section normal to the beam axis; l is the beam path within the disperse system in question; N is the particle concentration, N and l appear correlated. In holographical disdrometers l = 1-100mm. For particles with a 10 µm diameter an admissible concentration varies in a 10^6- $10^4 cm^{-3}$ range. But 10^6 particles with a 10µm diameter placed in one plane screen an illuminating beam completely. This corresponds to a 10^8 cm^{-3} concentration. Consequently, in the available disdrometers a disperse system length in the third dimension lowers the admissible particle concentration by about two orders of magnitude. In the singleray disdrometer developed by the authors /6/ a path length is 150mm, that is why the maximum concentration of 10µm particles is $7.10^{-3} cm^3$. Examples of large and small and small particle images are given in Fig.1; Fig. 2 shows real size distributions; a)-in a fuel nozzle spray; b)- in a natural fog of an advective type. The fuel nozzle type and the distribution of counted concentration along the direction of sounding at a water film resolution are shown in fig. 3. The illustrations give the notion about the potentials of the new metrological trend at present.

A problem of measuring errors is a question of principle in metrology. A systematic experimental error which includes :
1)- an error of disdrometer calibration by a bench mark scale;
2)- an error due to a spread function;
3)- the total error due to a noise level along the path;
4)- an error due to holographical image distortions, due to the path optics imperfectness, film distortion and contraction was about 40%.

The above instrument was calibrated by a direct method.
Random errors are as follow :
1)- an error in the determination of typical distances in recording and image restoration schemes;
2)- an error due to image central sharpness setting in a linear direction;
3)- an error of the direct counting of transverse dimensions.

Paper /6/ gives the above apparatus random error characterized by a relative error of about 10%.

Thus, the total measuring error is about 50% and at present it is stipulated basically by a systematic error. However, in our opinion, holographical disdrometry takes only the first steps. The potentials of the new metrological trend will grow with the development of holographical technique

Fig.1 Water aerosol holographical image :

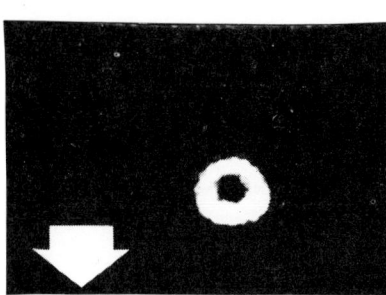

a)- 225 µm water droplet

b)- water aerosol droplets;
a 17.5µm diameter droplet

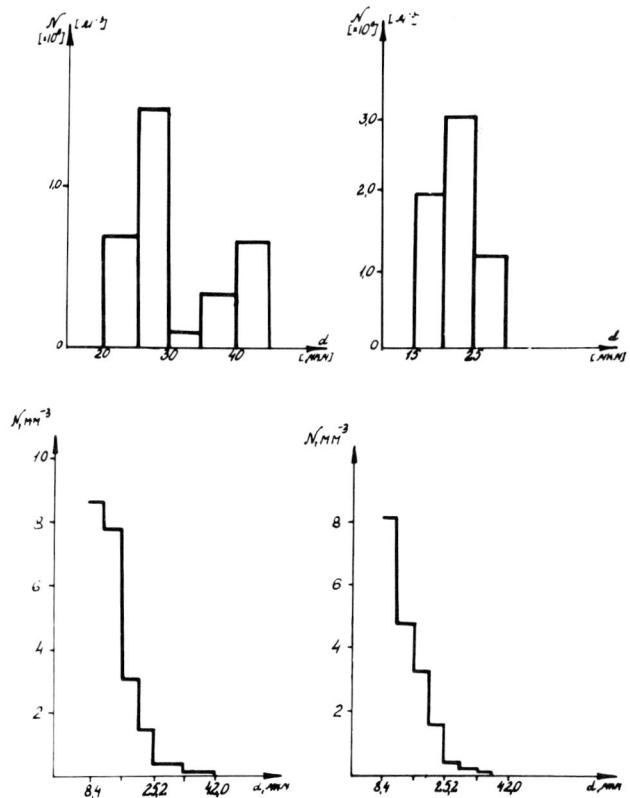
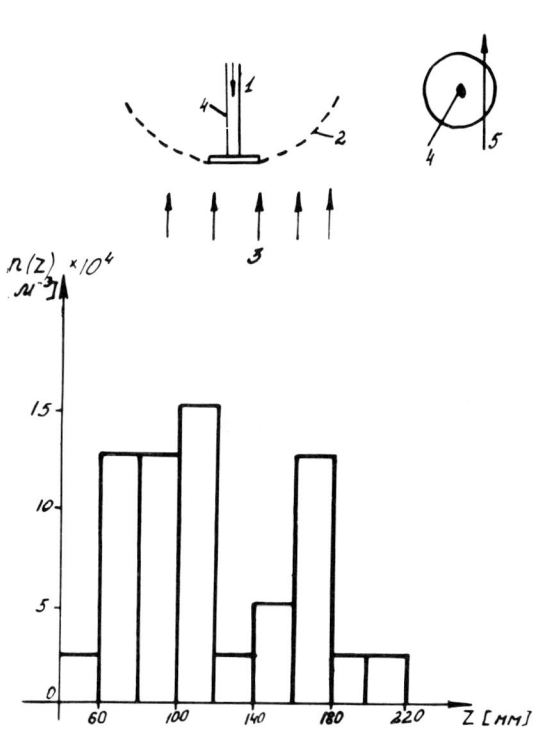

Fig.2 Size distribution
a)- in a disc nozzle fuel;
b)- in a natural fog of an advective type

Fig.3 Counted concentration distribution along the direction of sounding.

1)- water supply; 2)- film form; 3)- on-running air flow; 4)- cylindrical probe; 5)- sounding directin / coincides with the direction of holographical interometer signal beam/.

REFERENCES

1)- B.A. Silverman, B.J. Thompson and J.H. Ward. A laser Fog Disdrometer. J. Appl. Met. 3, 8, 792, 1964.
2)- D.J. Bitteto. A Holographic 3-P Movie with Constant-Velocity Film Transport. Laser Focus, 4, 17, 1968
3)- R.W. Meier. Magnification and Third-Order Aberrations in Holography. J. Opt. Soc. Am., 55, 8, 1965
4)- D.I. Stasselko, V.A. Kosnikovski. In Optics and Spectroscopy 34, N°2, p. 365, 1973
5)- E.M. Birger, V.M. Zakharov, S.P. Karlov, L.N. Rasumov. In Meteorology and Hydrology, N°1, p. 44-52, 1977
6)- E.M. Birger,.L.N. Rasumov. In "Sindai", September 3-6th, 1974

STUDY BY HOLOGRAPHIC INTERFEROMETRY OF DIMENSIONAL VARIABILITY
IN PRECISION-MOULDING MATERIALS USED IN ODONTOLOGY

M. Blandin, C. Durou, H. Soulet

Faculté de Chirurgie Dentaire et Laboratoire d'Optoélectronique
Université Paul Sabatier
118, route de Narbonne — 31077 Toulouse Cedex — France

Broadly speaking, every prosthesis is intended to replace a missing or deficient organ. In odontology, there are two main types of prosthetic reconstitution :
- movable devices resting on mucous membrane as support,
- fixed devices attached to a dental support.

The preparation of a fixed prosthesis demands, first of all, that a working model be prepared on which the prosthesis is to be formed. Consequently, the operation is performed in two basic phases :
    1 - The reproduction of a particular dental form, including, first of all, the taking of an impression in the mouth which gives a specific matrix mould, and secondly the preparation of a single positive casting in plaster of Paris, using the impression as a mould.
    2 - The formation of the prosthesis itself : the future prosthesis devices are modelled in wax or resin, then covered with a coating, the final casting being performed by the lost-wax process.

The considerable number of stapes briefly summarized here, combined with the large number of materials used and the wide range of possibilities offered by manufactures, can only multiply the sources of error, some of which are self-compensatory and some of which are additive.

Final success depends directly on achieving a perfect coincidence between the starting point, given by the drilled teeth, and the prosthesis device intended to fit them, and it seems absolutely essential to analyse each step of the process in detail. We have decided, however, to limit our investigation to the first step, the taking of the impression, and have tried to measure by holographic interferometry the volumetric variation of some materials used for taking impressions.

An impression is taken by using a special holder to insert into the mouth a soft plastic material which, after a certain setting time, becomes hard or elastic.

In usual practice, four types of materials are used : silicone rubbers, polysulphide rubbers, polyethers and hydrocolloids. However, only silicone rubbers have been examined in our study.

The classical methods [1,2,3,4] for studying moulding materials are based on the assumption that the materials are isotropic : measurement of the deformation of a sample in one direction only should give information on the three-dimensional variability of the impressions. In actual fact, the vulcanization of such materials takes place by cross-linking of a terminal hydroxyl group and elimination of water through the surface, so that the possibility of isotropic variation is definitely excluded. Furthermore, the imperfect mixing of the base and the hardener, the non-uniform temperature of the material at the time of mechanical constraints, the problem of altering elasticity (when the impression is withdrawn, slight deformations are produced), the influence of moisture, the reaction between the moulding material and the plaster used for casting are all factors which may also produce a definite anisotropy of the material.

We therefore thought it desirable to study, using an optical technique, the real deformation of an actual impression over a number of hours after its formation in the mouth. Holographic interferometry seemed well suited to this problem. To the best of our knowledge, holography has been used in odontology by two research teams : the American team of ALT-SCHULLER's [5], who tries to measure the strain induced by mastication, and WEDENDAL's [6] Swedish team, who were mainly interested in studying the deformation of prosthesis.

The problem which we have tackled is quite different and the difficulties we have encountered arise from the fact that the material studied undergoes, in the course of time, not only macroscopic deformations which we wish to measure, but also microscopic alterations in its surface. As a result of this latter effect, the degree of correlation between two successive holograms of the same object decreases with time (see the Appendix). Thus, in interferometry by double exposure of the hologram, the only method which does not create a difficulty in locating fringes with opaque objects, the two exposures must be sufficiently close in time, otherwise the contrast of the fringes is greatly diminished. So our very first attempts met with defeat, not because the material used remained undeformed, but because it was being deformed more rapidly than we had expected.

To overcome this lack of correlation, we first of all made use of a palliative : on the apparently flat surface of the test sample, we spread a very thin metal foil, either aluminium or gold. The metal surface conforms to the macroscopic deformations of the material, while at the same time giving inteference fringes with high contrast since its own microscopic structure obviously remains unaltered.

This indirect process, however, has the disadvantage of not being applicable to real impressions, which have an extremely uneven surface. In this case, on solution consists in decreasing the length of time between the two exposures of the hologram, so that a compromise might be found between the number of fringes observable and the quality of their contrast. In the case of the materials studied, experience has shown that it is difficult to observe more than ten fringes ; this corresponds to a time interval of the order of a few minutes during the first hour after the formation of the impression and a fews tens of minutes at later stages. It is therefore only by increasing the number of double-exposure holograms that the alteration in the material can be followed throughout the whole day that follows the taking of the impression.

It should also be noted that the exposure time should not be more than a fraction of a second.

Preliminary Results

As a preliminary to a systematic study which has not yet been undertaken, we have carried out two series of trials :
- firstly, using thin cylindrical test samples solidly fixed onto a metal support by means of retaining holes (Figure 1),
- secondly, using genuine impressions, prepared using a perforated holder.

The materials were of the silicone-rubber type : "Silaplast" and "Coltoflax".

The first sort of test sample allowed us to show up a non-uniform subsidence of the free surface of the material, this subsidence being most pronounced at the points where there were retaining holes (Figures 2, 3, 4).

The beam lighting the object was slightly divergent and oriented at an angle of 30° to the direction of observation, normal to the surface of the sample. Under these conditions, for $\lambda = 633$ nm, each fringe spacing represents a displacement of the surface by 0.33 µm. Thus we can estimate that, during the first hour, a layer of silicone rubber 2 mm thick undergoes a subsidence of 3 µm, i.e.[4] a relative contraction of 0.15 %, in good agreement with results given by another method.

On actual impressions, the fringes are found to be very irregular, which is not surprising, and the most important deformations take place around the marks left by the teeth (Figures 5, 6). This is particularly important, since it is in those zones that the realisation of the final prosthesis should be the most faithful. Thus, although these last tests cannot claim to be quantitative, we find that they are nonetheless most instructive, for they will allow us, firstly, by performing them systematically, to compare the numerous moulding materials offered to dental practitioners and, secondly, will allow us to suggest a reasonable method for treating theses impressions in the laboratory.

Conclusion

Holographic interferometry, though well suited to the study of deformations or displacements of solids, was only applicable in the case of plastic materials on condition that certain precautions be taken.

These preliminary tests have brought out two facts :
- polymerizable materials evolue very slowly in time,
- in general, silicone rubbers are relatively stable in dimensions, the precision required in odontology only being of the order of 1/100 mm.

References

1. Skinner, E.W., Philips, R.W., <u>Sciences des Matériaux Dentaires</u>, Traduction Navarro, Ed. J. Prélat, 1971.
2. Poggioli, <u>Précision des Empreintes Obtenues avec des Matériaux Elastiques</u>, Cahiers de Prothèse n°4, p. 43, 1973.
3. Caïtucoli, <u>Cours CES, Matériaux</u>, Bordeaux, 1975.

4. Gibert, Y., Joniot, B., Lubespere, A., Soulet, H., Cariou, J.M., Collette, G., Jalabert, M., Les Cahiers de Prothèse, 19, 124, 1977.
5. Altschuller, B.R., SPIE Proc., Vol. 89, p. 40, 1976.
6. Wedendal, P.R., Bjelkhagen, H.I., Appl. Opt., 13, 2481, 1974, and "Holography in Dentistry" in "Laser Applications in Medecine and Biology", Vol. 3, Wolbarsht M.L. Ed., Plenum Press, n°4, 1977.
7. Matsuda, K., Tsujiuchi, J., Proc. ICO Conf. on Optical Methods in Scientific and Industrial Measurements, Jap. J. Appl. Phys., 14, Suppl. 14-1, p. 265, 1975.

Appendix

Visibility of Double-Exposure Fringes

In theory, the visibility of fringes obtained by double exposure ought to be equal to 1, since the amplitudes are equal.

In reality, in the case of a substance undergoing polymerization, microdeformations appear on the surface of the material which profoundly alter the contrast of fringes. This phenomenon has been studied by several authors, who have even developed it into a method for measuring microscopic deformations [7]. It should be noted that practically all that varies is the contrast and not the position of the fringes, which is only dependent, as a first approximation, on the displacement of the "mean surface".

Addendum

After the reading of this paper, the Chairman, Prof. J. Ch. Vienot, drew our attention to the fact that a very similar study on the same subject had been prepared by Mr Ch. Pirel in collaboration with Messrs A. Lacourt, M. Spajer and J.P. Goedgebuer of the Laboratoire d'Optique of the Université de Besançon (France). This work is described in Mr Pirel's thesis : Thèse de 3ème Cycle de Sciences Odontologiques - "Contribution à l'étude de la stabilité dimensionnelle des élastomères silicones et recherche de protocoles expérimentaux" - Université Claude Bernard Lyon I, Th. n°04, 1976.

STUDY BY HOLOGRAPHIC INTERFEROMETRY OF DIMENSIONAL VARIABILITY IN PRECISION-MOULDING MATERIALS USED IN ODONTOLOGY

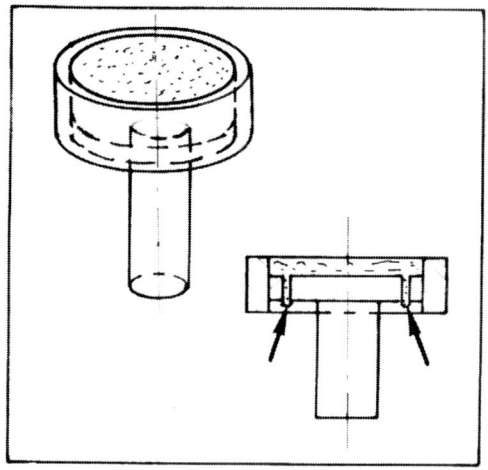

Fig. 1 : Diagram of the cylindrical sample-holder (the arrows indicate anchorage points), e is the thickness of the sample.

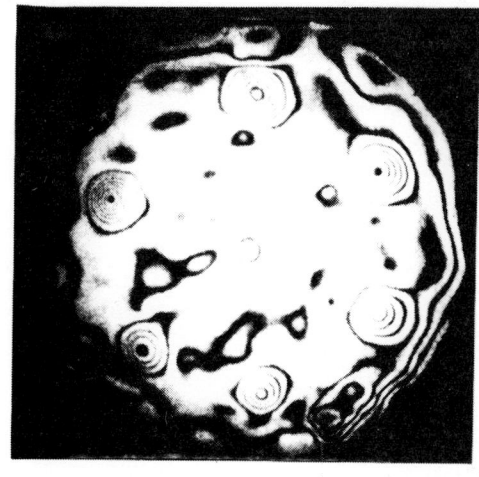

Fig. 2 : Cylindrical test sample of SILAPLAST, covered in aluminium, (e = 1.8 mm), t = 5 min., Δt = 1 min

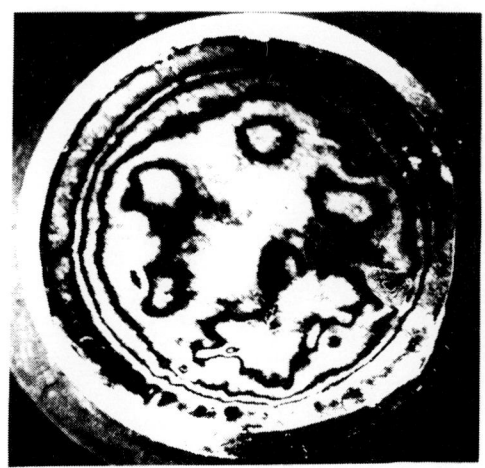

Fig. 3 : Test sample of COLTOFLAX covered in aluminium (e = 5 mm), t = 30 min. Δt = 5 min.

Fig. 4 : Test sample of COLTOFLAX covered in aluminium (e = 5 mm), t = 60 min., Δt = 30 min.

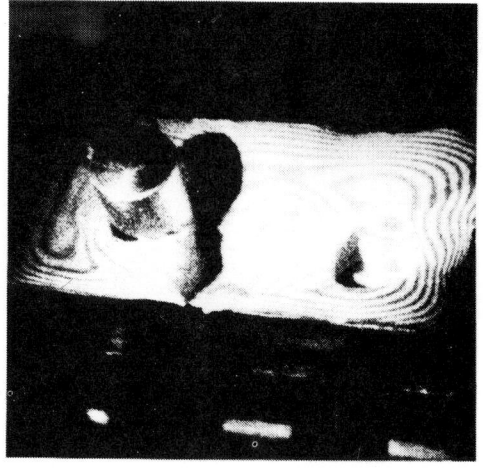

Fig. 5 : Sample of COLTOFLAX. Fringes surrounding two obstacles, t = 5 min., Δt = 1 min.

Fig. 6 : Sample of COLTOFLAX. Impressions of teeth, t = 5 min., Δt = 1 min.

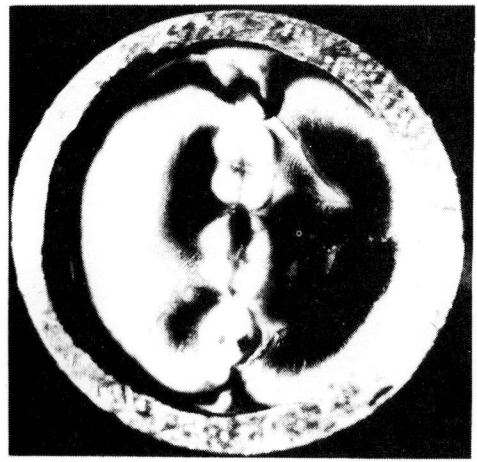
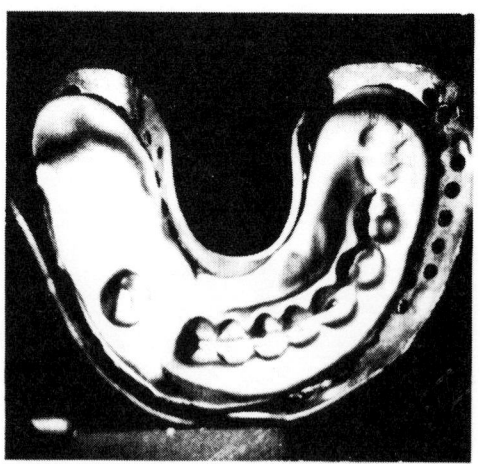

Fig. 7 : Impressions of teeth in a sample of COLTOFLAX (without metallic coating), t = 60 min., Δt = 2 min.

Fig. 8 : Complete impression in COLTOFLAX, held in a perforated impression-tray, t = 60 min., Δt = 2 min.

1st EUROPEAN CONGRESS ON OPTICS APPLIED TO METROLOGY

Volume 136

SESSION 5.2

HOLOGRAPHIC INTERFEROMETRY

Session Chairmen
Burch
Lowenthal

APPLICATION OF HOLOGRAPHIC INTERFEROMETRY TO THE STUDY OF STRUCTURAL DEFORMATIONS IN CIVIL ENGINEERING

J. M. Caussignac
Laboratoire Central des Ponts & Chaussees
58, Bd. Lefebvre — Paris Cédex 15

Evolving construction techniques in civil engineering during recent years have favoured the concomitant and growing extension of complex structures. The design of these structures most often calls for the introduction of basic assumptions which greatly condition the results. In certain cases difficult to analyse, it is particularly desirable to be able to infirm or confirm, on the basis of other methods, the validity of the assumptions made. It is thus possible to examine more closely the behaviour of these structures under load. In particular, the optical methods of stress analysis can help to facilitate the interpretation and utilization of design calculation data. Among the optical methods available to us, we endeavoured to develop in a quantitative manner the use of holographic interferometry for application to certain civil engineering problems.

The main studies conducted up to the present at the LCPC concerned essentially the fiels of the deformation of thin sections using reduced models of carriageways and bridges.

To illustrate this text, we have chosen a recent study relative to the behaviour of a "Y"-type slab bridge model under the action of vertical loads. At present, there are design calculation programs making it possible to estimate the stresses in the slab at least in the zones not having singular points.

Within the framework of this problem, holographic interferometry was used, firstly, for confirming the calculation results and, secondly, for dealing with certain characteristic magnitudes in particular in areas not easily handled by calculations. In fact, like all optical methods, the holographic method with back scattering gives an overall view of the phenomenon and thus makes it possible, by the continuity of the fringe system systems, to provide information on all the zones of the model. In general, a given study is carried out in several phases which we shall examine successively :

1. Recording of holograms. 2. Analysis of fringe patterns and automatic processing. 3. Results and comparison between different methods.

Experimentation

Design of model

The experimentation was carried out on a Y-type slab bridge model in plexiglas to 1/100 scale. This model rests on three simple supports made up of knife-edges. Consequently, this simple support boundary condition calls for zero vertical deflection along these lines (figure 1).

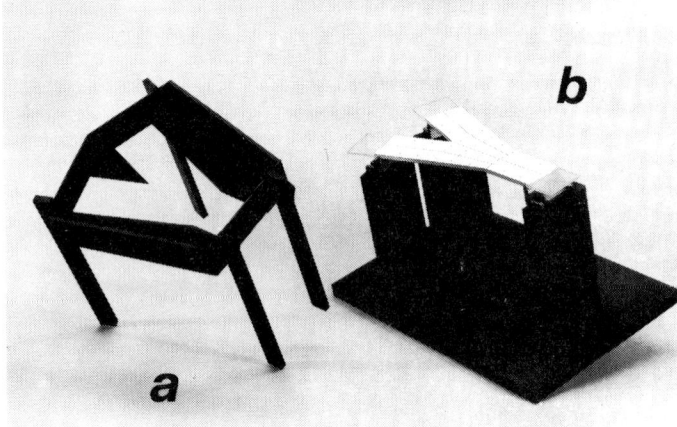

Figure 1

a) Tank used for obtaining uniformly distributed load

b) Model arranged on its supports.

This may appear simple to reconstitute experimentally. In reality, the great sensitivity of the holographic method presents an obstacle. There must in fact be a perfect contact to within $\lambda/2$ between the thin slab and the knife-edge, bearing in mind however that one cannot expect machining more precise than 1/100 mm. An upstream study of thin-slab support condition made it possible to define a device capable of correctly providing simple supports (Figure 2) (Réf. 15).

This problem of boundary conditions is especially crucial when the model has a skew geometry. A sligh fixing can in fact cause variations of about 50 percent in the results.

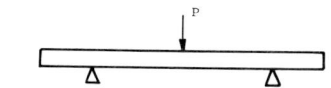
Simple supports with partial fixing

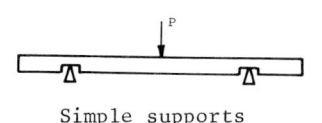
Simple supports with detachment possibilities

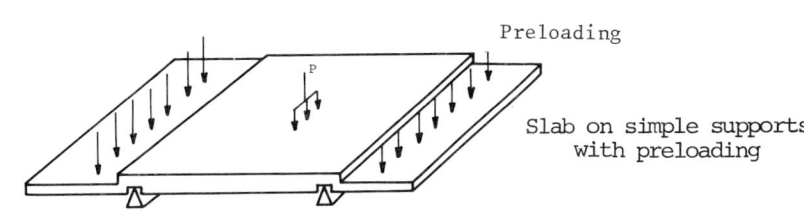
Slab on simple supports with preloading

Fulfilling the support conditions of thin slabs

Figure 2.

Experimental conditions

If one wishes to use holographic data for metrology purposes, it is essential that the vertically-stressed model should work in linear elasticity. This requires that, on the one hand, one should remain within the field of small deformations (mechanical condition) and that, on the other, the horizontal displacements of each point of the model should not introduce working differences greater than $\lambda/4$ (optical condition resulting from the hypothesis whereby a point of a diffuser cannot interfere with itself after deformation). With these conditions, and within the framework of a pure deformation (fringes localized in the vicinity of the structure and absence of fringes in the order 0), it may be assumed that the interference fringes represent the level lines of the deformed surface.

The vertical deflection is then given by the expression : $W = \dfrac{2n-1}{2} \dfrac{\lambda}{1 + \cos \theta}$

in which λ represents the lighting wavelength, n the order of interference and θ the angle formed between the direction of the incident object beam and the average direction defined by the object and the hologram.

Loading systems

Two types of loading were applied (Figure 3).

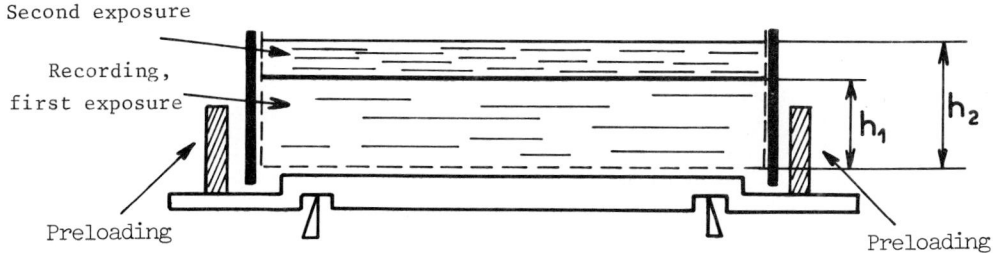

Continuous load applied between the two exposures : $P = \rho g(h_2 - h_1)$

Figure 3.

1. Localized loads at points M, N, P of the model (Figure 4)
2. Uniformly distributed loading achieved by uniform hydrostatic pressure.

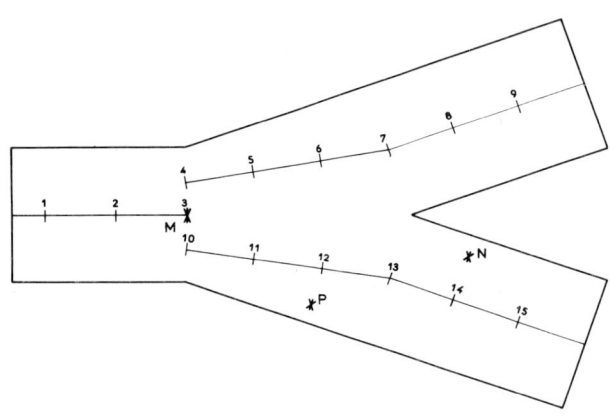

Figure 4.

Recording of holograms

In every case, the double exposure technique is used with the introduction of the stress between the two exposures. The interferograms obtained for two types of loading are given in Figs. 5 and 6.

Fringes of equal deflection

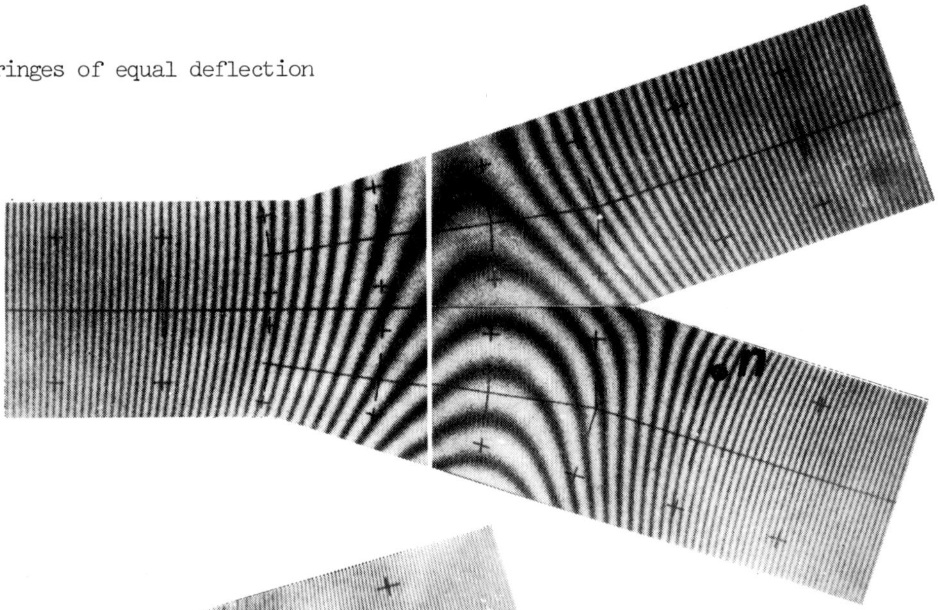

Point loading at N : 50 g

Figure 5.

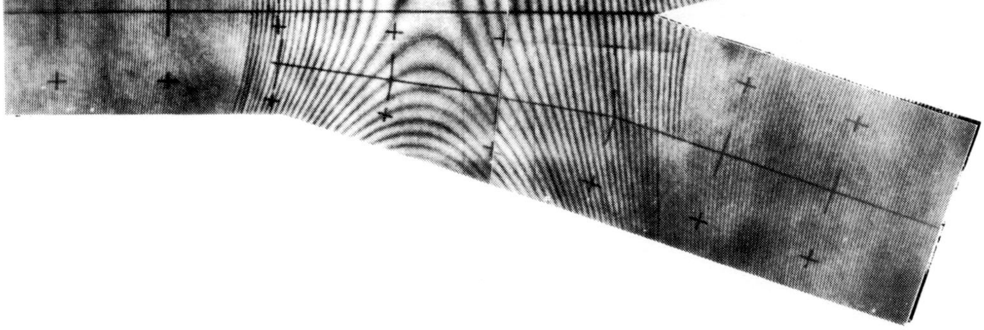

Uniformly distributed loading
(0.585 g/cm^2)

Figure 6.

Dimensional analysis

The purpose of dimensional analysis is to transpose the results obtained on a reduced model by holography to the real case. This correspondence takes into account the geometrical and mechanical characteristics of the model and of the real structure (ratio of moduli of elasticity, dimensions, and so forth). In the case of a thin slab undergoing deformation, Lagrange's equation makes it possible to relate the vertical deflection to the applied load :

$$\Delta^2 W = -\frac{P}{D}$$

in which $\Delta^2 W$ represents the bilaplacian of W and D the flexural rigidity, namely

$$D = \frac{E h^3}{12(1-\nu^2)}$$

E : modulus of elasticity,
h : thickness, ν Poisson's ratio
P : the applied load.

The bending moments of the slab at a point (x, y) of the model are furnished by the expressions :

$$M_{xx} = D \left(\frac{\partial^2 W}{\partial x^2} + \nu \frac{\partial^2 W}{\partial y^2} \right)$$

$$M_{yy} = D \left(\frac{\partial^2 W}{\partial y^2} + \nu \frac{\partial^2 W}{\partial x^2} \right)$$

$$M_{xy} = D (1-\nu) \frac{\partial^2 W}{\partial x \partial y}$$

The sign of the vertical deflection is chosen so that M_{xx} is positive when the top surface of the model is compressed.

The different expressions mentioned above are applicable to the real structure as well as to the model. It is merely necessary to establish the relationships in order to obtain the desired result. By setting the index b for the real structure in concrete and the index m for the model, we have the following relationships :

Deflections : $W_B = W_m \left(\frac{P_b}{P_m}\right) \left(\frac{L_b}{L_m}\right)^4 \left(\frac{E_m}{E_b}\right) \left(\frac{h_m}{h_b}\right)^3 \left(\frac{1-\nu_b^2}{1-\nu_m^2}\right)$

E : modulus of elasticity
P : load

Bending moments : $M_{xxb} = M_{xxm} \left(\frac{W_b}{W_m}\right) \left(\frac{E_b}{E_m}\right) \left(\frac{h_b}{h_m}\right)^3 \left(\frac{L_m}{L_b}\right)^2 \left(\frac{1-\nu_m^2}{1-\nu_b^2}\right)$

L : length
ν : Poisson's ratio

Strictly speaking, this dimensional analysis is valid only if Poisson's ratios are equal in the concrete structure and in the model. This condition is assumed to be fulfilled.

Data reduction

On the basis of the interferograms obtained, the reduction of data takes place in two phases :

1 - A manual review which consists in digitizing the interference fringes on a point reader. This first reduction makes it possible to plot the defection curves.
2 - Automatic processing of these data for calculating the bending moments by the double derivation of deflections.

Manual reduction

The principle of this operation is to assign, to each point digitized on the coordinatograph along an analysis line, a pair of values indicating, on the one hand, the position coordinates and, on the other, the value of the vertical deflection related to the considered fringe order. The difficulty in this reduction has to do to a great extent with the numbering of the fringes as this is particularly influenced by the exact determination of the fringe order 0. The mechanical boundary conditions make it possible to overcome this indetermination.

Automatic processing

Automatic processing consists in smooth ing a mathematical curve on the basis of experimental points initially digitized in order to carry out a certain number of calculations. In mechanics, the bending moments are the most interesting magnitudes which are obtained by the double derivation of the vertical deflection. It will consequently be essential to precess the fringe patterns by computer to reach the bending moments. The study of the Y-type slab bridge was dealt with by single-dimensional smooth ing along the previously chosen reduction lines. This smooth ing uses "cubic splines" for which there is a minimizing of the second derivative of the function (Ref. 14).

Results

The purpose of the study of the Y-type slab bridge was to confirm by experimentation the results obtained with a skew-slab computer program(EUGENE)* specially suited to the calculation of this type of structure. At the same time, comparisons were made by means of one finite-element program (ROSALIE). (Réf. 19)

These comparisons are made for given loads : a uniformly distributed load and an eccentric load. The results obtained on these two types of loading are given in Figures 7 and 8.

The deflection curve corresponds to the holographic method. As concerns automatic processing, we confined ourselves to a comparison between the longitudinal curvatures ($D(\partial^2 w/\partial y^2)$)

In fact, the appearance of the fringes (Figures 5 and 6) shows that the influence of the transverse curvature term is negligible in view of the limited number of experimental points, automatic smoothening being possible only when there is a sufficient number of data.

Remarks :

The analysis method used consist in calculating the curvatures along a straight line. This continuous line was obtained in the longitudinal direction by developing the broken line made up of the axes of the three local references.

The error bars mentioned on the experimental curves characterize the differences obtained when the analysis of the fringes is carried out along two symmetrical lines.

The study conducted by holography made it possible to arrive at several interesting observations :
1 - The influence of the point of the Y remains small and does not involve any significant discontinuities.
2 - Good agreement exists between the analytical method and the experimental method in zones not exhibiting any singularity.

* Program (EUGENE). (Ref. 18)

Conclusion

The example considered in this text illustrates clearly that it is possible in certain cases of structural deformation (field of thin slabs) to achieve quantitative results by the holographic method. Several similar civil engineering studies (pavements and bridges) have made it possible to better understand these phenomena which cannot be dealt with easily by calculation (Ref. 16 - 17). This experimental technique nevertheless requires an operating procedure which must be employed very carefully, considering the extreme sensitivity of the method.

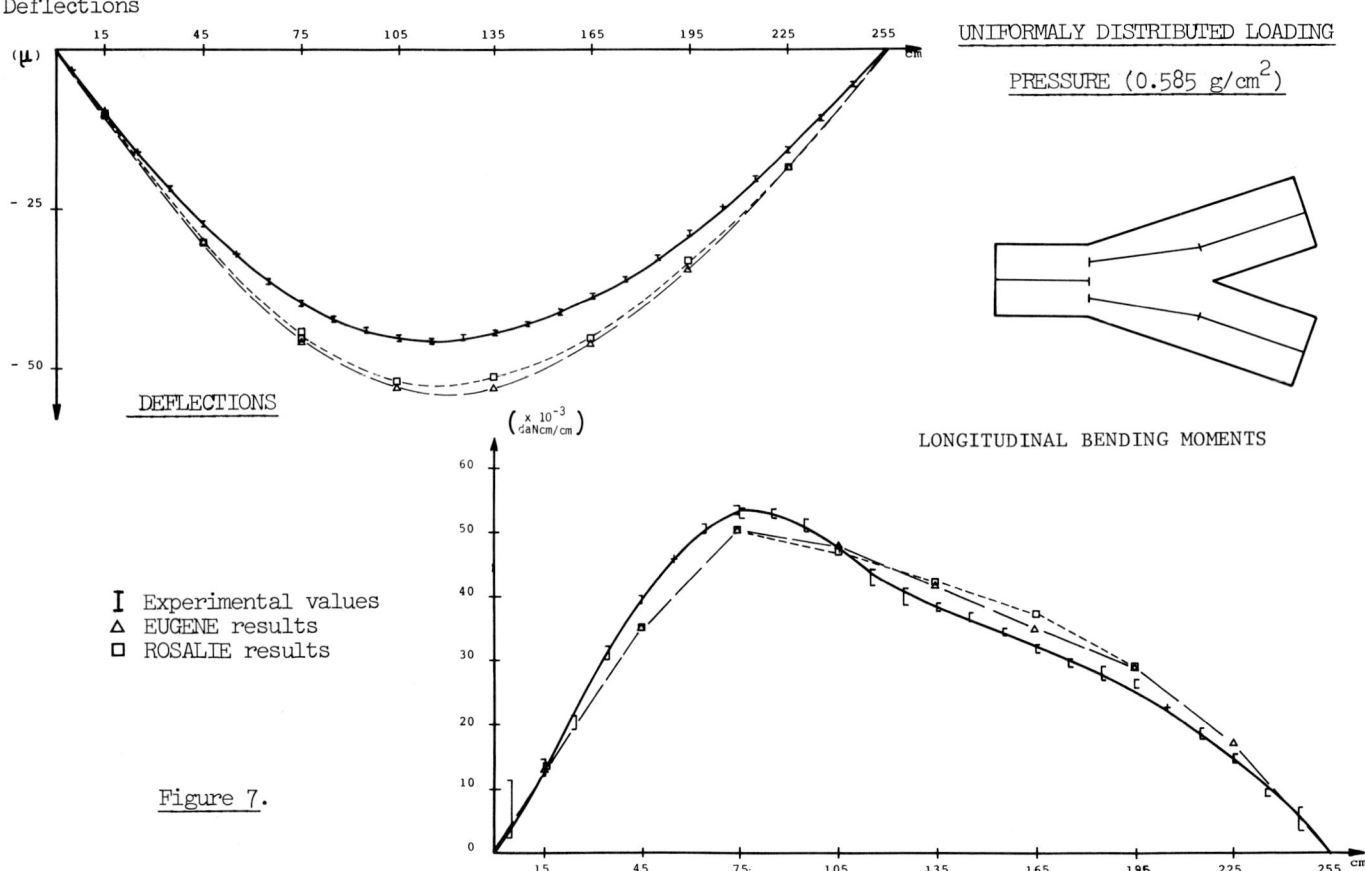

Figure 7.

APPLICATION OF HOLOGRAPHIC INTERFEROMETRY TO THE STUDY OF STRUCTURAL DEFORMATIONS IN CIVIL ENGINEERING

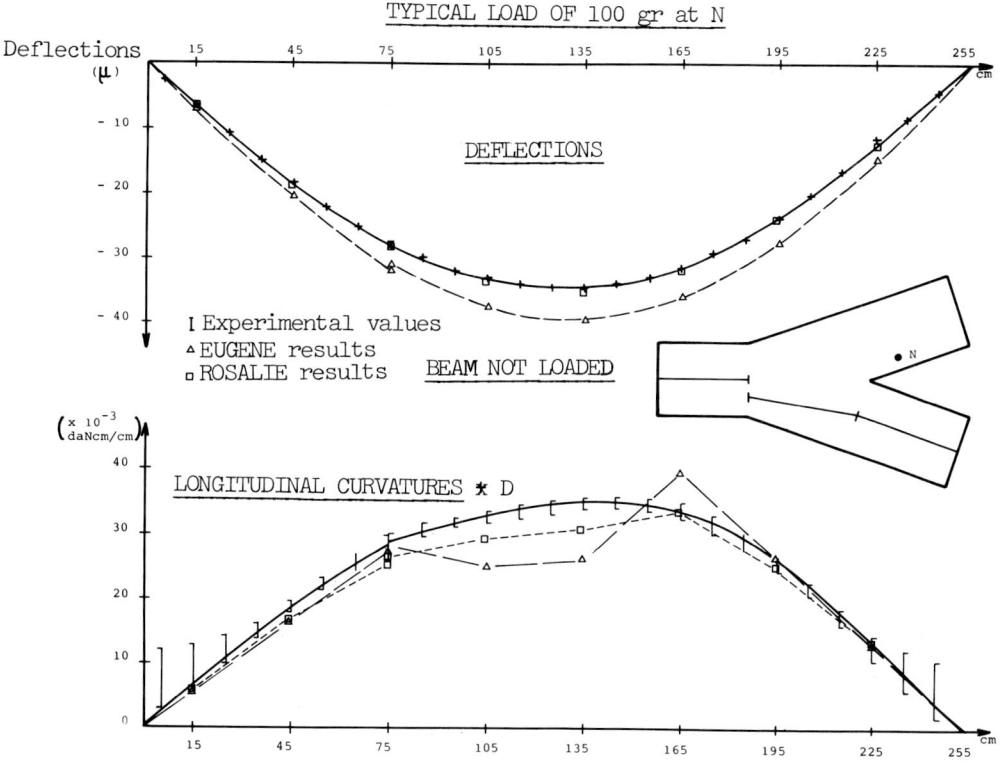

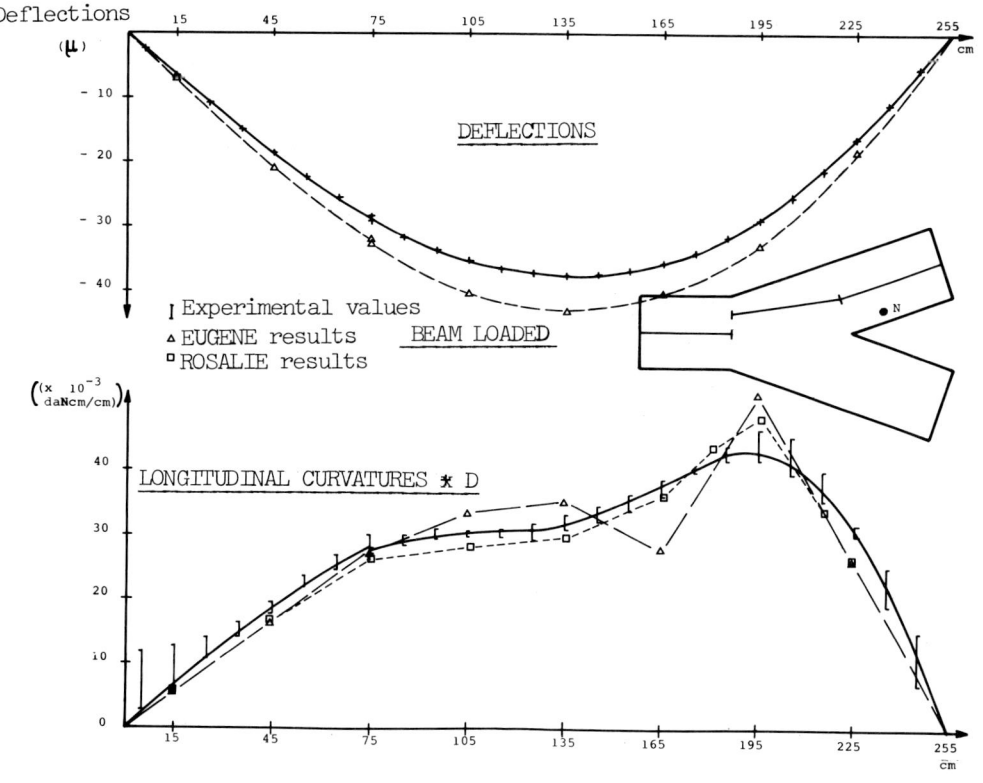

Bibliography

1. JOSEPH DER HOVANESIAM and JERRY VARNER - Méthods for determining the bending moments in Normally loaded thin plates by hologram interferometry. "The Engineering uses of Holography". Proceedings on a conference held at the University of Strathclyde Sept. 1968 - Edited by ROVERTSON and HARVEY.
2. JP DUNCAN and SABIN - An experimental Method for recording curvature contours in flexed elastic plates Experimental Mechanics (1965).
3. E. MIDDLETON - A reflection technique for the survey of the deflection of flat plates - Experimental Mechanics (1968).
4. K A HAINES and BP HILDEBRAND - Interferometric Measurements on diffuse surface by holographic techniques. IEEE Transactions on Instrumentation an measurement - Vol. IM 15 (4) pp 149 - 161 (1966).
5. SP TIMENSHENKO and S WOINOWSKY-KRIEGER - Theory of plates and shells New York Mc GRANW-HILL (1959)
6. JM BURCH - The 1965 viscount Nuffield Memorial paper - Prodt Eng. 36 431 (1965)
7. D. DANLIKER, B. ELLIASSON, B. INEICHEN and F.M. MOTTIER - Quantitative Determination of bending and torsion through holographic Interferometry - Proceedings uses of coherent Optics - April 1975 - Edited by ER ROBERTSON University of Strathclyde - Cambridge University Press.
8. AE. ENNOS - Measurement of in-plane surface strain by hologram interferometry - J. Phys. E. (Sci Inst) 1, 731-734, 1968
9. J.W.C. GATES - Holographic measurement of surface distorsion in three dimensions Optics technology I (5), 274-250, 1969.
10. J. EBBENI - Comparaison des différentes méthodes d'Interférométrie holographique Mécanique Matériaux Electricité - 1-7 Mars 1974.
11. P.M. BOONE - Holographic determination of in plane deformations Optics Technology, 2 (2) 94-98, 1970
12. J. Ch. VIENOT, C. FROEHLY, J. MONNERET and J. PASTEUR - Hologram interferometry : surface displacement f fringe analysis as an approach to the study of Mechanical strains and other applications to the determination of anisotropy in transparent objects. The Engineering Uses of holography - London Cambridge University Press 1979, pp. 133-150
13. CHF. VELZEL - Contours of Equal in-plane displacement in holographic interferometry - Optics Communications, 7 (4) 302-304 1973.
14. H. REINSCH - Smoothing by spline functions-numerische mathematik 10, 177-183 (1967)
15. JM. CAUSSIGNAC - Etude des conditions d'appui d'une plaque mince par holographie. Compte rendu interne Laboratoire Central des Ponts et Chaussées - 1974.
16. JM. CAUSSIGNAC - Study of rigid pavements by holographic method. Proceeding of Internation Commission for Optics 10 PRAGUE 1975 - p. 487-500. Recent Advances in Optical Physics Edited by BEDRICH HAVELKA and JAN BLABLA.
17. K. ABDUNNUR - Contribution à l'étude des ponts à deux poutres avec et sans entretoises sur appuis. Thèse de Doctorat d'Etat présentée à l'Université PARIS VI. - 12 Avril 1974.
18. Program (EUGENE): Catalogue SETRA des documents types des manuels du projecteur - document 4.75.
19. GUELLEC, IMBERT, RICARD - La méthode des éléments finis et le système ROSALIE - Bulletin de Liaison Ponts et Chaussées 81 - Janv.-Fév. 76 - Inf. 1801.

STUDY BY HOLOGRAPHIC INTERFEROMETRY OF MASS TRANSFER DURING ELECTROCHEMICAL PROCESSES AT SOLID-LIQUID INTERFACES

Michael Clifton, Victor Sanchez and Christian Durou

Laboratoire de Chimie-Physique et Electrochimie L.A. 192 and Laboratoire d'Optoélectronique
Université Paul Sabatier, 31077 — Toulouse, France

Abstract

In electrochemical processes at solid-liquid interfaces, one of the main factors controlling mass transfer is the formation of boundary layers. These layers have a thickness between 0.15 and 0.7 mm, depending on the flow velocity of the liquid. Classical electrochemical techniques only give overall measurements of mass transfer; the aim of this study was to measure local mass transfer using holographic interferometry in real time, a technique particularly well adapted to the use of cells with complex forms. Two processes were observed: electrolysis and electrodialysis. The refractive index profile in the boundary layer was observed as a function of the liquid flow-rate and the applied difference in potential. Because of high refractive-index gradients in the boundary layer, the observed profile may be considerably distorted by light-deflection effects. Calculations have shown, however, that light-deflection errors may be kept within tolerable limits by using sufficiently dilute solutions. These solutions of limited concentration still allow useful results to be obtained.

Introduction

Boundary layer phenomena play an important role in all electrochemical processes which involve a phase change. The reaction occurring at the interface between the two phases is generally faster than the mass-transfer processes, such as diffusion and migration, which bring the participant ions to the interface. A layer then forms next to the interface, in which the concentration of the ions involved is altered under the effect of the process at the interface.

The two processes which have been studied in this work are electrolysis and electrodialysis. In both of these processes, the presence of boundary layers is an important factor controlling the rate of the process.

It would be desirable, then, to have some method which would allow the formation of these boundary layers to be studied directly. The measurements involved must be highly localized since these boundary layers are normally less than one millimetre thick. Holographic interferometry in real time has an advantage over more classical optical techniques in that it may be used when the observation cell is not made with high-quality optical glass. Observation cells for electrochemical processes are generally fairly complex structures and the use of high-quality glass would make them much more difficult and much more expensive to prepare. So holographic interferometry has considerable advantages in this domain.

The Electrochemical Processes

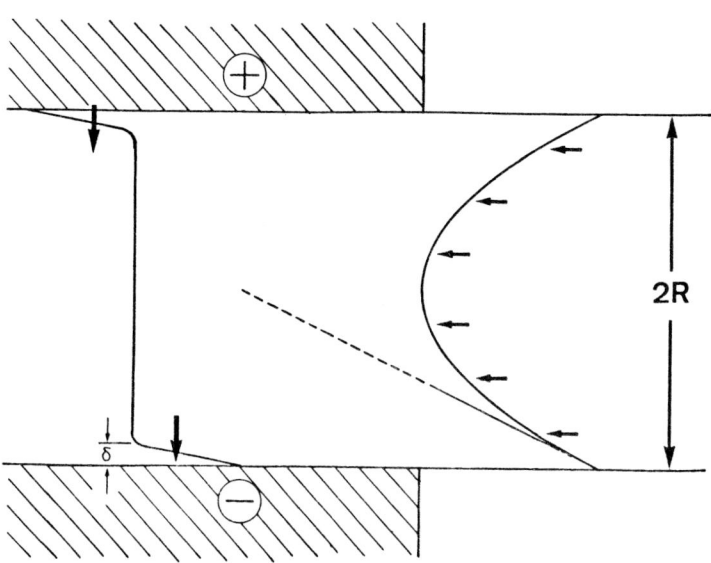

Fig. 1. The electrolysis cell, with concentration and velocity profiles

Figure 1 shows the electrolysis cell. The shaded areas are the two electrodes, and between them is the electrolyte solution, which flows from right to left in a laminar flow regime. The cell is made sufficiently long, so that by the time the solution reaches the electrodes, it already has a fully developed, parabolic velocity-profile. This velocity profile is shown as equation 1; equation 2 is a useful linear approximation.

$$v = v°(2 - y/R)y/R \qquad (1)$$

$$v = v° \, 2y/R \qquad (2)$$

The electrodes are of copper and the solution is a copper sulphate solution. At the anode the copper dissolves into solution and at the cathode the copper is deposited. These processes of dissolution and deposition bring about the formation of a boundary layer at the surface of each electrode. Its thickness is zero at the "point of attack", where the solution first meets the electrode, but the layer gets progressively thicker as the liquid moves down-stream from the point of attack. Fairly quickly, however, the thickening of the boundary layer becomes insignificant and the layer remains roughly the same thickness across most of the surface of the electrodes. A typical thickness would be 0.5 mm.

The thickness of the boundary layer varies with the rate of liquid flow: a fast-flowing solution causes a thinner boundary layer. The thickness also varies with the diffusion coefficient of the ions: slowly diffusing ions produce a wider boundary layer.

The concentration difference in the boundary layer profile is determined by the rate at which the ions are absorbed or released at the active surface. A rapid process at the active surface demands a rapid diffusion of the ions involved and so requires a greater concentration difference.

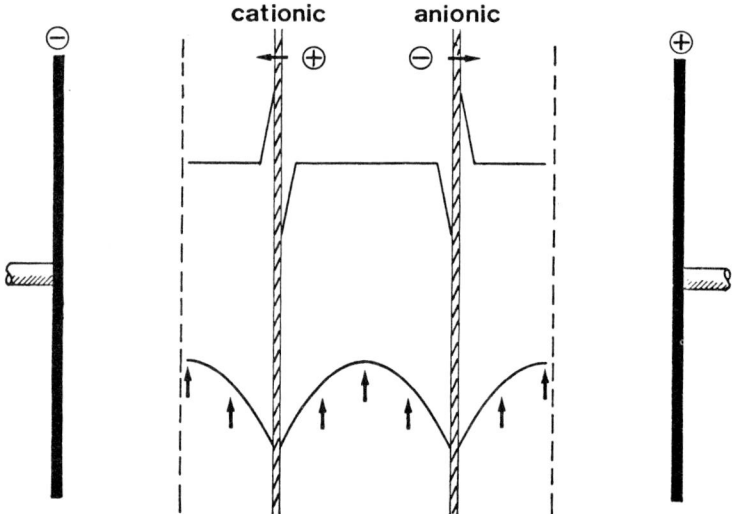

Fig. 2. Concentration and velocity profiles in an element of an electrodialysis cell.

Figure 2 shows an element of an electrodialysis cell. The solution enters with the same concentration in all three compartments. But the cell is subjected to an electric field which causes migration of the ions in solution. At the same time, the ion-selective membranes in the cell only allow the passage of ions of one type. The cationic membrane allows the cations to pass and the anionic membrane allows the passage of anions.

In this way, the concentration in the central compartment falls as the solution flows through the cell, while the concentration in the two outer compartments rises. A complete electrodialysis unit contains a number of such elements placed side by side, in a filter press arrangement. Electrodialysis has been applied to the desalination of sea water and brackish waters as well as to a number of more unusual operations, such as the deionization of cheese-whey.

The efficiency of the process is limited, however, by the fact that boundary layers, similar to those produced on electrode surfaces are also found to build up on the membrane surfaces. The transport of the ions within the membranes occurs more rapidly than the diffusion and migration of the ions in solution, so that the rate of the process is governed mainly by the transfer of the ions in the solution.

The aim in applying holographic interferometry to this sort of system is to study the variation of the thickness of the layer and the concentration difference within the boundary layer as a function of other parameters, such as liquid flow-rate and potential difference across the cell. It should then be possible to compare these results with the behaviour of the system which would be predicted by theory.

Mathematical Model

A mathematical description of the boundary layer can be built up on the basis of equations 3 and 4. Equation 3 is an equation for the conservation of mass; J_i is the flux vector of the ion i:

$$\text{div } J_i = 0 \qquad (3)$$

Equation 4 shows the flux vector as the sum of the three types of flux present in the system: flux due to convection, diffusion and migration respectively.

$$J_i = c_i v - D_i \text{ grad } c_i + \frac{D_i z_i F}{RT} E c_i \qquad (4)$$

These two equations can be combined to give an equation of the following form:

$$2 v^\circ y/R \; \frac{\partial c}{\partial z} = D \frac{\partial^2 c}{\partial y^2} \qquad (5)$$

In this partial differential equation, a linear approximation to the parabolic velocity profile (equation 2) is used. The z-axis lies along the active surface in the direction of flow; the y-axis is perpendicular to the active surface.

Equation 5 may be solved by substituting the quantity η:

$$\eta = y(2v^\circ/9DRz)^{1/3} = 0.893 \, y/\delta \qquad (6)$$

We must assume that the concentration at the active surface is everywhere a constant value c_e; we then arrive at this rather simple expression for the concentration at any point:

$$c = c_e + (c_o - c_e)G(\eta) \qquad (7)$$

The function $G(\eta)$ is a special function which is shown in graphical form in Figure 3. It is really a concentration profile of the boundary layer expressed in terms of a dimensionless concentration and a dimensionless distance, since η is, in fact, proportional to the distance from the active surface divided by δ, the boundary layer thickness. The distance δ has been marked off on the lower alternative scale.

Using this equation, the concentration and the concentration gradient can be predicted at any point in the boundary layer. This mathematical model may, of course, be tested by comparing its predictions directly with the interferograms produced by holographic interferometry. But it is also useful in considering certain problems of a purely optical kind which arise when interferometry is used to examine boundary layers. These are the so-called "mirage" effects.

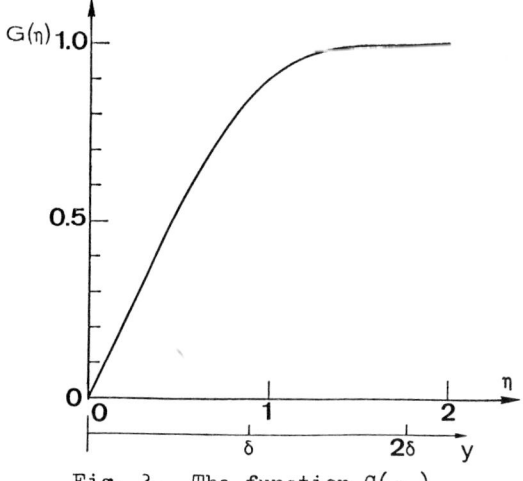

Fig. 3. The function $G(\eta)$

Light-Deflection Effects

Because the boundary layers are usually very narrow and often contain large differences in concentration, they also contain high refractive-index gradients which can cause serious distortion of the interference pattern, due to light deflection.

We have examined this problem using a modified version of the approach developed by Muller, Beach and others at the University of California.[1,2] In Figure 4, we have shown the boundary layer with its thickness somewhat exaggerated. A ray of light enters the solution at the point A and is deflected by the refractive index field. The effective focal plane of the photographic system is located at the point M, so that the deflected ray ABC appears to originate from M and is focussed at the same point on the interferogram as the ray LMN which comes from the holographic reconstruction of the cell in its reference state.

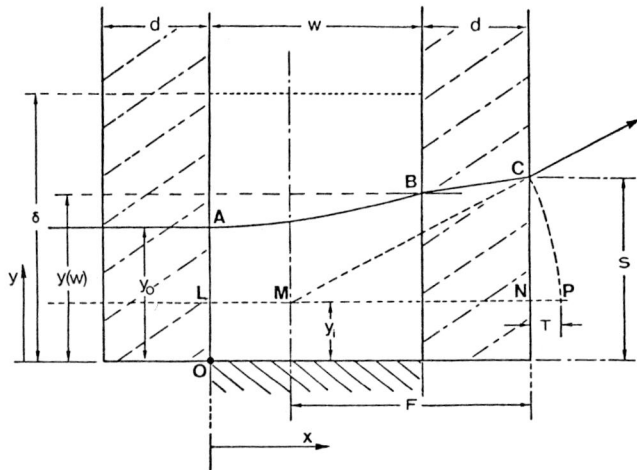

Fig. 4. Deflection of a light ray in the boundary layer

The refractive index field can be calculated from the equation for the concentration (equation 7) and, by carrying out a numerical integration, we can calculate the difference in optical path length between the deflected ray and the reference ray. When this calculation is performed for a number of rays, entering the cell at different distances from the active surface, the profile of the interference fringe can be predicted.

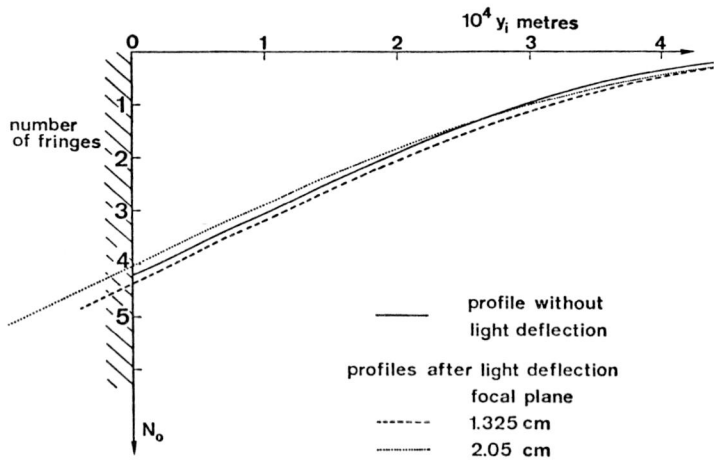

Fig. 5. Calculated fringe profiles

In Figure 5, we have shown some of the fringe profiles calculated for the dimensions of our observation cell. Similar profiles have been calculated for a wide range of solution concentrations; the concentration in this case is close to our experimental concentration.

The unbroken line is the profile of the interference fringe as it would be if there were no light deflection. The two other lines show profiles calculated by taking into account light-deflection effects. The two were calculated for different positions of the focal plane: one at the centre of the cell, one at the inner face of the cell on the side facing the light source.

At first sight, there seems to be little difference between the three profiles, but one important point is the fact that the profiles affected by deflection are longer. It is interesting to consider whether this lengthening would actually be seen in the case of holographic interferometry. It is true that it would be an important effect in the case of classical interferometry in which the part of the image lit up by the reference beam is not limited to any particular area. But in holographic interferograms, the area to the left of the N-axis receives no light at all from the hologram because it lies in the shadow of the electrode or of the membranes. We must remember that the reference beam coming from the hologram is, of course, unaffected by the refractive index gradient in the observation cell, so that in the reconstructed image of the cell the shadow of the active surface is not displaced. Since the shaded area only receives light from one source, there can be no interference there and the interference fringes must stop at the line $y_i = 0$. In this way, the errors due to light deflection are automatically reduced to some extent by using holographic interferometry instead of classical interferometry.

Our calculations of fringe profiles have shown that it is possible to keep distortion errors within reasonable limits by using solutions with moderate concentrations. These concentrations can still be high enough to allow useful work to be done, and no correction of the recorded interferograms is necessary.

Experimental Observations

Figure 6 is an example of some fringes recorded with the electrolysis cell. The two metal electrodes can be seen above and below and, near the cathode, the capillary of the reference electrode.

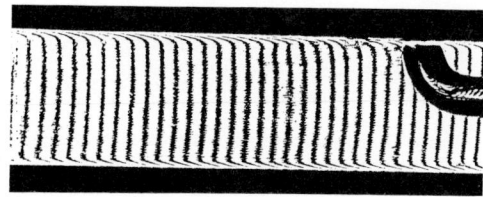

Fig. 6. Fringes recorded with the electrolysis cell.

Figure 7 shows the electrodialysis cell, with some typical interference fringes. The sodium chloride solution enters at the bottom and flows upwards. There is a flow of salt out of the central compartment into the two outer compartments.

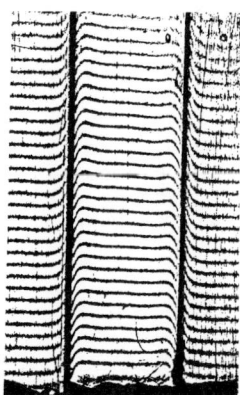

Fig. 7. Fringes recorded with the electrodialysis cell.

Luggin capillaries near the membrane surfaces are connected to calomel electrodes which give a reading of the potential difference applied across each membrane. It has been possible to establish some interesting correlations between measurements of the interferograms and purely electrochemical measurements, both in the case of electrolysis and in the case of electrodialysis.

To sum up our experiences with holographic interferometry, it would be fair to say that we have found it a useful tool for examining boundary layers, in spite of light-deflection effects. The light-deflection errors can, in fact, be regarded as negligible if moderate solution concentrations are used.

References

1. Beach, K. W., _A Laser Interferometer for Mass Transfer Studies_, M. S. Thesis, University of California, Berkeley, 1968.
2. McLarnon, F. R.; Muller, R. H.; Tobias, C. W., "Derivation of one-dimensional refractive-index profiles from interferograms," _J. Opt. Soc. Am._, Vol. 65, pp. 1011-8. 1975.

HOLOGRAPHIC ANALYSIS OF OSCILLATIONS IN SQUEALING DISK BRAKES

Armin Felske
Volkswagenwerk AG, Division of Research
and Development — Messtechnik/Optik
D-3180 Wolfsburg 1, West Germany

Abstract

On a brake test stand the vibration modes of the calliper frame, of the disk, and of the brake pads occuring simultaneously are made visible during squealing in an interference figure by means of double-exposure interferometric holography.

Introduction

Brake squeal has been a problem since the early days of motoring. It has generally been tackled empirically, and a number of ways of reducing squeal on existing brake systems have been developed. However, until today it has not been determined on what principles a brake should be designed in order to eliminate the unpleasant squeal noise.

Experiments suggest that squeal in disk brakes arises from the coupling of vibrations within assembly and that the two most important parameters are the coefficient with the friction material μ, the geometry of the brakes and in particular the location of the areas of contact between brake disk and brake pads, and between the backplate of the brake piston as well as the cantilever.

Very accurate measurements of acceleration on various parts of the brake system showed that during squeal the friction material of the inner or outer pad is deflected elasticaly along the disk surface by frictional force. This deflection causes a second deflection with a component vertical to the surface of the pad and the disk which reduces friction. Thus the stored energy returns into the system to the first configuration and the cycle is repeated. The phase difference between the coupled vibrations necessary to maintain oscillations arises from inertias in the system.

Holographic vibration analysis has become a remarkable tool used at the factory of the VOLKSWAGENWERK AG make to understand this hypothesis and to solve special vibration problems concerning the unpleasant squeal of disk brakes (1)(2)(3)(4).

In order to remove the tendency of squealing of a brake system the first step is to analyze the patterns of vibration of the different parts of the brakes. In Figure 1 I will show you our disk brakes with a frame-type calliper. With respect to the following examples of sliding callipers and fist-type brakes, these parts are the brake disk, the brake pads, and the frame or fist.

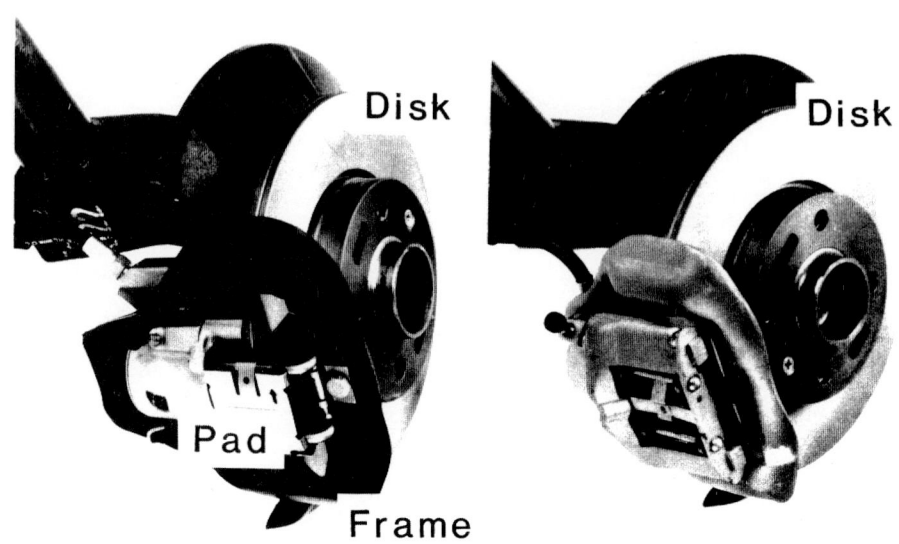

Fig. 1. Disk brakes with frame type callipers.

A large number of squeals has been recorded when pulling up and on a test stand, too. It has been seen that squeal (look at Figure 2) contains only a few high frequency vibrations during a single stop. Generally a single main frequency at 2 or 3 kHz appears in that one squeal, and this is accompanied by harmonics.

By means of the holographic interferometry it is possible to visualize the vibration patterns of the different parts during braking with squeal on a test stand.

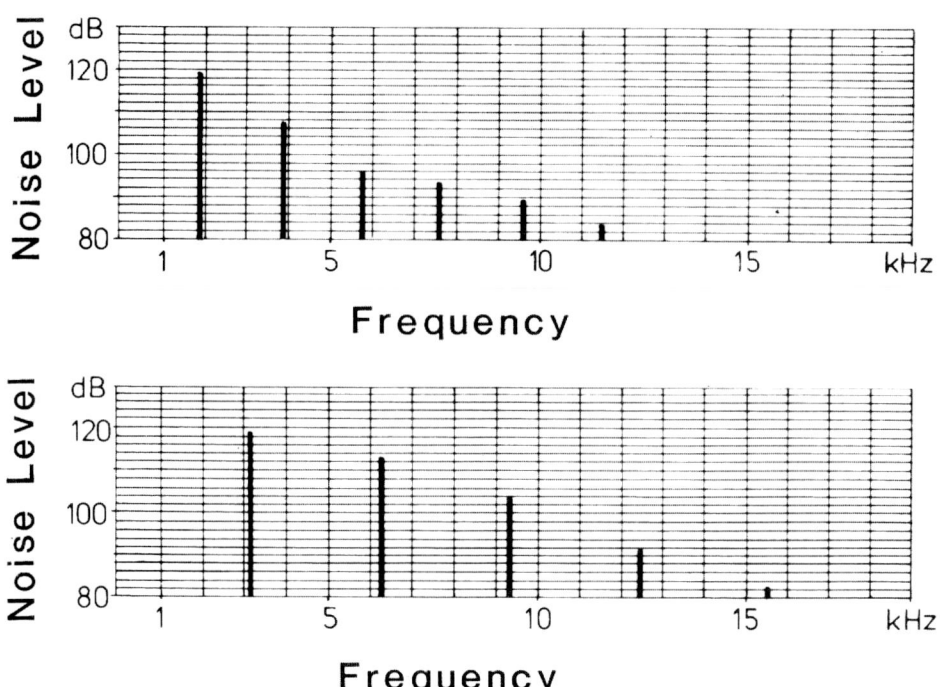

Fig. 2. Component vibration frequency of brake noise measured on an actual vehicle and on a test stand.

Holographic techniques

We used two variations of the basic holographic interferometric technique. Each possesses certain advantages of the other in particular test situations.

(1) Dynamic time-average interferometric holography:
Single holographic recordings are made of the non-rotating disk with pressed brake pads mounted on an isolated table undergoing resonant cyclic vibratory motion. Consequently, upon reconstruction of the hologram, the interference among the ensemble of images produces an interference pattern which is weighted towards the deformation extremes of the test brake. Therefore, regions having no motion or nodes in the vibratory pattern exhibit the greatest intensity and can readily be detected upon inspection of reconstructed image.

An example is presented in Figure 3. The images are those of disk brakes excited at resonant frequencies of 2.5 and 3.4 kHz. The nodal lines are clearly visible. In addition, the adjoining decrease of intensity of the successive bright fringes can be observed when we are going from a node to an antinode i. e. the maximum of displacements.

At 2.5 kHz (Fig. 3 a) the brim of the disk shows 6 nodal lines and 6 vibration centres. At 3.4 kHz (Fig. 3 b) there are 7 nodal lines which are in touch with 7 antinodes. The eighth vibration maximum is suppressed by the pad and here the vibratory energy of the pad runs over to the disk. According to vibrations energy the number of nodes and antinodes must be even.

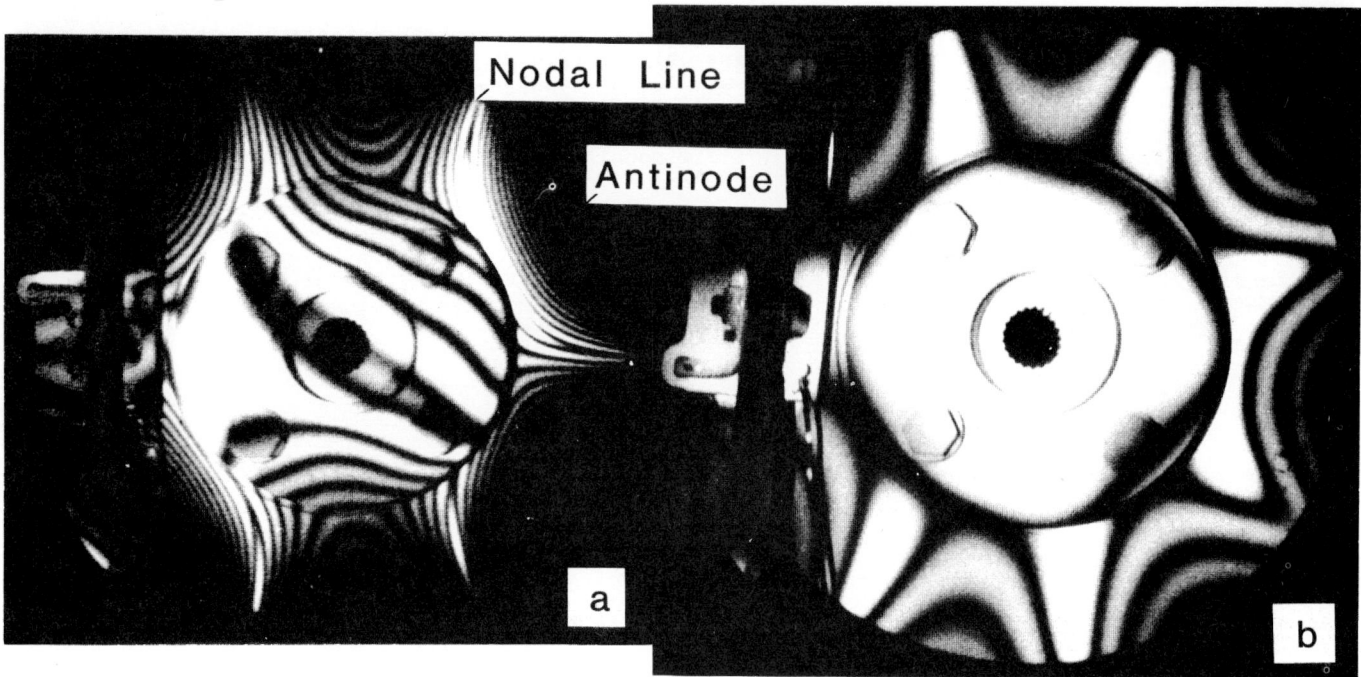

Fig. 3. Reconstruction from time-average holograms showing the vibration modes of a brake disk.
(a) Vibration frequency 2.5 kHz; the brake pad covers a nodal line of the disk.
(b) Vibration frequency 3.4 kHz; the brake pad suppresses an antinode.

The next Figure 4 presents the reconstructed image of another brake system excited at resonance frequency 4.8 kHz. Here the pad vibrates transversely with respect to the disk. For instance, the upper part of the pad swings up. At the same time the lower part swings down. The halves of the outer pad are excited by adjacent antinodes.

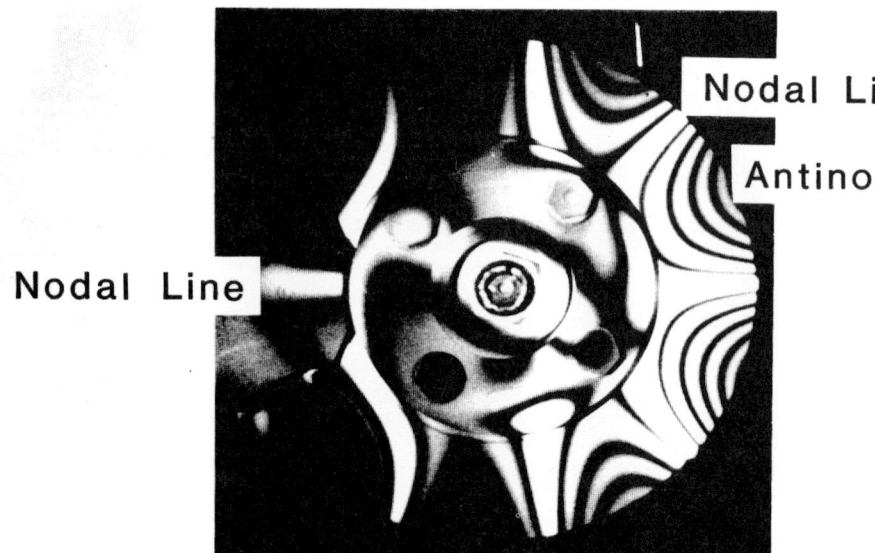

Fig. 4. Reconstruction from time-average hologram showing the vibration mode of a brake disk. Vibration frequency is 4.8 kHz; the halves of the brake pad are excited enforced by adjacent antinodes of the disk.

(2) <u>Double-exposure interferometric holography:</u>
Two holographic recordings of the brake system while squealing are made in two different positions of the vibrating pad, calliper frame, or disk. Pulsed holography can be utilized to eliminate the effects of environmental vibrations which occur on the test stand. Thus the practical application of holography is not limited.

I have used a self-made pulsed holographic camera. The entire ruby-Laser and the optical accessories are mounted on a tripod. The camera is shown in the next Figure 5. The camera generates two holographic Laser pulses of short duration (20 ns) within the same pump flash to eliminate the vibration isolation requirement. This is accomplished by the double pulsing of the Pockels cell. So pulse separations of a few 100 μs can be obtained.

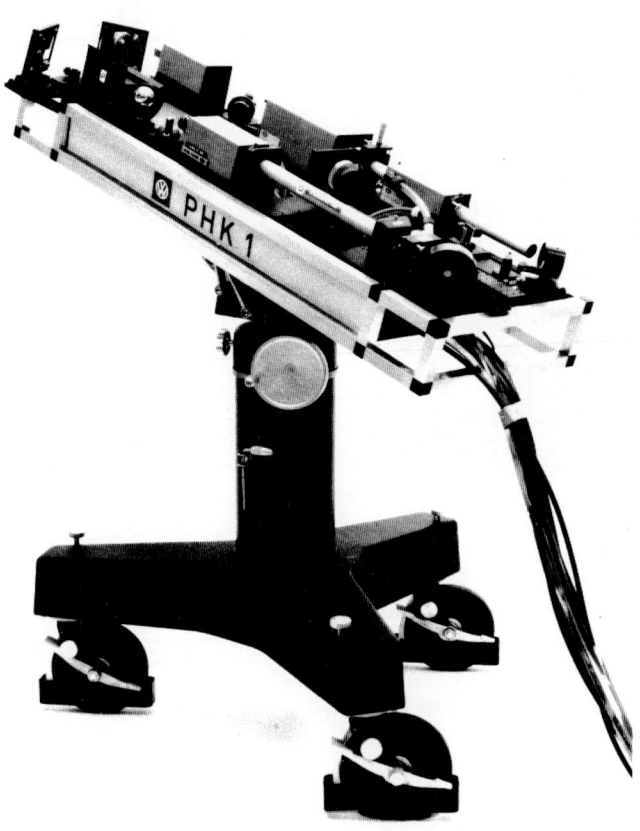

Fig. 5. Double Pulse Hologram Camera (Development of Volkswagenwerk AG) produced by Rottenkolber Holo-System, D-8201 Obing, Allertsham, West-Germany.

Vibration pattern of calliper frame and brake pads
You can imagine that the displacement of the vibrating brake system is as small as that of some micrometers. Thus you have to start triggering the first Laser pulse coinciding at a maximum of vibration amplitude.
Furthermore, the Laser pulses are set on fire only at the time when discriminated amplitudes of the vibration signal occur from a microphone or accelerometer.

The next Figure 6 shows a typical resonance vibration pattern of the calliper frame exposed at a noise level of 120 dB and 2.4 kHz. We see 6 antinodes spread over the curved and arched brim of the frame with amplitudes greater than 3 μm. The artificial excitement of this vibration pattern and the amplitudes onto the freely hanging frame generates a noise level of 120 dB within the immediate vicinity of the frame. This means that the calliper frame is the part which transmits the main part of the noise energy.

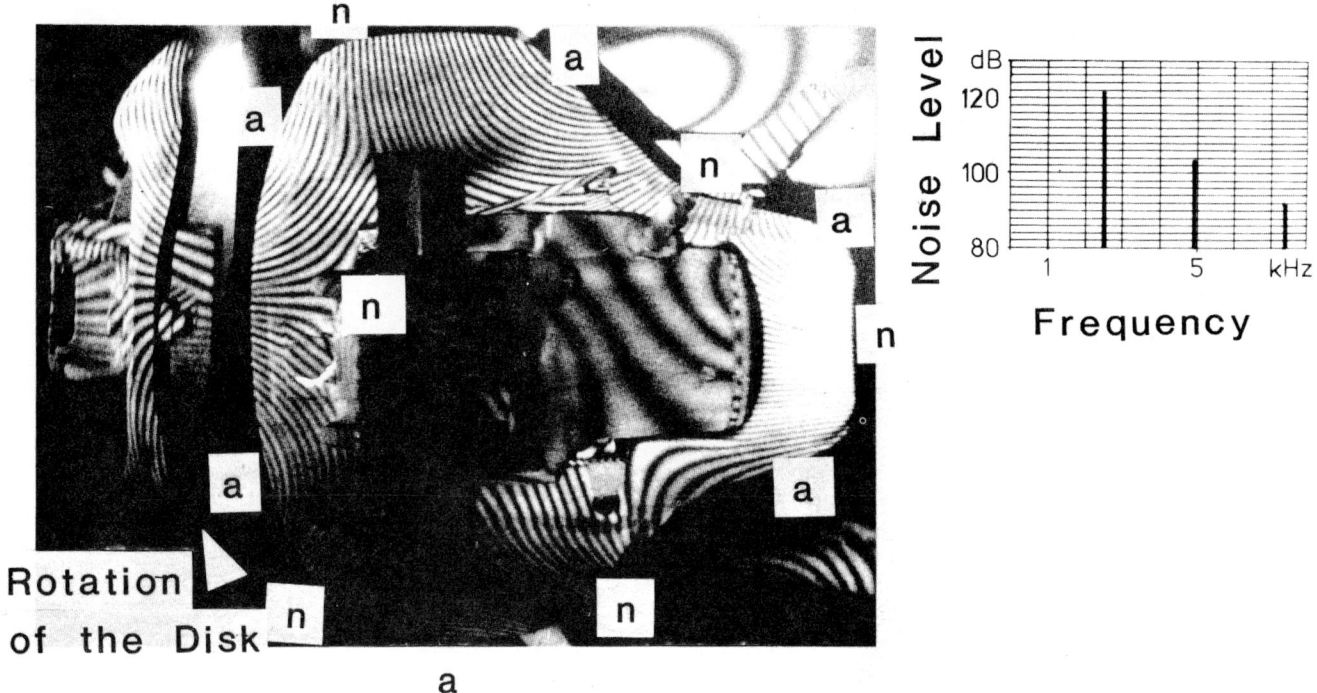

Fig. 6. Reconstruction from double-pulsed hologram of a squealing calliper frame exposed at a noise level of 120 dB and 2.4 kHz. The vibration amplitudes are greater than 3 µm. (a) antinodes, (n) nodes of vibration pattern.

For the first time holographic interference pictures demonstrate brake pad vibrations vertical to the disk, as for instance when the pad is not enough moved back after braking, the brake pad will be excited once when the disk is turning round. The hologram shows a free - free bending vibration of a plate at 6.5 kHz (Fig. 7 c)

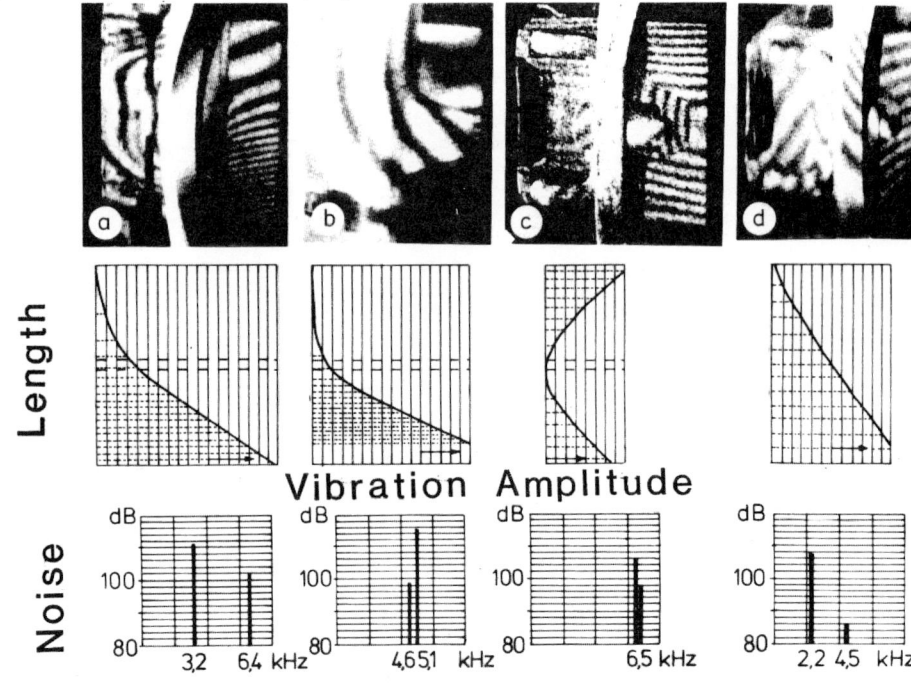

Fig. 7. Reconstructions from double-pulsed holograms of brake pads. (a) and (b) bending vibration mode of pads with a rain groove in the middle of the rubbing surface, (c) free-free bending vibration mode at 6.5 kHz, and (d) weak bending of the pad without a rain groove.

Reducing the squeal noise

One way of reducing the squeal derived from the results of holographic investigations is to remove the brake pads weak structure which is produced by cutting a rain groove into the rubbing surface. On the right side of the Fig.7 you can see the weak bending of the pad without a rain groove. And it is certain and has been controlled that the function on the disk brake is not at all impaired.

The second method is an alteration of the cutting-out of the calliper frame for the brake pad in a way so that the edges of the outer pad are pressed by the frame. This method works as an additional impediment of brake pads bending motion. The result is given by the noise level spectrum in the next Figure 8. It shows one peak at 9.4 kHz which occurs very rarely at a level up to 106 dB. Its origin is a high frequency vibration of the outer pad which excites on a reduced scale the calliper frame because of a strong coupling between frame and pad. The peaks at the other frequencies can no longer be found.

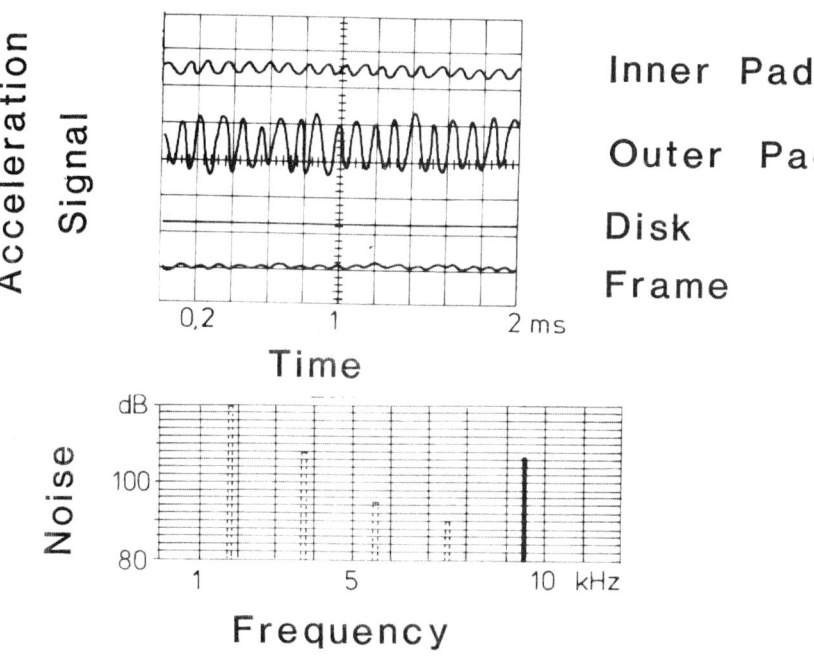

Fig. 8.
Noise level spectrum of the brake system with a cutting-out calliper frame for an additional impediment of brake pads bending motion. The outer pad vibrates at 9.4 kHz and gives the main part of the noise level.

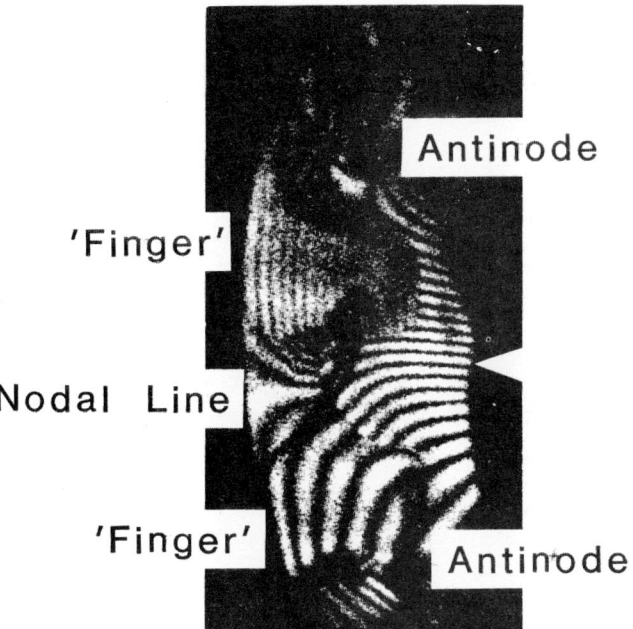

The interference picture of another type of brake system reveals a new mechanism of vibration (Fig. 9.). The longer brake pad vibrates enforced by adjacent antinodes of the disk. The excitation of the so-called "fingers" of this brake system is reactionary. You will be able to localize the nodal line between the fingers. The yoke itself executes its first harmonic vibration pattern at 4.9 kHz confronting the coupled braking system.

Fig. 9.
Reconstruction from a double pulsed hologram of a longer brake pad. As shown in Fig. 4. the edges of the brake pad vibrate enforced by adjacent antinodes of the disk and the so-called "fingers" of the yoke, too, because of the nodal line between the fingers.

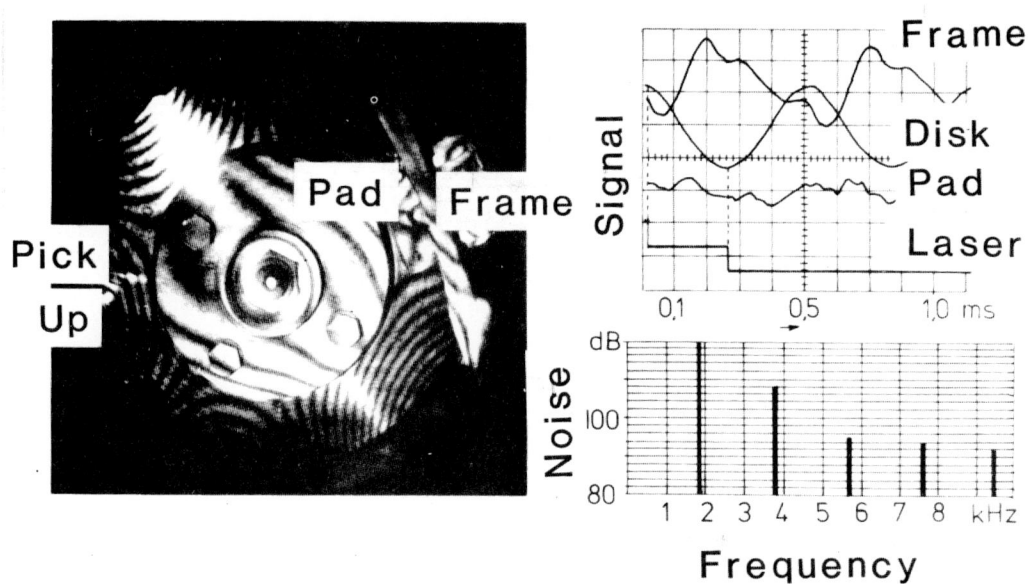

Fig. 10. Reconstruction from a double-pulsed hologram of a squealing disk and the coupled brake system. The oscillographic picture shows the displacement signal of the vibrating disk at the maximum of the noise level. The fringe pattern of the disk amounts to the double of amplitudes (2.8 µm)

Vibration pattern of the disk and its noise level

It is very important to explain the vibration modes of the disk and to calculate the noise level dissipated from the ring-shaped surface of the disk. In this case (Fig. 10) the Laser has been triggered at a maximum amplitude of the disk, measured by a magnetic pick-up and a minimum of speed rotation of the disk. The oscillographic picture which you see on the Fig. 10 shows the displacement signals of the disk. The first Laser pulse coincides with a maximum amplitude and the second one with a minimum. The steps at the lowest trace are the signals of the laser pulses. The 1.9 kHz noise note with an intensity of 120 dB corresponds with a six-node vibration. The sixth antinode is wholly suppressed by the brake pad. The evaluation of the fringe pattern of the disk amounts to the double of amplitudes. Now, we are measuring an amplitude of 1.4 µm.

Vibration patterns with 8 and 10 nodes have been found at frequencies of 3.2 and 6.0 kHz.

Now we have artificially excited the brake disks. The distribution of the amplitudes and the sound levels transmitted from the disk has been measured.

The Table 1 shows that for instance at 1.9 kHz the artificially produces sound level of 100 dB carries a vibration amplitude of about 1.5 µm which is an amount higher than can be exited by the squeal itself.

In this special case we can conclude that the sound level produced by the disk is minor than an intensity factor of one hundred in relevance to the calliper frame.

Although squeal is such a nuisance and so many ways of reducing have been tried, surprisingly, little has been published on the mechanism of squeal. It is as sure as anything that squeal again occurs with the construction of a new brake system. It is also clear that the intricate problem of brake squeal is not solved once at all, but steady progress has been made toward its understanding by holographic analysis.

Nevertheless, further work is very desirable and it is felt that the best hope of providing a permanent cure is by way of thorough understanding of the process through which squeal arises.

Final remarks

Holographic vibration analysis is a very useful method for looking at these small vibrations. If there is an immediate effect of the palliatives used to lessen squeal and chosen by holographic analysis it will be visible for any engineer in the **interference figur**. From the given picture he will see the working mechanism of the disk, pad, and frame vibrations. From that he will be able to develop some new practical antisqueal devices.

Table 1. Frequencies, Noise Levels, Vibration Modes, and Amplitudes of Disks While Squealing and at Artificial Excitement

Exitement while squealing

Frequency	Noise Level	Vibration Mode	Amplitude
1.9 kHz		6 Nodes	1,4 µm
3.2 kHz	104 dB	8 Nodes	0,87 µm
6.0 kHz		10 Nodes	0,63 µm

Artifical excitement

Frequency	Noise Level	Vibration Mode	Amplitude
1.9 kHz	90 dB	6 Nodes	0,31 µm
1.9 kHz	100 dB	6 Nodes	1,5 µm
1.9 kHz	103 dB	6 Nodes	1,7 µm
3.2 kHz	90 dB	not visible	0,15 µm
3.2 kHz	100 dB	8 Nodes	0,48 µm
6.0 kHz	90 dB	10 Nodes	0,15 µm
6.0 kHz	100 dB	10 Nodes	0,36 µm

Acknowledgements

The author wishes to thank A. Happe, G. Hoppe, and M. Matthäi for carrying out the experiments, and for encouragement and helpful discussions.

References

1. Felske, Happe, A, Schwingungsuntersuchungen an Karosserien und Aggregaten mit Hilfe der holografischen Interferometrie, A TZ 75, 1973, S. 96-102
2. Felske, A., Happe, A., Double Pulse Laser Holography as a diagnostic method in the Automotive Industry. E. Robertson, The Engineering Uses of Coherent Optics, Cambridge University Press 1976.
3. Felske, A., Happe, A., Vibration analysis by double pulsed Laser Holography, SAE Paper No. 770030, Detroit 1977.
4. Felske, A., Happe, A., Quietschen von Scheibenbremsen - Holografische Schwingungsanalyse und Abhilfemaßnahmen, ATZ 79, 1977, 7/8, S. 281-288.

SOME CONSIDERATIONS ON THE QUANTITATIVE INTERPRETATION OF HOLOGRAPHIC INTERFEROGRAMS

R. F. C. Kriens
Institute of Applied Physics TNO-TH
P. O. Box 155, 2600 AD Delft, Netherlands

Abstract

In a holographic recording and reconstruction configuration, in which the object illumination source and the center of projection of the observation system seem to coincide, the accurate measurement of the sensitivity vectors is simplified considerably. Therefore, quantitative interpretations of the holographic interferograms are obtained with relatively small effort. If a dual reference beam double-exposure set-up is used, the sense of displacement components can be found and extremes in the scalar product of displacement vector and sensitivity vector can be detected when a relative phase shift is introduced in the reconstruction beams.

Introduction

In literature several methods, either of dynamic or of static character have been proposed for the quantitative interpretation of double-exposure holographic interferograms. However, actual calculations on displacement vectors and strain have only been carried out in a few cases; often the measurements were performed for a single component of displacement or small objects were used of simple and sometimes analytically known geometry.

Typical problems that are encountered in the evaluation of interferograms taken with conventional holographic recording and reconstruction configurations are:

— the accurate determination of length and orientation of the sensitivity vectors
— the possible occurrence of shadows in the reconstructed image of the object, which intersect with the fringe system and therefore hamper its evaluation
— the detection of extremes in the scalar product of displacement and sensitivity vectors over the interferogram
— the determination of the sense of the displacement components.

To solve the first two problems a configuration is proposed in which the object illumination and observation seem to be performed from the same point. This is obtained by means of a beamsplitter in front of the holographic recording medium.
For the other problems fringe shifts are introduced in the interferograms. A dual reference beam double-exposure configuration is required for this purpose.
Finally, it is shown how displacement vectors can be composed from seperate displacement components obtained from multiple interferograms.

Conventional Recording and Reconstruction Configuration

In Fig. 1 the conventional recording and reconstruction configurations are shown for taking double-exposure holographic interferograms. S is the point-source from which the object is illuminated. H is the holographic recording medium and V is the "viewing point", i.e. the center of projection of the system with which the reconstructed images are observed (human eye, photographic or video camera). Also an arbitrary point P on the object's surface is shown, and a point of reference R for the quantitative interpretation.

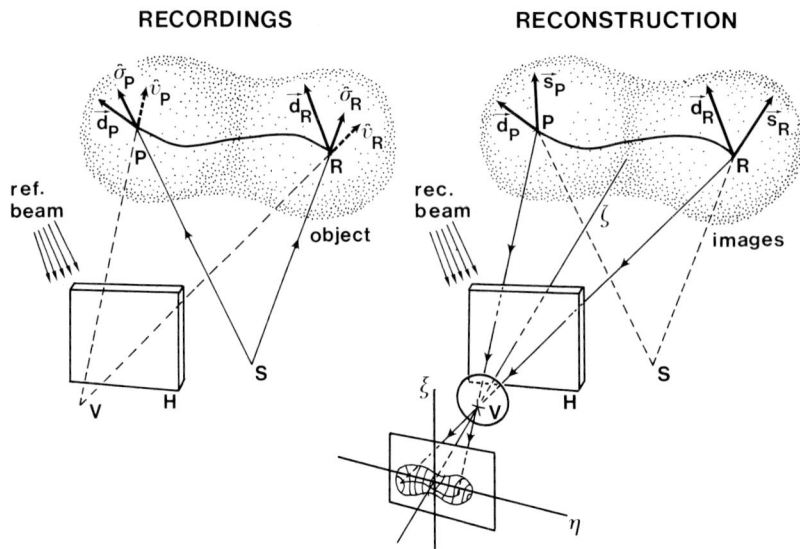

Fig. 1 Conventional recording and reconstruction configuration for a double-exposure holographic interferogram

In Fig. 2 a time scheme is given for the recording and reconstruction procedures of the conventional double-exposure holographic interferogram.

SOME CONSIDERATIONS ON THE QUANTITATIVE INTERPRETATION OF HOLOGRAPHIC INTERFEROGRAMS

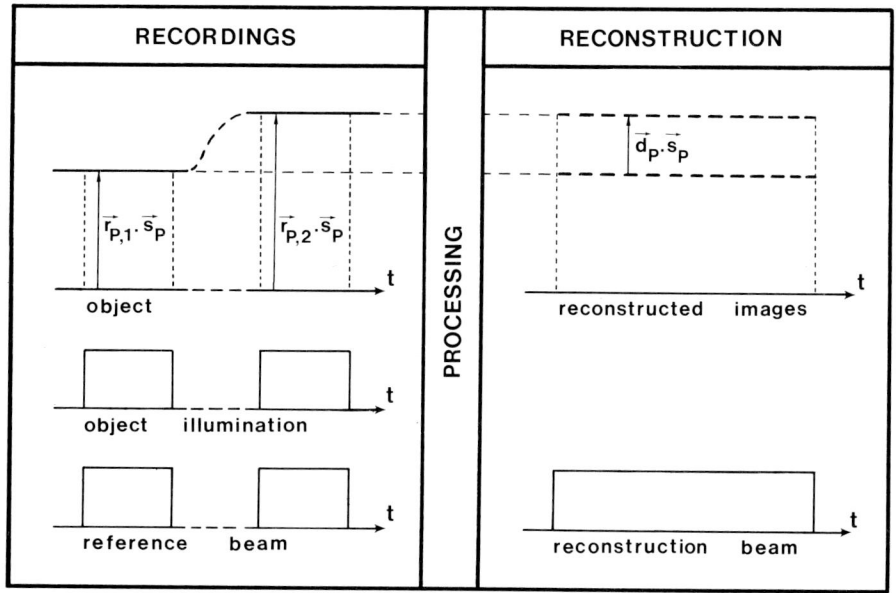

Fig. 2   Time scheme of recording and reconstruction procedures of a single reference beam double-exposure holographic interferogram

During the first holographic recording the object is in a static situation. The position of a point P on the surface of the object is given by the position vector $\vec{r}_{P,1}$. The holographic recording medium H is exposed to the interference of the diffusely reflected object beam and the reference beam.

After the object's position or shape (or both) have changed, a new static situation is reached and a second holographic recording is made on H. The position of point P is now represented by the vector $\vec{r}_{P,2}$. The displacement of point P is given by the displacement vector $\vec{d}_P$:

$$\vec{d}_P = \vec{r}_{P,2} - \vec{r}_{P,1} \tag{1}$$

After processing of the holographic recording medium a hologram is obtained from which with the aid of a reconstruction beam two virtual images are reconstructed corresponding to the two states of the object. Since both images are reconstructed in coherent light, there will be interference and fringes may be seen over the object if they can be resolved by the observation system.

The location of these fringes is determined by the difference in the length of the optical paths from the object illumination source S via the positions of an object surface point P to the viewing point V.

If the displacement of point P, i.e. $|\vec{d}_P|$, is small compared to the distances SP and VP, this path length difference δ_P is given by:

$$\delta_P = \vec{d}_P \cdot \left(\frac{\vec{SP}}{SP} + \frac{\vec{VP}}{VP} \right) = \vec{d}_P \cdot (\hat{o}_P + \hat{v}_P) \tag{2}$$

in which $\hat{o}_P$ and $\hat{v}_P$ are the unit vectors in object illumination and observation directions.

The vector sum of these unit vectors multiplied with $2\pi/\lambda$ is called the "sensitivity vector" $\vec{s}_P$:

$$\vec{s}_P = \frac{2\pi}{\lambda} (\hat{o}_P + \hat{v}_P) = \frac{2\pi}{\lambda} \left(\frac{\vec{r}_P - \vec{r}_S}{|\vec{r}_P - \vec{r}_S|} + \frac{\vec{r}_P - \vec{r}_V}{|\vec{r}_P - \vec{r}_V|} \right) \tag{3}$$

in which $\vec{r}_P$ is taken either equal to $\vec{r}_{P,1}$ or to $\vec{r}_{P,2}$.

Now, the phase difference Φ_P associated with a path length difference δ_P is given by:

$$\Phi_P = 2\pi\delta_P/\lambda = \vec{d}_P \cdot \vec{s}_P \tag{4}$$

The scalar product $\vec{d}_P \cdot \vec{s}_P$ was called the "fringe locus function" by Stetson[1] because the location of the fringes is governed by it.

For the quantitative interpretation of the holographic interferogram Φ_P and $\vec{s}_P$ must be measured in points P where the displacement is required.

One technique used for this purpose is given by Aleksandrov and Bonch-Bruevich[2]. Basically it consists of changing the direction along which the reconstruction of a point P is observed and counting the fringes that pass this point during this procedure.

If the viewing point V is shifted from a position $\vec{r}_{V,0}$ to a position $\vec{r}_{V,1}$ the difference in fringe locus function is:

$$\Phi_{P,1} - \Phi_{P,0} = \frac{2\pi}{\lambda} \vec{d}_P \cdot (\hat{v}_{P,1} - \hat{v}_{P,0}) = \frac{2\pi}{\lambda} \vec{d}_P \cdot \left(\frac{\vec{r}_P - \vec{r}_{V,1}}{|\vec{r}_P - \vec{r}_{V,1}|} - \frac{\vec{r}_P - \vec{r}_{V,0}}{|\vec{r}_P - \vec{r}_{V,0}|} \right) = 2n\pi \tag{5}$$

in which n is the number of fringes which pass the point P during the shift of V (n is a real number).
From equation (5) the component of $\vec{d_P}$ along $\hat{v}_{P,1} - \hat{v}_{P,0}$ is calculated.
An advantage of this technique is, that there is no need for knowing the position of the illumination source S and no point of reference R is required where the fringe locus function Φ_R is known.
However, if the unit vector in the observation direction, i.e. $\hat{v}_P$, passes a situation in which its scalar product with $\vec{d_P}$ reaches an extreme (e.g. a minimum in a situation of closest approach of the two vectors), the sense in which the fringes are counted must be reversed. Such situations are not uncommon in actual holographic interferograms and they are hard to detect. Also the sign for n to be used in equation (5) is not known, and thus two opposite values are found for the displacement component.
Another drawback is the limited angle over which the observation direction can be changed in practical configurations. If the number of counted fringes n is small, large relative errors may be introduced in the calculated component of $\vec{d_P}$.
Furthermore, the technique tends to be time-consuming, since only one point P can be evaluated at a time.
Fossati Bellani and Sona [3] [4] and later Ek and Biedermann [5] [6] presented automatic measurement systems that essentially are based on the same idea, but the real constructed images of the object are used, while the hologram is scanned by a thin conjugate reconstruction beam.
Also Boone and De Backer [7] and Steinbichler [8] have introduced methods in which the real reconstructed images are used. They scan the real images with a small aperture and analyze the intersection of the interference pattern consisting of hyperboloids with a plane. Major drawbacks of this technique are the fact that it is time-consuming and that it is difficult to obtain a high degree of accuracy.

Ewers, Fritzsch and Wachutka [9] and Peeck and Kreitlow [10] proposed the quantitative interpretation of double-exposure holographic interferograms by measurement of the relative values of the fringe locus function in points P with respect to its value in a point of reference R. The evaluation is based on the following equation:

$$\Phi_P - \Phi_P = \vec{d_P} \cdot \vec{s_P} - \vec{d_R} \cdot \vec{s_R} = 2 n \lambda \tag{6}$$

in which n is the number of fringes that is counted on a path over the surface of the object starting in the point of reference R and ending in P (n is a real number). See Fig. 2.
The fringes on a path can be counted by analyzing a photographic recording of the holographic interferogram or by using a scanning observation system such as is found in a video camera with associated electronics for image analysis.
The main feature that makes this method attractive is that for all points that lie on the path from R to P measurements can be carried out in a single traverse of this path. This implies that the technique is suited for automation.
Unfortunately, practical evaluation of interferograms is not simple. The sign of n is not known from the interferogram. Moreover, the sensitivity vector $\vec{s_P}$ has to be measured in all points P where the displacement is wanted; this applies also to the sensitivity vector $\vec{s_R}$ in the point of reference R in cases where only $\vec{d_R}$ is known and non-zero.
From equation (3) it is seen that measurement of these sensitivity vectors requires the composition of the unit vectors in object illumination and observation directions. At reconstruction the unit vectors in the observation directions are easily measured relative to a cartesian coordinate system ξ, η and ζ with its ζ-axis coinciding with the optical axis of the observation system (see Fig. 2).
Measurement of the unit vectors in the object illumination directions cannot be carried out along this way. Schönebeck [11] showed experiments in which in the holographic interferogram the object illumination direction was determined from the orientation and length of a shadow cast on a calibrated sun-dial. Naturally, this system is only applicable to near parallel object illumination beams.
To eliminate the problem of measuring the unit object illumination vectors, one could consider configurations for recording and reconstruction of holographic interferograms in which for any point P the unit vector $\hat{v}_P$ is equal to $\hat{o}_P$.

Configuration with Coincident Object Illumination and Observation

The condition of equal unit vectors $\hat{v}_P$ and $\hat{o}_P$ for all points P requires either configurations in which V and S coincide or configurations in which V is made to coincide with a virtual position of S.
Because the first configurations appeared to be impractical, a configuration of the second type was used (see Fig. 3).

Fig. 3   Holographic recording and reconstruction configuration with coincident object illumination and observation

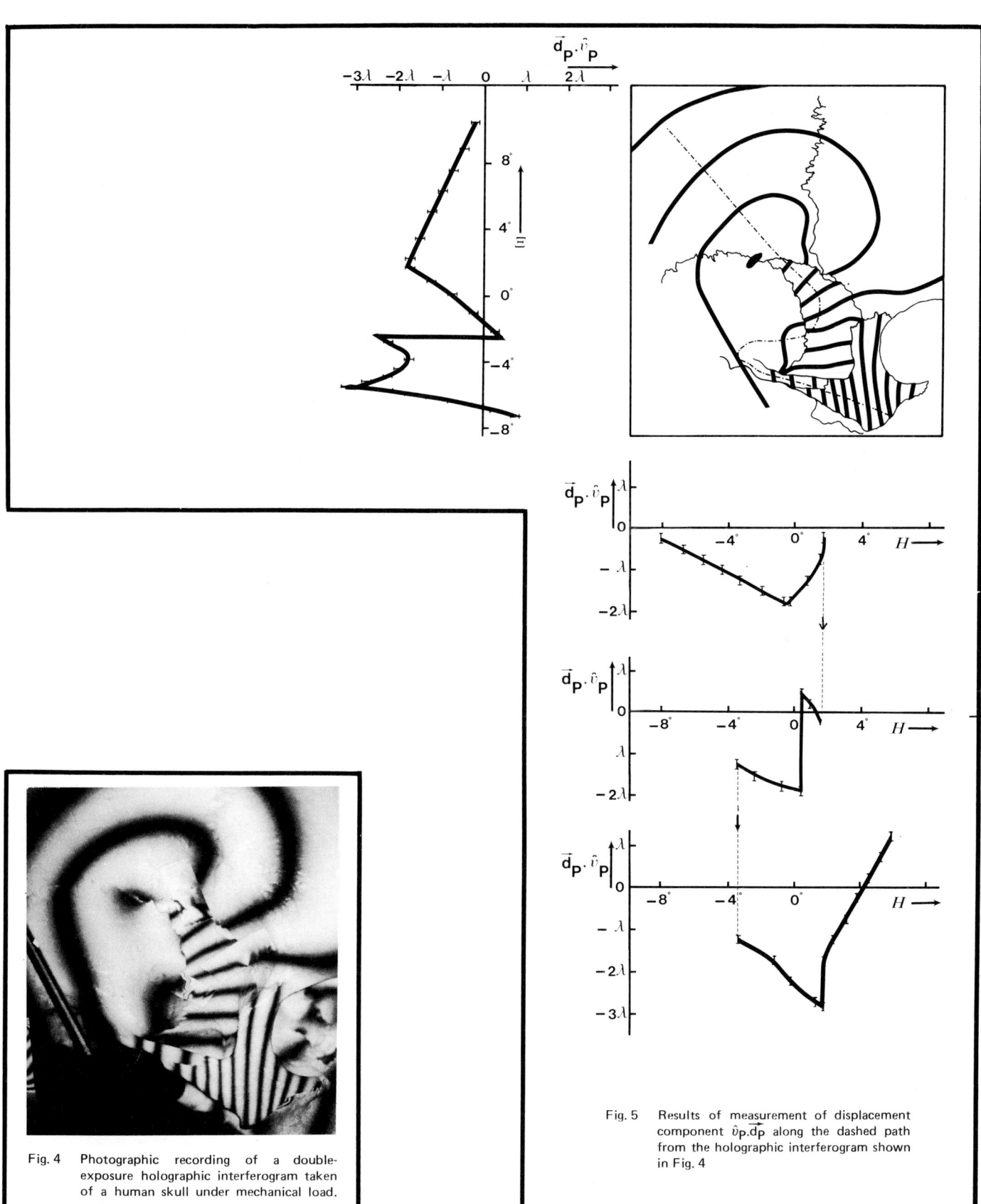

Fig. 4 Photographic recording of a double-exposure holographic interferogram taken of a human skull under mechanical load.

Fig. 5 Results of measurement of displacement component $\hat{v}_P \cdot \vec{d}_P$ along the dashed path from the holographic interferogram shown in Fig. 4

During both holographic recordings the object illumination beam originates from its source S, and is partially reflected towards the object by a large beamsplitter BS. Part of the diffusely reflected object beam is transmitted through the same beamsplitter and reaches the holographic recording medium H, where it is holographically recorded.

At reconstruction of the hologram the observation system is positioned in such a way that the viewing point V is located in the mirror position of S with respect to the beamsplitter.

Now, for any point P on the surface of the object the sensitivity vector $\vec{s}_P$ has uniform length $4\pi/\lambda$ and is radially directed from V:

$$\vec{s}_P = 4\pi \hat{v}_P/\lambda = 4\pi \hat{o}_P/\lambda \tag{7}$$

Unfortunately, no configurations are conceivable in which the sensitivity vector has uniform length and uniform direction, except for set-ups in which parallel object illumination is used in combination with telecentric observation. However, such set-ups are only applicable to small objects.

In Fig. 4 a photographic recording is shown of a double-exposure holographic interferogram taken of a human skull with the beamsplitter configuration. It was studied how various cranial bones and the joints in-between react to mechanical loading.[12].

Along a path of special interest the component of displacement was determined along the sensitivity vector, which is radially directed from the viewing point V. This point V is located in the center of projection of the photographic camera. The observation directions were easily measured in a cartesian coordinate system ξ, η and ζ in which the ζ-axis coincided with the axis of the photographic camera.

As point of reference R any point could be taken on top of the mechanical structure that was attached to the top of the skull and that was used for applying the mechanical load, since displacement is zero there. Accordingly, the fringe locus function Φ_R is zero.

In Fig. 5 the results are shown of the measurements on the holographic interferogram of Fig. 4. The component of the displacement vector $\vec{d}_P$ along the $\hat{v}_P$ direction is displayed as functions of the angles $\Xi = \arctan(\hat{v}_P \cdot \hat{i}_\xi / \hat{v}_P \cdot \hat{i}_\zeta)$ and $\P = \arctan(\hat{v}_P \cdot \hat{i}_\eta / \hat{v}_P \cdot \hat{i}_\zeta)$, with $\hat{i}_\xi$, $\hat{i}_\eta$ and $\hat{i}_\zeta$ as unit vectors of the ξ, η and ζ coordinate system.

The error made in the measurement of the fringe locus function is estimated to be $\pi/4$. The sense of the displacement displayed in the results was based on prior knowledge of the qualitative behaviour of the bone structure.

An additional advantageous feature of the beamsplitter configuration is seen from Fig. 4: now shadows are cast over the object, since each point is observed in exactly the same direction as in which it was illuminated. Shadows may provide difficulties in a quantitative interpretation of holographic interferograms for two reasons.

Firstly, the displacement of points on the surface that are located within a shadow cannot be evaluated. Secondly, the relation of the fringe orders of the interference patterns on either side of a shadow cannot always be deduced from the rest of the patterns. In other words, the continuity of the fringe locus functions is not guaranteed. Both types of difficulties are seen in the holographic interferogram of Fig. 6.

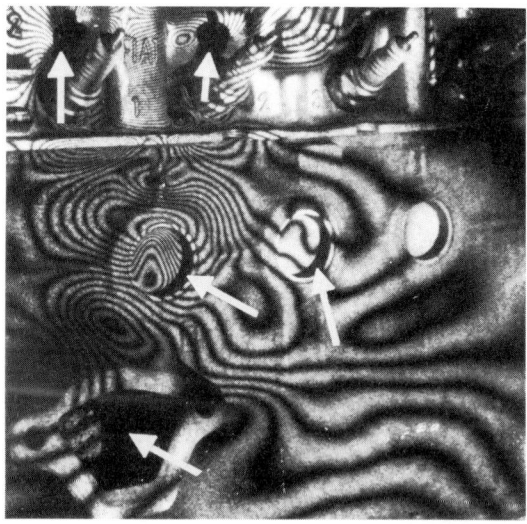

Fig. 6 Photographic recording of a double-exposure holographic interferogram taken of a car engine under mechanical load in a configuration with non-coincident object illumination and observation

Naturally, a disadvantage of the use of a beamsplitter is the loss of power available for hologram recording. The loss of power in the object beam is larger than 75 percent of the power available in this beam without the use of a beamsplitter. In cases that the diffusely reflected object beam is depolarized an improvement could be made by using a polarizing beamsplitter in conjunction with a polarization rotator.

For holographic interferometry of objects that are covered with retroreflective coatings Fagan, Waddell and McCracken[13] proposed similar beamsplitter configurations.

Dual Reference Beam Double-exposure Holographic Interferometry

In the above experiments the sense of the displacement component could not be evaluated from the double-exposure holographic interferogram.

Basically, the reason that such information cannot be gathered from the interferogram, is that neither exposure is "indexed". Both virtual images of the object in its two states are reconstructed with the aid of a single reference beam (see Fig. 2).

Two reference beams with different angles of incidence to the holographic recording material have been used for various reasons throughout the history of holography and holographic interferometry[14] [15] [16] [17] [18] [19]. Tsurata, Shiotake and Itoh[20] applied dual reference beams to holographic interferometry of diffusely reflecting objects. Most recently, Dändliker et al.[21][22][23][24] have used dual reference beam configurations for heterodyne detection of interference fringes in holographic interferograms enabling highly accurate interpolations. Here some other features will be shown that are also available from the dual reference beam technique.

In Fig. 7 a time scheme is given of the recording and reconstruction procedure of a dual reference beam double-exposure hologram.

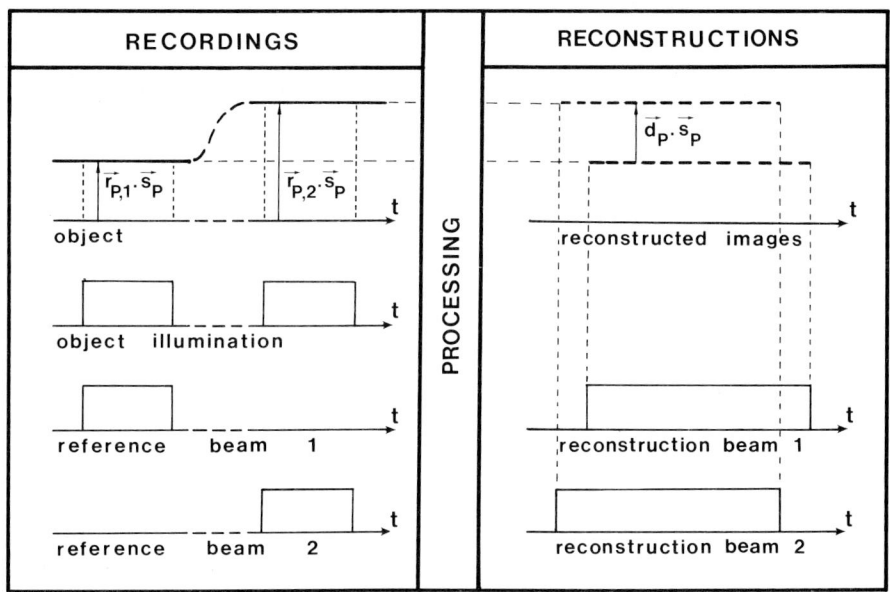

Fig. 7 Time scheme of recording and reconstruction procedures of a dual reference beam double-exposure holographic interferogram

The beam that is diffusely reflected by the object in its first state is recorded in a holographic exposure with the aid of a reference beam 1. The object beam and the reference beam are represented by the wave-functions $U_{obj,1}$ and $U_{ref,1}$ respectively, while the position of a point P on the surface of the object is given by the vector $\vec{r}_{P,1}$ (cf. Fig. 2).

After changes in the position or shape of the object, a new static situation is reached. Again a holographic recording is made. This time with the aid of a separate reference beam 2 which is incident to the holographic recording material from a different direction. Now, the object beam, the reference beam and the position of P are given by $U_{obj,2}$, $U_{ref,2}$ and $\vec{r}_{P,2}$.

The holographic recording material H is thus exposed to the two exposures E_1 and E_2. The exposure times are $t_{exp,1}$ and $t_{exp,2}$

$$E_1 = t_{exp,1} \left(I_{ref,1} + I_{obj,1} + U_{ref,1} U^*_{obj,1} + U^*_{ref,1} U_{obj,1} \right) \tag{9}$$

$$E_2 = t_{exp,2} \left(I_{ref,2} + I_{obj,2} + U_{ref,2} U^*_{obj,2} + U^*_{ref,2} U_{obj,2} \right) \tag{10}$$

If processing into a linear amplitude hologram is assumed, the amplitude transmittance of the hologram τ will be given by:

$$\tau = \tau_0 + \tau_1 (E_1 + E_2) \tag{11}$$

in which τ_0 and τ_1 are in general complex constants (the real part of τ_1 will be negative for conventional photosensitive materials). Reconstruction of the hologram is carried out by illumination with two reconstruction beams $U_{rec,1}$ and $U_{rec,2}$ that are proportional to the two reference beams 1 and 2:

$$U_{rec,1} = a_1 U_{ref,1} \quad \text{and} \tag{12}$$

$$U_{rec,2} = a_2 U_{ref,2} \tag{13}$$

in which a_1 and a_2 are complex constants.
The reconstructed beams from the hologram are given by:

$$\begin{aligned}
U = \tau(U_{rec,1} + U_{rec,2}) = \\
a_1 \left\{ \tau_0 + \tau_1 t_{exp,1}(I_{ref,1}+I_{obj,1}) + \tau_1 t_{exp,2}(I_{ref,2}+I_{obj,2}) \right\} U_{ref,1} + \\
+a_2 \left\{ \tau_0 + \tau_1 t_{exp,2}(I_{ref,1}+I_{obj,1}) + \tau_1 t_{exp,2}(I_{ref,2}+I_{obj,2}) \right\} U_{ref,2} + \\
+a_1 \tau_1 t_{exp,1} U_{ref,1} U_{ref,1} U^*_{obj,1} \quad + a_2 \tau_1 t_{exp,2} U_{ref,2} U_{ref,2} U^*_{obj,2} + \\
+a_1 \tau_1 t_{exp,1} I_{ref,1} U_{obj,1} \quad + a_2 \tau_1 t_{exp,2} I_{ref,2} U_{obj,2} + \\
+a_1 \tau_1 t_{exp,2} U_{ref,1} U^*_{ref,2} U_{obj,2} \quad + a_2 \tau_1 t_{exp,1} U_{ref,1} U^*_{ref,2} U_{obj,1} + \\
+a_1 \tau_1 t_{exp,2} U_{ref,1} U_{ref,2} U^*_{obj,2} \quad + a_2 \tau_1 t_{exp,1} U^*_{ref,1} U_{ref,2} U_{obj,1}
\end{aligned} \tag{14}$$

On the top two lines are the transmitted reconstruction beams. On the third line are the real pseudoscopic reconstructions corresponding to both holographic recordings.
The fourth line displays the virtual orthoscopic reconstructions of the object in its two states. It are these reconstructed object beams that can yield interferograms of the changes in position and shape of the object. On the bottom two lines are the cross-reconstructions, which are formed by the reconstruction of one holographic recording by the reconstruction beam belonging to the other recording. The two beams on the fifth line will also produce an interferogram. This interferogram is related to the interferogram in the virtual holographic reconstructions.
Not all of the beams in equation (14) are spatially separated. Actually, overlap of the virtual orthoscopic reconstructions by the cross-reconstructions may occur depending on the geometry of the various beams used in the recordings. Dändliker, Marom and Mottier[23] have given a rigorous treatment of this effect and they have outlined how to avoid the overlapping reconstructions.

The object beams can be considered as composed of beams originating from all the points in the illuminated part of the surface of the object. Beams coming from a point P at the first and the second holographic exposures are represented by the wave-functions $U_{P,1}$ and $U_{P,2}$, respectively.
It is assumed, that the object illumination is the same for both holographic recordings, and that the reflectivity of the around P is not changed due to its displacement or deformation.
In that case the wave-functions $U_{P,1}$ and $U_{P,2}$ are related by:

$$U_{P,2} = U_{P,1} \exp(i\vec{d}_P \cdot \vec{s}_P) \qquad (15)$$

The wave-function corresponding to the virtual orthoscopic reconstruction of point P is given by (cf. fourth line of equation (14))

$$\tau_1 U_{P,1} \left\{ a_1 t_{exp,1} I_{ref,1} + a_2 t_{exp,2} I_{ref,2} \exp(i\vec{d}_P \cdot \vec{s}_P) \right\} \qquad (16)$$

And the intensity becomes, accordingly:

$$|\tau_1|^2 I_{P,1} \left[|a_1|^2 t_{exp,1}^2 I_{ref,1}^2 + |a_2|^2 t_{exp,2}^2 I_{ref,2}^2 + \right.$$
$$\left. + 2|a_1 a_2| t_{exp,1} t_{exp,2} I_{ref,1} I_{ref,2} \cos\{\vec{d}_P \cdot \vec{s}_P + \arg(a_2/a_1)\} \right] \qquad (17)$$

Now a generalized fringe locus function is introduced for dual reference beam double-exposure holographic interferograms:

$$\Phi_P = \vec{d}_P \cdot \vec{s}_P + \arg(a_2/a_1) \qquad (18)$$

The virtual orthoscopic reconstruction of the object is spatially modulated due to variations of Φ_P over the surface of the object. Full contrast of the interference fringes is accomplished by controlling the ratio of the reconstruction beam intensities in such a way that:

$$\frac{I_{rec,2}}{I_{rec,1}} = \frac{|a_2|^2 I_{ref,2}}{|a_1|^2 I_{ref,1}} = \frac{t_{exp,1}^2 I_{ref,1}}{t_{exp,2}^2 I_{ref,2}} \qquad (19)$$

From equation (17) it follows that fringes shift to positions of smaller $\vec{d}_P \cdot \vec{s}_P$, if the relative phase of reconstruction beam 2 is increased with respect to the phase of reconstruction beam 1, i.e. $\arg(a_2/a_1)$ is increased. Fig. 8 illustrates how through shifts in the relative phase of the reconstruction beams the difference $\Phi_P - \Phi_P$ can be evaluated as point P is moved along a certain path over the object starting in R.
Two interferograms are shown. The fringes in these interferograms are drawn as solid lines. No effort was made to display the actual modulation of the interferogram.

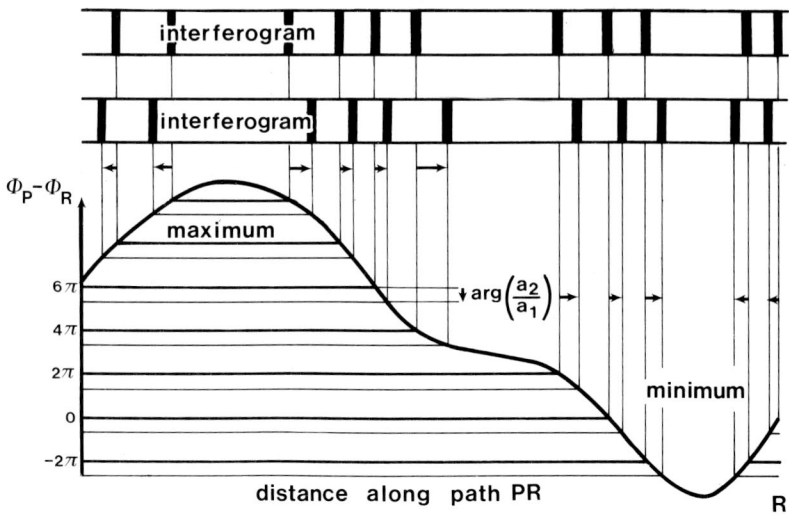

Fig. 8 Detection of extremes in the scalar product $\vec{d}_P \cdot \vec{s}_P$ and determination of the proper sense of $\vec{d}_P \cdot \vec{s}_P$ by introduction of relative phase shifts in the dual reconstruction beams

The upper interferogram corresponds to a situation in which no shift was introduced in the relative phase of the reconstruction beams: $\arg(a_2/a_1) = 0$. This interferogram is identical to the interferogram that would be obtained, if a single reference beam double-exposure hologram was taken of the same deformation.

If this interferogram would be evaluated using an automatic fringe reading system that scans the interferogram along a certain path, solving equation (6) would require additional data on the proper values of n to be substituted.

In such a system extremes in $\vec{d_P}.\vec{s_P} - \vec{d_R}.\vec{s_R}$ cannot be detected. At the locations of extremes the sense of fringe counting must be reversed. Also the sign of n in equation (6) cannot be evaluated from a static fringe pattern. In Fig. 8 the lower interferogram illustrates the effect of a positive shift in the relative phase $\arg(a_2/a_1)$ of the reconstruction beams. As is seen from equation (18) fringes will move to positions of smaller $\vec{d_P}.\vec{s_P}$ in this case. Extremes in $\vec{d_P}.\vec{s_P}$ are detected by the reversal of the direction of fringe movement along the scan path, fringes move away from a maximum and move towards a minimal if a positive shift is applied.

It is evident, that from the way in which the fringes shift one can conclude upon the proper profile for $\vec{d_P}.\vec{s_P}$, or in other words on the proper sign and value of n.

The above theory has been verified with experiments. In Fig. 9 results are presented of a dual reference beam double-exposure holographic interferogram taken of a flat plate in a frame. The plate was pulled to the rear at its center with the aid of an electromagnet. Both plate and frame were deformed.

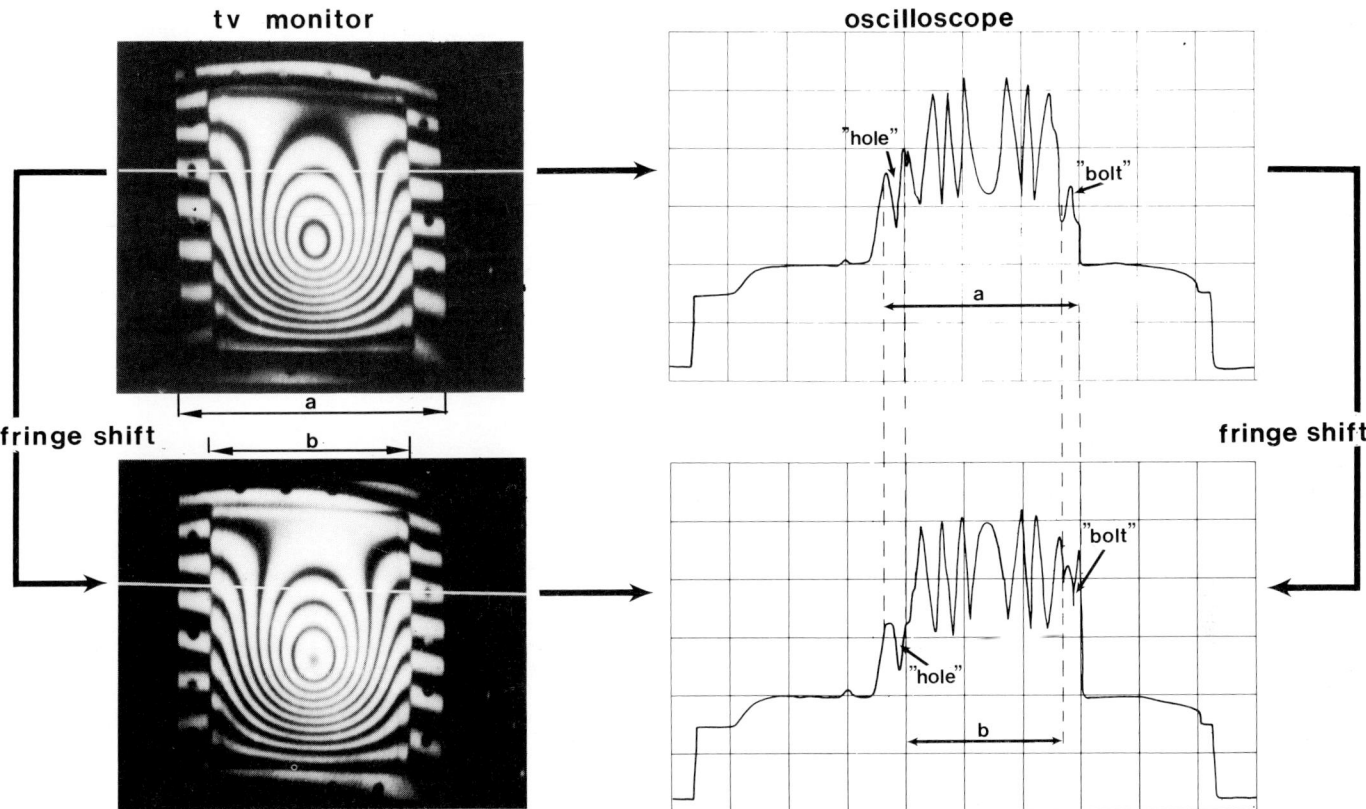

Fig. 9 TV monitor displays of the reconstructions from a dual reference beam double-exposure holographic interferogram under different conditions for the relative phase between the reconstruction beams. The video signal of a selected TV scan-line is also shown in both cases.

The two photographs of Fig. 9 show the dual reference beam double-exposure hologram under different conditions for the relative phase as displayed on a TV monitor. A two-channel oscilloscope was used to analyze the video-signal of one particular TV scan-line. The shifts in the relative phase were introduced by a piezo-electrically driven mirror in one of the reconstruction beam paths. The mirror drive and the oscilloscope were both synchronized on the frame synchronisation of the TV channel.

In this way a situation was obtained in which the channels of the oscilloscope displayed intensity profiles of the video signal along one selected TV scan-line under different relative phases of the reconstruction beams. The interference fringes in the two signals are shifted just less than one half of a period. An extreme (maximum) in $\vec{d_P}.\vec{s_P}$ is observed near the center of the plate.

Furthermore, another feature of the fringe-shift technique is noted: dark holes in the object field which easily could be held for dark fringes in a stationary interferogram may be identified as such.

The TV scan-line in Fig. 9 ran through a hole located in the left hand side of the frame. In the right hand side of the frame it ran over a dark bolt. In the video signals the dip corresponding to the hole remained stationary when the fringes were shifted, while the reconstruction of the bolt changed from bright to dark.

Compositions of Displacement Vectors from Separate Displacement Components

In the above theory and experiments it is discussed how in any point P on the surface of an object a single component of displacement can be measured. Through the use of a configuration with coincident object illumination and observation the components measured are radially directed from the viewing point.

To measure vectorial displacements a set-up must be used in which in each point P three independent components are determined. An example of such a configuration is shown in Fig. 10.

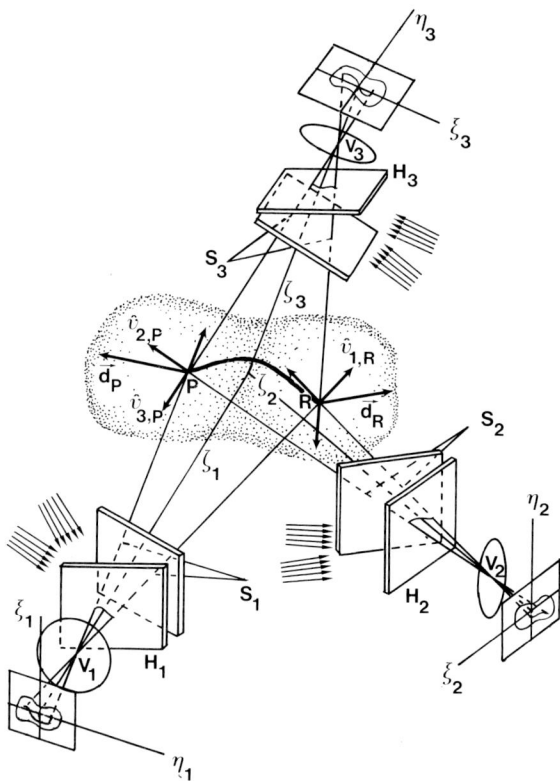

Fig. 10 Triple dual reference beam double-exposure holographic interferometer for measurement of vectorial displacements of object surface points.

The measurement system consists of three independent dual reference beam double-exposure holographic interferometers, each built in a beamsplitter configuration. Three separate object illumination sources S_1, S_2 and S_3, and three separate "viewing points" V_1, V_2 and V_3 are used.
During the holographic recordings of the interferograms only one object illumination can be performed at a time, in order to avoid overlapping fringe patterns. Thus, three successive holographic exposures are required during both states of the object.
Quantitative interpretation is performed by solving the following set of linear equations:

$$\Phi_{i,P} - \Phi_{i,R} = 4\pi(\vec{d}_P \cdot \hat{v}_{i,R})/\lambda = 2n_i\pi \quad (i=1,2,3) \tag{20}$$

in which n_i (i = 1,2,3) are the numbers of fringes counted along three paths from R to P.
In order to measure the vectors $\hat{v}_{i,P}$ and $\hat{v}_{i,R}$ (i = 1,2,3) the points P and R must be identified in each of the observation systems regardless under which perspective the object is seen.
The identification is relatively simple done, if both points are illuminated on the object itself by a small laser spot. This eliminates the need for knowing data on the object's shape and the perspectives under which it is seen.
If the three paths from R to P, along which the fringes are counted in the observation systems, were made to coincide, vectorial displacement data could be obtained for all points on this path. For this purpose, a laser spot could be moved stepwise over the surface of the object. The path over which the spot moves, would then be time-encoded in all three observation systems.
Experiments to verify the above theory have not yet been carried out.

Conclusion

It was shown, how with the aid of holographic configurations with coincident object illumination and observation sensitivity vectors in double-exposure holographic interferograms can be measured accurately and relatively easy.
Furthermore, with the introduction of dual reference beams and a shift in the relative phase of these beams during reconstruction, the detection of extremes in the fringe locus function and the determination of the sense of displacement components appears to be possible.
Without the above options being implemented in double-exposure holographic interferometers, measurement of displacement components would require data on the shape and orientation of the objects under study and additional analysis of the interferograms by skilled technicians.

Acknowledgements

The author wishes to thank H.J. Frankena and H.J. Raterink for their support and M. Lodewijk for his technical assistance and suggestions. Part of this work was carried out under contract with Fiat Industries, Italy.

References

1 K.A. Stetson, "Homogeneous deformations: determination by fringe vectors in hologram interferometry", Appl. Opt. 14, 2256 (1975)

2. E.B. Aleksandrov and A.M. Bonch-Bruevich, "Investigation of surface strains by the hologram technique", Sov. Phys. – Techn. Phys, 12, 258 (1967)

3. V. Fossati Bellani and A. Sona, "Measurement of three-dimensional displacements by scanning a double-exposure hologram", Appl. Opt. 13, 1337 (1974)

4. V. Fossati Bellani, "Automatic measurement of 3-D displacements by using the scanning technique in double exposure holograms", Applications of holography and optical data processing, Pergamon Press, Oxford, 1977, p. 225-231

5. L. Ek and K. Biedermann, "Hologram interferometry with a continuously scanning reconstruction beam", Applications of holography and optical data processing, Pergamon Press, Oxford, 1977, p. 233-239

6. L. Ek and K. Biedermann, "Analysis of a system for hologram interferometry with a continuously scanning reconstruction beam", Appl. Opt. 16, 2535 (1977)

7. P.M. Boone and L.C. de Backer, "Determination of three orthogonal displacement components from one double-exposure hologram", Optik 37, 61 (1973)

8. H. Steinbichler, "Beitrag zur quantitativen Auswertung von holografischen Interferogrammen", dissertation University of Technology Munich January 1973

9. W.M. Ewers, W. Fritzsch and H. Wachutka, "Bestimmung dreidimensionaler Verformungsfelder mit Hilfe der holographischen Interferometrie", Symposium "Aktuelle Probleme der holographischen Interferometrie in der zerstörungsfreien Werkstoffuntersuchung", Meersburg, September 1973

10. A. Peeck and H. Kreitlow, "Quantitative Bestimmung der vollständigen Veränderungsvektoren von Objektpunkten in der Schwingungs- und Verformungsanalyse mit Hilfe der holographischen Interferometrie", Messen + Prüfen/automatik (October 1973)

11. G. Schönebeck, "Anwendungen der holographischen Interferometrie im Stömungsmaschinenbau", paper presented at "Frühjahrsschule für holographische Interferometrie" Hannover (March 1975)

12. R.F.C. Kriens and M. Lodewijk, "Meting van de vervorming van een schedel onder orthodontische belasting met behulp van kwantitatieve holografische interferometrie", internal report 705.220 (in Dutch; September 1977)

13. W.F. Fagan, P. Waddell and W. Mc Cracken, "The study of vibration patterns using real-time hologram interferometry". Optics and laser Technology 4, 167 (1972)

14. A.W. Lohmann, "Reconstructions of vectorial wavefronts", Appl. Opt. 4 1667 (1965)

15. O. Bryngdahl, "Polarizing holography", J. Opt. Soc. Am. 57, 545 (1967)

16. M. De and L. Sévigny, "Three beam holography", Appl. Phys. Letters 10, 79 (1967)

17. M. De and L. Sévigny, "Three beam holographic interferometry", Appl. Opt. 6, 1665 (1967)

18. G.S. Ballard, "Double-exposure holographic interferometry with separate reference beams", J. Appl. Phys. 39, 4846 (1968)

19. J. Surget, "Two reference beam holographic interferometry for aerodynamic flow studies", Applications of holography and optical data processing, Pergamon Press, Oxford, 1977, p. 183-192

20. T. Tsurata, N. Shiotake and Y. Itoh, "Hologram interferometry using two reference beams", Jap. J. Appl. Phys. 7, 1092 (1968)

21. R. Dändliker, B. Ineichen and F.M. Mottier, "High resolution interferometry by electronic phase measurement", Opt. Commun. 9, 412 (1973)

22. R. Dändliker, B. Eliasson, B. Ineichen and F.M. Mottier, "Quantitative determination of bending and torsion through holographic interferometry", The engineering uses of coherent optics, Cambridge U.P., Cambridge, 1976, p. 99-117

23. R. Dändliker, E. Marom and F.M. Mottier, "Two-reference-beam holographic interferometry", J. Opt. Soc. Am. 66, 23 (1976)

24. R. Dändliker and B. Ineichen "Quantitative strain measurement through holographic interferometry", Proc. Third European Electro-Optics Conference (1976), SPIE Vol. 99, p. 90-98

VIDEO-ELECTRONIC ANALYSIS OF HOLOGRAPHIC INTERFEROGRAMS

Franz Lanzl[+] and Michael Schlüter
Institut für Angewandte Physik, Universität Hamburg
2000 Hamburg 36, Germany

Abstract

A hybrid system involving analogue and digital video techniques is presented, which enables the user to determine the loci of interference fringes with high precision in double-exposure holography. Experimental results for interferometric applications are given.

Introduction

One of the most commonly used techniques in holographic interferometry is the double exposure technique. Two object waves associated with two different states of an object are recorded holographically on a photographic plate. Upon reconstruction, the two waves interfere to produce an interferogram covering the actual image of the object.

In order to perform a quantitative analysis on a holographic interferogram, one has to count interference orders up to the image point of interest (static method) [1] or count the number of fringes passing over the object point during a change of the observation angle (dynamic method) [2].

Both methods yield information from which the three orthogonal displacement components of an object point can be calculated. To reach high precision in these results it is of utmost importance to determine the interference orders or the number of fringes with high accuracy.

To reduce measurement errors in the dynamic method, a technique was proposed [3] which relates the interference pattern to an overdetermined set of linearized equations for each object point. The three displacement components are then calculated by the least squares method.

Starting from holographic interferograms where the two different states of an object are recorded with different reference beams, an electronic processing system was developed [4], which enables the user to reconstruct the two interfering images at different optical frequencies. The image is scanned by one (or two) photodetectors yielding a high resolution for the displacement (or its derivative) of the image point under investigation.

In this paper we also use a hologram recorded with two reference beams. A video-electronic analysing system is presented which yields a two-dimensional map of fringe loci, based on the principle of intensity subtraction of two successively reconstructed images before and after changing the relative phase difference of the reference beams by 180°.

Image Subtraction

The first reconstructed image intensity from a double exposure hologram is basically (reference beam intensity set to unity)

$$I_1(\vec{r}) = 2|A(\vec{r})|^2 (1 + \cos\Delta\phi(\vec{r})) \tag{1}$$

with $A(\vec{r})$ being the light amplitude of the object wave and $\Delta\phi(\vec{r})$ the phase difference originating from the object deformation. Figure 1a shows such an intensity pattern, with a Gaussian distribution of $|A(\vec{r})|^2$.

After introduction of a 180° phase shift between the reference beams an image intensity

$$I_2(\vec{r}) = 2|A(\vec{r})|^2 (1 - \cos\Delta\phi(\vec{r})) \tag{2}$$

results, where the maxima and minima are interchanged as shown in Figure 1b. Subtraction of the two intensity patterns yields

$$I(\vec{r}) = I_1(\vec{r}) - I_2(\vec{r}) = 4|A(\vec{r})|^2 \cos\Delta\phi(\vec{r}) \tag{3}$$

[+] This author is coworker of the Gesellschaft für Strahlen- und Umweltforschung, 8042 Neuherberg, Germany.

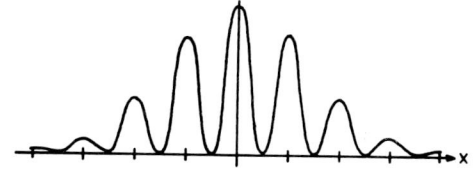

a) Intensity distribution of a Michelson interferogram

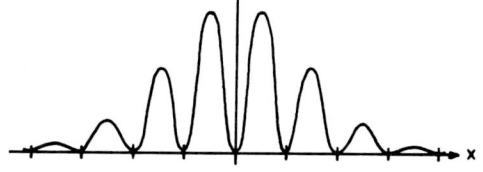

b) Same; with a 180° phase shift

c) Subtraction pattern on constant bias

Fig. 1. Intensity Distributions of Interferograms

as indicated in Figure 1c.

The zeros of the cosine function occur for phase differences

$$\Delta\phi = \frac{2n + 1}{2} \pi \quad (n = 0, \pm 1, \pm 2, \ldots).$$

It is essential for our method that the zeros can be detected with higher accuracy than the extrema, as will be shown in the following discussion.

In order to discuss the local variation of the intensity difference I(x) in eq. (3), we assume that $|A(x)|^2$ varies to a lesser extent than $\Delta\phi(x)$ with respect to x, and get the result

$$I(x+\Delta x) = I(x) + \Delta I$$
$$= 4 |A(x)|^2 \cos\{\Delta\phi(x) + \frac{d\Delta\phi}{dx}\Delta x\}$$
$$= 4 |A(x)|^2 \{\cos\Delta\phi(x)\cos\frac{d\Delta\phi}{dx}\Delta x$$
$$- \sin\Delta\phi(x)\sin\frac{d\Delta\phi}{dx}\Delta x\} \quad (4)$$

Thus the local variation ΔI of I(x) becomes

$$\Delta I = 4 |A(x)|^2 \{\cos\Delta\phi(x)\{\cos\frac{d\Delta\phi}{dx}\Delta x - 1\}$$
$$- \sin\Delta\phi(x)\sin\frac{d\Delta\phi}{dx}\Delta x\} \quad (5)$$

A second order expansion in Δx yields

$$\Delta I \simeq 4 |A(x)|^2 \{-\frac{1}{2}\left(\frac{d\Delta\phi}{dx}\Delta x\right)^2 \cos\Delta\phi(x) - \frac{d\Delta\phi}{dx}\Delta x \sin\Delta\phi(x)\} \quad (6)$$

At the extrema of the interference fringes $\cos\Delta\phi(x)=1$ and $\sin\Delta\phi(x)=0$ hold true, so we get

$$|\Delta I| = 2 |A(x)|^2 \left|\frac{d\Delta\phi}{dx}\right|^2 |\Delta x_e|^2. \quad (7)$$

At the zeros of the cosine function $\cos\Delta\phi(x)=0$ and $\sin\Delta\phi(x)=1$, so

$$|\Delta I| = 4 |A(x)|^2 \left|\frac{d\Delta\phi}{dx}\right| |\Delta x_o|. \quad (8)$$

Assuming equal minimal values of ΔI being resolved, a comparison of eq. (7) and (8) yields

$$\frac{|\Delta x_o|}{|\Delta x_e|} = \frac{1}{2}\sqrt{\frac{|\Delta I|}{|A(x)|^2}} \quad (9)$$

To give a numerical example, we take $\frac{|\Delta I|}{|A(x)|^2} = \frac{1}{100}$ which is easily verified in our experimental set-up, and get

$$\frac{|\Delta x_o|}{|\Delta x_e|} = \frac{1}{20}.$$

So the detection of zeros is more accurate by a factor of 20 than the detection of extrema.

From these considerations we see the two advantages of the subtraction method:

1) The fringe loci can be determined with higher accuracy from the zeros than from the extrema.

2) These loci are independent of the amplitude variation $|A(x)|^2$ of the object including the factor $|R(x)|^2$ i.e. the intensity distribution of the reference beam, which was assumed unity in the formulae. This is valid if $|A(x)|^2$ and $|R(x)|^2$ vary slowly compared to $\Delta\phi(x)$. Only the width of the lines of the loci of the zeros is influenced by $|A(x)|^2$ and $|R(x)|^2$ according to eq. 9.

General Description of the Processing System

In order to perform the image subtraction experimentally, the holographic interferograms are fed into an electronic processing circuit by means of an extremely sensitive high-resolution TV camera (Figure 2).

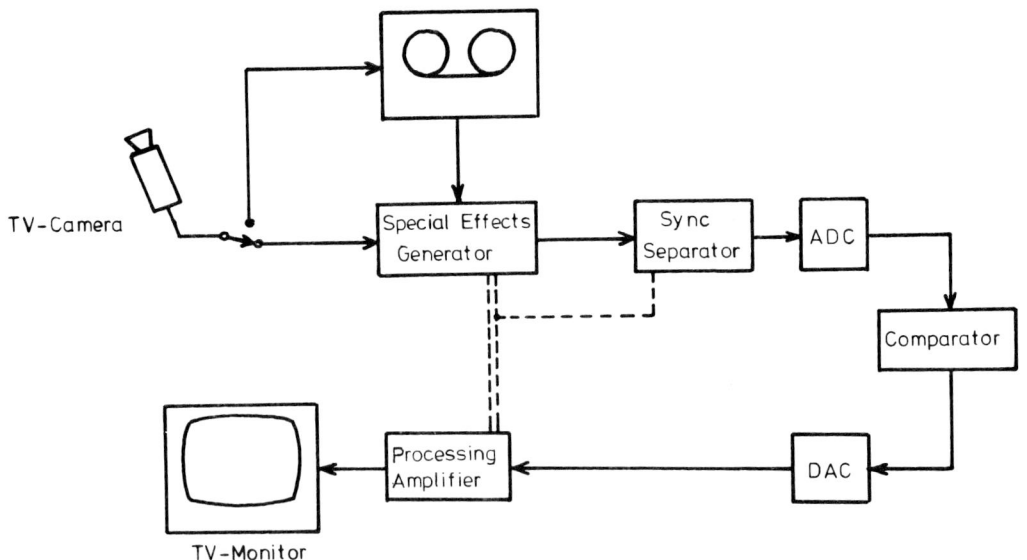

Fig. 2. Experimental Set-Up for the Analysis of Interferograms

In a first step the interferogram to be analysed is stored onto a one-inch video tape recorder (VTR). Now a 180° degree phase shift is introduced into the interferogram which is then fed directly together with the original interferogram played back from the VTR into the video special effects generator (SEG). The SEG performs contrast inversion and superposition of both interferograms onto each other thus yielding a subtraction pattern on a constant bias. The subtraction interferogram is then further processed.

Following the separation of the synchronizing and blanking pulses from the composite analogue video signal, the video data are digitised into 8 bit words (giving 256 possible grey levels) at a 10 MHz sampling frequency in the video analogue to digital converter (ADC).

A digital comparator circuit selects any chosen range of grey levels down to a single grey level between 0 and 255 to be written white on the TV monitor screen, while the remaining grey levels are written black.

The processed data are passed through a video digital to analogue converter (DAC) and a processing amplifier to generate composite video signals for display on the monitor. The resulting TV pictures are binary images displaying lines or areas of constant grey level in white.

Experimental Results

Preliminary experiments were performed using a common Michelson interferometer with a rectangular glass plate in one of the arms. The interference pattern due to a small inclination of one of the mirrors was projected onto a screen and stored onto the VTR by means of the TV camera.

Figure 3a shows a monitor photograph of the pattern, Figure 3b depicts the corresponding oscillogram of a single TV line (roughly along the centre of the interferogram; synchronizing pulses to the left and to the right of the actual TV line).

a) Monitor Photograph

c) Monitor Photograph

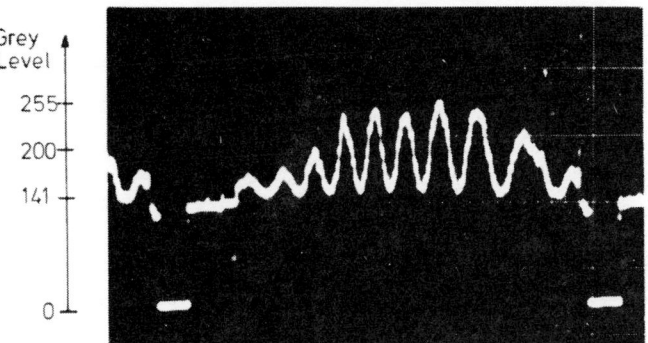
b) TV Line Oscillogram

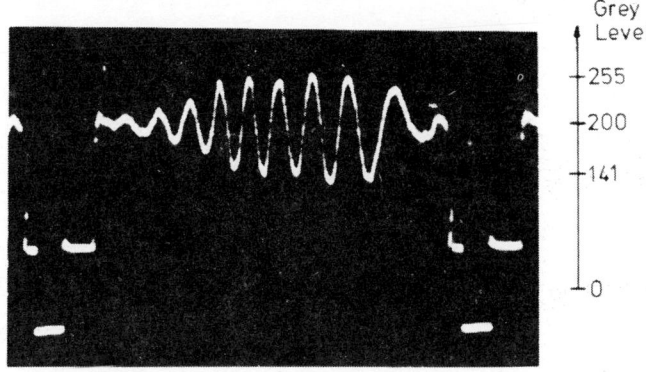
d) TV Line Oscillogram

MICHELSON INTERFEROGRAM   SUBTRACTION INTERFEROGRAM

Figure 3

By rotating the glass plate, a 180° phase shift is introduced into the interference pattern which is then subtracted from the original pattern in the above mentioned manner. The resulting subtraction pattern is shown in Figure 3c, while Figure 3d depicts the corresponding TV line oscillogram.

Figure 3d shows that in image subtraction via contrast inversion a constant bias is added to the subtraction pattern, since no negative intensities may appear in a TV signal. Thus the zeros of the cosine function are shifted to a positive value.

The subtraction pattern is then converted into digital data. The comparator circuit selects the grey levels 141, 200 and 255 to be written in white on the monitor screen (see Figures 4d, e, f). These grey levels correspond to the minima, shifted zero points and maxima of the cosine function, respectively.

It is clearly seen that the determination of the extrema is less accurate by a factor of the order of 10 than the determination of the shifted zero points.

Furthermore, we see from Figure 4e (grey level 200) that the width of the lines of constant phase difference increases towards the left and right edges of the TV picture. This effect is also predicted by theory (eq. (8)). The local line width Δx_0 increases with decreasing illumination intensity $|A(x)|^2$ for a fixed intensity interval ΔI (which is given by the digitising process). Because of the Gaussian distribution of the illuminating laser beam there is less light intensity in the edges of the object than in the centre, which gives rise to broader lines of constant phase difference in the edges.

If we process the original interference pattern (see Figure 3a) in the described manner without performing the image subtraction, binary images like the ones shown in the monitor

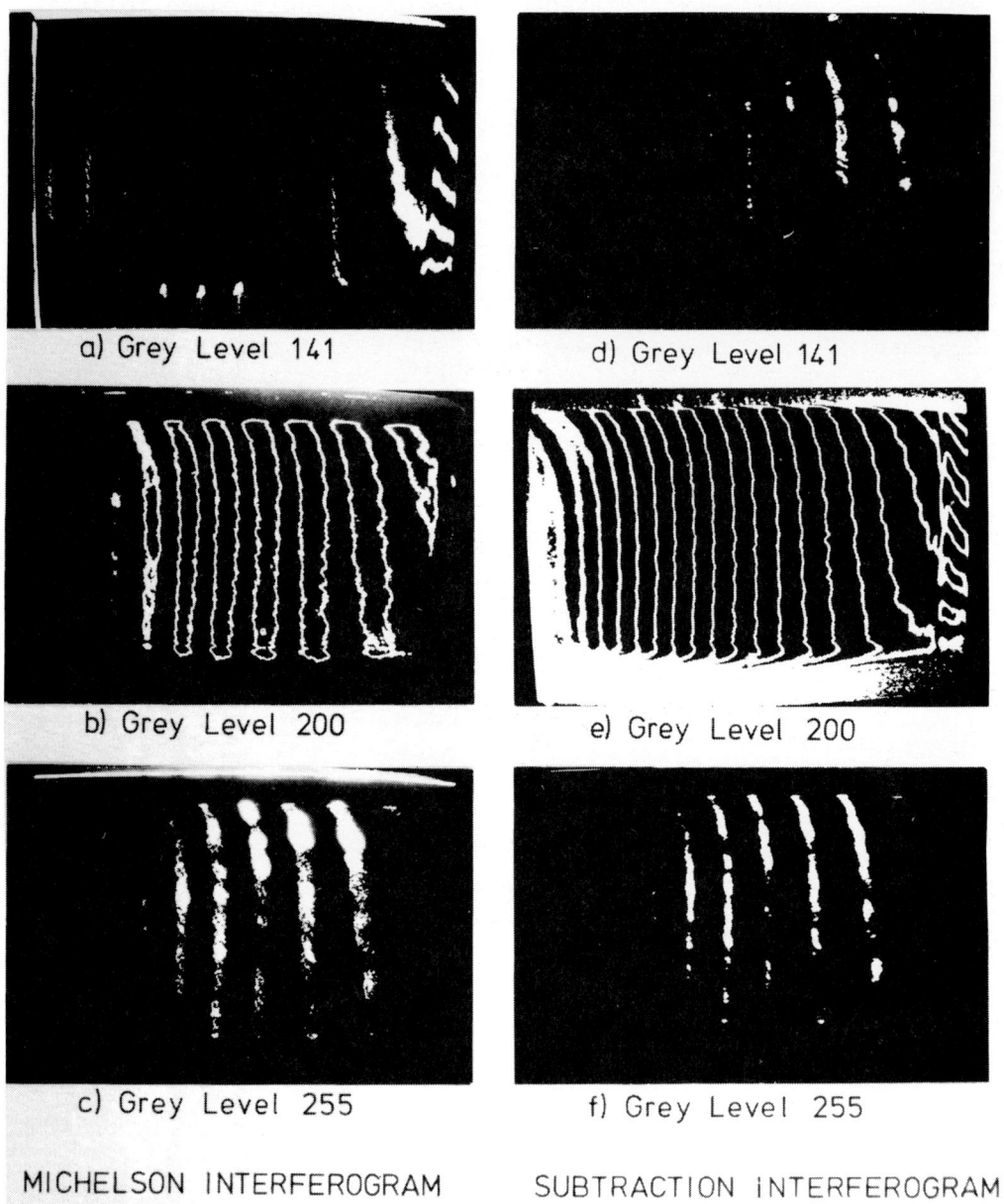

Figure 4

photographs of Figures 4a, b, c will result. They correspond to the selected grey levels 141 (Figure 4a), 200 (Figure 4b) and 255 (Figure 4c).

A comparison of Figures 4e and 4b shows that it is possible by image subtraction to detect the shifted zero points over the whole interferogram by selecting just one grey level. (In addition, equal noise contributions are suppressed in the subtraction method). It is not possible, however, to detect all cosine function inflection points of Figure 4b, which roughly correspond to the shifted zero points of Figure 4e, in one single step.

It is desirable to superimpose the lines of constant phase difference of Figure 4e onto the original interferogram of Figure 3a. This can only be done by using two synchronous special effects generators, and experiments to accomplish this task are under way.

Holographic Interferograms

The image subtraction principle calls for a means to introduce a 180° phase shift into the interferogram under investigation. In holographic interferometry we therefore have to use an experimental set-up with two reference beams of variable phase relation.

Using the double exposure technique, the two states of the object will be recorded each with a separate reference beam. Upon reconstruction with both beams the regular holographic interferogram will appear. A change in the relative phase of the reference beams will cause a corresponding phase shift in the interferogram which has to be adjusted to 180°.

Experiments in which the image subtraction method is applied to holographic interferograms are also being performed at the moment.

Conclusions

In this paper we discuss the following fringe analysis method:

1) Two interferograms with a relative phase shift of 180° of the fringes are created.

2) The intensity distributions of these interferograms are subtracted.

3) Only the zeros of this difference are recorded as fringe loci.

The advantages of this method are:

1) A far more exact positioning of the fringes compared to positioning by the extrema of of the fringes.

2) Independence of the amplitudes of object and reference beam as well as varying background.

Application is possible in the following fields:

1) Classical interferometry

2) Holographic interferometry

3) Moiré analysis.

References

1. I.E. Sollid: Appl. Opt. $\underline{8}$, 1587 (1969)
2. E.B. Aleksandrov, A.M. Bonch-Bruevich: Sov. Phys. $\underline{12}$, 258 (1967)
3. S.K. Dhir, I.P. Sikora: Proc. Symp. Engineering Applications of Holography (SPIE) Los Angeles 1972
4. R. Dändliker, B. Ineichen, F.M. Mottier: Opt. Commun. $\underline{9}$, 247 (1973)

1st EUROPEAN CONGRESS ON OPTICS APPLIED TO METROLOGY

Volume 136

SESSION 5.3

HOLOGRAPHIC INTERFEROMETRY

Session Chairmen
Lanzel
Grosmann

HOLOGRAPHIC INTERFEROMETRY WITH THE POSSIBILITY
OF MODIFYING THE FRINGES DURING RECONSTRUCTION

W. Schumann and M. Dubas

Laboratory of Photoelasticity
Swiss Federal Institute of Technology, CH-8092 Zurich

Abstract

In holographic interferometry, fringe control may be achieved in several manners. In this paper, the general way of reasoning is applied to the case where the two interfering wavefronts are recorded on separate holograms, one of which is moved during the reconstruction. In the first part, we investigate the image reconstructed by a hologram displaced with regard to the recording. We determine especially the position of the aberrated image of a point source. In the second part we examine the fringes produced by the interference of the wavefront coming (virtually) from an undeformed object and the wavefront coming from the same object which has been mechanically deformed and moreover "optically modified" by the movement of the hologram. We first determine the optical path difference which does depend upon the mechanical displacement of the object point and upon the motion of the hologram, but not upon the position of the images. We then calculate the derivative of this path difference and express so the fringe interspace and direction. The derivative contains in addition to the previously mentioned displacement vectors, the strain and rotation tensors of the object and the rotation tensor of the hologram. In conclusion we show how such a modification of the position of a hologram may be useful for the deformation measurement.

1. Introduction

In order to increase the flexibility of holographic interferometry, modification of the fringes during the reconstruction proves useful. If both wavefronts are reconstructed ones, one can act separately on them in several manners. For instance, when the wavefronts coming from the undeformed and from the deformed object have been recorded with different reference sources, one can move either one of the reconstruction sources (see e.g. references [1-4]) or the hologram [5]. Or, when the wavefronts have been recorded on different holograms [6], one can displace both "sandwiched" holograms [7,8], or only one of them. As an illustration of the general theory, the latter case of two holograms, one of which is movable, will be discussed here.

In the first part, we investigate the image of a point source as it is reconstructed by a hologram which has been moved with respect to its position during the recording. In the second part, we examine the fringes produced by the addition of two wavefronts. One comes (virtually) from the undeformed object and the other, from the same object which has been mechanically deformed and moreover optically modified. We determine here the optical path difference of the interfering rays and the derivatives of this quantity, i.e. the fringe interspace and direction. These functions depend only on the mechanical object deformation and on the hologram position. In conclusion we indicate how this method of fringe control may help to measure the deformation.

Let us add that we do not take into account here a change of the wavelength of the light, the thickness of the emulsion, its shrinkage and its change of refractive index during the development. We also neglect the influence of the glass plate. This is warrantable if both plates are located "out of the interferometer", that is to say if the emulsions face each other. On the other hand the optical arrangement may have any geometry.

2. Image of a point source reconstructed by a moved hologram

Figure 1 describes the problem we are going to deal with. The wavefront emitted by a point source P' has been recorded on a hologram by means of a reference source Q. In fact P' is a point on the diffusely reflecting surface of an object in the deformed state, P being the position of this point before the deformation. For the recording the object was illuminated by a source located in S, at the distance L'_s in front of it. For the reconstruction the hologram is placed in front of Q but in a position different from the recording one, so that a certain point H' on it now lies in $\tilde{H}$. As a consequence, we may expect that the image of P' has also been shifted to a certain point $\tilde{P}$. Similarly, we must assume the phase in $\tilde{P}$ to be given by the distance $\tilde{L}_s$ from a fictitious source $\tilde{S}$.

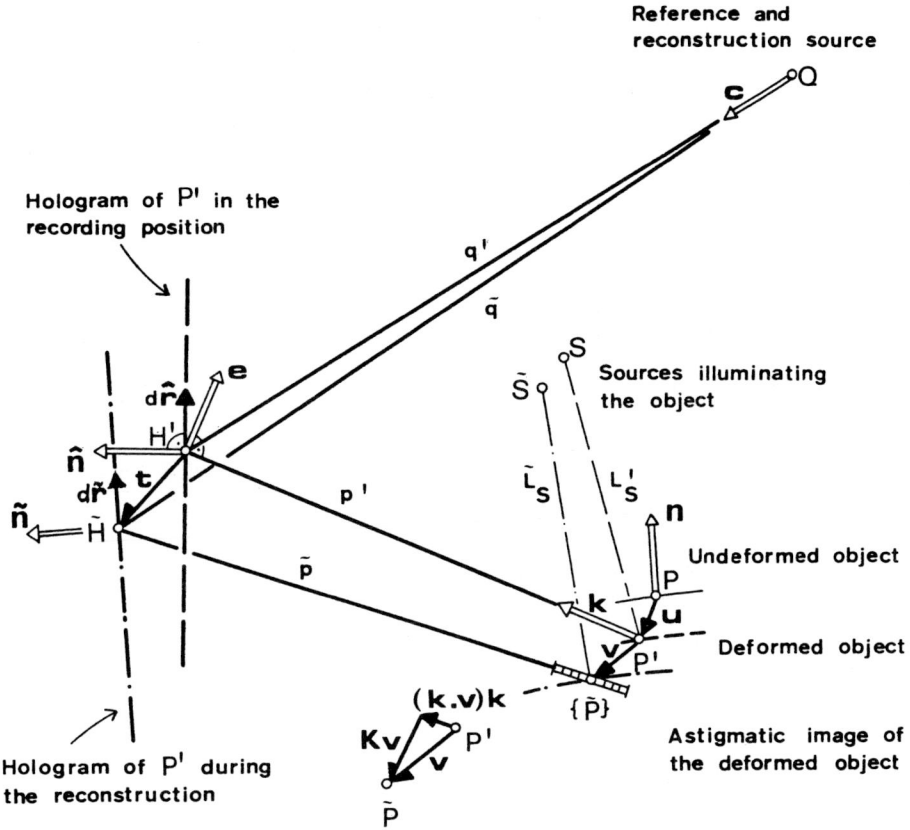

Fig. 1. Formation of the aberrated image $\tilde{P}$ of the point P'.

If these assumptions were correct, the interference produced in $\tilde{H}$ by Q and $\tilde{P}$ ought to be the same as that produced in H' by Q and P'. So the following *condition of interference identity* in H' and $\tilde{H}$ should be fulfilled

$$(\tilde{L}_s + \tilde{p} - \tilde{q}) - (L'_s + p' - q') = 0 , \qquad (1)$$

where $\tilde{p}$, p', $\tilde{q}$ and q' are the distances $\tilde{P}\tilde{H}$, P'H', Q$\tilde{H}$ and QH' respectively. Since the points P' and $\tilde{P}$ are fixed, the *optical path difference* (heavy lines in figure 1)

$$\Theta = (\tilde{p} - \tilde{q}) - (p' - q') = (\tilde{p} - p') - (\tilde{q} - q') \qquad (2)$$

will be a function of the position vectors $\hat{r}$ and $\tilde{r}$ of H' and $\tilde{H}$. Obviously, it is not possible to define a point $\tilde{P}$ such that Θ remains constant over the whole hologram surface. Let us then see if we may at least keep it "constant" in the vicinity of a hologram point. First we have thus the *stationarity condition* of Θ round $\tilde{H}$ and H'

$$d\Theta = d\tilde{r} \cdot \nabla(\tilde{p} - \tilde{q}) - d\hat{r} \cdot \nabla(p' - q') = 0 . \qquad (3)$$

∇ is the three-dimensional operator of derivation. As for $d\tilde{r}$ and $d\hat{r}$, they are increments situated on the hologram planes respectively in the reconstruction and in the recording position. These two vectors determine the position of corresponding points in the vicinity of $\tilde{H}$ and H' and are therefore related by the linear transformation

$$d\tilde{r} = R d\hat{r} , \qquad (4)$$

where **R** is the *orthogonal tensor of the hologram rotation*. (If the hologram had also been deformed we should have had a general deformation gradient instead of **R**.) Further, the derivatives of the distances between a fixed point and a variable point constitute *unit vectors*: $\nabla p' = \mathbf{k}, \nabla q' = \mathbf{c}, \ldots$ (3,4,8-10,12). Thus equation (3) reads now

$$d\Theta = d\hat{r} \cdot R^T(\tilde{\mathbf{k}} - \tilde{\mathbf{c}}) - d\hat{r} \cdot (\mathbf{k} - \mathbf{c}) = 0 . \qquad (5)$$

This equation must be valid for an arbitrary $d\hat{r}$, but staying always in the hologram plane, i.e. staying normal to the unit vector $\hat{n}$. Using the *normal projection* onto this plane, $\hat{N} = I - \hat{n} \otimes \hat{n}$ (applied as a linear transformation onto a vector, f.i. **k**, it gives

$\hat{N}k = k - \hat{n}(\hat{n}\cdot k))$, we conclude that

$$\hat{N}R^T(\tilde{k} - \tilde{c}) - \hat{N}(k - c) = 0 \;. \tag{6}$$

This equation, which is similar to that obtained in the case of a movement of the source[3,9,10], allows one to find, together with the auxiliary condition $|\tilde{k}| = 1$, the direction $\tilde{k}$ of the image $\tilde{P}$.

Similarly to the investigation of the *small* mechanical deformation of an object, we may express the preceding results in function of the displacement vectors **v** between P' and $\tilde{P}$ and **t** between H' and $\tilde{H}$ if we assume v,t << p',q'. We may then write the *linearized* relation

$$\Theta = (t - v)\cdot k - t\cdot c \;. \tag{7}$$

As for the derivative of this function, we must remember that **v** is considered as a constant vector and that the derivative of a unit direction vector contains a normal projection along this direction[8,9]: $\nabla \otimes k = K/p'$, $\nabla \otimes c = C/q'$, where $K = I - k \otimes k$, $C = I - c \otimes c$ are normal projections along **k** and **c** (figure 2). Furthermore, the derivative of **t**

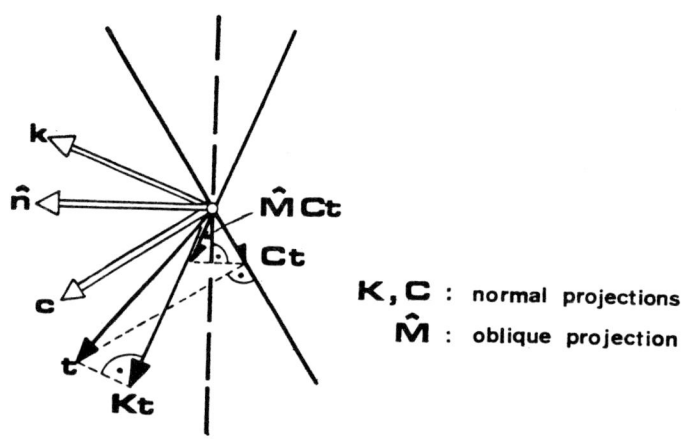

Fig. 2. Normal and oblique projections of the vector **t**.

consists in the *skew-symmetric tensor* Ψ of the *small hologram rotation*: $\Psi \simeq R^T - I = \nabla \otimes t$. Instead of (6) we obtain now

$$\hat{N}\left[\frac{1}{p'} K(t - v) - \frac{1}{q'} Ct + \Psi(k - c)\right] = 0 \;. \tag{8}$$

The *lateral part* **Kv** of the displacement **v** may be isolated if one uses the *oblique projection* $\hat{M} = I - \hat{n} \otimes k/\hat{n}\cdot k$ (**I** denotes still the identity) which projects along $\hat{n}$ onto the plane perpendicular to **k** (figure 2). As a matter of fact, because of $\hat{M}\hat{N} = \hat{M}$ and $\hat{M}K = K$, we are able to write

$$Kv = Kt - \frac{p'}{q'} \hat{M}Ct + p'\hat{M}\Psi(k - c) \;. \tag{9}$$

Thus, the stationarity condition $d\Theta = 0$ around H' and $\tilde{H}$ leads to the lateral shift of the image P', either by (6) or by (9). In order to obtain more information about this image, we might state that $d\Theta = 0$ not only in H' and $\tilde{H}$ but also in any other pair of points around, and so define a whole set $\{\tilde{k}\}$ of directions. We might then verify that all such rays do in fact not pass through one point $\tilde{P}$ and thus that the image $\tilde{P}$ suffers *aberrations*[9,11,12,13]. This constitutes the *ray tracing method* (see e.g. references). Alternatively we may examine the *higher order increments* of Θ and proceed in a way similar to that of the geometrical theory of lens aberrations (see e.g.[10]). Here we shall restrict our attention to the study of $d^2\Theta = 0$ which we shall express in function of the displacements **t** and **v**[3].

To calculate $d^2\Theta$, we derive once more the bracket in equation (8), i.e. $\nabla\Theta$. In addition to vector derivatives we already encountered, second order tensors must be derived, so for example[3,8] $\nabla \otimes (K/p') = -[k \otimes K + K \otimes k + K \otimes k)^T]/p'^2$ which constitutes a third order tensor. In analogy with the preceding we may call the latter bracket a *superprojection*. So, the developments lead to ($d^2\tilde{r} = 0$)

$$d^2\Theta = d\hat{r} \cdot (\nabla \otimes \nabla \Theta) d\hat{r} = d\hat{r} \cdot \{ - \frac{1}{p'^2}[k \otimes K(t-v) + K(k \cdot t - k \cdot v) + K(t-v) \otimes k] +$$
$$+ \frac{1}{q'^2}[c \otimes Ct + C(c \cdot t) + Ct \otimes c] + \frac{1}{p'}[\Psi K - K\Psi] - \frac{1}{q'}[\Psi C - C\Psi] \} d\hat{r}. \quad (10)$$

Using (9), we make the lateral part Kv disappear. Further, again by the aid of the oblique projection $\hat{M}$ and of its *transpose* $\hat{M}^T$, which projects along k onto the plane perpendicular to $\hat{n}$, we may isolate the *longitudinal part* $k \cdot v$. $d\hat{r}$ is now replaced by a corresponding unit vector e perpendicular to k. Because of $\hat{M}K\hat{M}^T = K$ and $e \cdot Ke = 1$, we obtain

$$k \cdot v = e \cdot Te , \quad (11)$$

where

$$T = \hat{M}\{k \otimes [\frac{p'}{q'}Ct - p'\Psi(k-c)] + K(k \cdot t) + [\frac{p'}{q'}Ct - p'\Psi(k-c)] \otimes k -$$
$$- \frac{p'^2}{q'^2}[c \otimes Ct + C(c \cdot t) + Ct \otimes c] - p'[\Psi K - K\Psi] + \frac{p'^2}{q'^2}[\Psi C - C\Psi]\}\hat{M}^T. \quad (12)$$

The relation (11) let thus appear the *astigmatism* of the image $\tilde{P}$: as a matter of fact, to each vector e, i.e. to each dr corresponding to this e, belongs a longitudinal shift $k \cdot v$. The limits of the *interval* in which the set of images $\{\tilde{P}\}$ is located may be found in the following manner: considering $|e| = 1$ as an auxiliary condition, $k \cdot v$ as a Lagrange multiplicator and K as the two-dimensional identity in the interesting plane, one may write the characteristic eigenvalue equation

$$|T - (k \cdot v)K| = 0 . \quad (13)$$

By means of (13) and (9) it is thus possible to determine the position of $\tilde{P}$, or, more precisely, of the set of images $\{\tilde{P}\}$.

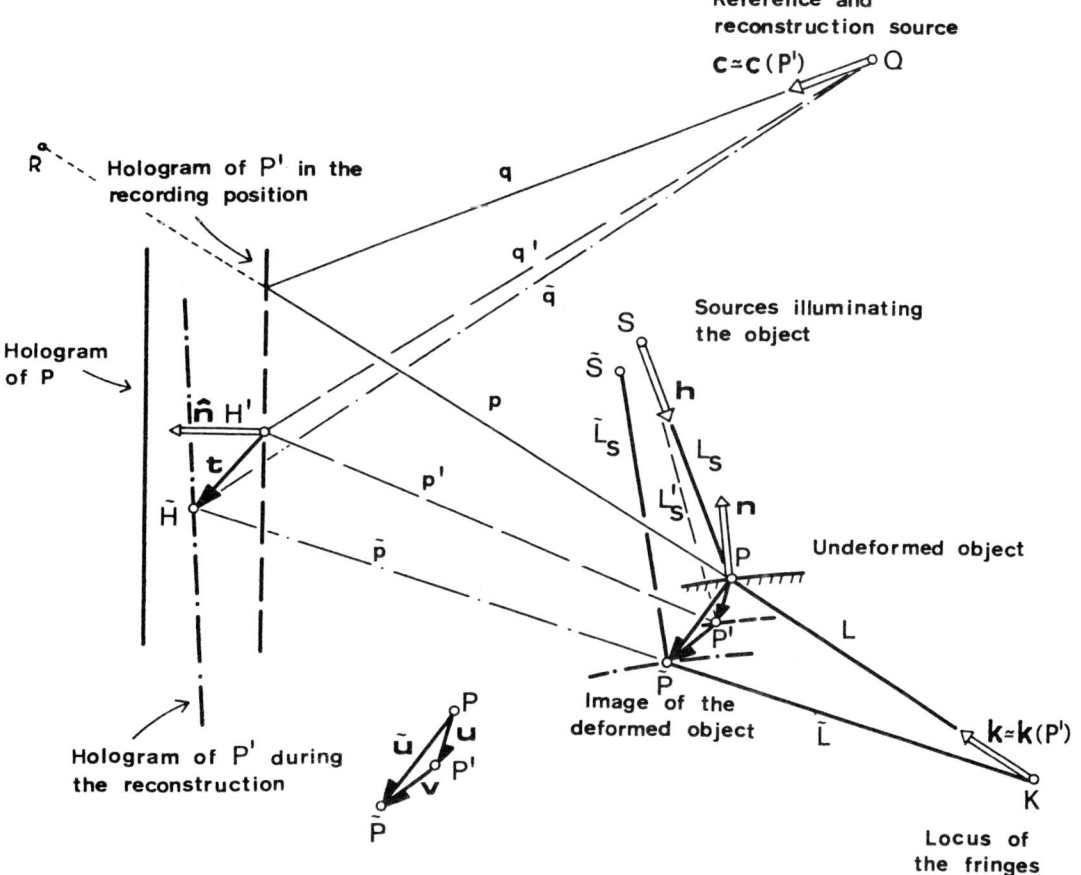

Fig. 3. Formation of the fringes.

3. Formation of the fringes

Let us go over now to the problem of interference fringes. These are formed by the superposition of two wavefronts. Firstly, the image of the undeformed body (set of points $\{P\}$) is reconstructed by means of a fixed hologram (figure 3). Secondly, the other, movable hologram produces the image of the mechanically and moreover "optically deformed" body, that is to say of $\{\tilde{P}\}$. We assume that the interference phenomenon is formed in a point K situated at the distance L behind (in this case L > 0) or in front of (L < 0) the object surface. The *optical path difference* (heavy lines in figure 3) of the rays intersecting in K is then

$$D = (L_S - L) - (\tilde{L}_S - \tilde{L}) \ . \tag{14}$$

The unknown distance $\tilde{L}_S$ may be eliminated thanks to the condition of interference identity (1). D becomes then

$$D = (L_S - L'_S) + (\tilde{p} - p') - (\tilde{q} - q') - (L - \tilde{L}) \ . \tag{15}$$

Assuming a *small deformation*, i.e. $u,v,t \ll L_S,L,p,q,\ldots$ permits us to write the *linearized* relation

$$D = -\mathbf{u}\cdot\mathbf{h} + (\mathbf{t} - \mathbf{v})\cdot\mathbf{k} - \mathbf{t}\cdot\mathbf{c} + \tilde{\mathbf{u}}\cdot\mathbf{k} \ ,$$
$$D = \mathbf{u}\cdot(\mathbf{k} - \mathbf{h}) + \mathbf{t}\cdot(\mathbf{k} - \mathbf{c}) = \mathbf{u}\cdot\mathbf{g} + \mathbf{t}\cdot\mathbf{f} \ . \tag{16}$$

u denotes here the *displacement vector* of the mechanical deformation, i.e. the vector from P to P', whereas $\tilde{\mathbf{u}} = \mathbf{u} + \mathbf{v}$ is the *total displacement*, i.e. from P to $\tilde{P}$. It may be mentioned that the observation direction **k**, the object illumination direction **h** and the hologram illumination direction **c** are unit vectors considered here in the reference configuration, that is to say along lines such as SP and KP. The first term in (16), where $\mathbf{k} - \mathbf{h} = \mathbf{g}$ is usually called the sensitivity vector, is the one occurring in conventional holographic interferometry. The second one, where $\mathbf{k} - \mathbf{c} = \mathbf{f}$ could similarly be called the *sensitivity vector of the fringe control*, takes into account the modification of the optical arrangement. Furthermore, it is important to note that the exact location of $\tilde{P}$ does not matter at all since it appears neither in (15) (if one considers $\tilde{p} + \tilde{L}$ as a whole) nor in (16). So, in this paragraph, we have only spoken of a "point" $\tilde{P}$ or of a "point" $\tilde{S}$ for convenience, but they do not arise in the results. In other words, fringes are formed whatever the image looks like.

If we now desire to know the direction and the interspace of the fringes, we must derive expression (16). An observer located in R at the distance L_R from P (figure 3), looks at K in the direction **k**. If he now looks in a neighbouring direction $\mathbf{k} + d\mathbf{k}$ differing by an infinitesimal angle $d\phi$ from the previous direction, the *change of the fringe order* in the direction $d\mathbf{k}/d\phi = \mathbf{m}$ is

$$\frac{dD}{d\phi} = L_R \mathbf{m}\cdot\mathbf{M}\left\{\left[\boldsymbol{\gamma} + \Omega\mathbf{E} + \boldsymbol{\omega}\otimes\mathbf{n}\right](\mathbf{k} - \mathbf{h}) - \frac{1}{L_S}\mathbf{H}\mathbf{u} - \frac{1}{L_R}\mathbf{K}(\mathbf{u} + \mathbf{t}) + \frac{L_R - p}{L_R}\hat{\mathbf{M}}\left[\boldsymbol{\Psi}(\mathbf{k} - \mathbf{c}) - \frac{1}{q}\mathbf{C}\mathbf{t}\right]\right\} \ . \tag{17}$$

$\mathbf{M} = \mathbf{I} - \mathbf{n}\otimes\mathbf{k}/\mathbf{n}\cdot\mathbf{k}$ is an oblique projection (not to be confused with $\hat{\mathbf{M}}$) along the normal **n** of the undeformed object onto the plane perpendicular to **k**. At first sight one would expect the normal projection $\mathbf{N} = \mathbf{I} - \mathbf{n}\otimes\mathbf{n}$ to be present. But the tensor **M** allows us (with $\mathbf{MN} = \mathbf{M}$) to relate all derivatives on the object surface to the increment $d\mathbf{k}$ or to **m**. The derivative of the displacement vector **u** is contained in the first bracket and consists in (figure 4): the *surface strain tensor* $\boldsymbol{\gamma}$, the *pivot rotation* $\Omega\mathbf{E}$ with the scalar Ω and the two-dimensional permutation tensor **E**, the vector $\boldsymbol{\omega}$ describing the *inclination* of **n**. As in the preceding paragraph, the derivatives of unit vectors yield normal projections; in addition to the already encountered **K** and **C**, $\mathbf{H} = \mathbf{I} - \mathbf{h}\otimes\mathbf{h}$, which projects onto a plane normal to **h**, is present. Similarly to (16), equation (17) does not depend on the position of the image $\tilde{P}$ but, by use of (9), one could make **Kv** appear again.

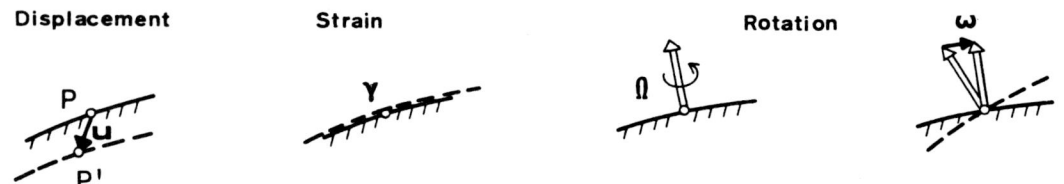

Fig. 4. The displacement and the elements of the displacement gradient.

The relation (17) gives thus the fringe interspace in any direction **m**. In particular, when **m** is parallel to a fringe, $dD/d\phi = 0$ and **m** must be perpendicular to the vector **M**{...}. By means of a calculus similar to the preceding one, we might also express the condition of complete localization of the fringe in K[4,8]. This condition states that D is *stationary* on the object surface in the vicinity of P and is thus analogous to the stationarity condition (5) for Θ in the vicinity of H'. If the fringes are completely localized in K, (17) may now conveniently be written in terms of the image shift

$$\frac{dD}{d\phi} = \mathbf{m} \cdot \left[-\frac{L_R + L}{L} (\mathbf{Ku} + \mathbf{Kv}) \right] . \tag{18}$$

This means that in this case (and only in this case) *the fringes are perpendicular to the apparent total displacement*, i.e. to $\mathbf{K}(\mathbf{u}+\mathbf{v}) = \mathbf{K}\tilde{\mathbf{u}}$. This theorem is analogous to the one of conventional holography.

4. Conclusion

In the first part of this paper, we investigated the aberrated image of a point source reconstructed by a hologram which has been moved with respect to its position during the recording. In the second part, devoted to the description of the fringes formed by two holograms, one of them movable, we used the results of the first part in two ways: on the one hand, we spoke provisionally of the image point of the mechanically and optically deformed object; on the other hand, we were able to calculate the phase of the light ray concerning this image by means of the condition of interference identity on the hologram and to make so the fringe problem independent of the image problem. Nevertheless it seems interesting to emphasize also the likeness of the reasonings and of the formulae in these two fields.

The results we have obtained as for the optical path difference (16) and its derivative (17) may be used to measure the deformation of an object in the same way as in conventional holographic interferometry, with the additional freedom of being able to modify adequately the fringe pattern. Moreover, since the relation (17) contains on the one hand the elements of the mechanical deformation (**u**, **γ**, Ω, **ω**) and on the other hand those of the movement of the hologram (**t**, **Ψ**), it could allow to "*translate*" the object's deformation into the hologram's motion; this corresponds to the concepts propounded by Abramson[14]. For instance, if $q = L_s = \infty$ (collimated beams) and **h** = **c** (parallel illumination of the object and of the hologram) one could first translate the movable hologram (without rotating it) until the zero order fringe appears on the investigated point P independently of the observation direction **k**. Then, as shown by (16) **t** = -**u**. In a second step, still maintaining **t** = -**u** in the point H' defined by a given observation direction **k**, one could rotate the hologram until the fringes disappear round P, i.e. until *the fringe order becomes stationary*. Then $dD/d\phi = 0$ for any direction **m** and (17) would lead to, since $\mathbf{M}\hat{\mathbf{M}} = \hat{\mathbf{M}}$,

$$\mathbf{M}(\boldsymbol{\gamma} + \Omega \mathbf{E} + \boldsymbol{\omega} \otimes \mathbf{n})\mathbf{g} = \frac{p - L_R}{L_R} \hat{\mathbf{M}} \boldsymbol{\Psi} \mathbf{g} . \tag{19}$$

The strain and the rotation could so be obtained without measuring explicitly the displacement vector. This would be an important advantage due to the fringe control. Nevertheless numerous practical problems are brought up too: the holograms must be exactly repositioned and the movements of the shifted one must be precisely known; the zero order fringe must be identified and one must be able to recognize when the fringe order is stationary. Furthermore, it must be added that, if one desires to measure all the components of the strain and rotation tensors, three linearly independent sensitivity vectors $\mathbf{g}_i$ are necessary in any case.

(See the references on the following page).

References

1. Ballard, G. S.,"Double-exposure holographic interferometry with separate reference beams", *J. Appl. Phys.*, Vol. 39, pp. 4846-4849. 1968.
2. Tsuruta, T., Shiotake, N., Itoh, Y., "Hologram interferometry using two reference beams", *Japan. J. Appl. Phys.*, Vol. 7, pp. 1092-1100. 1968.
3. Schumann, W., Dubas, M., "On the motion of holographic images caused by movements of the reconstruction light source, with the aim of application to deformation analysis", *Optik*, Vol. 46, pp. 377-392. 1976.
4. Schumann, W., Dubas, M., "On the holographic interferometry used for deformation analysis, with one fixed and one movable reconstruction source", *Optik*, Vol. 47, pp. 391-404. 1977.
5. Dändliker, R., Marom, E., Mottier, F. M., "Two-reference-beam holographic interferometry", *J. Opt. Soc. Am.*, Vol. 66, pp. 23-30. 1976.
6. Gates, J. W. C., "Holographic phase recording by interference between reconstructed wavefronts from separate holograms", *Nature Lond.*, Vol. 220, pp. 473-474. 1968.
7. Abramson, N., "Sandwich hologram interferometry: a new dimension in holographic comparison", *Appl. Optics*, Vol. 13, pp. 2019-2025. 1974.
8. Dubas, M., Schumann, W., "Contribution à l'étude théorique des images et des franges produites par deux hologrammes en sandwich", *Opt. Acta*, Vol. 24, to appear. 1977.
9. Helstrom, C. W., "Image luminance and ray tracing in holography", *J. Opt. Soc. Am.*, Vol. 56, pp. 433-441. 1966.
10. Champagne, E. B., "Nonparaxial imaging, magnification, and aberration properties in holography", *J. Opt. Soc. Am.*, Vol. 57, pp. 51-55. 1967.
11. Latta, J. N., "Computer-based analysis of holography using ray tracing", *Appl. Optics*, Vol. 10, pp. 2698-2710. 1971.
12. Miles, J. F., "Imaging and magnification properties in holography", *Opt. Acta*, Vol. 19, pp. 165-186. 1972.
13. Přikryl, I., "Studying hologram imagery by a ray-tracing method", *Opt. Acta*, Vol. 19, pp. 623-631. 1972.
14. Abramson, N., "Sandwich hologram interferometry. 2: some practical calculations", *Appl. Optics*, Vol. 14, pp. 981-984. 1975.

TESTING BY HOLOGRAPHIC INTERFEROMETRY OF SOLID PROPERGOL ENGINES

Paul Smigielski
Franco-German Research Institute
68300 Saint-Louis, France

Daniel Cesario
Central Technical Establishment for Armament
94114 Arcueil, France

Claude Patanchon
National Society for Powders and Explosives
33160 St-Medard-en-Jalles, France

Abstract

We present an application of holographic interferometry by double exposure to non-destructive testing of solid propergol engines 350 mm in diameter and 1000 mm in length at a maximum. The apparatus installed for industrial use is described. It permits us to easily visualize defects in the adhesive of 10 mm x 7 mm under 2.5 mm of thermic protection by application of a weak pneumatic stress (around 10 mmHg). The engine is placed in a vacuum chamber provided with an observation port. The parasitical contour fringes due to the variations in refractive index of the air with the partial vacuum can be eliminated by the use of a different gas with each of the two exposures of the hologram. Results concerning real engines are shown.

Introduction

We present a prototype of an apparatus for non-destructive testing of solid propergol engines, produced in the Central Technical Establishment for Armament (ETCA). This prototype, adapted to the search for defects in adhesive in engines manufactured by the National Society for Powders and Explosives (SNPE), uses the method of holographic interferometry by double exposure[1].
This method presents, particularly in comparison to tangential radiography, the following advantages:
- considerable reduction in number of plates,
- global view of the object
- possible detection of defects of smaller dimensions.

The study and production of this prototype were financed by the Direction of Research, Study and Techniques (DRET). The goal was to construct an autonomous device functioning at the production site and able to be serviced by non-specialist personnel.

Previous Studies

In 1973[2] the Franco-German Institute of Research in Saint Louis (ISL) was assigned by the DRET to study the feasibility of the method. To this end, the SNPE produced samples of adhesives including calibrated defects. These defects, of different natures, (inclusion of air, Teflon, adhesive without hardener, leaks) were situated between powder and thermic protection or between metal and thermic protection.
The studies undertaken at the ISL permitted us to test diverse methods of loading and to determine the method of holographic interferometry which was best adapted to the problem[3].

Determination of the Method of Stress

Various defects have been visualized by the use of thermic and pneumatic stresses. The pneumatic stress causes defects involving air to appear clearly. Thermic stress also permits visualization of inclusions of air, while demonstrating other types of defects (inclusions of Teflon, for example, or leaks). Nonetheless, the difficulties met upon application of this method (non-uniformity of heating, overall expansion of the material) make it more delicate to use than the pneumatic method.
Other stresses (shocks, vibrations) were tested without contributing significant improvements compared to the stresses mentioned above, except in the case of the detection of leaking defects, or non-leaking defects not containing gaseous inclusions.

Determination of the Interferometric Method

The search for defects took place in real time as well as in deferred time. In the latter case, we used holographic interferometry by double exposure. For an industrial application it is simpler to use the double exposure method, which is easier to apply than the real time method, and gives more highly contrasted interferograms.

Definition of the Prototype

The results obtained at the ISL demonstrating the feasibility of the method permitted us to define the prototype which the ETCA was ordered to construct.

It was decided to search for defects containing air, by application of a pneumatic stress. The engines to be tested are cylindrical blocks having a diameter of 350 mm and a length of about a meter, with a weight of about 100 kilograms. The thickness of the thermic protection which covers the defects is on the order of 2.5 mm. We look for defects in adhesive of dimensions at least equal to 20 mm x 20 mm. These engines were chosen because they are of average size in the production of the SNPE. The apparatus may be used to find defects in blocks of smaller dimensions.

Since this apparatus will be installed in an industrial environment (SNPE) after its production, it will need to be serviced by non-specialized personnel, thus proving that holography is in fact useful in industrial non-destructive testing. After a trial period of several months, it should be possible to know the sensitivity of the method, its performances and its limitations.

Description of the Prototype [4]

This prototype is composed of (figure 1):
- a vacuum chamber in which a cradle holds the engine,
- a granite bench supporting the optical setup,
- a mechanically welded framework supporting the aforementioned elements.

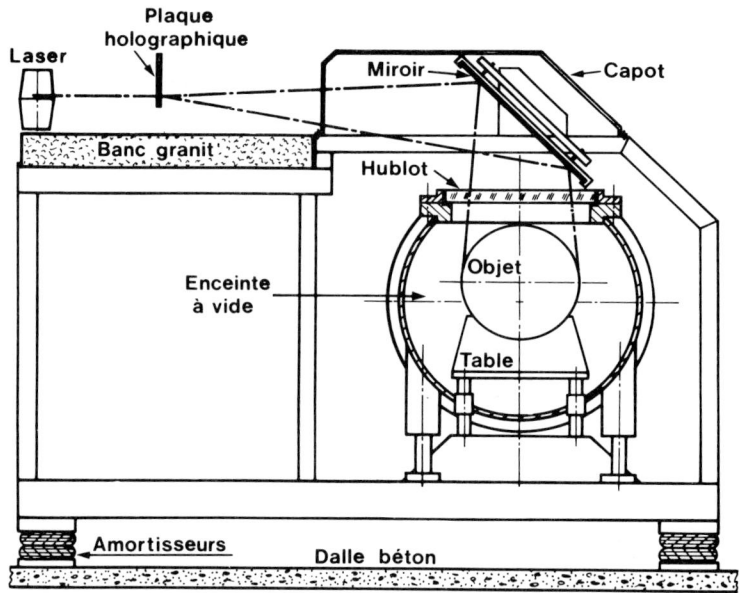

Fig. 1. Simplified diagram of the holographic bench

Several points have been more particularly studied:

1) Chamber-cradle separation: The drop in pressure created inside the chamber entails an overall displacement of the engine by the intermediary of its support. This displacement is expressed by a network of parasitical fringes which makes the detection of the defects difficult, even impossible. We thus produced a flexible cradle-chamber connection, practically eliminating the parasitical movements of the support of the engine.

2) Rigidity: The chamber was specially reinforced by several ribs. The frame of the port situated at the upper part of the chamber is a frame of equal resistance. The stresses due to the deformation of the chamber are thus applied in a uniform manner.

3) Insulation: Special pneumatic screws sustain the prototype and insulate it from the parasitical vibrations which are propagated through the ground.

Other Characteristics

We use an argon laser whose power is 1W on the 5145 Å ray. The coherence of this laser, increased by placing a Fabry-Perot etalon in the cavity, is on the order of 10 m, which permits the exploration of objects of large dimensions without the necessity of equalizing the optical paths. The relatively great power of this laser permits relatively short exposure times (on the order of a second).

The holograms are made on holographic plates 9 x 12 cm. We observe 1/3 of the engine, which we record on 1/3 of the plate. We thus conserve the information corresponding to the entire engine on a single plate 9 x 12 cm.

The prototype is entirely hooded (figure 2). There is thus no parasitical light near the sensitive plate. These hoods also assure protection of the optical system against dust and protect the holographic setup from acoustical and thermic disturbances.

Fig. 3. Prototype with engine on cradle

The reconstruction of the holograms is made on an adjacent bench using a low-power He-Ne laser as light source.

Results

Once the prototype was produced, we proceeded to tests in two stages:

1) On an inert engine, already laboratory tested, we again visualized typical defects.

2) On a real engine, with the prototype in place at the SNPE, we caused calibrated defects to appear. The decreases in pressure necessary to visualize inclusions of air are generally very small. We demonstrate defects of 10 x 7 mm under 2.5 mm of thermic protection, with a pressure of about 10 mm of mercury (figure 3).

Despite all the precautions taken, parasitical fringes are present, sometimes interfering with the interpretation of the holograms. These fringes have two origins:

1. <u>Flow of the propellant</u>: The propellants tested are relatively soft and settle under the effect of their own weight. This flow is a slow phenomena and can last for several hours. Care was taken to adapt the cradle to the diameter of the propellants. Nonetheless, the overall displacement of the block between two exposures causes a network of parasitical fringes to appear.

2. <u>Contouring</u>: The partial vacuum created in the chamber entails a variation of the refractive index of the air. For each point of the object, the variation in index Δn entails a variation of optical path $2e\Delta n$ (e being the distance from one point of the engine to the port). This variation of optical path creates a state of variable interference. Upon reconstruction of the hologram we observe a network of interference fringes (figure 4). The ISL has developed a method permitting the elimination of these parasitical fringes. We propose to operate in the following manner:
    1. The first exposure is made with helium at the pressure p_1 (refractive index n_1);
    2. The second is made with air at pressure p_2 (refractive index n_2) inferior to p_1 and such that $n_2 = n_1$.
Figure 5 illustrates the procedure which permits us, by adapting the pressure p_2, to compensate the flow equally and in large measure. The indicated order of the exposures can be reversed.

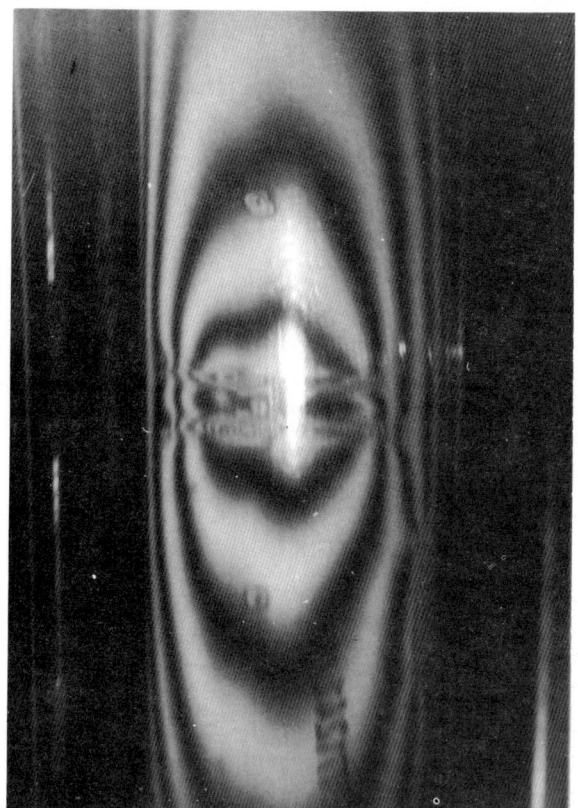 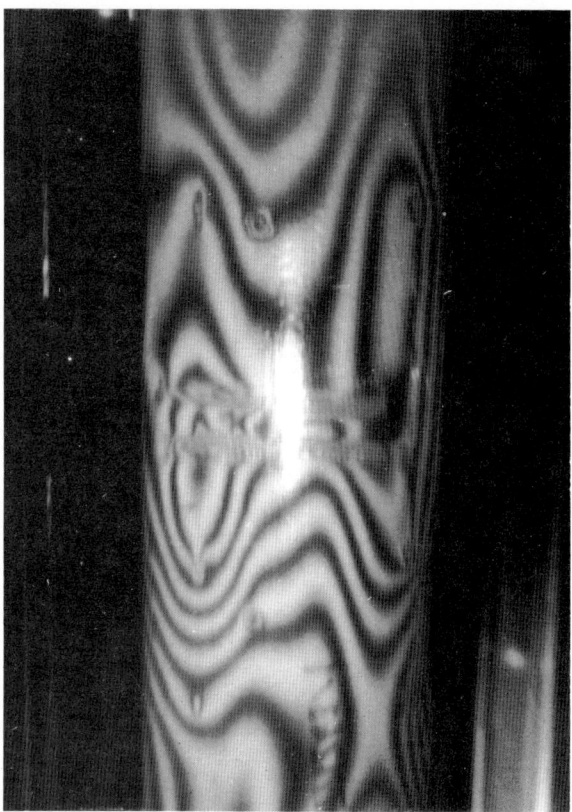

Fig. 3. Holographic images of a real engine showing artificial defects in adhesive for two different pressures: 10 mmHg (left) and 30 mmHg (right).

Conclusion

The goal which we fixed for ourselves at the beginning of this study has thus been attained. We have produced an apparatus for non-destructive testing by holography, adapted to the search for inclusions of gas in propellants. It functions in an industrial environment and can be used by non-specialists. It permits the testing of cylindrical engines whose maximal dimensions are 1 m in length, 350 mm in diameter and a weight of about a hundred kilograms. The examination of a block of this size can be made in about an hour. The future use of photopolymers for recording the holograms will considerably reduce this time.

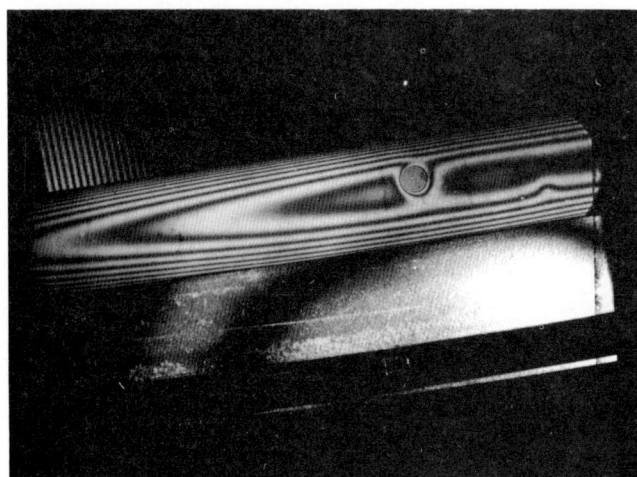 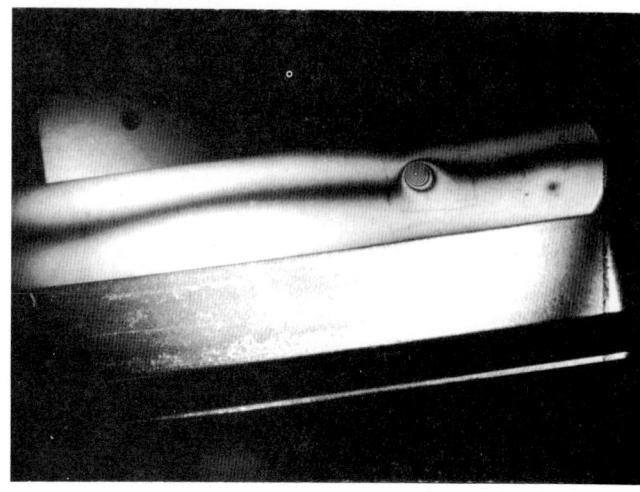

Fig. 4. Contour fringes for a partial vacuum of 340 mmHg. The two exposures of the hologram were made with air in the chamber.

Fig. 5. Elimination of contour fringes for a partial vacuum of 340 mmHg. The first exposure was made with helium, the second with air.

References

1. Waters, J.P., Applied Optics 10, 2364 (1971).
2. Smigielski, P., Albe, F., Stimpfling, A., Mouchain, J.L., "Application de l'interférométrie holographique à la détection des déformations d'une maquette de propulseur", Notice ISL 1/73, 1973.
3. Smigielski, P., Stimpfling, A., "Contrôle des propulseurs par holographie", Rapport ISL R 134/75, 1975.

   Smigielski, P., Albe, F., Fagot, H., Stimpfling, A., "Contrôle non déstructif par holographie. Application a la détection de défauts de collage". 3rd International Colloquium on Methods of Non-Destructive Testing, Toulouse, May 1974. Rapport ISL CO 14/74, 1974.

   Smigielski, P., Stimpfling, A., Patanchon, C., Lamarque, P., "Interférométrie holographique appliquée au contrôle non destructif des propulseurs a propergol solide". 8th World Conference on Non-Destructive Tests, Cannes, September, 1976. Rapport ISL CO 218/76, 1976.
4. Cesario, D., "Contrôle holographique de maquette de propulseur". Rapport E.T.C.A., n° 77 RT 55, 1977.

APPLICATION OF HOLOGRAPHIC INTERFEROMETRY TO TESTING OF SPUN STRUCTURES

J. D. Dubourg
SNIAS Aquitaine
33160 Isaac, France

Abstract

Up to now, holography has been employed industrially only in the testing of thin shells (helicopter or turbine blades, for example). The studies undertaken at SNIAS Aquitaine since 1972 have been oriented toward the testing in series of casings of ballistic engines in spun fibers. This paper presents the different stages and the results obtained.

These studies permit us to favorably envisage the introduction of this technique into the testing of our products. Several conclusions are drawn as to this method.

Introduction

This article presents the principal results of a study whose goal was the application of holographic interferometry to the testing of propellant casings of ballistic engines. This work, done at the Aerospatial Establishment of Aquitaine, was financed by the D.T.E. as well as by the company.

The Problem

The Product to be Tested

Figure 1 shows an engine; it is composed of a resistant fiberglass shell saturated with resin, spun and polymerized, and lined with rubber protecting the structure upon combustion of the powder.

It is provided with two openings, one in front holding the priming system, the other in back holding the nozzle. Figure 1 indicates the presence of powder, but all our tests were made on the casing alone before loading of the powder.

Types of Defects

The principal defects encountered on this type of product are de-laminations in the spun glass and detachments between the external shell and the thermic protection (figure 2).

Tests Used

At present these defects are searched for globally by infrared thermography and the point-by point verifications are made in X-radiography.

Thermography has, however, limitations which caused us to look for another global method of testing for these anomalies. We oriented ourselves toward holographic interferometry, of which we had discovered the potential at the 6th Conference on Non-Destructive Testing in Hanover in 1970.

Tests

The tests were made on a marble bench 2.40 x 1.60 m, equipped with a 15 mW He-Ne laser.

Objects Tested

The samples studied had the following dimensions:
- ⌀ 380 mm
- length: 400 mm
- thickness of spun glass: 8 mm
- thickness of rubber lining: 6mm.

The programmed defects were:
- de-laminations
- detachments

concerning surfaces of ⌀ 60, 30, 15 mm.

Application of Stress

The following stresses were studied:
- internal partial vacuum
- internal or external heating
- mechanical stresses
- vibrations

Results

Figure 3 shows the visualization of a de-lamination by placement in a partial vacuum (a) and by external heating (b). An increase in the number of fringes or a local disturbance of them permits us to locate the defect. In figure 4, a detachment between the thermic protection and the "roving" is detected by placement in partial vacuum. In this case, it is a local disturbance in the form of the fringes which allows us to locate the defect. Tangential radiography permits us to situate it within the thickness of the shell.

From these tests on samples, we concluded that placement in partial vacuum and heating were the types of stresses which offered the best compromise: simplicity of application, quality of visualization of defects.

The positive results obtained led us to search for the means permitting the application of holography to industrial testing on real structures.

Evaluation of Pulsed Lasers

Pulsed lasers seemed to be able to solve certain particular problems posed by the application of testing by holographic interferometry at the industrial stage:
- problems of insulation
- necessity of working in a darkroom

Insulation

The use of pulsed lasers, delivering two pulses of very short duration (30 ns), very close together (adjustable time between pulses: 100 μs to 1 ms), permits the elimination of all ambient disturbance but imposes the use of dynamic stresses: vibrations or shocks.

We thus began by evaluating these types of stresses on the samples in our possession. This did not give conclusive enough results for us to consider extending this technique to the testing of real structures. Also, we oriented ourselves toward static stresses but without insulation.

Figure 5 shows the practical realization of the test. The sample $\emptyset$ 380 mm rests on a cradle fixed to a laboratory table; the ruby laser delivering 1J per pulse is also set on a table; between the two is situated the optical assembly, independent of the laser and the sample.

The results obtained (figure 6b) show, and this is a first point which it is important to retain, that the vibrations contributed by the ambient medium have practically no influence on the interferograms.

One may think that these vibrations are of very small amplitude; thus, a rigid displacement between the object, the optics and the laser can cause a change of the order of the fringes, but this is not bothersome in non-destructive testing since the holographic examination is solely qualitative.

The second important point is that the stability is indispensable only during the exposure time and the pulse of the laser is sufficiently short to fulfill this condition without any precautions.

Holography in Daylight

The second inconvenience which the use of a pulsed laser eliminates is the need to work in reduced ambient lighting.

The solution of the problem is simple: the photographic plate is protected by a shutter, then exposed to the laser pulses during its aperture. The plate is not exposed to the ambient light long enough to be spoiled, while the high-energy pulse, at the right wavelength, is recorded and constitutes the hologram.

Results

Figure 6 compares a hologram made with a continuous laser in a darkroom on a pneumatically insulated marble bench (figure 6a) with a hologram made with a pulsed laser without insulation and in broad daylight, in a workshop situation (figure 6b). These holograms are made at the same level of stress, and we can see the analogous quality of the results obtained.

Tests on Real Structures

Remark

In figure 2 we defined the different defects which may occur in this type of product (de-laminations, detachments). Figure 7 shows a cross-section of the engine and indicates the possible position of these anomalies.

In figure 7, at the rear of the engine, a rubber skin is shown, called detached skin since it is not glued to the structure, whose role is to compensate for the differential dilatations between the powder block and the exterior casing when the powder is poured or upon combustion. (The powder block and the shell do not have the same dilatation coefficients.) It is in this zone that we find the most difficult defects to test for.

In fact, infrared thermography is a rapid and efficient solution for testing of detachments in most cases, and for de-laminations in general; the choice between this method and holographic interferometry can be made only according to criteria of sensitivity or cost.

Thermography, on the other hand, is ineffective in the zone of the detached skin, because of the presence of the unglued interface and the great thickness of the rubber in this part of the engine.

It was consequently in this zone that testing by holographic interferometry was examined with priority in order to replace the systematic radiography used at present, which is a very long and expensive method of testing.

Test

Figure 8 shows the proceeding. In order to detect a detachment between the thermic protection and the "roving", we first create the vacuum in the detached skin so as to press it against the lining.

Also, we examine by holography the interior of the engine; in order to do this the optical assembly is placed inside it.

The external or internal heating presents no difficulty; however, the creation of the vacuum requires the sealing of the openings of the tank and we thus caused the laser beam to pass through a glass provided in one of the stoppers.

With Pulsed Laser

Figure 9 shows the assembly of the test run with a pulsed laser (ruby: 600 mJ/pulse).

With Continuous Laser

For the evaluation of the continuous laser and the testing of real structures, we constructed a holographic bench in concrete resting on insulating feet. The .4W ionized argon laser was then placed on the bench, the test assembly remaining in all respects identical to that used with the ruby laser.

Results

Figure 10 recapitulates the results obtained for testing for a defect (detachment under the detached skin).

The interpretation of these plates requires a certain amount of practice and good knowledge of the fabrication of the product. But this is facilitated by the comparison of holograms made in the same conditions of stress, of different zones presenting the same geometry.

From this we can retain that the thermic stress permits an easier interpretation of the interferograms obtained.

From these trials on real objects the following lessons were learned:
- the analogous quality of the holograms obtained with either type of laser;
- the advantage in focusing of the adjustments of the continuous laser;
- the advantages in testing in a workshop situation of the pulsed laser.

The choice between them should be made on the basis of the investments and the type of testing considered: on-site or in a prepared location.

It should be remarked that the results obtained from a study with a continuous laser are directly transposable if we use a pulsed laser.

Conclusion

The theme of this report, which is a very particular problem solved by holographic interferometry, should not restrain personal conclusions as to the possible applications of this method; used in this case on pieces of large dimensions and in difficult configurations, it gave positive results where other global methods remained ineffective.

Concerning its ultimate interest, the following aspects should be considered:
- decrease in cost of testing
- improvement of quality of tests.

The decrease in the cost appears when holographic interferometry can replace local tests such as radiography or ultrasound.

The improvement of the quality is an intrinsic advantage of this method, which bases the detection of defects on differences in the mechanical response of the piece tested. This can permit us to better understand the influence of an anomaly, whatever it may be, on the future performances under stress of the product.

The definition of an industrial testing station using the holographic technique is the last stage of this study. It should be completed by the end of this year and we hope to see its installation for the testing of our new products sometime in 1978.

Thus, this technique seems to lead to industrial service for several particular subjects, principally aeronautic: helicopter propellor blades, propellant casings, and turbine blades. Its application to series testing will permit us, after exploitation, to draw the true conclusions as to its reliability and interest to industry.

Bibliography

1. Françon, Holographie Masson, publisher.
2. Erf, Robert K., Holographic Non-Destructive Testing, Academic Press, New York.
3. Smigielski, Stimpfling, "Controle des propulseurs par holographie", ISL-R 134/75.

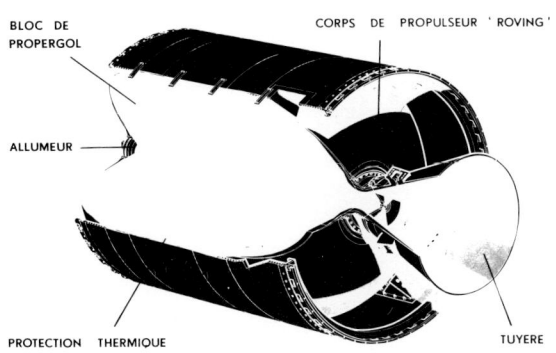

Fig. 1

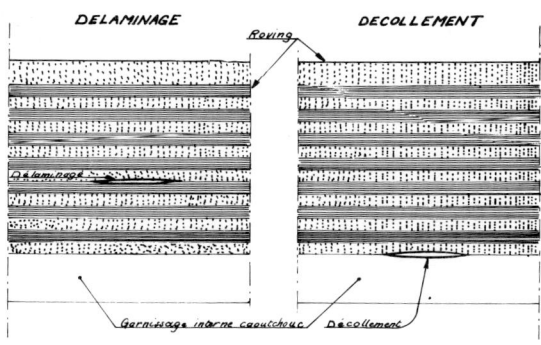

Fig. 2

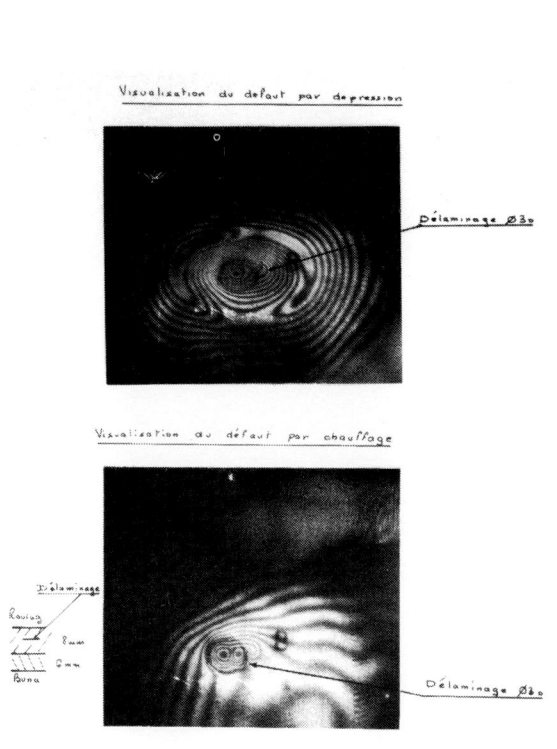

Fig. 3

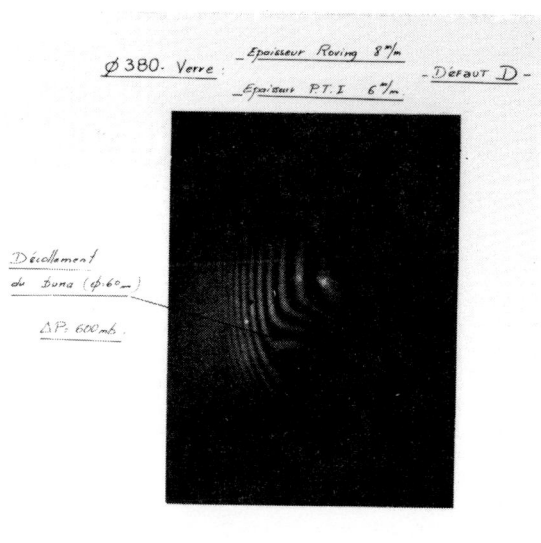
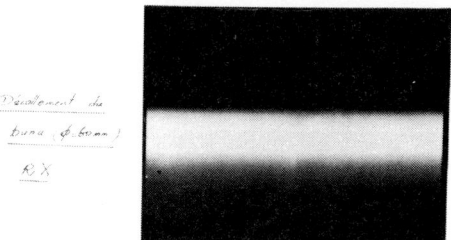

Fig. 4

Fig. 5

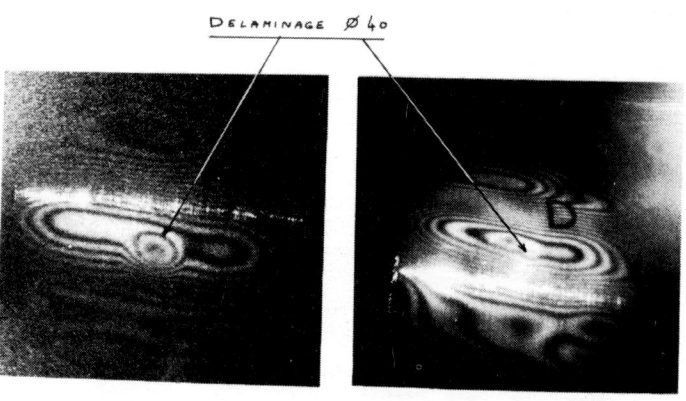

Fig. 6

(a) LASER CONTINU   LASER DECLENCHE (b)

MISE SOUS CONTRAINTE PAR DEPRESSION

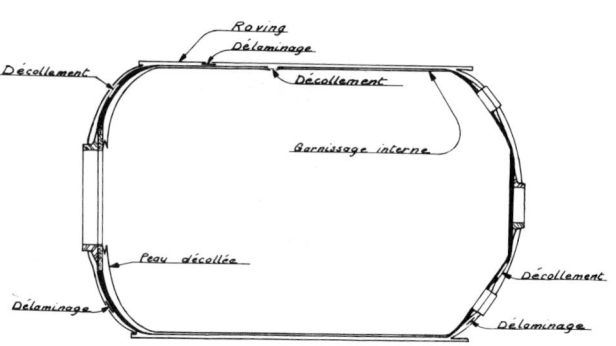

Fig. 7

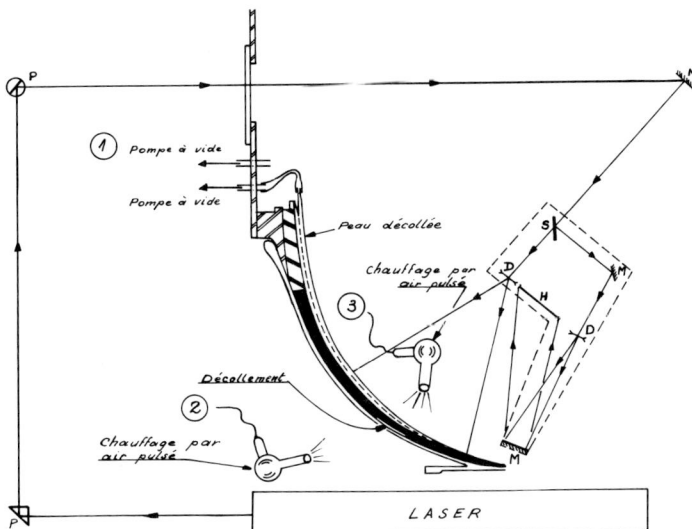

Fig. 8

Fig. 9

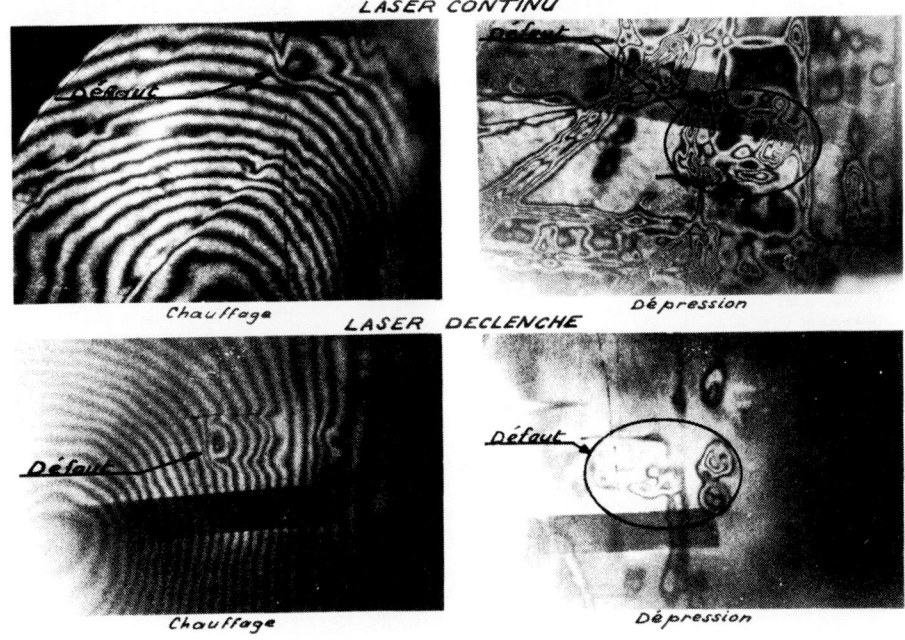

Fig. 10

HOLOGRAPHIC INTERFEROMETRY APPLIED TO THE METROLOGY OF GASEOUS FLOWS

Jean Surget, Jean Délery and Jean-Paul Lacharme
Office National d'Etudes et de Recherches Aérospatiales (ONERA)
92320 Chatillon, France

Abstract

Holographic interferometry is widely used at ONERA for the detailed analysis of transonic flows, whose great sensitivity to perturbating effects of material probes is well known. The implementation of holographic techniques made it possible to build an interferometer presenting many advantages over the classical Mach-Zehnder instrument as regards the possible extent of the field, but also ease of use and cost of fabrication.

Interferograms corresponding to the infinite fringe adjustment provide a visualization of the aerodynamic phenomena. But the quantitative study is preferably carried out from interferograms presenting an initial finite fringe pattern, in order to multiply the measuring points.

Processing, which makes use of an entirely automatic process, is ensured on photographic reproductions of the interferograms restituted by double exposure holograms. The negatives are explored by means of a high precision microdensitometer coupled with a data acquisition unit, comprising a minicomputer and a disc memory where the programme and intermediary data are stored. The computer ensures two functions : 1) it pilots the displacements of the plate holder of the densitometer according to a pre-programmed path, and 2) it ensures the acquisition and processing of the densitometric data. The results are edited on a printer and on a tracing table. Thanks to this procedure, the processing time is considerably reduced : the exploitation time of an interferogram along 25 lines comprising each 60 gas density measurements (i.e. 1500 measuring points) is less than 7 hours.

The holographic bench and its acquisition unit are at present used for the systematic study of the shock wave-turbulent boundary layer interaction in transonic flow at the S8A wind tunnel of the ONERA Fluid Mechanics Laboratory of Chalais-Meudon (near Paris).

1. Introduction

Interferometry, long used by ONERA for experiments in wind tunnels[1], makes it possible to obtain, in addition to the visualization, a great number of simultaneous measurements in a two-dimensional or axisymmetrical aerodynamic flow. In spite of the interest of this method, notably as regards accuracy, the generalization of the use of interferometry was impeded, until the last few years, by two main obstacles : the apparatus cost (usually a Mach-Zehnder interferometer) and the size of the workload imposed by interferogram processing. At the present time, these difficulties have been considerably reduced, thanks to the use of holography to make the interferograms and, for their exploitation, the utilization of a data acquisition and processing system taking advantage of modern informatics methods.

Because of the success of previous experiments[2][3], a holographic interferometry apparatus, connected to a small computer, has been built in the S8A wind tunnel of the Chalais ONERA Fluid Mechanics Laboratory. It is widely used for the systematic study of the interaction between shock wave and turbulent boundary layer in transonic plane flow, permitting a great number of measurements whose validity has been verified. Specifically non intrusive, because it is an investigation process without contact, interferometry is particularly suited to the analysis of this type of flow, whose great sensitivity to perturbating effects of material probes, such as Pitot tubes, static pressure probes or hot films, is well known.

Quantitative interferometry is carried out an finite fringe holographic interferograms so as to increase the measurement resolving power. An entirely automatic processing procedure has been developed[4] ; it makes use of a high accuracy microdensitometer connected to the minicomputer. The programme ensures the identification of the fringe centre lines and the calculation of the difference between their respective interference order. Then it determines the local value of the density relative to that in a reference point, where this parameter is known by means of pressure and temperature measurements. Processing times have thus been considerably reduced.

2. Holographic interferometer

2.1. Description

The holographic bench used for single pass aerodynamic flow interferometric examination is shown in Figure 1 and its optical scheme in Figure 2.

Fig. 1 — Holography bench.

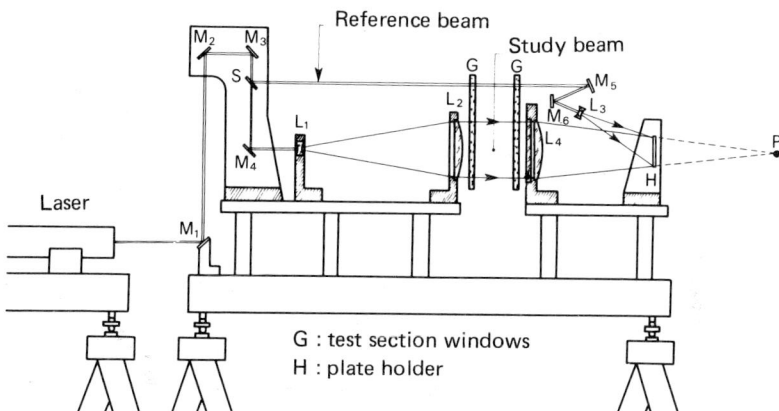

Fig. 2 — Layout of the holography bench.

It is an inexpensive piece of equipment, designed with a view to make the best use of the advantage of simplicity and economy offered by holography. It is not cumbersome, as it occupies only two tables of flanged cast aluminium of standard fabrication.

The first table (0.75 x 0.50 m) carries only the laser ; the second one (1.50 x 0.50 m) ensures the assembly of the emission and reception units of the holographic mounting itself, while preserving between them the minimum possible space for the passage without contact of the wind tunnel test section.

The light source is a 15 mW helium-neon laser. The laser beam is directed to the beam splitter S by means of the plane mirrors M_1, M_2 and M_3 (fig. 2). The beam crossing S is reflected by the plane mirror M_4 and expanded by the afocal system constituted by the lenses L_1 and L_2. The lens L_2, 200 mm in diameter, provides the test beam composed of parallel light rays in order to permit the quantitative study of the flow, which is two dimensional. The wind tunnel test chamber closed by the windows G (fig. 2), limits the test beam section to 140 x 100 mm. The light rays passing through the test chamber are collected by the collimating lens L_4 and converge at its focal point P behind the holographic plate holder H.

Moreover, the part of the laser beam reflected by the beam splitter S constitutes the reference beam of the set-up. This beam, and the test beam, pass through the test chamber, but outside of the aerodynamic channel. It is reflected by the two plane mirrors M_5, M_6 and it is expanded by the diverging lens L_3, and illuminates H in order to create, by superimposition with the test beam, the necessary interference to record the hologram of the wind tunnel test chamber.

The intensity ratio test-beam/reference-beam is about 0.2. The value of the angle made by the two beam axes abutting on H is 27°.

The distance between mirrors M_5 and M_6 can be adjusted to equalize approximately the two optical path lengths of the apparatus.

2.2. Mode of operation

The two conventional holographic techniques of real-time interferometry and double exposure interferometry can be performed with this device.

With the former, a reference hologram of the test chamber, without flow, is recorded in a suitable photographic plate which is then processed and put back exactly in its previous location. The special ONERA plate holder[5] permits one to ensure this operation very easily and with a sufficient accuracy. Observing through this holographic plate illuminated simultaneously by the two beams, the interferogram appears in real time as with a classical interferometer, changing with the flow inside the test chamber. For that reason, the real time holographic interferometry is mainly used for examination of evolutive phenomena, such as transient aerodynamic flow, with possibility of motion picture recording[6].

With the double exposure technique, a first exposure is taken without flow, and then a second exposure, on the same photographic plate, when the wind tunnel is running. The processed plate is then replaced in its holder and illuminated by the reference beam alone. The reconstructed interferogram of the flow observed is recordable by means of a photographic camera whose objective is placed at the point P of Figure 2.

The double exposure method provides only one interferogram of the phenomenon and only in deferred time. But it is the best when the interferogram has to be processed for quantitative measurement, for two main reasons : on the one hand, because the time necessary to make the two exposures is too short for casual mechanical bench deformation to appear ; on the other hand, because there is no risk, in this case, of introducing an error in the interferogram, due to an imperfect positioning when replacing the plate in its holder.

By whatever technique they are obtained (real time or double exposure) the interferograms correspond to two types of adjustment : in infinite fringe or finite fringe pattern (Fig. 3).

Infinite fringe interferograms are mostly used as a means of visualizing the phenomenon since, in two-dimensional flow, the interference fringes trace lines of equal density, which are also equal velocity lines in the isentropic flow regions.

For quantitative analysis, finite fringe adjustment is used in order to improve the accuracy by increasing the number of point measurements. The additional fringe grating is obtained by translation of the lens L_1 of the bench (fig. 2), in its plane, between the two exposures of the holographic plate. The fringe spacing is inversely proportional to the lens displacement (fig. 4). The phenomenon is indicated by the deformations it brings to the initial pattern of equidistant fringes.

Consequently, for each aerodynamic experiment, the analysis is carried out on the photographic reproductions of two interferograms : the original pattern and the same pattern modulated by the aerodynamic flow field information.

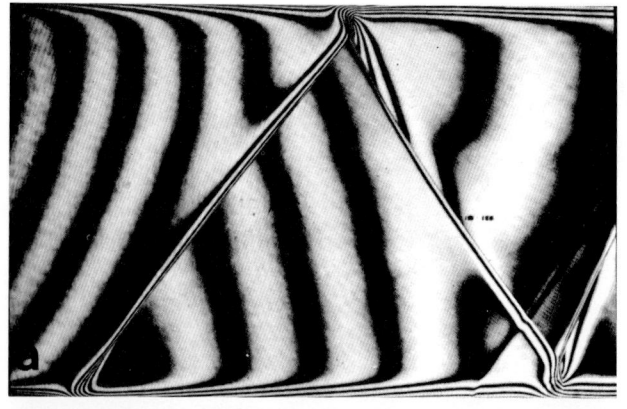

Fig. 3 — Interferograms of the flow behind a profile.

   a) infinite fringe pattern
   b) horizontal finite fringe pattern
   c) vertical finite fringe pattern.

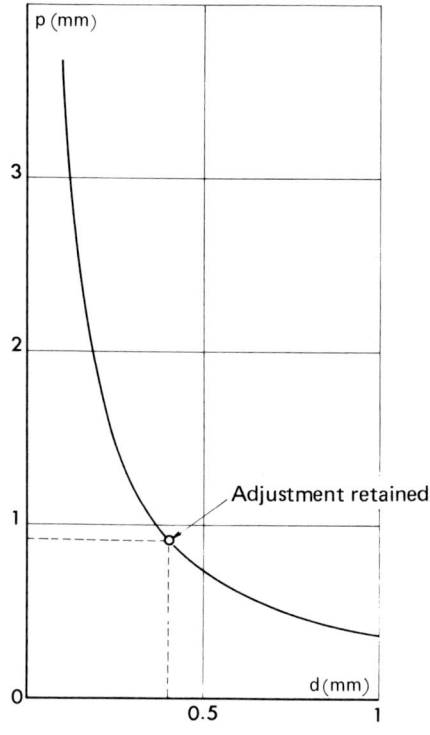

Fig. 4 — Evolution of the fringe width p as a function of lens L_1 displacement d.

3. Interferogram processing method

3.1. Picture analysis

3.1.1. General — Let us first briefly recall how, with the double exposure method, interferometry provides a density measurement in a gas.

Let R be a so-called "reference" point, where the density ρ_R is known. The reference state in R is characterized by the fringe number N_R on the interferogram taken when the wind tunnel is running, and N'_R on that taken when it is stopped, these two numbers being taken arbitrarily. Let M be a point of the aerodynamic field, of unknown density ρ_M, charactirized by fringe numbers N_M and N'_M. The difference in the lengths of the light paths having crossed the flow in R and M respectively can be written, for the interferogram taken during the run :

$$\Delta = (N_M - N_R) \lambda \qquad (1)$$

This quantity represents the superposition of the effects of the aerodynamic phenomenon Δ_1 and of that of the finite fringe adjustment Δ_2 ; hence : $\Delta = \Delta_1 + \Delta_2$. If ν_M and ν_R are respectively the refraction index values in M and R, e the length of the medium crossed, here the test section, we write :

$$\Delta_1 = (\nu_M - \nu_R) e \qquad (2)$$

In air, ν is related to density by the Glodstone-Dale law :

$$\nu - 1 = B\rho, \qquad (3)$$

B being a constant depending on the light source used ; hence

$$\rho_M = \frac{\Delta_1}{eB} + \rho_R. \qquad (4)$$

Δ_2 is given by the difference of light path lengths in M and R on the interferogram at rest :

$$\Delta_2 = (N'_M - N'_R) \lambda \qquad (5)$$

$$\Delta_1 = \Delta - \Delta_2 = [N_M - N_R - (N'_M - N'_R)] \lambda$$

or, recalling eq. (4) :

$$\rho_M = \frac{\lambda}{eB} [N_M - N_R - (N'_M - N'_R)] + \rho_R. \qquad (6)$$

The study of this relation shows that the fringe numberings on the pictures during the run and at rest are independent, as only the differences between their values at the test and the reference points have to be considered.

In practice, only the fringe centres can be precisely pinpointed, and thus be assigned with a number N. The minimum distance between two measuring points cannot thus be smaller than half the local fringe pitch on the picture taken during the run. We shall show later how the corresponding numbers can be determined on the picture taken at rest.

3.1.2. Automatic processing procedure — As mentioned in Section 2.2, processing is carried out on the photographic pictures of the interferograms provided by the holograms.

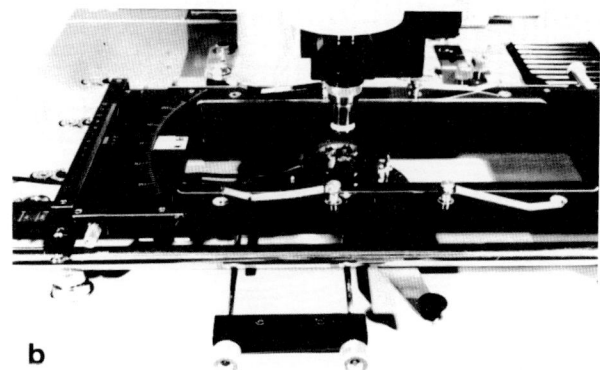

Fig. 5 — Joyce Loebl 3 CS microdensitometer.
a) General view ; b) Plate holding chariot.

The negatives are explored by means of a Joyce Loebel, type 3 CS, microdensitometer whose Figure 5a gives a general view. The main parts of the instrument are :
— a plate-holding chariot (fig. 5b) able to move along two perpendicular directions OX, OY, that can be adjusted at an angular position in the XOY plane ; movements along OX and OY can be ensured by 5 μm increments, and are controlled by a computer ;
— a device for measuring optical density by comparison of a small part of the picture with standard reference wedge ; the observation window can be adjusted in width and length so as to adapt at best the area over which the measurement is taken, as a function of the minimum fringe dimension and the grain of the sensitive surface. The value of optical density is converted in numerical form into a integer between 1 and 1000.

The microdensitometer is connected to a data acquisition unit comprising a Hewlett-Packard 2100 minicomputer, a disc unit where the programme and intermediary data are stored, a printer and a tracing table.

The computer ensures two functions : on the one hand it pilots the movements of the plate-holding chariot of the densitometer according to a programmed path, and on the other hand it ensures the acquisition and processing of densitometric data.

The fineness of analysis of which the microdensitometer coupled with the computer is capable is illustrated by the example presented in Figure 6. Figure 6a shows an interferogram with tight fringes, taken in a transonic flow where a quasi-normal shock is formed. A small, 1 mm^2 square area, centered on the shock image, has been explored. Its densitometric analysis is represented, with expansion of the vertical scale, in figure 6b where the contour lines are separated by an optical density step of about 0.07. It should be noticed that because, on the one hand, of the three-dimensional effects resulting from the interaction of the shock wave with the boundary layers along the windows, and on the other hand of the strong diffraction taking place near the shock, the observed fringes, especially on the upstream side, are not representative of the density variation within a supposedly plane flow. The example presented here has thus a purely optical interest and should not be interpreted on the aerodynamic viewpoint.

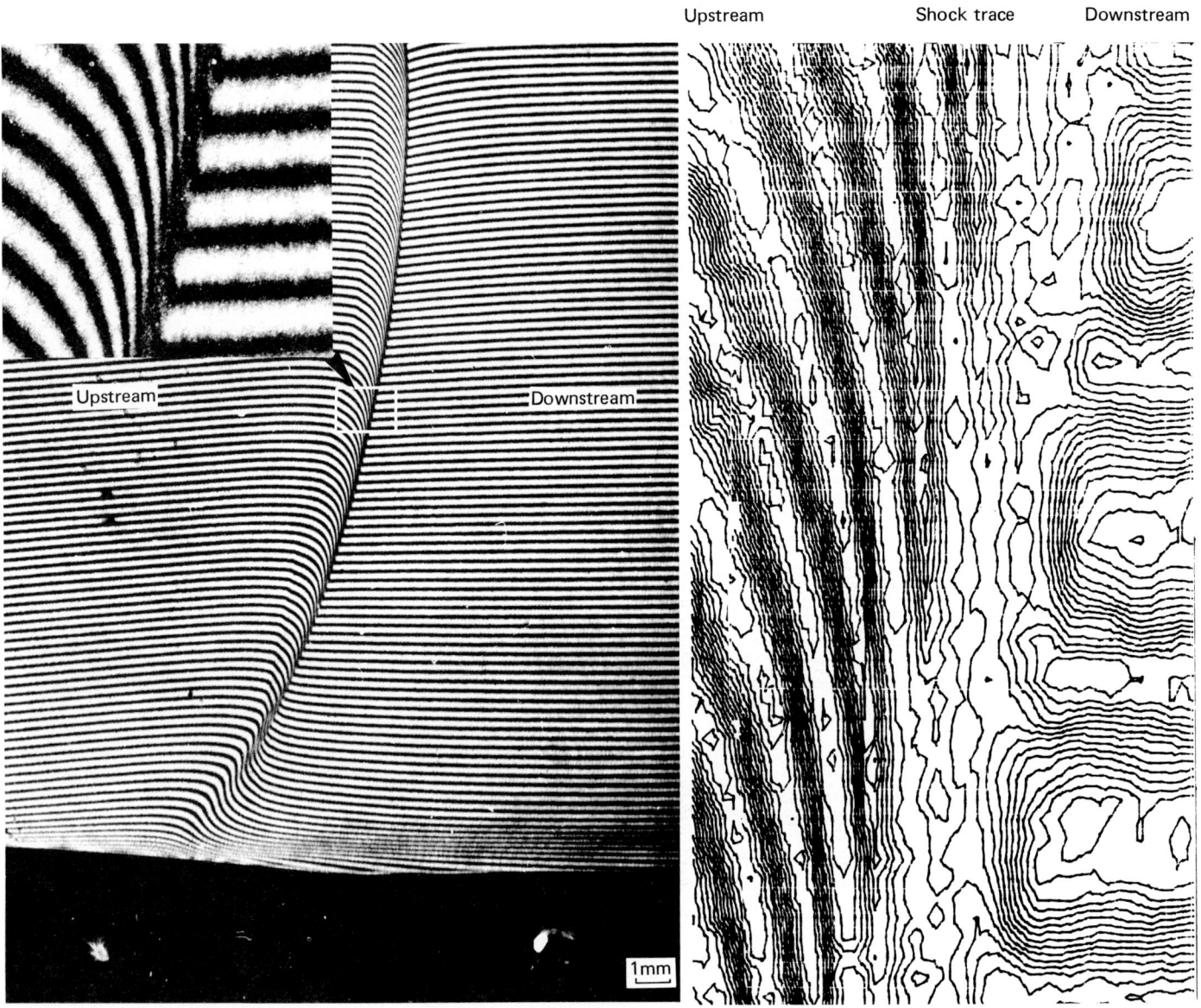

Fig. 6 — Densitometric analysis of a small area of an interferogram.

Figure 7 gives a representation with perspective effect of the fringe structure in the same region.

We shall describe the exploitation process of an interferogram by considering a concrete case such as the one shown on Figure 8. The pictures have been taken at rest and during the run concerning a transonic flow in a duct carrying on its bottom surface xx' a half profile whose only the downstream part is visible on the photographs.

We can notice the presence of a quasi-normal shock wave of small intensity, crossing the field of view.

The main purpose of the study being the boundary layer that develops along the bottom surface, the finite fringes have been adjusted to be almost parallel to xx' so that the exploration lines on the pictures cross them almost normally. The abscissae x_k of these lines (k = 5 on figure 8) are determined as a function of the phenomenon to be analyzed, one of them passing through the reference point R, where density ρ_R is known. For convenience, R is chosen at the centre of a fringe, in a region where the flow is practically uniform, so that ρ_R can be calculated from a local measurement of the wall pressure p_R by using the relation of isentropic expansion. The "run" and "rest" interferograms are successively explored along the lines x_k, each picture being placed on the densitometer table so that the displacement axis XX' be exactly parallel to xx'. This is obtained through a line of small diameter (0.8 mm) holes bored in the wall, one of them being the origin Ω of the densitometer coordinates. The positions of these holes, in an xoy frame related to the profile, is precisely known by construction, as well as the picture magnification (here, G = 0.37). Thus it is easy to determine the position of the measuring points relative to the wall, whose shape is defined by its equation in the xoy frame.

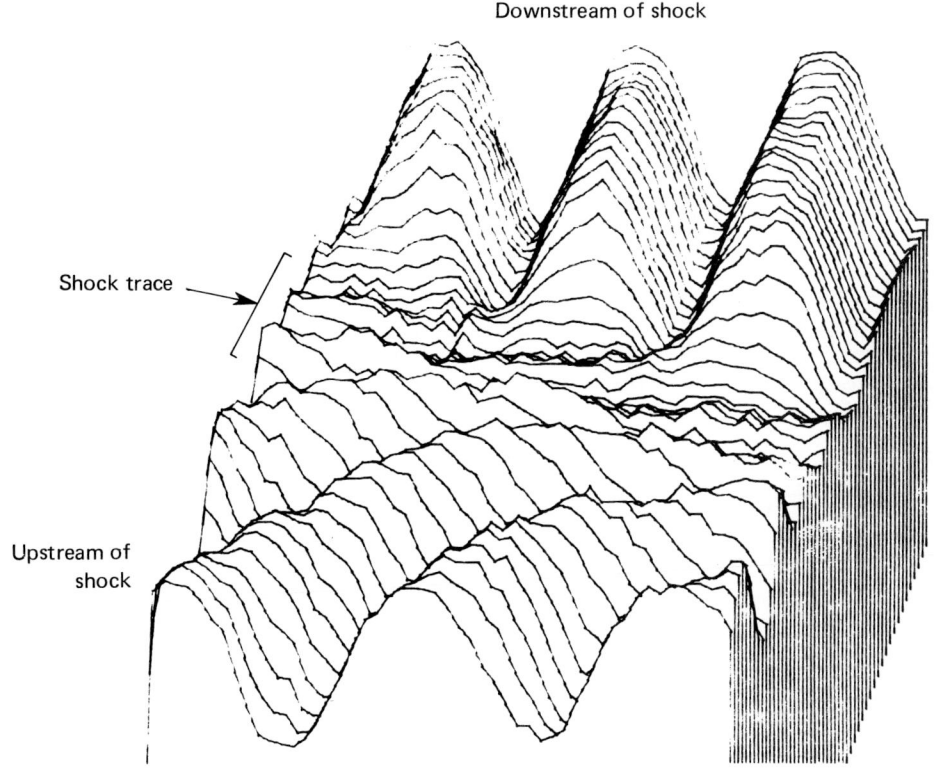

Fig. 7 — Fringe structure around a shock.

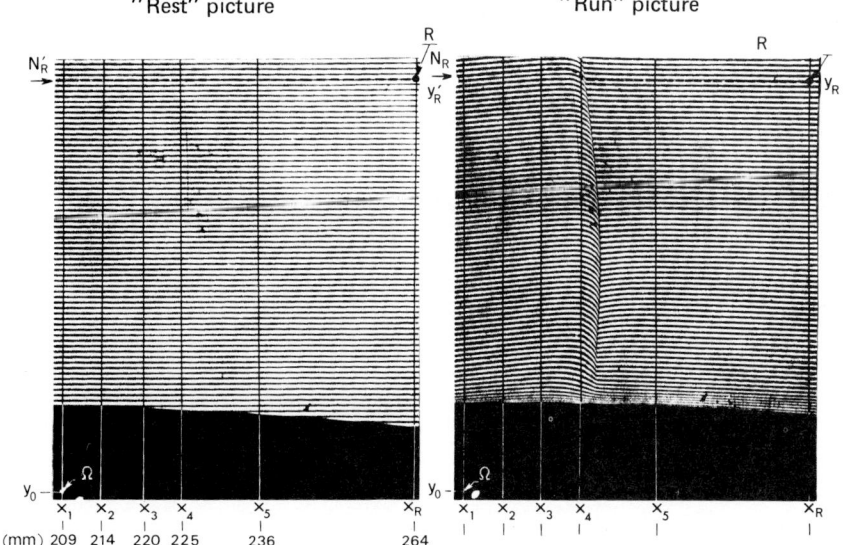

Fig. 8 — Interferogram exploration procedure.

The densitometer, piloted by the computer, explores the picture by incremental steps, starting, for each line k, from the fringe passing through the reference point R whose number N_R is arbitrarily fixed. When a fringe i (black or white) is crossed, the coordinates $y_{k,i}$, $Y_{k,i}$ of its centre, and its number $N_{k,i}$, are memorized. This operation stops when the wall is reached. The programme then repositions the microdensitometer on the departure point of exploration k, then orders a displacement along the fringe number N_R, up to abscissa x_{k+1}. The optical analysis process is then repeated until all four lines are scanned.

The method used to determine the ordinates $y_{k,i}$ of the extrema is explained on figure 9. The fringes are defined by discrete measurement of optical density $d(y)$, $d(y + \Delta y)$... $d(y + n\Delta y)$ picked up at ordinates (on the picture) every 10 μm in the boundary layer and every 20 μm in the external field. A mean density level d_m being fixed with a view to discriminate approximately two adjacent fringes, all the values of d lying on the same side of d_m are smoothed by a parabola $\mathcal{P}$ (y) passing as close as possible to the experimental points according to the least mean square rule. The apex of $\mathcal{S}$ is then taken as the centre of the fringe considered. At the next alternation, the mean density d_m is recalculated as the half sum of the extrema densities measured at the previous step; this process enables one to follow at best an overall variation of the negative picture transparency due to a non uniformity of the field illumination.

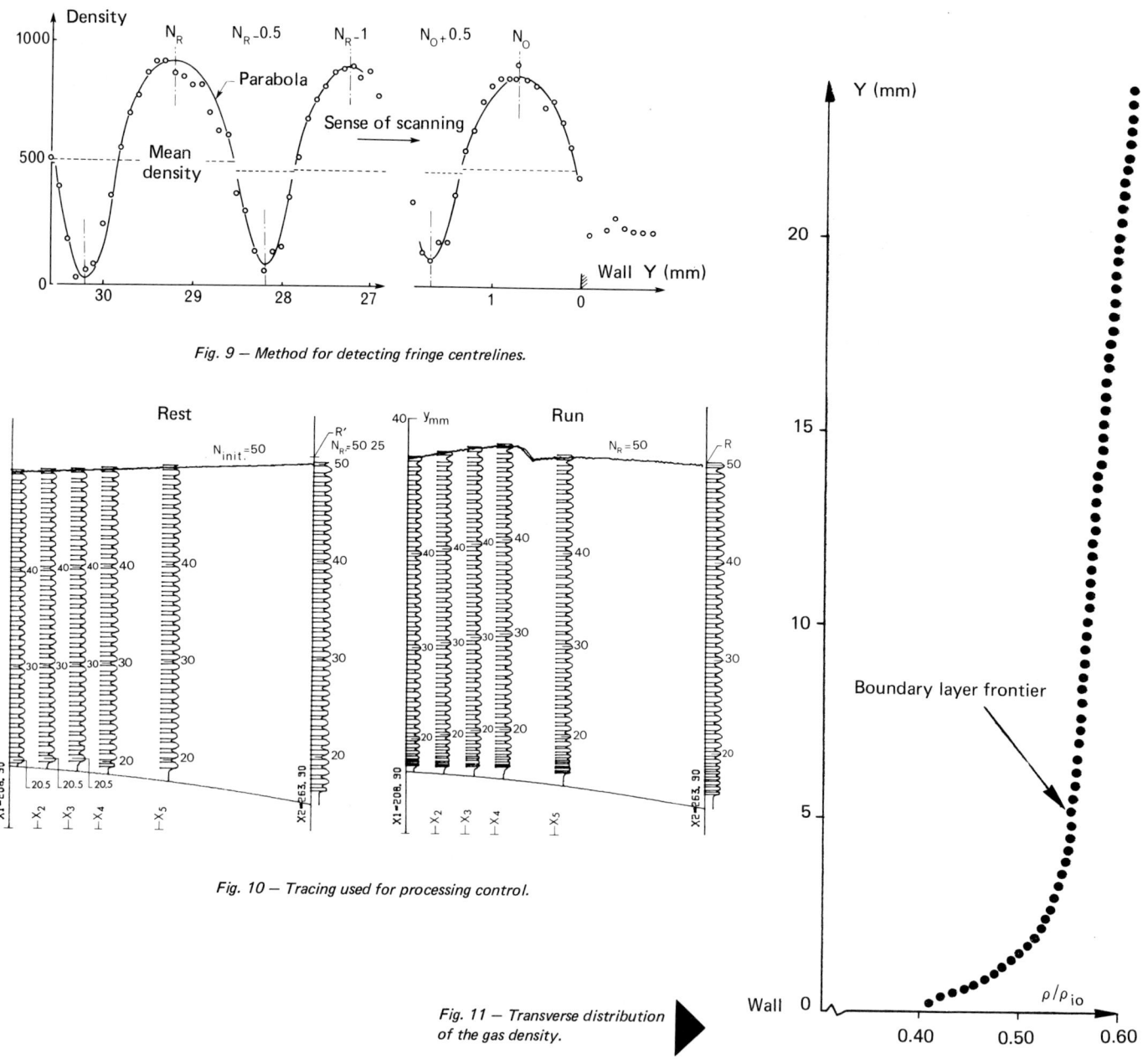

Fig. 9 — Method for detecting fringe centrelines.

Fig. 10 — Tracing used for processing control.

Fig. 11 — Transverse distribution of the gas density.

Figure 10 shows a tracing describing the procedure of fringe localization for six exploration lines on the two pictures, "rest" and "run". Such a document allows one to make sure that all operations took place normally. If it is not the case, a manual intervention on the disc file permits in most cases the reintroduction of "missed" fringes or the elimination of parasitic fringes unduly accounted for and due to picture defects.

3.1.3. Density calculation — The programme defines the gas density ρ_M at every point M, of coordinates $(x_k, y_{k,i})$ placed at the centreline of a fringe of a "run" picture. According to eq. (6), the calculation of ρ_M requires the knowledge of the fringe number N'_M at M on the "rest" interferogram, where M has no reason to be located on a fringe centreline. As, without the flow, the fringes are very nearly parallel and evenly spaced, N'_M is interpolated by a linear relation.

$$N'_M = A_k y_{k,i} + B_k$$

where the A_k and B_k coefficients are calculated for each abscissa x_k by the mean square method over the whole set of sightings along x_k. The distributions of $(\rho/\rho_{io})(y_{k,i})$ for each x_k constitute the "rough" results of the processing (ρ_{io} is the gas density in the stagnation state). They are edited on the printer and automatically traced. Figure 11 gives an example of such a distribution.

3.2. Exploitation of the results

Before any further step, the gas density profiles obtained this way are smoothed by a least square method with a view to eliminate a slight dispersion, and the recalculated values of (ρ/ρ_{io}) are stored on the disc. They may be exploited with a view to obtain a very complete description of the aerodynamic field. As an example, we are going to show the procedure adopted to deduce the characteristics of the boundary layer that develops along the profile.

The flow not being isentropic in the dissipative zone, a second measurement (temperature or pressure) is, in principle, necessary to determine the velocity U. However, in the present case where the wall is practically athermane and the Mach numbers moderate, we can avoid this second measurement either by assuming that the stagnation temperature T_{io} remains constant within the boundary layers or by adopting, to relate temperature and velocity, the modified Crocco law[7] :

$$\frac{T}{T_e} = \frac{T_p}{T_e} + \frac{T_f - T_p}{T_e} \frac{U}{U_e} - \left(\frac{T_f - T_e}{T_e}\right)\left(\frac{U}{U_e}\right)^2$$

where T : local temperature ; T_e : temperature at the boundary layer frontier δ ; T_p : wall temperature ; T_f : friction temperature, and U_e : velocity at the frontier δ.

In permanent regime, without heat flux, $T_p = T_f$ the friction temperature being given by :

$$\frac{T_f}{T_e} = 1 + 0.9 \frac{\gamma - 1}{2} M_e^2$$

where M_e is the Mach number at the frontier δ.

In these conditions, the reduced velocity is calculated by :

$$\frac{U}{U_e} = \sqrt{\frac{1 + 0.9 \frac{\gamma-1}{2} M_e^2 - \frac{T}{T_e}}{0.9 \frac{\gamma-1}{2} M_e^2}}$$

If pressure p is transversally constant, a condition very often verified in a boundary layer, T/T_e is given by :

$$\frac{T}{T_e} \rightleftharpoons \frac{\rho_e}{\rho} \quad \text{(equation of state)}$$

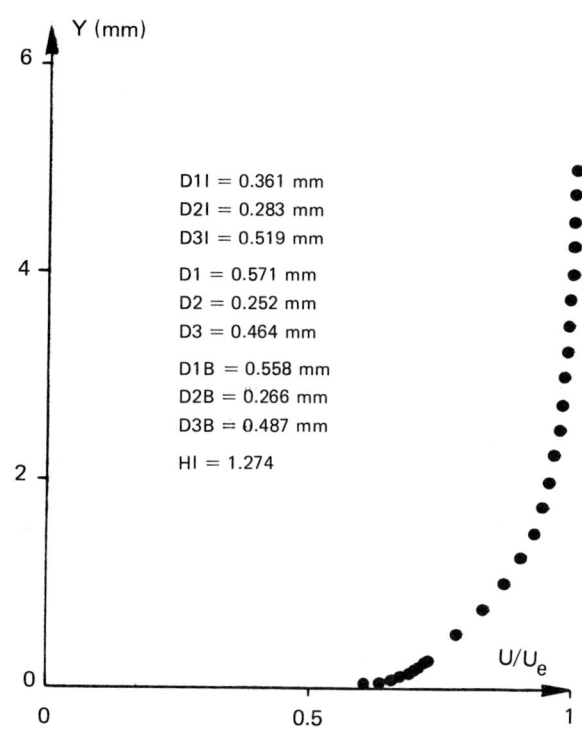

D1I = 0.361 mm
D2I = 0.283 mm
D3I = 0.519 mm
D1 = 0.571 mm
D2 = 0.252 mm
D3 = 0.464 mm
D1B = 0.558 mm
D2B = 0.266 mm
D3B = 0.487 mm
HI = 1.274

Fig. 12 — Velocity profile within the boundary layer. $M_e = 1.426$.

where ρ/ρ_e is directly deduced from the measurements.

The velocity profile U/U_e calculated in this manner is automatically traced (fig. 12), then exploited to deduce the integral thicknesses and the shape parameters characterizing the boundary layer : displacement thickness D1, momentum thickness D2, kinetic energy thickness D3, incompressible shape parameter HI, among others. Their values appear in the insert of figure 12. In order to improve the precision in the calculation of the characteristic thicknesses, 10 values of U/U_e are interpolated between the wall, y = 0, and the first measuring point, usually located around y = 0.25 mm.

3.3. Examples of results obtained

The S8A wind tunnel of the ONERA Fluid Mechanics Laboratory, where holographic interferometry is routinely used, is of continuous run type. It is supplied by dissicated atmospheric air, and its stagnation conditions are close to atmospheric ones, i.e. $p_{io} \simeq 1$ bar, $T_{io} \simeq 300$ K. The experimental set-up is schematized on figure 13. It comprises a transonic section 100 mm high and 120 mm wide. On the lower wall is installed a lenghtened half profile (its chord is 280 mm for a maximum thickness of 6 mm) which provokes an acceleration of the flow up to transonic regimes. The velocity gradient can be adjusted by action on the slope of the upper wall, which rotates around point A, at the section entrance. Downstream, a second throat of variable aperture makes it possible to adjust the shock wave location and intensity.

A presentation of the complete set of tests performed[8] would be outside the scope of this paper, which is devoted to the measuring and processing techniques. We shall only give here a sampling of results that emphasize the possibilities and the precision of the interferometric method.

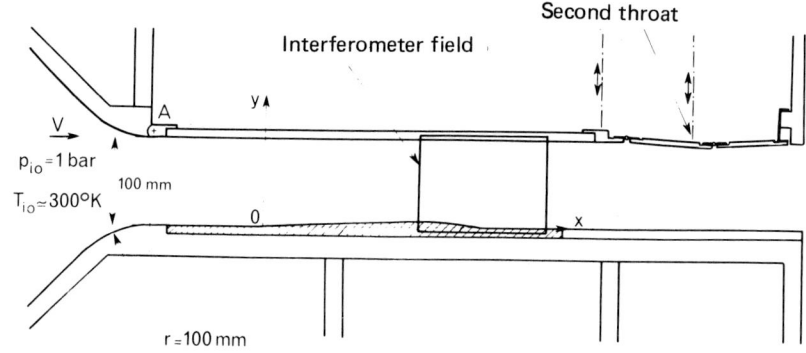

Fig. 13 — Transonic test section of the S8A wind tunnel.

Figure 14 concerns three flow configurations, visualized by the corresponding infinite fringe interferograms. In the first case a quasi-normal, quite strong shock forms upstream of the profile trailing edge ; it provokes a noticeable thickening of the boundary layer which, in the interaction, reaches conditions close to separation. The second example shows a "lambda" shock structure : an oblique shock is created by the change of flow direction at the trailing edge ; the zone of slightly supersonic flow that follows it is limited by a weak straight shock becoming evanescent near the wall ; these two shocks converge into a triple point from which a single resulting shock is initiated. In the third configuration, the trailing edge oblique shock reflects on the upper wall, thus starting the so-called "Mach phenomenon" ; in this case, the flow is supersonic in almost the whole field. The measurements presented on this figure give the evolutions of the Mach number M at the frontier of the boundary layer that develops on the lower wall (abscissa x is counted from the profile leading edge). The values of M_e deduced from the local gas density resulting from the interferometric processing are compared to the distributions $M_e(x)$ calculated from the wall pressure readings : the agreement between the two measuring methods can be considered as very good.

Another example concerns explorations of the boundary layer performed in front of and behind the shock, in the first configuration. The profiles of reduced velocity U/U_e are shown on figure 15. The adoption of a 0.91 mm fringe pitch in the "rest" adjustment (see Section 2.2) makes it possible to obtain about 25 measuring points inside the boundary layer, whose thickness behind the shock is about 9 mm. The measuring point closest to the wall has a minimum altitude of 0.3 mm. Interferometric measurements are compared, for a few velocity profiles, to probings by laser anemometry[9] taken at the centre of the observed zone. The two techniques reveal here slight differences, whose main cause is a rather great uncertainty (0.2 to 0.3 mm) in the localization of the measuring volume during the laser anemometry tests.

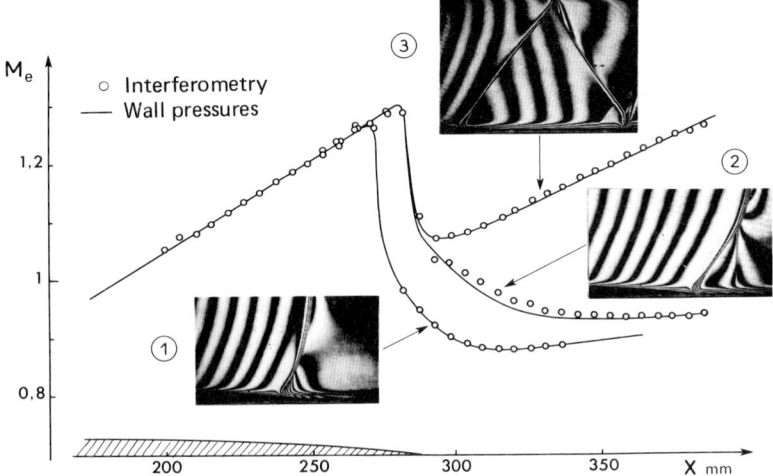

Fig. 14 — Mach number conditions at the boundary layer frontier.
1 - Separation in front of the trailing edge.
2 - λ-shock at the trailing edge.
3 - Shock reflection on the upper wall.

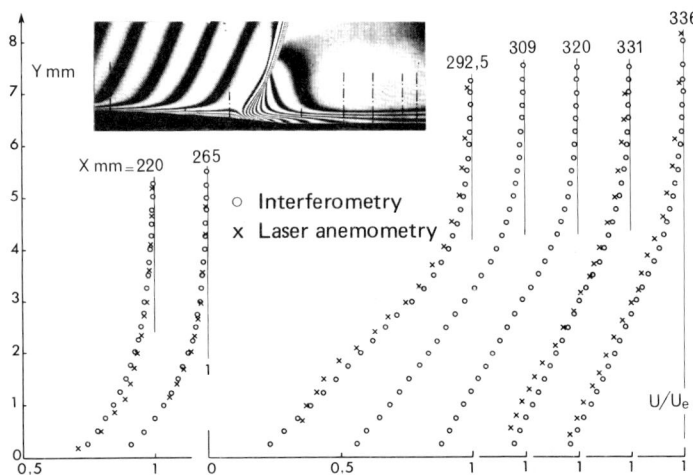

Fig. 15 — Velocity profiles within the boundary layer (configuration 1 of figure 14).

4. Conclusion

Interferometry is a measuring means perfectly adapted to the investigation of compressible flows very sensitive to the perturbating effect of material probes, either because of their exiguity or their nature (e.g. transonic). From another view point, interferometry permits limitation of the test time to that of the recording of a very samll number of interferograms, which make up a true memorization of the phenomena. The processing, in deferred time, of the photographs taken during the test can provide a large number of informations whose acquisition by classical methods (pitot tube, static probe, etc.) would require very long testing times.

In spite of these undeniable advantages, this process is but rarely used because, principally, of the high cost of fabrication and the difficulty of use of the Mach-Zehnder type interferometers, and also of the lengthy processing it involves. The recent development of holographic techniques allowed a considerable reduction of both the instrumentation cost and its operational constraints. Now, it is possible to build hardy interferometric benches that can be easily carried from one wind tunnel to another, and easy to set in operation. Moreover, the processing procedure is much more rapid and easy thanks to a high degree of automatization based on a microdensitometer connected to a minicomputer. Thus, the exploitation of an interferogram along 25 lines providing each 60 values of the gas density (i.e. 1500 measuring points) is now feasible within 7 hours, while the same work done manually would require a whole week for a very experienced technician. The precision of the results is usually better than 1%.

References

1. Véret, C., Philbert, M., Surget, J., Fertin, G., Aerodynamic flow visualization in the ONERA facilities. Proceedings of the International Symposium on Flow Visualization, Tokyo, Oct. 1977, p. 255-260.
2. Philbert, M., Surget, J., Application de l'interférométrie holographique en soufflerie. Rech. Aérosp. No 122 (1968), p. 55-60.
3. Surget, J., Etude quantitative d'un écoulement aérodynamique par interférométrie holographique. Rech. Aérosp., No 1973-3, p. 161-171.
4. Délery, J., Surget, J., Lacharme, J.P., Interférométrie holographique quantitative en écoulement bidimensionnel. Rech. Aérosp., No 1977-2, p. 89-101.
5. Dispositif de positionnement de plaque. French patent No 72-22592, 22nd June 1972.
6. Surget, J., Chatriot, J., Cinématographie ultra-rapide d'interférogrammes holographiques. Rech. Aérosp., No 132, Sept-Oct. 1969, p. 51-55.
7. Michel, R., Couches limites, frottement et transfert de chaleur. ENSAE Lecture Course (ed. 1972).
8. Délery, J., Recherches sur l'interaction onde de choc-couche limite turbulente. Rech. Aérosp. No 1977-6.
9. Boutier, A., Etude et réalisation d'un vélocimètre compact. Application à des mesures de vitesses en écoulements supersoniques et transsoniques très turbulents. NT ONERA No 237 (1974).

HOLOGRAPHIC INTERFEROMETRY IN OSTEOSYNTHESIS

P. Meyrueis, M. Pharok, J. Fontaine
Louis Pasteur University
67000 Strasbourg, France

Abstract

The bony structure of the human face is particularly exposed to the risk of fractures. Repair by conventional methods imposes a long and constraining immobilization of the patient. New methods, using holographic interferometry, are presently being developed in Strasbourg.

Introduction

One method presently used to repair broken jaws consists in the use of an adapted metal clamp of small dimensions, which replaces the bony structure in the transmission of stresses. For this it is necessary to know with precision the behavior of the jaw when submitted to a force, to characterize the lines of force and to study their dynamic behavior. It is then possible to determine the mechanical resistance of the bone.

On the microscopic scale, the configuration of this material is particularly complex. Furthermore, studies on samples of small dimensions have shown that the properties vary from one donor to another. It is also to be noted that, the bone being a living material, the mechanical properties are altered by any process of conservation. It is thus necessary to perform the tests on bones which have not yet lost their natural mechanical properties.

Preliminary tests were made by photoelasticimetry: a transparent model of the jaw was made. This model was submitted to stresses and analyzed in polarized light. This technique gives only approximate results since plastic and bone behave differently. Nonetheless, this permits us to find an optimal arrangement for the definitive setup and to determine the zones to be studied in greater detail.

Another technique consists in coating the bone to be studied in a special varnish which, in polarized light, causes the distribution of stresses to appear. The inconvenience is that this varnish alters the mechanical properties of the bony material and adheres badly to an inadequately prepared bone.

More precise results are obtained by holographic interferometry. The general principle is well known. The principal difficulty resides in the qualitative analysis of the fringes characterizing the deformations undergone by the object.

Theory

The treatment of the deformations is made as follows:

(1) mechanical formulation of surface displacements
(2) determination of the relations between the derivatives of the surface displacement and the interference fringes in the image plane
(3) interpolation of the network of fringes

It is necessary first to define a mathematical model. The interpolation of the fringes is facilitated by a phase modulation permitting, in theory, measurements with a precision of λ. Different techniques based on holographic interferometry permit us to measure the displacement vector $\vec{u}$ and its derivatives. Double exposure was used in the case of the jaw. After applying a stress to the object during the second exposure, at reconstruction we observe a network of interference fringes which can be located on the surface of the object. The study of this system of fringes provides information about the rotation and flexion of the different parts of the jaw.

For the mathematical approach to the problem, we consider a system of local coordinates x,y and z, with z normal to the surface of the object. Let $\vec{u}(x,y,z)$ be the displacement vector of the object, $\mathcal{E}$ the flexion and Ω the rotation to be determined.

The surface of the object being submitted to a force, the components of the tensor of the efforts distributed on the surface are obtained by the expression:

$$\mathcal{E}_{ij} = \frac{1}{2}\left(\frac{\partial u_i}{\partial x_j} + \frac{\partial u_j}{\partial x_i}\right) \tag{1}$$

i and j are the indices of the coordinates of the (x,y) plane.
The vector Ω characterizing the rotation has as its component:

$$\Omega_x = \frac{\partial u_z}{\partial y} \quad \Omega_y = -\frac{\partial u_z}{\partial x} \quad \Omega_z = \frac{1}{2}\left(\frac{\partial u_y}{\partial x} - \frac{\partial u_x}{\partial y}\right) \tag{2}$$

The flexion and the torsion result from the change in curvature of the surface under the action of the force applied. The original curvature of the surface of the object is described by the tensor K which has the components:

$$K_{\alpha,\beta} = \frac{\partial^2 F}{\partial x_\alpha \partial x_\beta} \qquad \gamma = x,y \tag{3}$$

where $z = F(x,y)$ defines the surface of the object. The change of curvature is defined by the tensor of components:

$$\Delta K_{\alpha,\beta} = \frac{\partial^2 u_z}{\partial x_\alpha \partial x_\beta} - K_{\alpha\beta} \mathcal{E}_{\gamma\beta} - \mathcal{E}_{\alpha\beta} K_{\alpha\beta} \tag{4}$$

The summation is made on the repetitive indices.
This last expression is simplified in the case of small displacements.
The above calculation supposes the knowledge of the displacement vector. The following considerations permit us to determine it. The phase of the interference phenomenon is written:

$$\varphi_M(\xi,\eta) = (\vec{K}_i - \vec{K}_e) \cdot \vec{u} = 2K(\vec{E} \cdot \vec{U}) = 2K \cdot U_M(\xi,\eta) \tag{5}$$

$\vec{K}_i$ is the wave vector of the incident ray, and $\vec{K}_e$ that of the ray in the direction of observation. The vector $\vec{E}$, equal to the difference $\vec{K}_i - \vec{K}_e$, is called the sensitivity vector.
(ξ,η) is the system of image coordinates. U_M is the component of the displacement $\vec{u}$ in the direction of the sensitivity vector. In considering the displacement vector by its components U_K in the coordinate system of the image plane we can write:

$$U_M = e_{nK} U_K$$

e_{nK} being the components of $\vec{E}$ in this system.
Three different sensitivity vectors permit us to calculate the three components of the displacement vector, thus to determine the deformation of the object studied, here the jaw.

Setup

We used a conventional setup with a single reference beam (Figure 1). The jaw is placed in a concrete support resting on a metal receptacle. The assembly must have maximum stability and rigidity.
We used a depression plate holder with micro-control allowing six degrees of free play with a precision of adjustment on the order of a micron. This precision is necessary for the localization of the fringes.
The assembly rests on an anti-vibration table composed of a metal sand box on which rests, by the intermediary of a shock absorbing system, the work platform. The laser is of the Spectra Physics 170 type. A very good coherence is obtained by means of the temperature controlled etalon. The exposures are from 1/10 to 1/2 second. Oblique lighting permits us to reduce the sensitivity of the fringes.

Trials and Results

The stresses are applied to the jaw by means of 2 to 80 gram weights hung at various points. A local stress is thus simulated at the point of application of the force. The hooks used permit the precise location of the force and this simple method has proved to be very effective. A more complicated dynamometric system did not give better results.
The study consisted at first in analyzing the principal lines of force of the jaw, then making a crown and applying a miniaturized metallic plate to it. This plate should be as thin and narrow as possible. It should also be malleable to permit adaptation while retaining sufficient solidity. We then analyzed the behavior of the repaired jaw to deduce the best means of fixing the plate.
The principal difficulties are mechanical. The jaw must be fixed on a support remaining perfectly unchanged after the application of the stress.
It is equally important to wisely choose the angle of lighting to obtain an optimal spacing of the fringes for a determined stress.
The qualitative analysis which was made permits us to verify the limits of osteosynthesis by miniaturized plates. The study is continuing and will permit a surer and more precise use of this technique, representing real progress in the treatment of fractures of the jaw.

Conclusion

Holographic interferometry is a tool well adapted to the study of bony structures like the jaw. This method does not require a model and can supply interesting information to surgeons applying plates to fractured regions of the jaw.
This study was made with the collaboration of a team of surgeons specializing in osteosynthesis from the hospitals of Strasbourg.

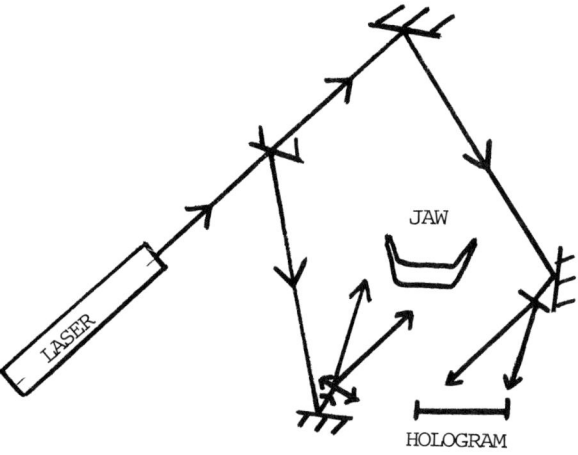

Fig. 1. Setup

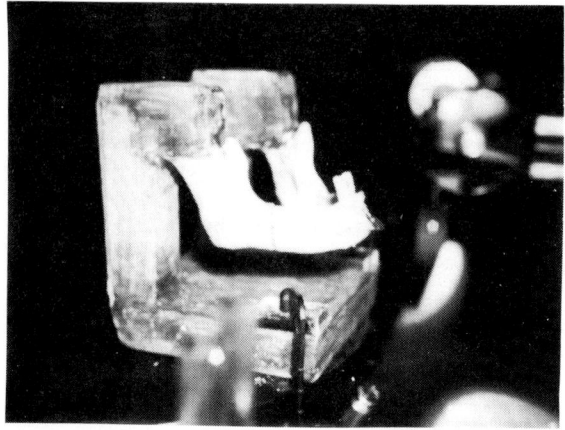

Fig. 2. View of the jaw setup in state of rest

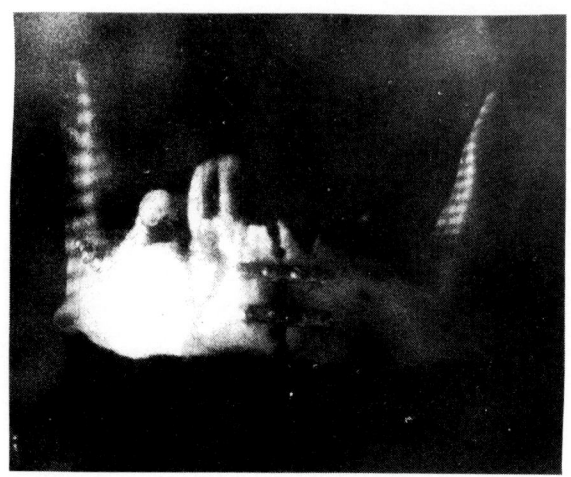

(a)   (b)
Fig. 3. (a) Front view of the hologram of the jaw after double exposure (b) Setup

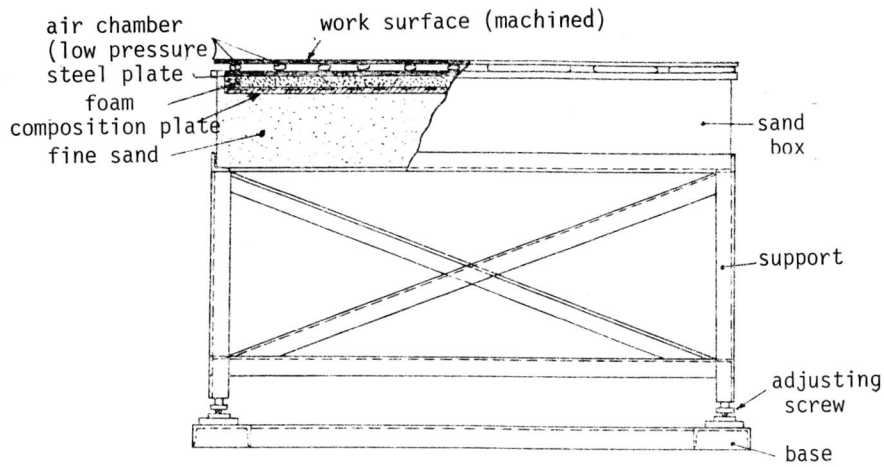

Fig. 4. Sketch of anti-vibration table

Bibliography

1. Gabor, D., "A New Microscopic Principle", Nature, vol. 161, 1948.
2. Collier, R., Burckardt, C. and Lin, L., Optical Holography, Academic Press, 1971.
3. Bremble, G.R., Lalor, M.J. and Hardinge, K., "A Preliminary Study of Fracture Fixation in Holography in Medicine".
4. Meyrueis, P. Thesis, University of Strasbourg, 1975.
5. Vienot, J.C., Smigielski, P. and Royer, H., Holographie Optique, Dunod (Paris), 1971.
6. Wilk, A. "Memory: l'Osthéosynthese mandibulaire par plaques miniaturisées vissées."
7. Stetson, K.A., "Homogeneous deformation by fringe vectors in holographic interferometry", Applied Optics, vol. 14, Sept., 1975.

1st EUROPEAN CONGRESS ON OPTICS APPLIED TO METROLOGY

Volume 136

SESSION 5.4

HOLOGRAPHIC INTERFEROMETRY

Session Chairmen
Smigielski
Sokolov

HOLOGRAPHIC TESTING OF ASPHERICAL SURFACES

R. Mercier
Institute of Optics, University of Paris — South
B.P. 43, 91406 Orsay Cedex, France

Abstract

The use of a hologram in an interferometer in order to test an aspherical surface poses the problem of the choice between holograms with the carrier on the axis and holograms with inclined carrier. We compare the performances which we can expect from these two types of holograms in the case of an aspherical deformation of the 4th degree.

Introduction

The present development of machines to manufacture aspherical surfaces poses in more and more critical fashion the problem of their checking and verification, whether at the level of adjustment of the machines, or of control of production.

Interferometry using a synthetic hologram[1] presents the interest of combining three advantages:
  -The use of a synthetic hologram contributes a total flexibility as to the surface forms. The only limitations are quantitative.
  -We can expect to obtain the usual precision of interferometric methods.
  -This is a global test of the surface, and not solely the analysis of a meridian.

Interferometer

Figure 1 shows the scheme of the setup which we used. It can be used with any type of hologram. The role of the hologram is to transform the aspherical wave issued from the mirror being tested into a plane wave which will be compared to the reference plane wave. The aspherical setup should permit us to generate a wave which is not greatly deformed in comparison to a plane wave.

Two important points should be noted: in the first place, the use of commercial emulsions led us to adopt the solution which consists in causing the two paths to pass through the hologram, which minimizes the effects of the defects of the plate serving as support of the hologram. Furthermore, the separator is traversed only by parallel rays, and it does not disturb the propagation of the aspherical wave, which is merely reflected in remaining outside the separator.

Hologram

For the hologram, we have our choice of two solutions: in-line holograms, that is, with the carrier on the axis, and holograms with inclined carrier. To compare these two types of holograms, we will examine the following four points:
  -the ease of production and use
  -the maximal deformation which the hologram is capable of testing
  -the spatial resolution of testing
  -the precision of the testing of the form of the surface.

The first criterion is clearly in favor of the in-line hologram, at least for the surfaces of revolution. The calculation of the hologram then becomes a problem in one dimension and the interferometer becomes the equivalent of a centered system.

We can thus use conventional methods of optical adjustments.

Problem of the Filtering of In-Line Holograms

For the filtering of the orders of the hologram, it has been proposed to use a series of masks designed to eliminate the spurious orders.

Lohmann and Ichioka[2] have shown that in the case of mirrors with a central obturation, for example a primary Cassegrain mirror, it was possible to completely eliminate the spurious orders by use of a simple diaphragm. We will analyse the conditions in which such a simplified filtering can be used for any optics.

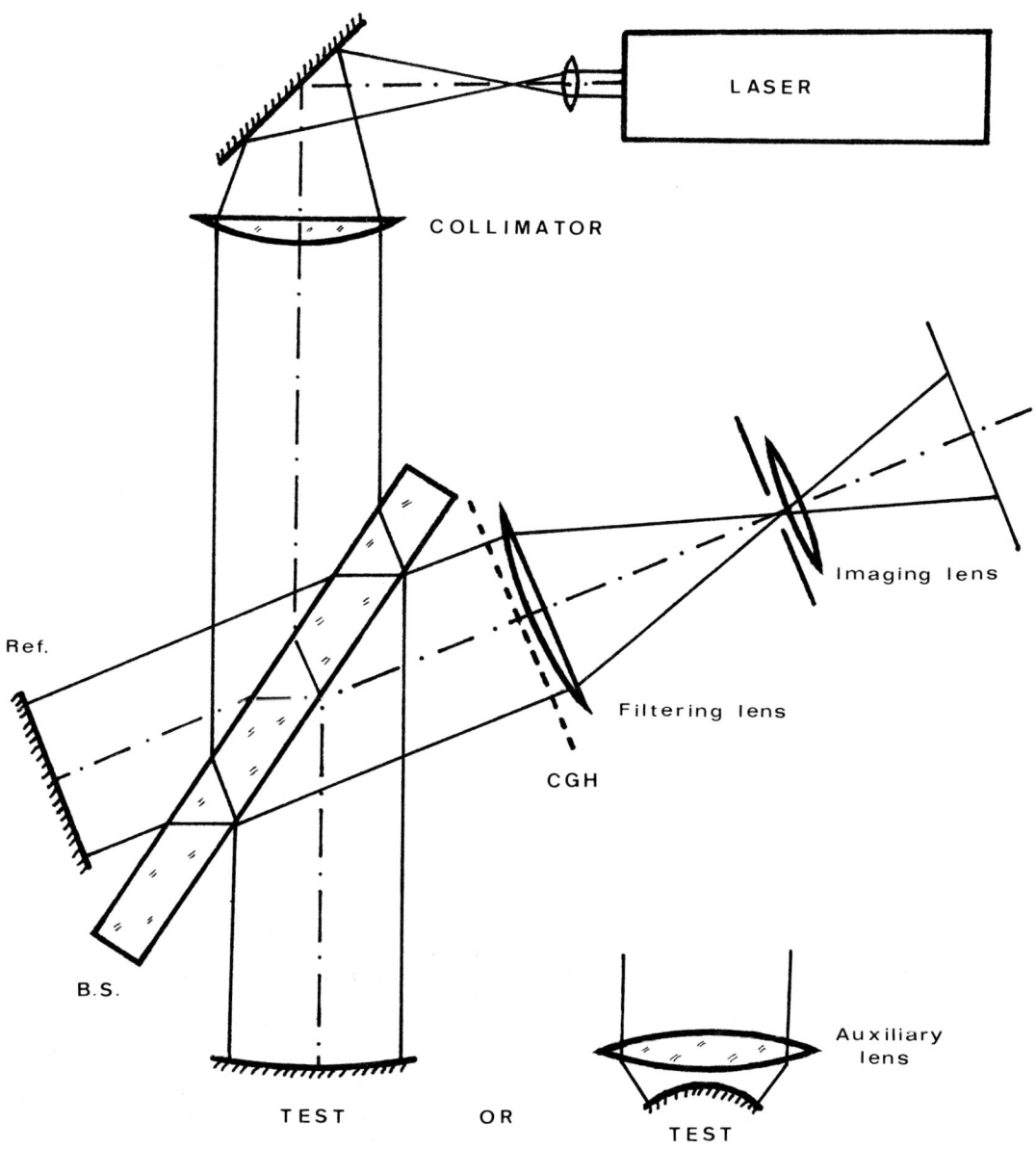

Fig. 1. Interferometer Setups

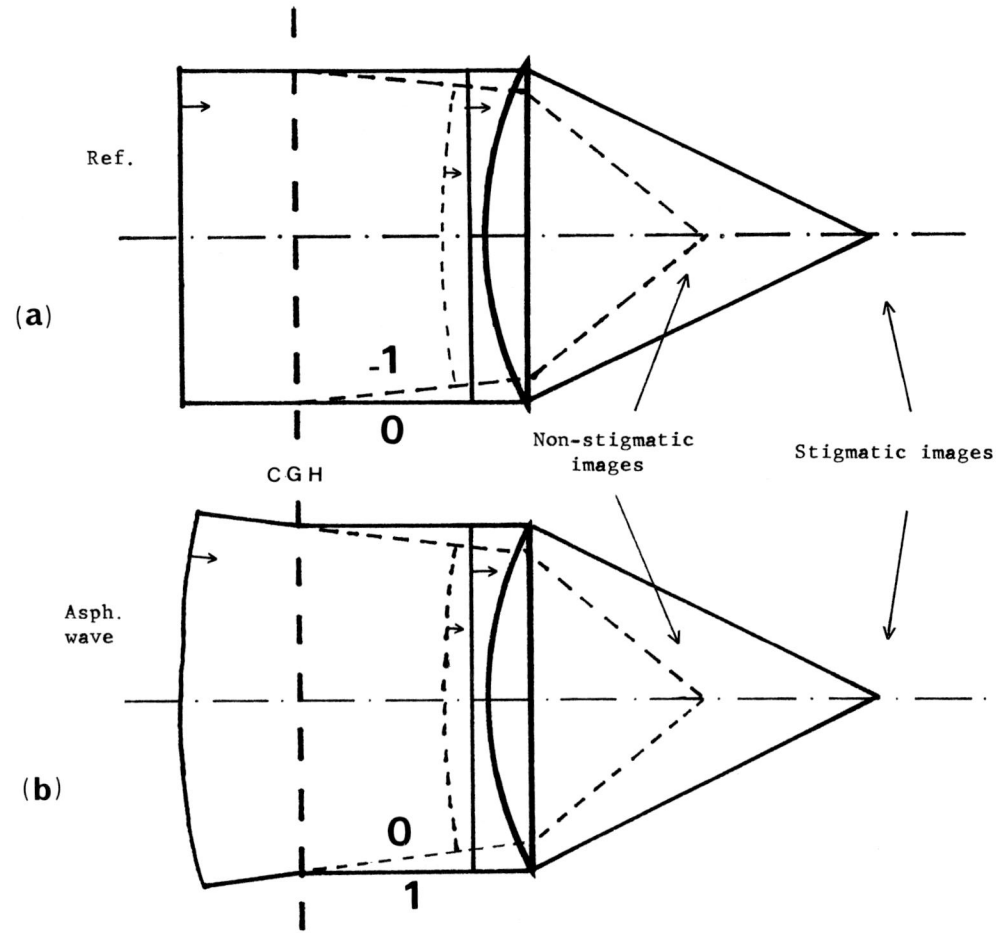

Fig. 2. Position and nature of the orders

In figure 2, we have represented the principal orders generated by the hologram. In 2a the reference, which is a plane wave, traverses the hologram in zero order and converges in a stigmatic image at the focus of the filtering lens. This reference also traverses the hologram in other orders which converge on the axis but are defocused in comparison to the preceeding one.

In 2b, the aspherical wave surface proceeding from the element being tested traverses the hologram in order 1 and gives a new stigmatic image coinciding with the zero order of the reference path. It is in this plane that we place the filtering diaphragm.

In figure 3, we find the same elements as before, as well as the filtering hole and an objective lens realizing the conjugation between the surface tested and the plane of the interferogram. The spurious orders project on the plane of the interferogram the shadow of the filtering hole, which is translated by a luminous disk at the center of the interferogram. This zone must be considered as lost for the testing and its extension will be an important parameter when compared with the carrier holograms.

HOLOGRAPHIC TESTING OF ASPHERICAL SURFACES

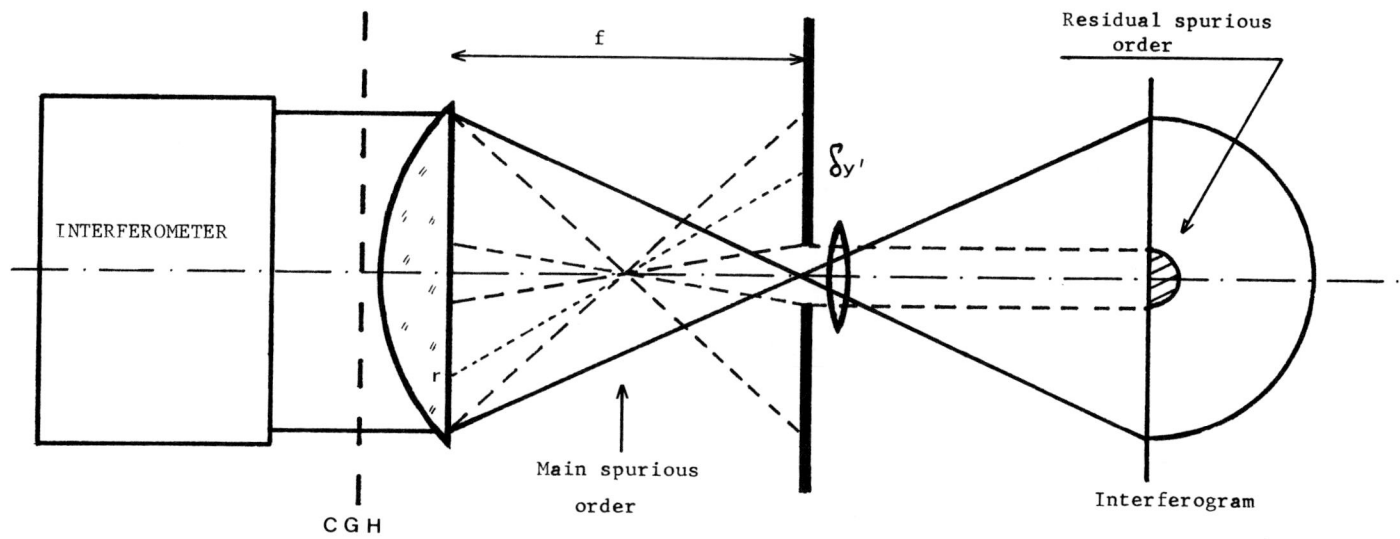

Fig. 3. Filtering of the orders

Calculation of Performances of In-Line Holograms

Non-controlled zone To calculate the size of the non- tested (non-controlled) zone, we may use certain results of the theory of aberrations: the spurious order of order k corresponds to an aberrant deviation $k\Delta$:

$$ \tag{1} $$

with: Δ_4^{Max} amplitude of the deformation of the 4th degree
     r   common radius
     R   radius of the hologram
     b   defocusing term

A given ray cuts the filtering plane at an incidence height which is none other than the transversal aberration $\delta y'$:

$$ \tag{2} $$

The extension of the non-controlled zone is deduced from the expression of $\delta y'$ as a function of r, in solving the inequality:

$$ \tag{3} $$

d = diameter of the filtering hole
r_1 = non-controlled radius
D where the non-controlled fraction: $u_1 = \dfrac{r_1}{R}$.

Effect of the diaphragm on the test The diaphragm carries out a filtering of two wave surfaces from which we wish to record the interferogram. The reference wave surface, a plane, is not greatly modified by this filtering, except at the edge where it will be deformed by fluctuations. On the contrary, the wave surface being tested contains *a priori* phase variations (information about the surface being tested) and it is this wave surface which will be affected by the filtering.

This effect is characterized by its cutoff frequency, whose expression is conventional. The lighting is, of course, coherent.

Results

In-line holograms: The mirror being tested is characterized by its radius R and by the deformation Δ_4^{Max} which it induces on the wave surface and which we suppose to be of the 4th degree.

The device serving to produce the hologram as well as the precision of control desired fix for us the minimum spacing of the hologram, and thus its inverse, which we call ν_{max}.

Finally, we accept that a fraction u_1 of the mirror not be controlled because of the spurious orders.

Calculation shows that the reduced variable ν_{cutoff}/ν_{max} depends only on the non-controlled fraction and on the variable without dimension X:

$$\tag{4}$$

Before comparing the two types of holograms, we will briefly examine the case of holograms with inclined carrier.

Holograms With Inclined Carrier

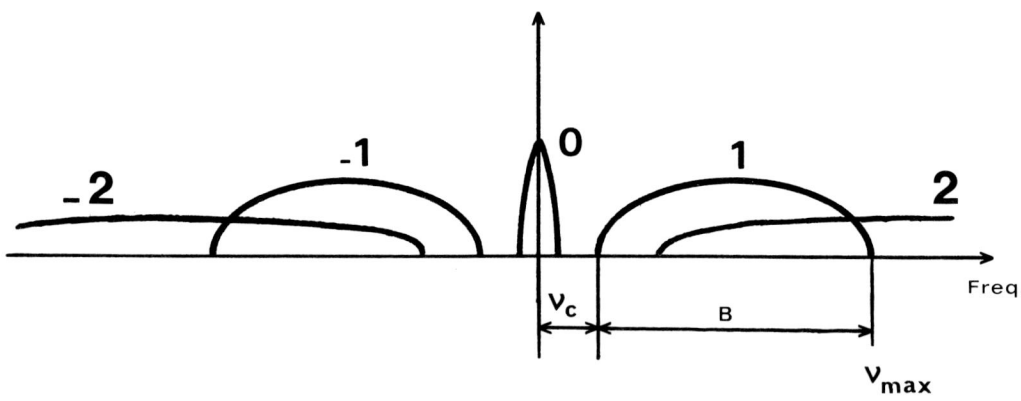

Fig. 4. Spectrum of frequencies of carrier holograms

In figure 4 we have represented the frequency spectrum of a hologram with inclined carrier (carrier frequency hologram). A spectral band of width B due to the wave surface to be compensated is displaced on both sides of the zero order by the carrier and gives two lateral bands. ν_{max} is the maximal frequency in the order 1 and always represents the inverse of the minimum spacing of the hologram. The spatial filtering is made in the plane of the spectrum and to totally eliminate the spurious orders, it will have a maximal cutoff frequency ν_2 corresponding to the space included between the orders -1 and +1. We thus obtain the expression of the cutoff frequency

(5)

Comparison of Holograms

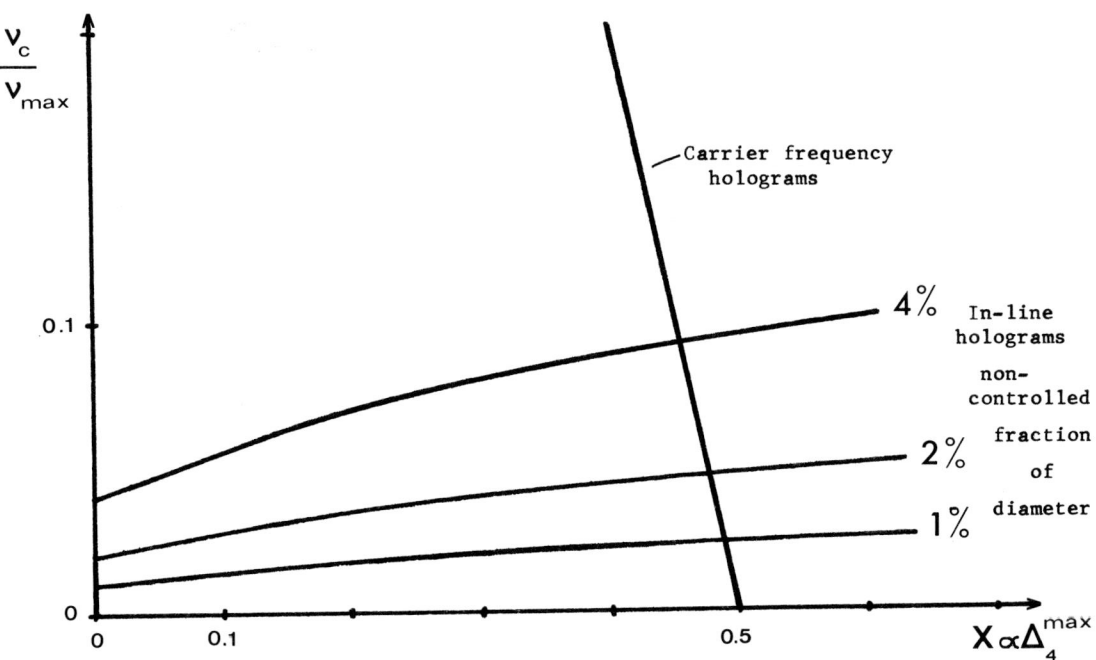

Fig. 5. Comparison of carrier frequency holograms with in-line holograms

In figure 5 we have represented the variation of the cutoff frequency on the mirror as a function of the variable X which we defined previously and which is proportional to Δ_4^{Max}.

Curve 1 concerns the carrier frequency holograms. The other curves correspond to the in-line holograms, for the different values of the non-controlled fraction u_1.

We see that, contrary to the case of carrier frequency holograms, the cutoff frequency increases with the deformation in the case of in-line holograms.

The examination of these curves shows that in-line holograms are superior to carrier holograms for large deformations. For small deformations, we should not lose sight of the specific advantages of in-line holograms.

It should also be noted that, given the large values which we will be led to give to ν_{max}, a relative cutoff frequency of 0.05 corresponds to satisfactory values for the spatial resolution, in general between 2/10 mm and 2 mm.

We ran trials testing a spherical mirror of small curvature which, over a diameter of 40 mm, is considered a surface deformed by 50 μ with regard to a plane. The hologram was traced large-scale on a Benson tracer and then photoreduced 18 times to obtain a diameter of 40 mm. The non-controlled fraction is 7% and the cutoff frequency at the level of the surface is 1 mm^{-1}.

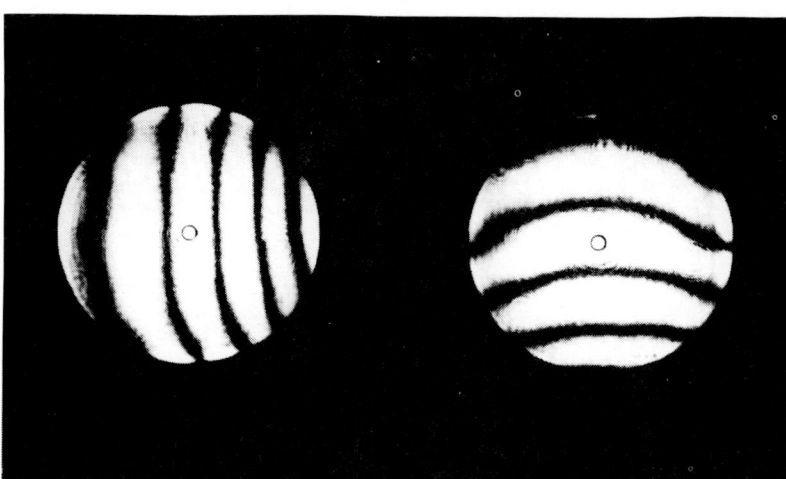

Fig. 6. Interferogram

Mechanical deformations of the paper when the hologram was traced introduced an astigmatism on the wave surface generated. The interferograms in figure 6 were taken in two perpendicular directions, for a single position of the hologram, which compensated on the average the wave surface emanating from the mirror.

This astigmatism would lead to an error of measurement of $\pm \lambda/4$ on the surface being controlled. A better choice of material should permit us to reduce it to acceptable values. The local fluctuations of the fringes are on the order of $\lambda/10$ (that is, $\pm \lambda/20$). These trials lead us to think that with the material cited we should be able to check, with reasonable precision, an aspherically deformed surface of around a hundred μ with regard to the best sphere, over a diameter of 80 mm^{-1} and a non-controlled fraction of 2%. Over a diameter of 40 mm, we could check a deformation of 57 μ with regard to the best sphere.

The difficulties of the tracing, the very sensitive character of the adjustments at the photoreduction and the desire to increase the amplitude of the controllable deformations lead us to envisage the use of a synthetisizer microdensitometer, which would permit us to produce the hologram directly with magnification unity.

References

1. Wolf, E., "Comparing a Surface With a Hologram", Progress in Optics XIII, ch. IV, #5.
2. Lohmann, A.W., Ichioka, Y., 1972, Appl. Opt. 11, 2597.

HETERODYNE HOLOGRAPHIC INTERFEROMETRY: A REVIEW

René Dändliker
Brown Boveri Research Center
CH-5405 Baden, Switzerland

Heterodyne holographic interferometry is a combination of holographic interferometry and opto-electronic fringe evaluation with the following outstanding properties[1,2]:

- fringe interpolation to better than 10^{-3} of a fringe ($\pm$ 0.3° for the interference phase);
- measurement with the same accuracy at any desired position in the image, therefore high spatial resolution (> 100x100 points);
- independent of brightness variations across the image;
- inherently direction sensitive, i.e. increase and decrease of interference phase can be distinguished;
- computer readable output both for position and phase easily obtained (allows on-line data processing);
- inherently less sensitive to speckle noise than fringe intensity measurements.

The heterodyne technique can be applied to nearly all known kinds of holographic interferometry, except to time average holograms. The list includes:

- real time holographic interferometry with high temporal resolution and accuracy;
- double exposure holography (recorded with two reference beams);
- vibration analysis from multiple exposure holographic interferometry recorded with stroboscopic illumination;
- depth contouring using dual illumination source or dual wavelength recording.

More detailed informations on both the experimental technique of heterodyne holographic interferometry and its application to strain and stress determination through double exposure holography are given in the following publications.

1. Dändliker, R., Ineichen, B., Mottier, F. M., "High resolution hologram interferometry by electronic phase measurement", Opt. Commun., Vol. 9, pp. 412-416. 1973.
2. Dändliker, R., Ineichen, B., Mottier, F. M., "Electronic processing of holographic interferograms", Proc. International Computing Conference, IEEE Inc., New York, 1974, pp. 69-72.
3. Dändliker, R., Marom, E., Mottier, F. M., "Two-reference-beam holographic interferometry", J. Opt. Soc. Am., Vol. 66, pp. 23-30. 1976.
4. Dändliker, R., Ineichen, B., "Nonlinear cross-talk in two-reference-beam holographic interferometry", Optics Commun., Vol. 19, pp. 365-369. 1976.
5. Dändliker, R., Eliasson, B., Ineichen, B., Mottier, F. M., "Quantitative determination of bending and torsion through holographic interferometry", The Engineering Uses of Coherent Optics, Cambridge University Press, Cambridge, 1976, pp. 99-117.
6. Dändliker, R., "Quantitative strain measurement through holographic interferometry", Applications of Holography and Optical Data Processing, Pergamon Press, Oxford, 1977, pp. 169-181.
7. Ineichen, B., Dändliker, R., Mastner, J., "Accuracy and reproducibility of heterodyne holographic interferometry", Applications of Holography and Optical Data Processing, Pergamon Press, Oxford, 1977, pp. 207-212.
8. Dändliker, R., Ineichen, B., "Quantitative strain measurement through holographic interferometry', SPIE vol. 99 - 3rd European Electro-Optics Conf., Soc. Photo-Optical Instr. Eng., Washington, 1977, pp. 90-98.

1st EUROPEAN CONGRESS ON OPTICS APPLIED TO METROLOGY

Volume 136

SESSION 6.1

SPECKLE

Session Chairmen
Ennos
Patanchon

DETERMINING THE INCLINATION OF A DIFFUSING SURFACE WITH REGARD TO VIEWING DIRECTION BY SPECKLE PHOTOGRAPHY

J. Montilla, R. Hernández
Instituto de Optica, Madrid-6, Spain

and

M. García
Facultad de Físicas, Universidad Complutense, Madrid-3, Spain

Abstract

A speckle-pattern image method is used to determine the inclination with regard to viewing direction of a diffusing plane surface. We recorded on a photographic plate a defocused double exposure image, illuminating along the viewing direction with two neighbouring wavelengths. In the experiments we have used, too, diffusing plane dihedrons illuminated in this way. Good agreement with the equation founded is observed.

Introduction

In the present work an optical method, based on recording speckle-pattern image was used. The method permits the measurement of a diffusing surface's inclination with respect to viewing direction.

Illuminating along the viewing direction with two wavelengths λ_1 and λ_2, two images of a diffusing plane are successively recorded on a photographic plate. Apart from the decorrelation effect, reported by several authors (1), (2), (3), it can be show that, under certain approximations, the two recorded structures are homothetic.

Spectral analysis of the plate, after it has been processed under usual linearity conditions, exhibits a set of Young's fringes from whose spacing the inclination of the diffusing surface with the viewing direction may be determined.

Bases of the method

The phenomenon is schematized in figure 1. A change of wavelenght of incident light gives rise to both a decorrelation and a radial shift between the two structures; it can be demonstrated that the latter shift is responsible for the homotethy between the two intensity distributions. As a consequence the homologues single speckles undergo a displacement, $a = x \Delta\lambda / \lambda_2$, which is directly proportional to the relative variation of the wavelength and depends on the distance "x". Therefore the projection of "a" over the plate plane is a function of wavelenght variation, the de-focus distance and the inclination "i" that the surface has respect to the illuminating direction. Simple geometrical considerations, for unity magnification, lead to

$$i = \frac{1}{4} \text{arc sin} \frac{2b}{d} \frac{\lambda_2}{\Delta\lambda} \qquad \{1\}$$

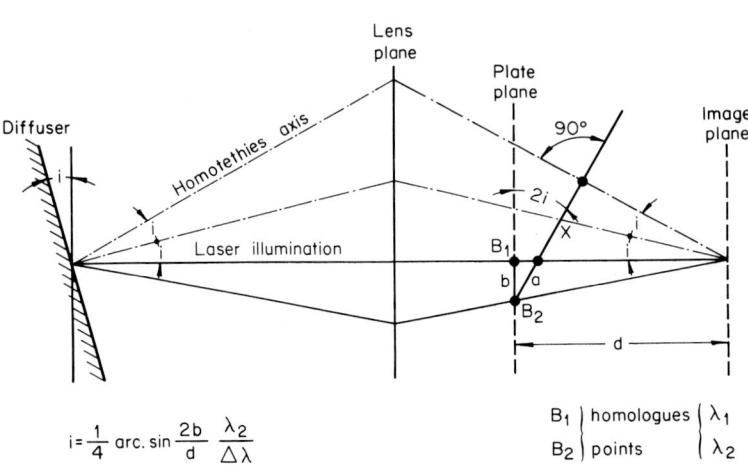

Fig. 1. Geometrial diagram for explaining the method.

where "i" is referred only to points neighbouring the center of the plate, "b" is the homologues single speckles displacement on the plane of the plate and "d" is the de-focus distance.

Recording of the speckle patterns

A plane diffusing surface (fig. 2) is illuminated by a coherent and monochromatic beam, produced by an argon ion laser, by means of a beam-splitter orientated at 45°. This permits the observation of back-scattering light. In order to obtain selected inclinations, the diffuser may be rotated about an in-plane axis perpendicular to the incident direction. The surface is imaged by a aberration free camera lens so that it is focused on a plane a predetermined distance from the surface. A defocused double exposure image of the surface is recorded. Between exposures the wavelength is changed without modifying the inclination of the object.

The diffusers are gypsum-model covered with aluminium diffusing paint. An aberration free camera lens at f/11 of 170 mm focal length was used. Magnification was, in every case, close to unity.

A double exposed speckle-pattern image of the surface is recorded, for a given value of de-focus distance, by illuminating a beam of wavelength $\lambda_1 = 496.5$ nm and $\lambda_2 = 501.7$ nm, from an argon ion laser, without modifying the diffuser position. Finally the plate is processed under linearity conditions.

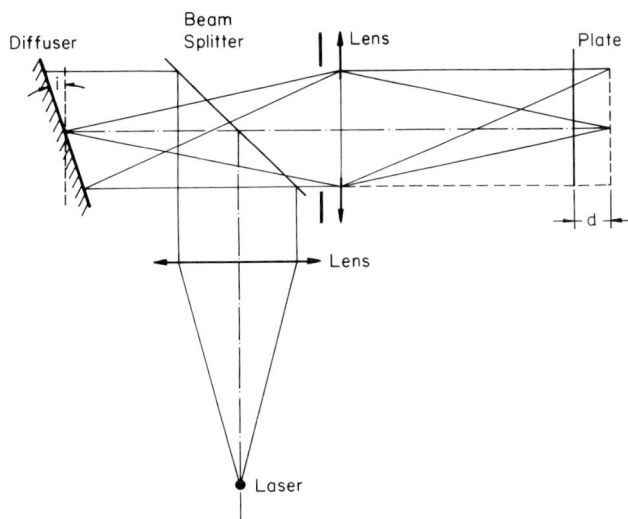

Fig. 2. Schematic arrangement for recording.

Spectral Analysis

The processed film is illuminated by a collimated beam of coherent light. The autocorrelation halo of the optical system, modulated by Young's fringes, is observed on the focal plane of a lens. The parameter "b" can then be calculated from the measurement of the fringe spacing. Finally, the inclination "i" can be then also obtained for given values of spectral interval and de-focus distance "d".

If the diffuser object presents different inclinations, a pointwise filtering on the Fourier plane allows that fringe patterns representing the loci of points of equal displacement, say, the loci of points having the equal inclination are manifestly evidenced.

In order to obtain acceptable fringe visibility, the decorrelation effect ocurring between speckles must be taken into account since the roughness of the object determines the useful spectral interval. As our optical system is aberration free and has a small aperture, the decorrelation effect of this system is not considered.

Experimental results

Measurements were made fitting degree to degree inclinations, in an interval of -15° and +15°. Recordings were carried out at spectral intervals of 5.2 nm and 8.5 nm, de-focus distance values were in the 5-20 mm range. From measured fringe spacing values, the application of Eq. {1} yielded values for the inclination "i" in closed agreement with those being applied.

Photograph 3 (a) shows the fringe pattern obtained by illuminating the whole plate for the plane diffuser at an inclination of 8° with regard to the incident direction and a de-focus distance of 10 mm.

Photograph 3 (b) shows the fringe pattern obtained where only inclination is varied (4°).

Photograph 3 (c) shows the fringe pattern obtained when the diffuser is oriented approximately along incident direction (1°).

 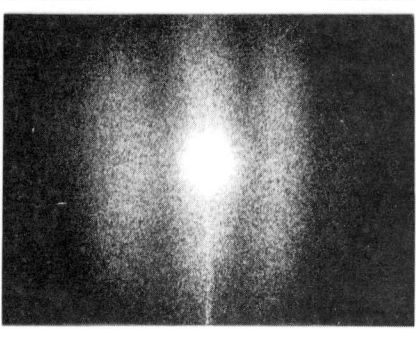 

(a) 8° (b) 4° (c) near incident direction, 1°

Fig. 3. Fringe patterns from whole field filtering for plane diffuser for three inclinations with regard to viewing direction. $\Delta\lambda$ = 5.2 nm. d = 10 mm.

We now turn to diffusers that present two inclinations. Photographs 4 (a) and 4 (c) show the fringe patterns corresponding, respectively, to a convex dihedron of 174° and to a concave dihedron of 174°. The two faces are symmetrically oriented with respect to the incident direction. It can be observed that while the curvature of the fringes are reversed in the two cases, the spacing of the fringe patterns are identical.

The relationship between the fringe spacing and inclination is apparent in photograph 4(b) obtained from the recording of a convex dihedron with an asymmetrical orientation. When this asymmetry is sufficiently acute, the orientation of the two faces became of the same kind, giving rise to an inversion of the curvature of one of the two patterns involved. This is the situation represented in photograph 4(d), which shows the recording of a concave dihedron.

Sensitivity of the Method

An improvement in sensitivity may be obtained by amplifying the spectral interval and the de-focus distance. The maximum spectral interval which can be used is determined by the roughness of the object which causes a certain degree of the decorrelation between the speckles. With our models fringe contrast is still acceptable for a relative wavelength variation of 8.5 nm. Maximum de-focus distance is limited by object-image correspondance; i.e., by amplification of the object surface integration area. Measuring techniques used insured precision of better than 30 minutes of arc.

Conclusion

This method, we believe, permits determination of diffuser object form and calculation of level curves by means of adequate computation procedures.

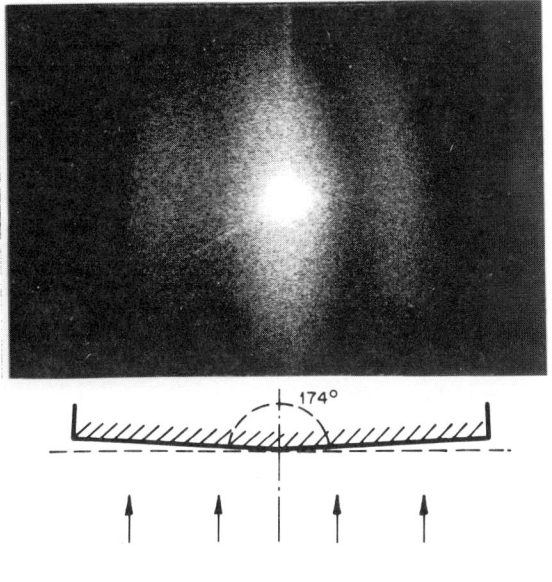

(a) Convex dihedron with a symmetrical orientation

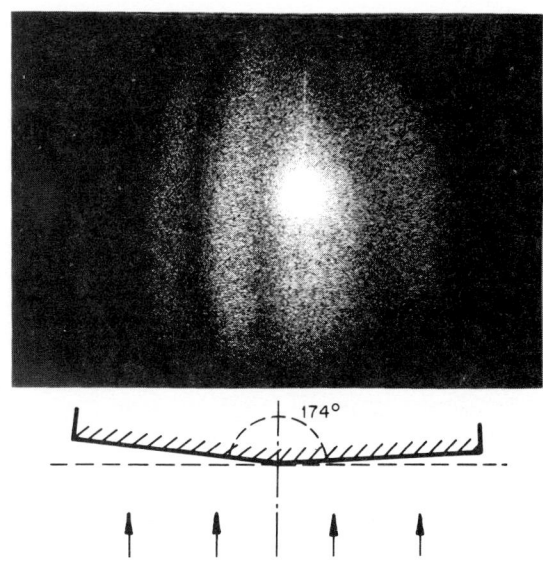

(b) Convex dihedron with an asymmetrical orientation

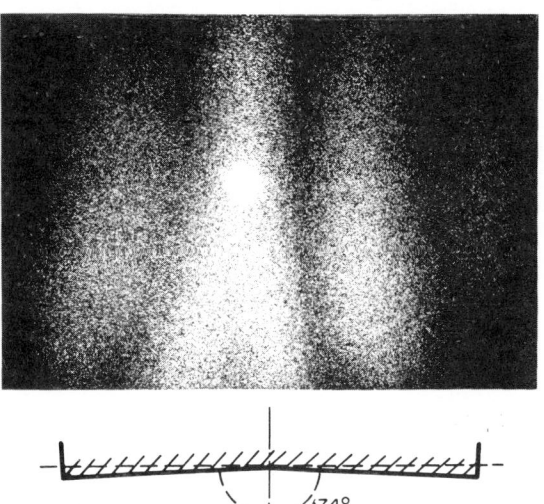

(c) Concave dihedron with a symmetrical orientation

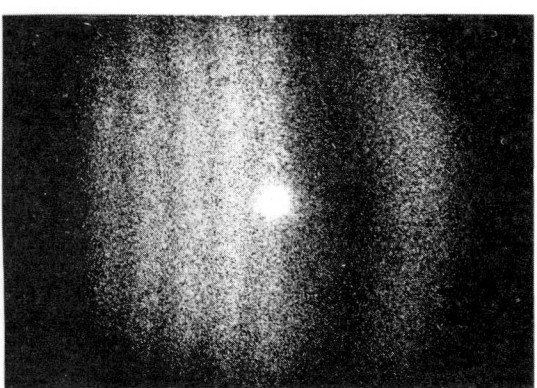

(d) Concave dihedron with acute asymmetry

Fig. 4. Fringe patterns from whole field filtering for diffusing dihedrons of 174°. $\Delta\lambda$ = 5.2 nm. d = 10 mm.

References

(1). G. Parry, Opt. Commun., 11, 172 (1974).
(2). G. Tribillon, Opt. Commun., 11, 172 (1974)
(3). J.A. Méndez and M.L. Roblin, Opt. Commun., 13, 2 (1975)

HOLOGRAPHIC METHODS MADE USEFUL BY PHASE MODULATED ESPI

Ole J. Løkberg and Kåre Høgmoen
Physics Department, The Norwegian Institute of Technology
N-7034 Trondheim-NTH, Norway

Electronic speckle pattern interferometry - ESPI

The use of speckle pattern interferometry to detect vibration modes was reported by Archbold et al. [1], and by several others [2,3] to detect vibration amplitudes. TV detection and electronic filtering have been added [4-6] to give a more flexible system: ESPI.

The main components of our ESPI setup are shown in Fig. 1. The object is imaged by two lenses onto the target of the TV camera. The reference wave is reflected by two mirrors, filtered, expanded, and superimposed in-line with the object wave. The video signal from the TV camera is filtered and rectified, and the processed image interferogram is presented on the TV monitor. The modulating mirror provides phase modulation of the reference wave. The supporting electronics facilitate control of the object excitation, amplitude and phase of the reference modulation, and it also includes a system for photoelectric measurement of vibrations directly from the monitor using lock-in technique [7].

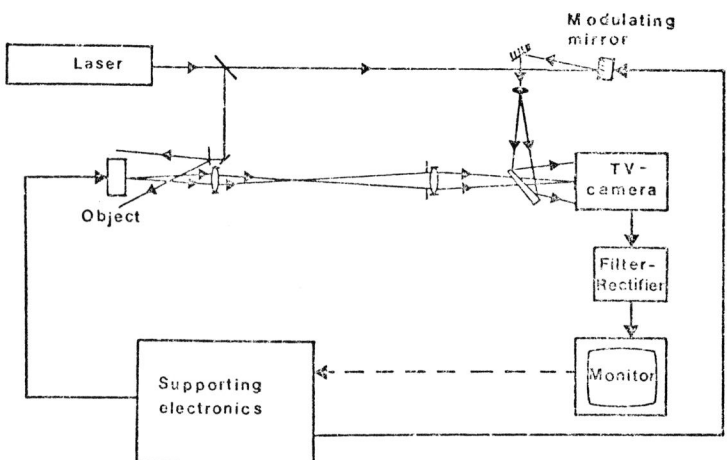

Fig. 1. ESPI setup

The ESPI has two major advantages as holographic system: (1) Short exposure time, 1/25 sec. (2) High repetition rate, 25 Hz. Therefore ESPI is very insensitive towards vibrational noise and object instabilities. The main disadvantage compared to ordinary holographic interferometry is the reduced picture quality, which is mainly due to the low resolution of the TV target, only about 25 lines/mm, while hologram films can resolve 2-3000 lines/mm. This and the generally high noise level have resulted in a rather narrow measuring range of ESPI in its ordinary time-average mode, only about one decade from 0.06 μm to 0.6 μm.

We have worked on extending this range by means of sinusoidal phase modulation of the reference wave in the ESPI system [7,8]. The reference wave modulation technique is well known in holographic literature [9], however, for practical uses of the technique the instantaneous presentation of interferograms as by ESPI is of great importance. We normally use sinusoidal phase modulation of the reference wave at the vibration frequency of the object, and we can adjust the modulation amplitude and the phase relative to the object vibration. To obtain a continuous phase variation we use a technique with a slightly shifted modulation frequency.

The reference modulation shifts the zero level for our vibration measurement, since we always detect the relative movement between the object and reference waves. The effect of the modulation is illustrated in Fig. 2, which shows fringe function (intensity versus vibration amplitude) without (fully drawn) and with phase modulation (dotted).

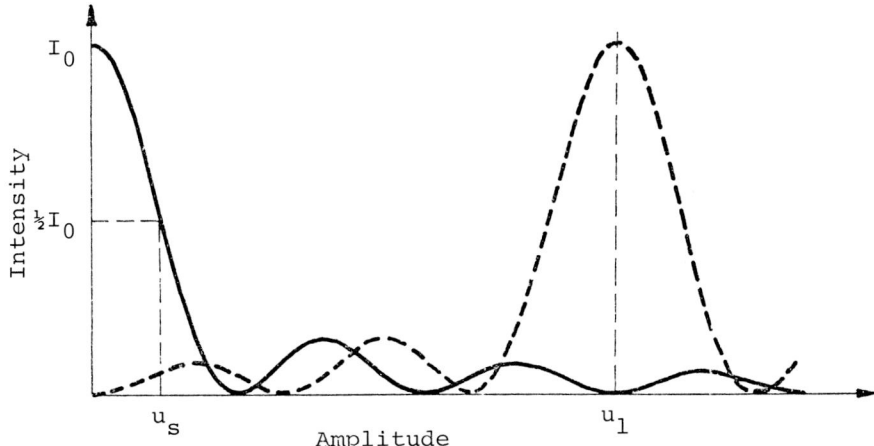

Fig. 2. Fringe functions by ESPI. Modulation amplitude $u_r = u_s, u_l$ by detection of smaller and larger amplitudes, respectively.

Extending the upper amplitude range. Phase measurement

By time-average ESPI the zero order fringe is easily detected as it is much brighter than the other fringes. By detection of the shifted zero order fringe we have extended the upper amplitude limit by ESPI up to 10 μm [8]. The technique is also useful for obtaining detailed vibration pictures at lower amplitude levels, especially on objects that give much optical noise with resulting low contrast in the fringe pattern. A variation of this technique can be used to separate the vibration phase distribution from the amplitude distribution [10]. We obtain the phase information as either lines or areas of constant vibration phase. The phase and amplitude contours together present the object's motion in a way which is very useful both for the exact analysis and for the intuitive understanding of the dynamics of the motion.

Extending the lower amplitude range

In many applications of ESPI, the detection and measurement of very small amplitudes are either required or desired. An example is vibrations in the ear, where the amplitudes are typically less than 0.1 μm at physiological sound pressure levels. To detect and measure vibrations in this range, we use a modulation amplitude that shifts the zero point for the vibration measurement to the steepest part of the fringe function, as indicated in Fig. 2. Here we have maximum intensity change as function of amplitude, and thus we have maximum sensitivity for small amplitude detection. The use of a modulation frequency that is slightly different from the vibration frequency results in a time-varying intensity beat in the displayed interferograms. The amplitude and phase of the beat signal are directly related to the object vibration. By visual observation we have detected amplitudes down to 20 Å, while we can study the deformation of a vibrating object in slow-motion when the amplitudes are in the range above 80 Å. The latter feature is especially valuable when we make a frequency scan to detect vibration anomalies, and to evaluate complex vibration modes.

By measuring the intensity beat at a selected object point on the monitor by photodetection and lock-in technique, we can get absolute amplitude and phase values. On a favourable test object we have measured amplitudes down to 0.1 Å. By this method for small vibration measurement there are two beneficial averaging processes of great importance: (1) The photodetector averages over many speckle grains. (2) The lock-in amplifier averages over many interferograms, and over many beat periods. These, and the other features of the ESPI, remove the need for fringe stabilization systems by this interferometer, and make possible mesurements on unstable objects where effective fringe stabilization cannot be performed. By this method we have the possibility of using a video tape recorder to store the interferograms and a corresponding reference signal for the lock-in amplifier. Then, at a later processing stage, we can select different object points and analyse their motions. This feature is especially valuable when we examine objects like human ear preparations, where aging effects limit the time available for the experiments.

Applications

We have found time-average ESPI, combined with the techniques described in this paper, extremely useful for a wide range of vibration measurement problems.

The main part of the work so far has been done on vibration analysis of human ear preparations, which are very difficult objects to use in any interferometer setup.

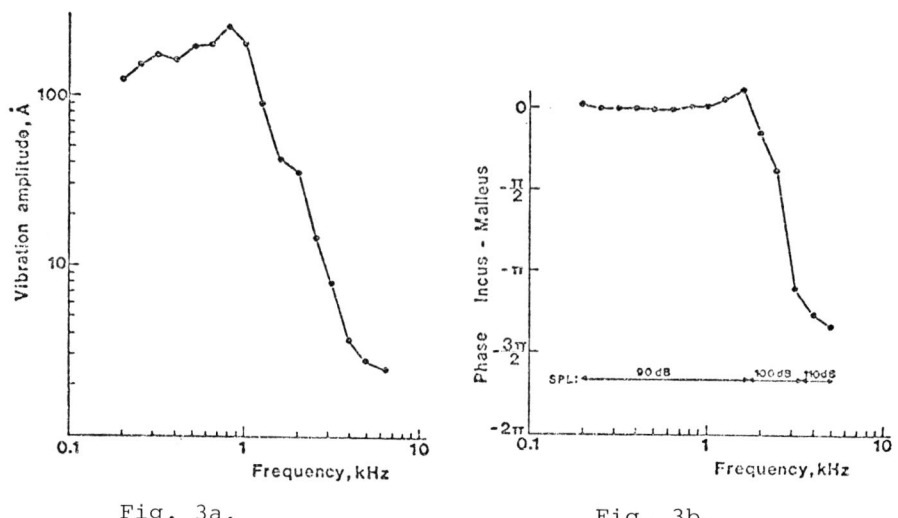

Fig. 3a. Fig. 3b.

Figures 3a,b show amplitude and phase versus frequency of the hammer (the phase is refered to the motion of the anvil). The measurements were done by first recording the vibrations on videotape and thereafter by measuring on the stored pictures. The photoelectric measurement could be performed even when we sprayed water on the preparation which made the individual speckles boil heavily. Visually we have also observed vibration anomalies in the vibrations of the drum membrane which were not detectable by ordinary time-average ESPI.

Complete vibration analysis of loudspeakers by use of holography usually presents problems due to their complex vibratory behaviour and their mechanical instability. We have found it more convenient to use ESPI with the slow-motion presentation at low amplitude levels to examine the overall performance of the loudspeakers, and thereafter plot the exact phase and amplitude distributions at interesting frequencies.

We have used our ESPI setup to detect and evaluate critical resonances in gasturbine wheels. By use of an Ar-laser the movements of the entire wheel could be seen on the monitor. However, for exact vibration mapping we examined single turbine blades and then it was sufficient to use a 20 mW HeNe-laser. In one set of experiments the blades were

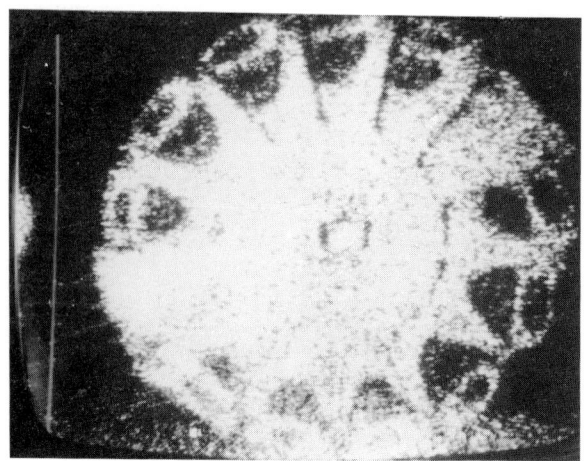 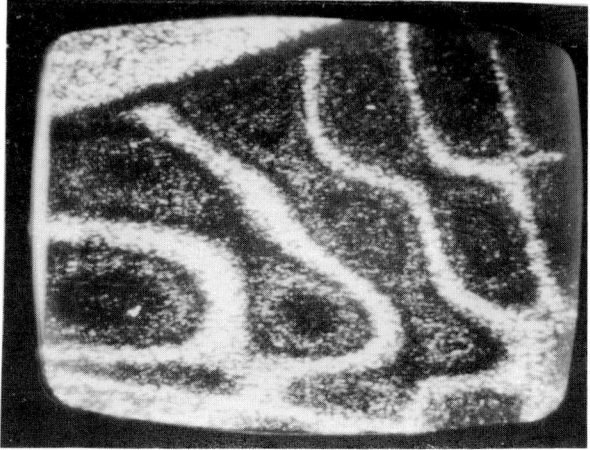

Fig. 4a. Turbine wheel vibrating at 6.5 kHz   Fig. 4b. Turbine blade vibrating at 14.5 kHz

heated to above 500 °C by a gas burner to simulate working conditions. Routine testing of gasturbine wheels is now done in factory environments. Figure 4a shows a vibratory pattern of an entire wheel. Here, it is difficult to resolve the fringes in the antinodes because of the limited resolution of the TV system. However, by using the slow-motion technique we get valuable information about the coupling between different blades. Figure 4b gives an example from a more detailed measurement of a single blade.

Another interesting project is concerned with objects vibrating under water like, e.g., sonar transducers. Contrary to holography, we did not have to make the reference beam traverse the same path in water as the object beam. Acceptable picture quality was obtained with the object submerged into 1 m of rather muddy water; see Figs. 5a,b.

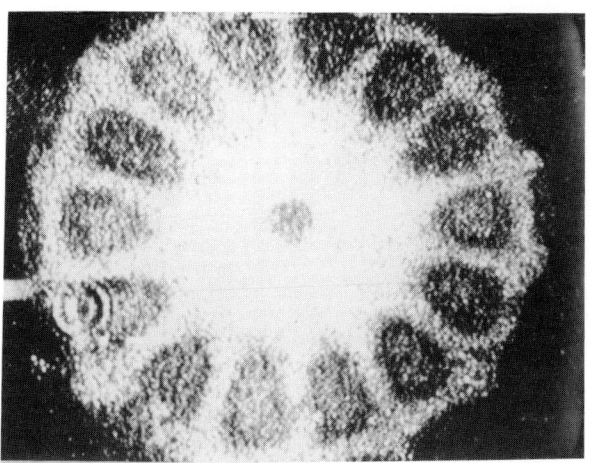
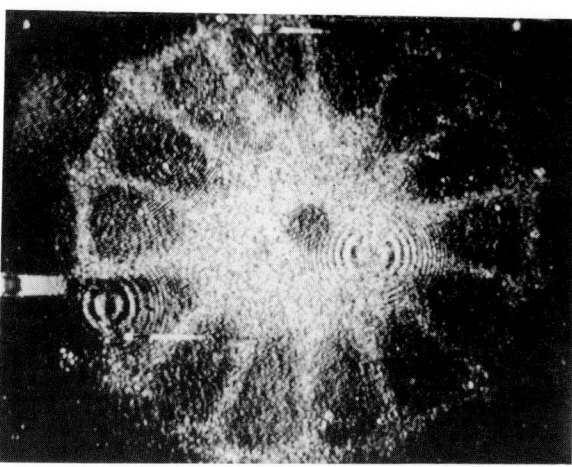

Fig. 5a. Plate vibrating at 20 kHz in clear water.

Fig. 5b. Same as 5a, but in muddy water.

Right now we are working in the high frequency region to measure surface wave vibrations. That means that we must extend our modulation technique to the MHz-region where the ability to detect and measure very small vibrations is really needed.

Conclusion

By combining sinusoidal phase modulation with ESPI, we have obtained an amplitude measuring range of 0.1 Å to 10 µm, which represents a major improvement compared to ordinary time-average ESPI. Essentially, the same modulation technique has been used to map phase distributions across vibrating objects. The favourable characteristics of ESPI — high sampling rate and short exposure time — are vital by use of the modulation technique and have made this instrument very convenient for practical vibration analysis.

References

1. E. Archbold et al., Nature 222, 263 (1969).
2. K.A. Stetson, Opt. Laser Technol. 2, 179 (1970).
3. L. Ek and N.-E. Molin, Opt. Commun. 2, 419 (1971).
4. J.N. Butters and J.A. Leendertz, Opt. Laser Technol. 3, 26 (1971).
5. A. Macovski et al., Appl. Opt. 10, 2722 (1971).
6. O. Schwomma, Austrian pat. no. 298830 (1972).
7. K. Høgmoen and O.J. Løkberg, Appl. Opt. 16, 1869 (1977).
8. O.J. Løkberg and K. Høgmoen, J. Phys. E: Sci. Instrum. 9, 847 (1976).
9. C.C. Aleksoff, in Holographic Nondestructive Testing, R.K. Erf, Ed. (Academic, New York, 1974), pp. 247.
10. O.J. Løkberg and K. Høgmoen, Appl. Opt. 15, 2701 (1976).

NEW POSSIBILITIES OF REAL-TIME INTERFEROMETRY WITH PHOTOCONDUCTIVE ELECTRO-OPTIC CRYSTALS $Bi_{12} SiO_{20}$

J. P. Huignard, J. P. Herriau
Thomson-CSF
Laboratoire Central de Recherches
BP.10 — 91401 . Orsay ; France

Abstract

Holographic interferometry is a nondestructive testing technique with a growing number of industrial applications. Photosensitive plates used for coherent wavefront recording generally require processing time, and it seems of particular interest to have new reusable materials allowing in situ interferogram writing and erasure with a recording energy comparable with high resolution photographic plates. We report in this conference a particular application of highly sensitive photoconductive cubic crystals $Bi_{12} SiO_{20}$ (BSO) to double-exposure real-time interferometry.

Development

BSO crystals provide the best known photorefractive sensitivity for phase volume hologram recording ; and these crystals exhibit no spatial resolution limitation when holographic fringes are perpendicular to the external applied field direction (transverse electrooptic configuration) (Ref.1). Phase hologram recording arises from the drift of photocarriers under fringe illumination and, after charge migration and trapping, the photoinduced space charge field modulates the refractive index via the electrooptic effect.

The required energy for hologram recording at argon laser lines with external applied field E_0 in the (110) direction is found equivalent to holographic silver halide plates (BSO : $S^{-1} \simeq 130 \mu J/cm^{-2}$ for 10^{-2} ; $E_0 = 6kV/cm^{-1}$). In accordance with theoretical models already proposed for hologram storage by space charge modulation in electrooptic crystals, [2-5] the efficient charge transport mechanism in these photoconductive crystals is consistent with an improvement of the recording sensitivity of about three orders of magnitude over other materials such as iron-doped $LiNbO_3$: $S^{-1} \simeq 200 mJ/cm^{-2}$ for $\eta \simeq 10^{-2}$). Ultimate photorefractive sensitivity is reached as already reported, in KTN,[8] but BSO crystals offer the advantage of linear electro-optic effect and optical quality in large size crystals. Hologram erasure is achieved by space charge relaxation under uniform illumination with the reference beam. Typical recording-erasure cycles in BSO crystals used for the present application at a given value of external applied field is given in Fig.1. These cycles are nearly symmetrical ; and, in agreement with Young's treatment in Ref.2, the photoinduced space charge at the initial stage of hologram recording becomes field independent for $E_0 = 6kV/cm^{-1}$.

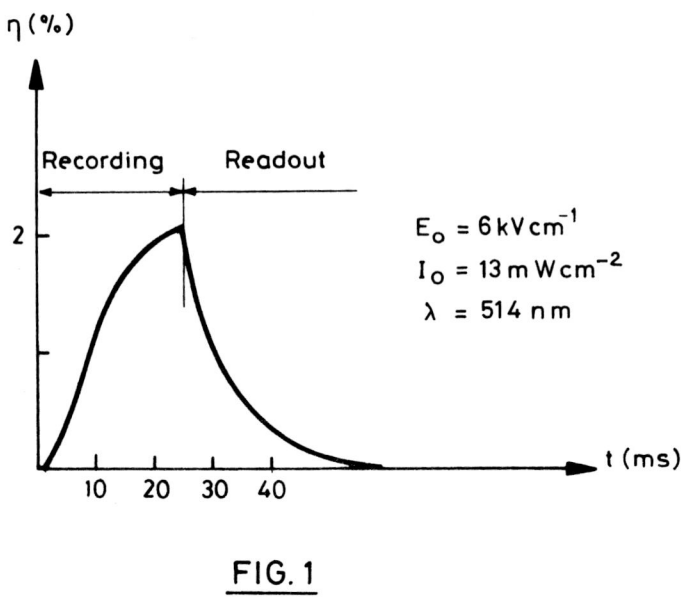

FIG. 1

Fig.1 - Typical experimental recording-erasure cycle with $Bi_{12} SiO_{20}$ (incident power = 13mW cm^{-2} λ = 488 nm, crystal size = 10x10x3mm, and fringe spacing = 2.5 μ) -

Since no fixing process has been found up to now in these crystals and no wavelength change is allowed for image reconstruction from a complex volume hologram, optical readout is destructive. Nevertheless, quasi-permanent observation of the stored information is achieved by projecting the temporarily reconstructed image on a vidicon memory tube (model TH-CSF . TH.7501).

Application of the photosensitive crystal

Real-time double-exposure interferometry is demonstrated with the holographic setup in Fig.2. The experimental procedure for interferogram observation with an incident power of $13mW/cm^{-2}$ and permanent applied field $E_o = 6kV\ cm^{-1}$ is as follows :

Phase 1 - Holographic recording of object A in BSO crystal. Recording time for saturation efficiency : $t_o \simeq 30msec$.

Phase 2 - Holographic recording of deformed wavefront from object A. Recording time : $t_o/2 \simeq 15$ msec. This sequence only provides a partial erasure of original wavefront A.

Phase 3 - Readout and erasure with reference beam. Projection of the reconstructed interferometric image on the vidicon memory tube and display on a television monitor. Readout time $\simeq 40$ msec. Storage time constant on the memory tube is about 20 min.

After this sequence the storage crystal can be recycled, and no fatigue effect to reversibility is noted.

Experimental results are given in Fig.3. Object under test is a thermal index gradient induced by a transistor in a radiator (Fig.3).

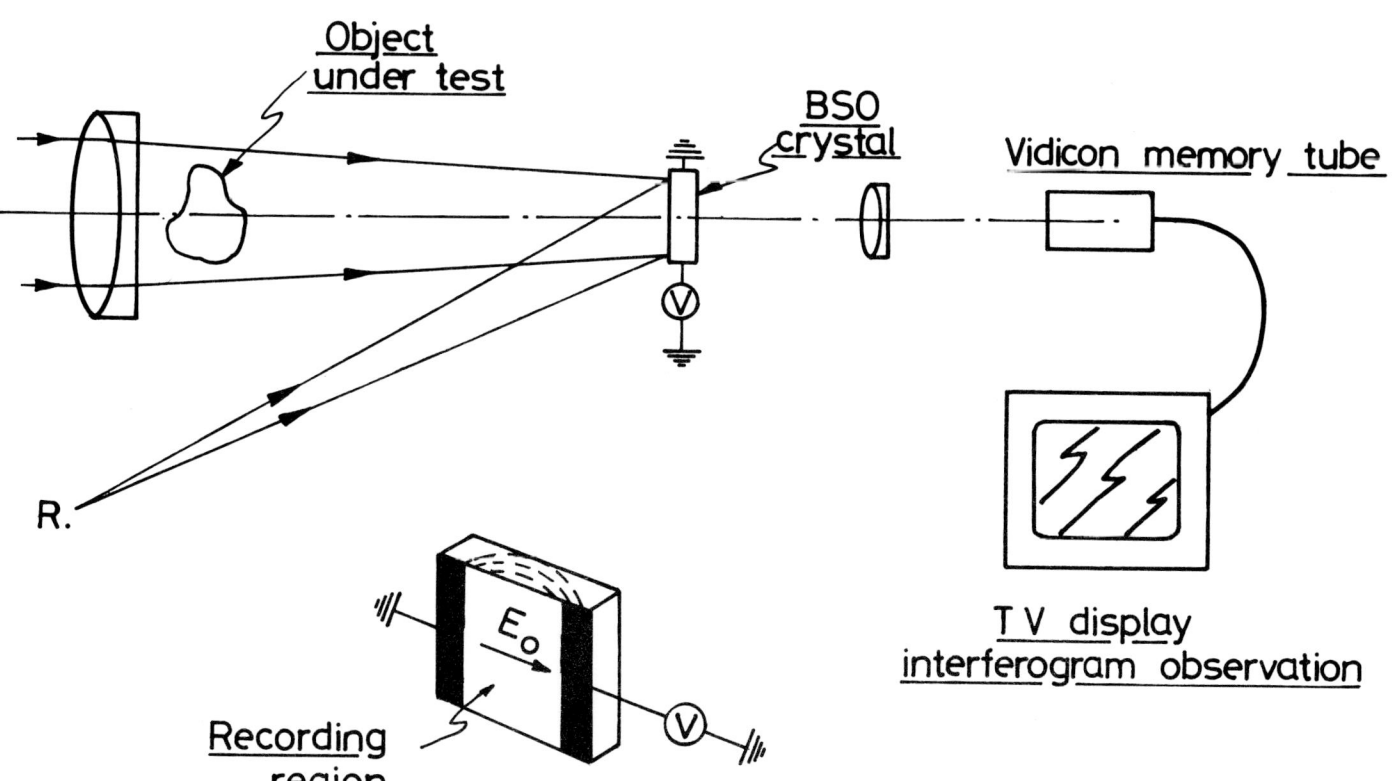

Fig.2 - Holographic setup for real-time double-exposure interferometry with image projection on vidicon memory tube. Incident power = $13mW/cm^{-2}$; applied voltage = 6kV ; crystal size = $10 \times 10 \times 3\ mm^3$.

Conclusion

The basic characteristics of the interferometric system may be summarized as follows :

Recording and erasure sensitivity equivalent to high resolution Kodak 649 F photographic plate[9] with about 1cm² crystal area.

Reusable material with no fatigue effect.

High quality of reconstructed images and control of saturation efficiency with external applied field E_0.

No displacement and processing of the storage medium.

Interferogram observation with a vidicon memory tube on TV monitor.

All these features make of $Bi_{12}SiO_{20}$ crystals and isomorphous $Bi_{12}GeO_{20}$ competitive materials for real-time double-exposure interferometric experiments of transparent or diffuse objects[9] ; consequently they should offer new possibilities in the field of non-destructive testing applications[10-11].

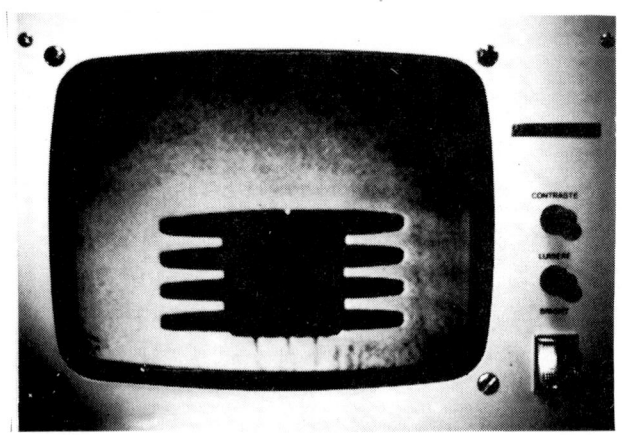

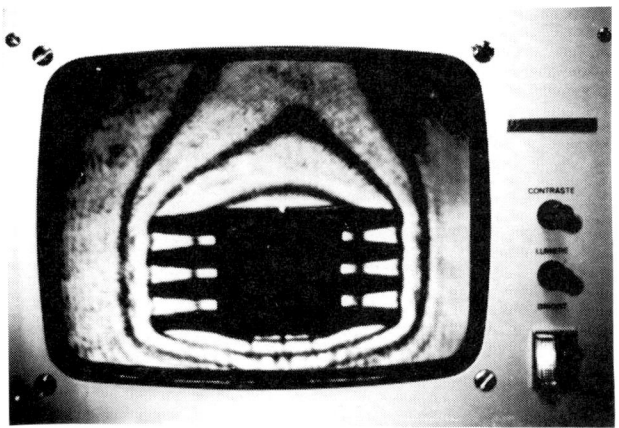

Fig.3 - Interferogram stored on vidicon memory tube and displayed on TV monitor. Initial reconstructed image and double-exposure interferogram of thermal index gradient induced by a transistor mounted in a radiator.

References

1. J.P. Huignard and F. Micheron, Appl. Phys. Lett. 29, 591 (1976)
2. L. Young, W.K.Y. Wong, M.L.W. Thewalt and W.D. Cornish, Appl. Phys. Lett. 24, 264 (1974)
3. J.J. Amodei and D.L. Staebler, RCA Rev. 33, 71 (1972)
4. D. Von Der Linde and A.M. Glass, Appl. Phys. 8, 85 (1975)
5. S.F. Su and T.K. Gaylord, J. Appl. Phys. 46, 5208 (1975)
6. R.R. Shah, D.M. Kim, T.A. Rabson and F.K. Tittel, J. Appl. Phys. 47, 5421 (1976)
7. J.P. Huignard, J.P. Herriau and F. Micheron, Ferroelectrics 11, 393 (1976)
8. D. Von Der Linde, A.M. Glass and K.F. Rodgers, Appl. Phys. Lett. 26, 22 (1975)
9. R.J. Collier, C.B. Burckhardt and L.H. Lin, in Optical Holography (Academic Press, New-York 1971) pp.280-309, 423-426
10. J.P. Huignard, J.P. Herriau, Applied Optics, Vol 16, 1809 (1977)
11. J.P. Huignard, J.P. Herriau, Th. Valentin, Applied Optics, Vol 16, 2796 (1977)

AUTOPROCESSOR MATERIALS FOR THE RECORDING OF PHASE HOLOGRAMS: PHOTOPOLYMERS AND ORGANO-METALLIC SEMICONDUCTORS

M. Jeudy
Laboratory of Applied Physical Chemistry ISSEC Inc.
13, Chemin du Levant, 01210 Ferney-Voltaire, France

Abstract

The principal attraction of autoprocessor materials for holographic recording is their simplicity of use by comparison to classical silver emulsions; their major disadvantage resides in the great sensitivity of the latter, which is difficult to equal with other photosensitive materials.

Two types of material are examined:
-Photopolymers, which take advantage of the change of refractive index caused by the polymerization induced by the absorption of light energy. Excellent diffraction efficiencies are obtained with satisfying sensitivity. We will present all their properties.
-Organometallic materials in which the variation of index is the result of an intrinsic and reversible chemical amplification of a photoelectric phenomenon. These materials are being studied and have already shown perfect reversibility as well as a comfortable spectral sensitization.

Introduction

The most commonly used materials, in holography as in photography, are silver materials. They possess, of course, numerous good qualities:
-good energy sensitivity
-spectral sensitivity extended to all commonly used visible wavelengths
-excellent resolution
-perfect stability of recording.
But they also present inconveniences which motivate the search for replacement materials. These inconveniences are, principally:
-weak diffraction efficiency
and above all,
-the necessity of a chemical treatment to develop and fix the modulation recorded.

Besides silver emulsions, there exists a whole range of materials the majority of which are materials of phase and not of amplitude; the recording is made in this case by modulation of the refractive index and not by modulation of the transparence; phase materials thus permit us to attain very high diffraction efficiencies.

In this category, photopolymers appear to complement silver emulsions very well; they are in effect autoprocessor phase materials, that is, the recording of the modulation is effected without treatment, either chemical or physical.

The process takes advantage of the difference in refractive index existing between the two forms, monomer and polymer, of the base constituent of the emulsion; the functioning process is sketched in figure 1: the light energy absorbed by the dye is transfered to the initiator, which provokes the local polymerization of the monomer.

The sensitivity depends on the initiation threshold of the polymerization chain reaction; the quantum efficiency is proportional to the number of molecules polymerized under the action of the light, while the resolution varies in the opposite direction.

Characteristic Properties of the MIV Photopolymer Developed at the ISSEC Under Contract from the DRME

-Spectral sensitivity: It is typically realized in red (He-Ne laser) but can be obtained in green or blue with other sensitizers. Figure 2 shows the curves corresponding to two different sensitizers for red and green.
-Energy sensitivity: Figure 3 shows the diffraction efficiencies obtained as a function of the illumination at 6328 Å for two densities of light power: 100 and 0.5 mJ/mm^2 or 50 mJ/cm^2 in the first case and only 20 mJ/cm^2 in the second.
Trials run with a photopolymer sensitized for the green of the argon laser (5145 Å) have shown a still better sensitivity: 5mJ/cm^2.
It should be noted that in the case of low power it is advisable, in order to obtain good diffraction efficiencies, to use a sufficiently intense readout beam, on the order of 20 to 100 mW/cm^2: this is due, on one hand, to the fact that the de-coloration of the sensitizer at the moment of absorption is a reversible reaction which means that the material becomes transparent only above a certain power; and on the other hand to the fact that the setting off of the polymerization reaction is effected only above a certain threshold which the readout beam can help to cross. Figure 4 shows this amplification effect in the evolution of the diffraction efficiency at the moment of reading: the curves represented by the solid lines both correspond to a readout beam of 50mW/cm^2; the curve represented by a dotted line corresponds to a readout beam equal to the recording beam, or 0.5 mW/cm^2.

Transfer of Spatial Frequencies

Figure 5 represents the holographic efficiency as a function of the spatial frequency: this efficiency is equal to the ratio of the energy reconstructed in the +1 order to the incident energy; it is measured on the diffraction by gratings recorded with equal energy beams of 50 mW/cm^2 each and at an angle of 2θ to each other, symmetrical with regard to the normal to the emulsion: the resolution surpasses 3000 mm^{-1}.

Stability of the Recording

Figure 6 shows the decrease in diffraction efficiency as a function of readout time under a beam of 50 mW/cm^2: the efficiency falls by half in about 3 hours. Nevertheless, the hologram is conserved no longer in darkness; the recoloration of the sensitizer seems to accelerate the erasure of the modulation by continuation of the polymerization. Different fixation techniques have been studied, but up to now the result is obtained to the detriment of the diffraction efficiency.

Conservation of the Material Before Use

The photosensitive layer can be conserved for two to three months before use. The best conditions of conservation are 20° C. and 50% relative humidity. The holographic efficiency obtained at recording can depend on the ambient conditions; if the plates are stored in a room where the conditions are different from those of the room where the recording takes place, a certain length of time may be necessary for an equilibrium to be established between the material and the surrounding medium.

Materials Based on Organometallic Compounds

A characteristic which we have not talked about up to now but which is somewhat attractive to the user is reversibility.

The ISSEC, in the framework of the same DRME contract as the one concerning MIV photopolymers, envisages the development of reversible materials based on organometallic compounds. Reversiblility is inherent in the fact that the medium undergoes no physical change at recording, but only variations of electronic state, such as the change of valence of the metallic atom and ruptures of organometallic bonds.

Unlike electro-optical materials such as niobate or lithium, with which a variation of index is photo-induced by simple redistribution of the electronic changes in the medium, the organometallic materials possess an intrinsic amplification which should confer on them a satisfying sensitivity: the photo-electronic effect is situated in fact at the level of the metallic atom and the amplification resides in the reversible modifications on the bonds of this atom with the organic medium which are thus entailed.

These materials are still being studied, but have already shown perfect reversibility and a convenient spectral sensitization.

AUTOPROCESSOR MATERIALS FOR THE RECORDING OF PHASE HOLOGRAMS: PHOTOPOLYMERS AND ORGANO-METALLIC SEMICONDUCTORS

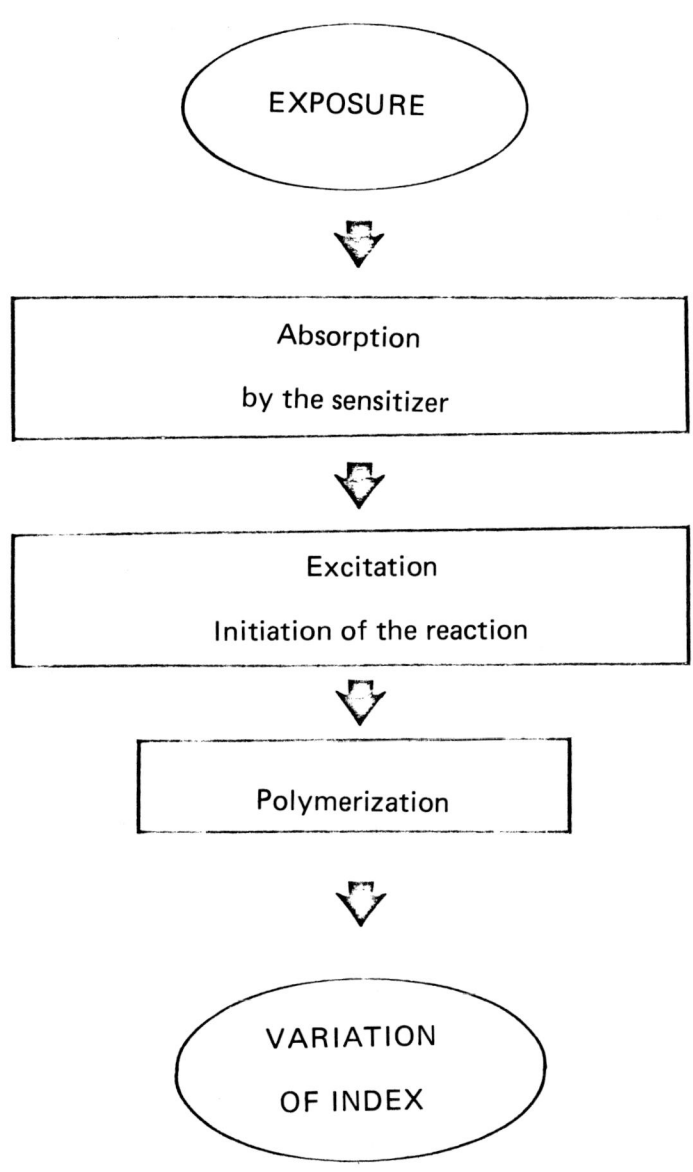

fig 1

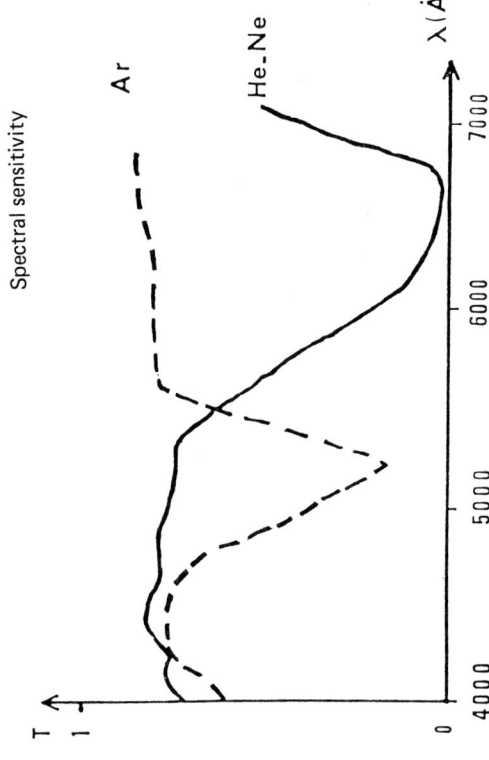

fig 2

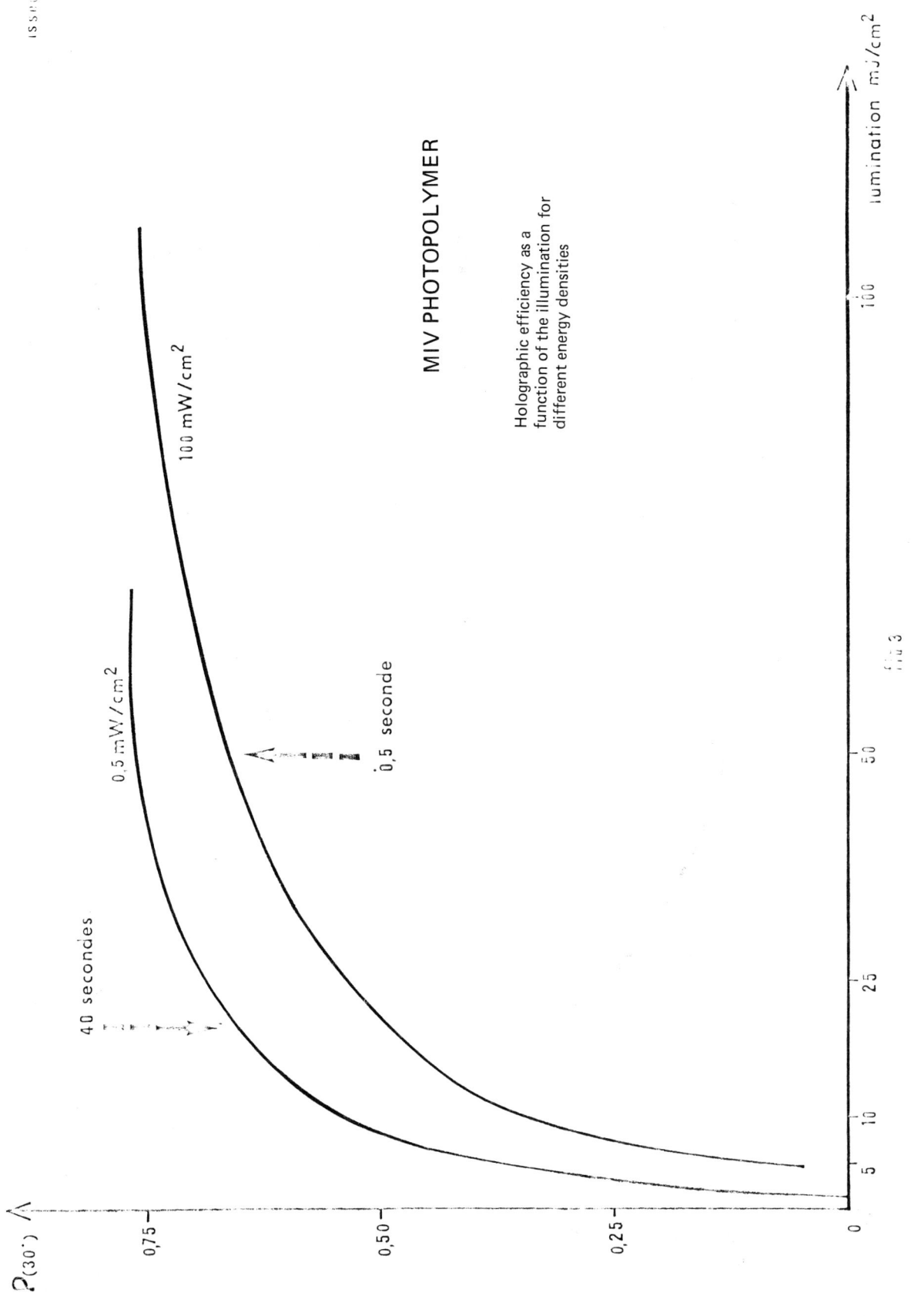

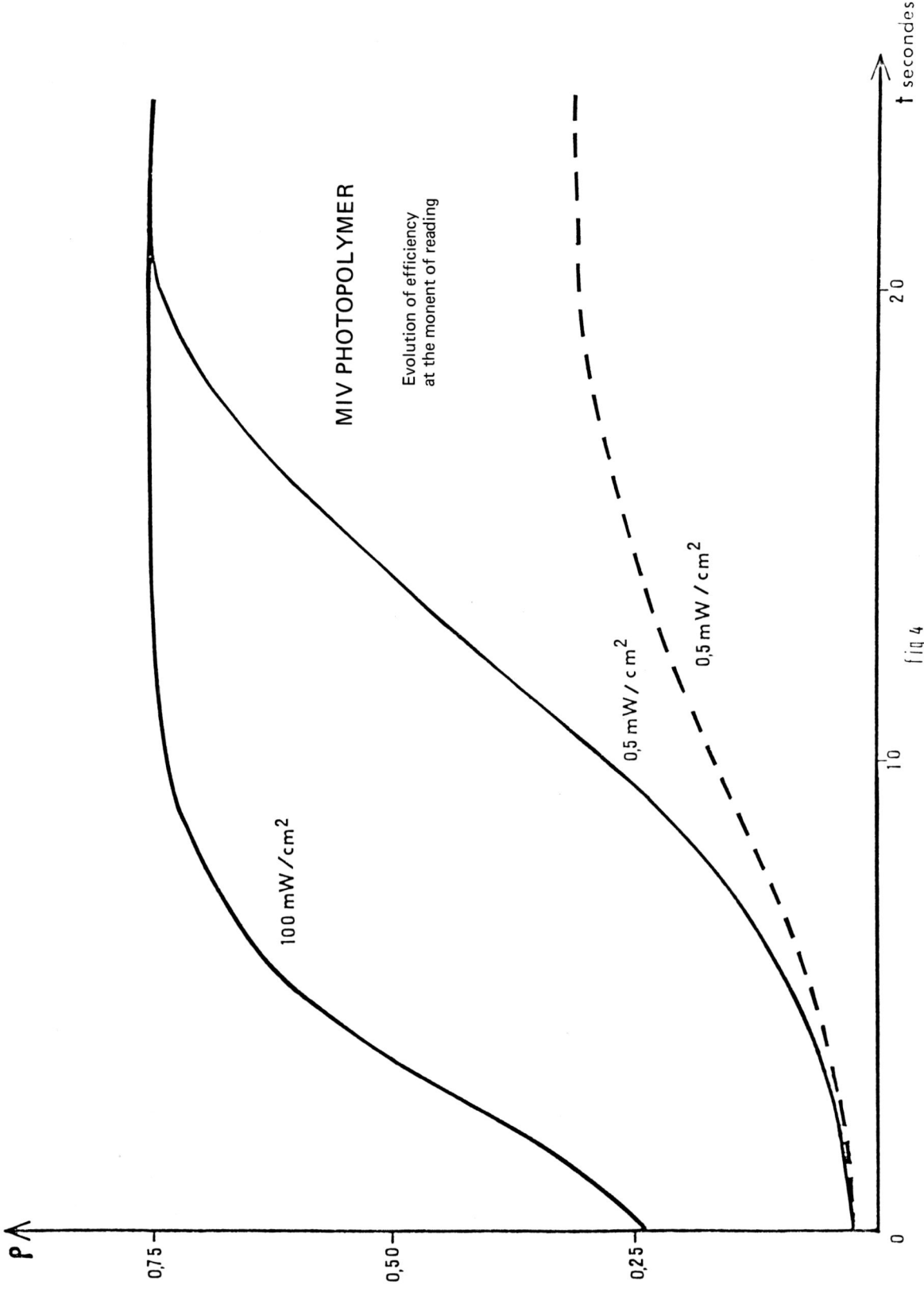

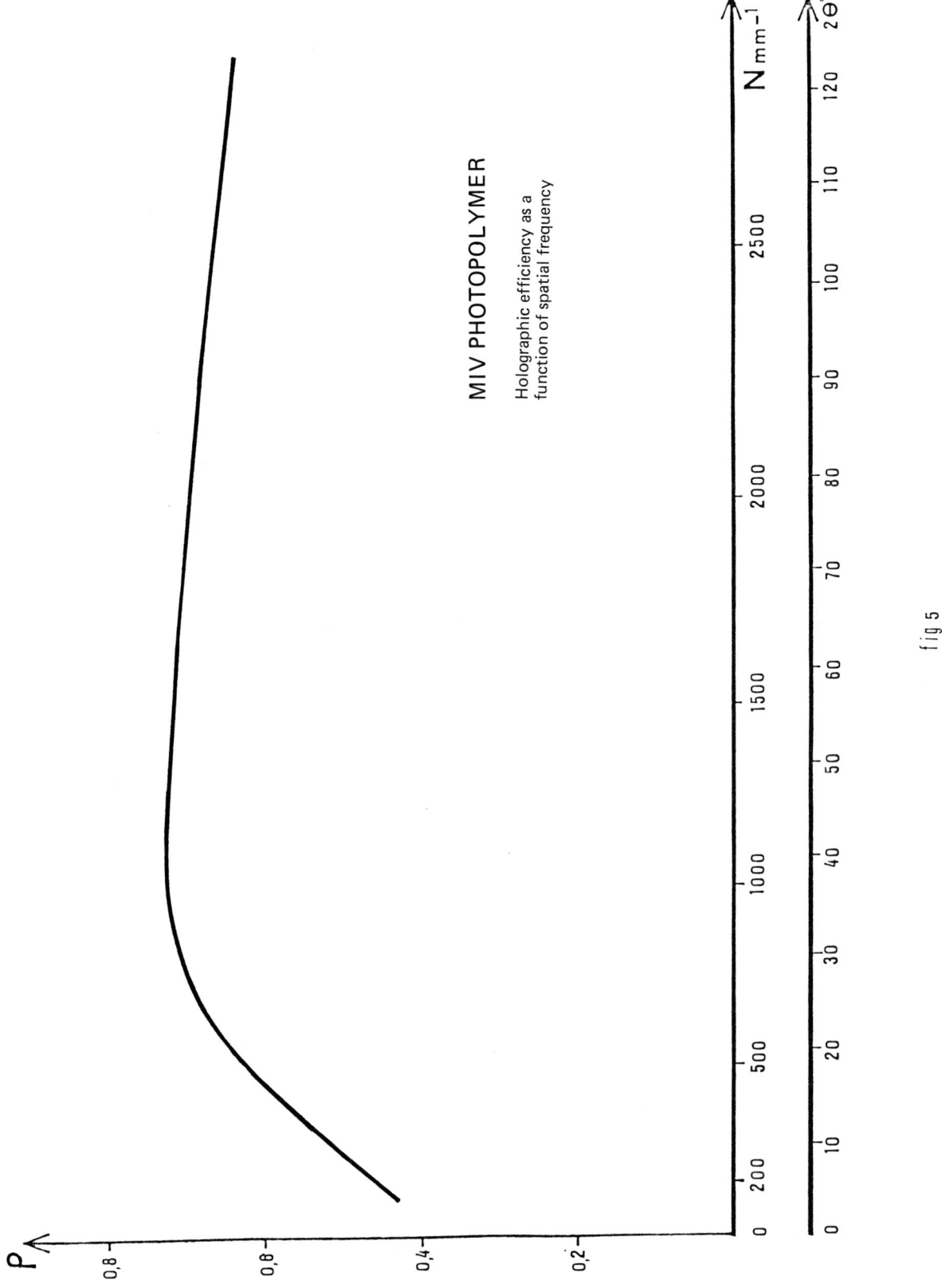

fig 5

MIV PHOTOPOLYMER

Holographic efficiency as a function of spatial frequency

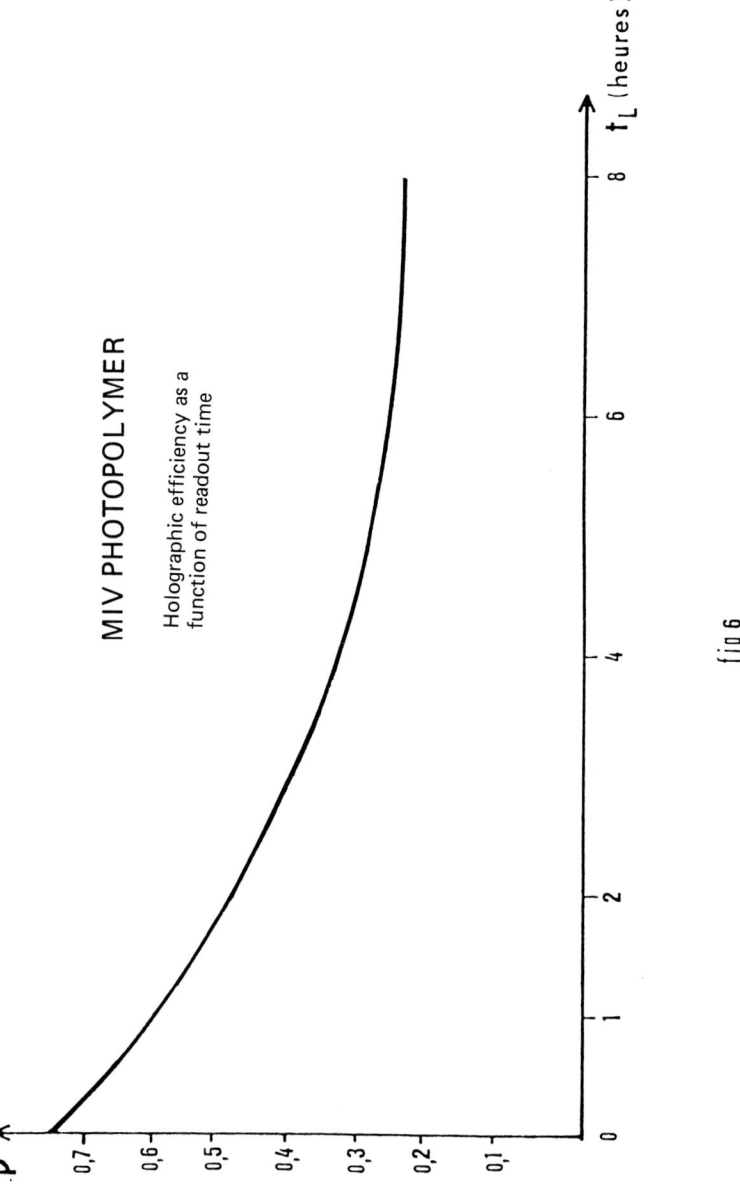

fig 6

IMPROVEMENT OF THE SIGNAL/NOISE RATIO IN THE METHOD OF SUBTRACTION OF IMAGES BY SPECKLE INTERFEROMETRY

S. Debrus and V. Sokolov

Optical Laboratory, Institute of Optics
Curie University Tour 13 3eme
4, Place Jussieu 75230 Paris Cedex 05, France

Abstract

The use of birefringent systems permits the recording, in a single exposure, of several different granularity figures. In two exposures, one with the image A, the other with the image B, we can obtain, in the focal plane of the recording plate of A and B exactly superposed, the profile of the interference fringes which would be given in fact by three exposures, made by alternating the images (A, B, A), with the respective exposure times 1/2, 1, 1/2. The signal/noise ratio is thus ten times greater than that which is given in the case of two poses in the absence of birefringent crystal

Introduction

The method of extraction of the differences between two images by speckle interferometry which we have already proposed (1) was initially effected in two exposures. We recorded on a single photographic plate P the two images to be compared, A and B, the identical parts being exactly superposed. During the recording the intensity of each image A or B is multiplied by that of a single granularity figure G, provided by an unpolished glass D, lit by a laser beam.

Each image is thus coded by very fine grains of 3 to 4 m. Furthermore, between the two superposed recordings, we caused the photographic plate P to undergo a translation x_o in its plane, superior to the diameter of the grains of the recorded granularity figure.

To obtain a good signal/noise ratio, while still using a relatively wide slit to filter the difference, it was necessary in fact to make three to five exposures instead of two (2,3). The two images are then placed successively on the photographic plate P, with exposure times respectively proportional to the binomial coefficients. Furthermore, between each exposure the plate undergoes the same translation x_o in its plane. For example, in the case of three exposures, the different exposure times are 1, 2, 1 with the images A, B, A respectively. In the case of five exposures, the exposure times are respectively proportional to the coefficients 1, 4, 6, 4, 1.

Nonetheless, it may be advantageous to have to make only two exposures instead of three or five, above all when the system is evolving.

We present here a method permitting us to make only two exposures while still obtaining a good signal/noise ratio, by using birefringent systems. We thus improve the performances in the initial case of two exposures.

Principle of the Method of Subtraction Between Two Images

Recording: The images A and B to be compared are on photographic plates presenting the transparence functions A(x,y) and B(x,y) respectively. Their difference is noted:

$$b(x,y) = B(x,y) - A(x,y) \tag{1}$$

Let G(x,y) be the intensity of the granularity figure formed in the plane of the recording plate by the unpolished glass D lit by the laser beam $\mathcal{L}$ (figure 1).

Between exposures, the granularity figure G(x,y) undergoes a translation x_o in the plane of the plate P. The translation is produced by the rotation of the parallel-face glass plate LV. The translation x_o is very much smaller than the dimension of the small details of the images to be treated. In this case, we can write approximately that an image A coded by a granularity figure displaced by x_i will present the following intensity on the plate P:

$$A(x,y)\, G(x-x_i,y) \simeq [A(x,y)\, G(x,y)] \otimes [\delta(x-x_i,y)] \tag{2}$$

For the two successive exposures, we can thus write the intensities as follows:

$$I_1(x,y) = [G(x,y)\, A(x,y)] \otimes \delta\left(x - \frac{x_o}{2}, y\right) \tag{3}$$

$$I_2(x,y) = [G(x,y)\, B(x,y)] \otimes \delta\left(x + \frac{x_o}{2}, y\right) \tag{4}$$

Mr. V. Sokolov's permanent address: Ioffe Physico-Technical Institute, Academy of Sciences of the USSR, 194021 LENINGRAD

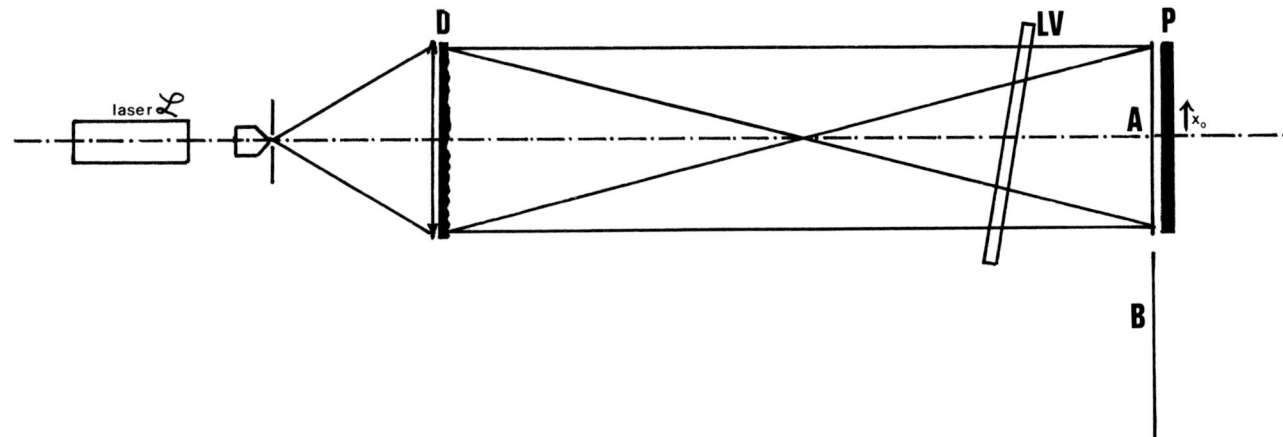

Fig. 1. Recording in two exposures

In total, the plate has received the intensity:

$$I(x,y) = I_1(x,y) + I_2(x,y) \tag{5}$$

We replace $B(x,y)$ by $A(x,y) + b(x,y)$ and we regroup the terms including the common parts $A(x,y)$. We thus isolate the difference $b(x,y)$ as follows, in the expression of $I(x,y)$:

$$I(x,y) = [G(x,y)\, b(x,y)] \otimes \delta(x + \tfrac{x_0}{2}, y) +$$

$$[G(x,y)\, A(x,y)] \otimes [\delta(x - \tfrac{x_0}{2}, y) + \delta(x + \tfrac{x_0}{2}, y)] \tag{6}$$

<u>Filtering</u>: The plate is developed in the linear conditions where the transparence in amplitude is a linear function of the intensity recorded. It is lit by a parallel beam of monochromatic light (figure 2). In the focal plane of a lens, besides the light directly transmitted by the plate, the light diffused by the speckle grains is distributed in amplitude according to the law:

$$a(u,v) = \text{F.T.}\,[I(x,y)] \tag{7}$$

The symbol F.T. represents the Fourier transform. We note $U(u,v)$, $\tilde{b}(u,v)$, $\tilde{\mathcal{A}}(u,v)$, the respective Fourier transforms of $G(x,y)$, $b(x,y)$, and $A(x,y)$.

The Fourier transform of $I(x,y)$ is obtained by taking the Fourier transform of each term of expression (6). For the diffused amplitude $a(u,v)$ we thus obtain:

$$a(u,v) = [U(u,v) \otimes \tilde{b}(u,v)]\, e^{jku\tfrac{x_0}{2}} + [U(u,v) \otimes \tilde{\mathcal{A}}(u,v)]\, 2\cos ku\tfrac{x_0}{2} \tag{8}$$

where $k = 2\pi/\lambda$.

The first term includes the spectrum $\tilde{b}(u,v)$ of the difference, which constitutes the signal. The second term includes the spectrum $\tilde{\mathcal{A}}(u,v)$ of the part common to both images, which constitutes the noise. This second term is modulated by the sinusoidal function $2\cos(ku\,x_0/2)$, where x_0 is the translation of the speckle between the two exposures. A diffusion halo $U(u,v)$ is extended widely around the image S' of the monochromatic and punctual source S which lights the plate P; this halo is modulated by the sinusoidal interference fringes perpendicular to the direction of translation x_0. When the images A and B are rigorously identical, $b(x,y) = 0$ and the fringe contrast is equal to 1. This means that there is no light on the minima. After filtering by a slit F placed on such a black fringe, the image P' of the plate P is entirely black (figure 3). When the two images are not completely identical, the fringe contrast is inferior to 1, since there is a little light on the minima, light which proceeds from the regions which are different on the two images. After filtering by a slit placed on a dark fringe (where $\cos ku\, x_0/2 = 0$), only the differences $b(x,y)$ appear on the image P' of the plate P, modulated by the granularity figure $G(x,y)$.

Improvement of the Signal/Noise Ratio

<u>Principle</u>: In the Fourier plane ϕ (figure 2) in the vicinity of a minimum, the light amplitude proceeding from the identical regions $A(x,y)$ on the two images is not strictly zero for the entire width of the filtering slit F. This small contribution to the formation of the image P' constitutes a noise (figure 3). For a slit of a given width, this noise is much weaker if we make three exposures instead of two, with the exposure times proportional to 1/2, 1, 1/2 (figure 4).

IMPROVEMENT OF THE SIGNAL/NOISE RATIO IN THE METHOD OF SUBTRACTION OF IMAGES BY SPECKLE INTERFEROMETRY

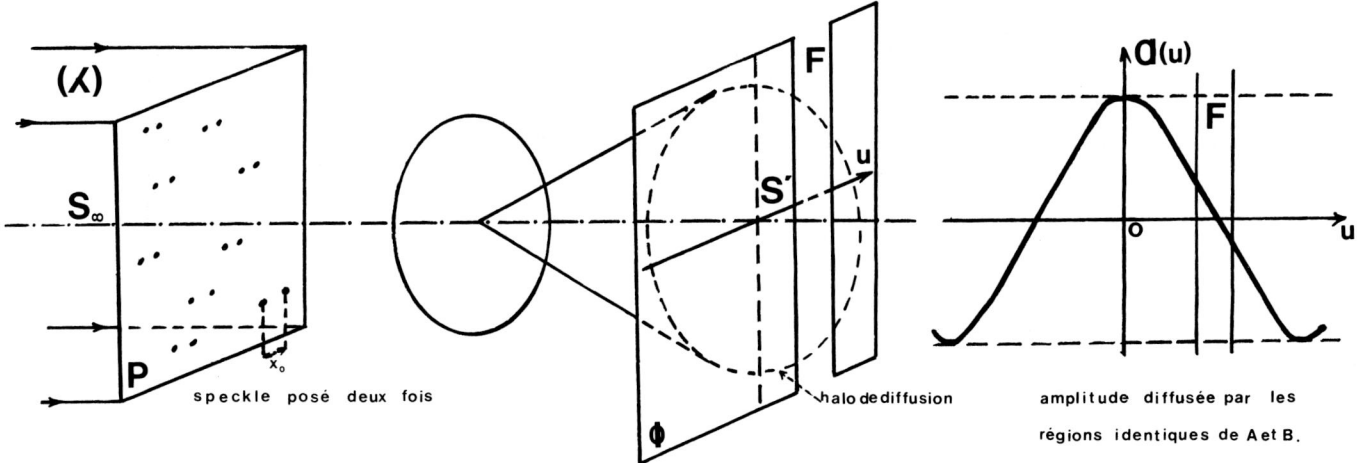

Fig. 2. Filtering in the case of two exposures

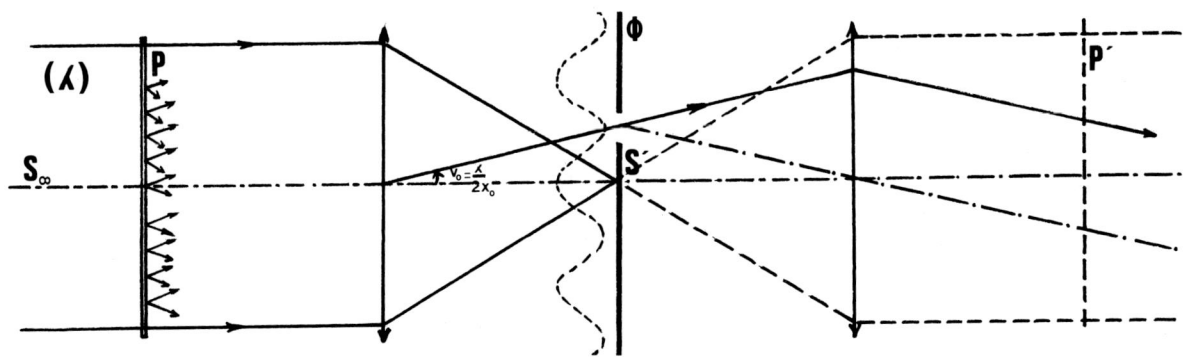

Fig. 3. Filtering and obtaining of the difference

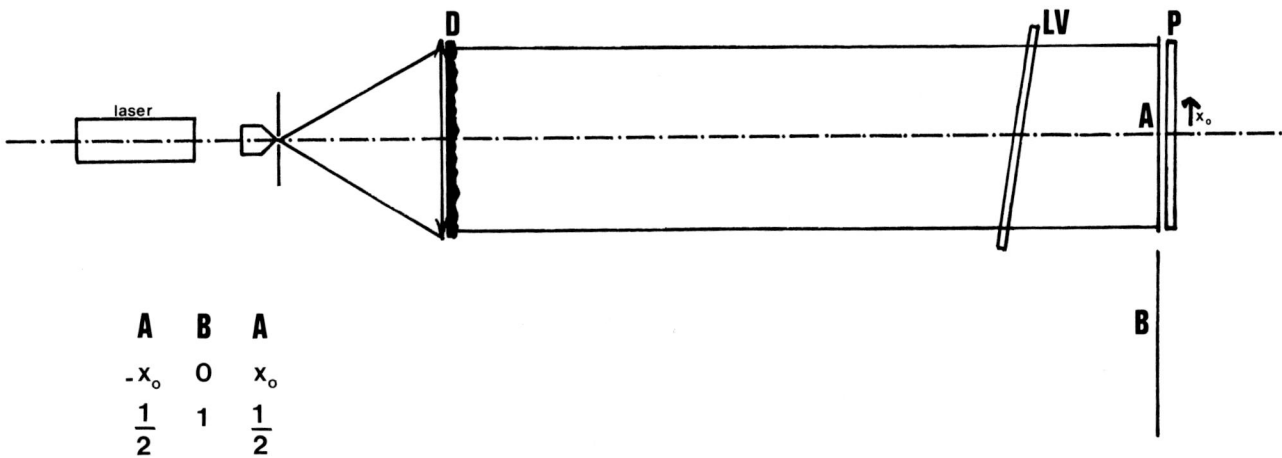

Fig. 4. Recording in three exposures

We alternately expose the images A and B and the recording plate P undergoes the same translation x_o between the different exposures.

The intensity received in the course of the three successive exposures is written:

$$I(x,y) = [G(x,y) \; A(x,y)] \otimes \tfrac{1}{2} \delta(x-x_o,y) + [G(x,y) \; B(x,y)] \otimes \delta(x,y) + [G(x,y) \; A(x,y)] \otimes \tfrac{1}{2} \delta(x+x_o,y) \quad (9)$$

In causing the difference $b(x,y)$ to appear, $I(x,y)$ is written;

$$I(x,y) = [G(x,y)\, b(x,y)] \otimes \delta(x,y) + [G(x,y)\, A(x,y)] \otimes [\tfrac{1}{2}\delta(x-x_o, y) + \delta(x,y) + \tfrac{1}{2}\delta(x+x_o, y)] \qquad (10)$$

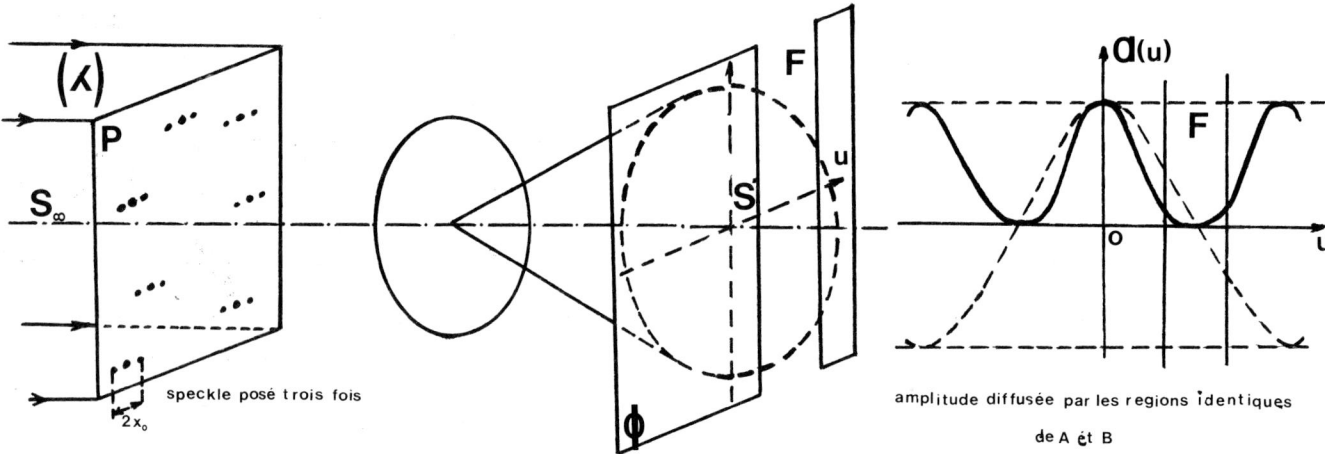

Fig. 5. Filtering in the case of three exposures

Upon filtering, the amplitude diffused by this plate exposed three times is written:

$$a(u,v) = U(u,v) \otimes \tilde{b}(u,v) + [U(u,v) \otimes \tilde{\mathcal{A}}(u,v)]\,[\tfrac{1}{2} e^{-jkux_o} + 1 + \tfrac{1}{2} e^{jkux_o}] \qquad (11)$$

that is:

$$a(u,v) = U(u,v) \otimes \tilde{b}(u,v) + [U(u,v) \otimes \tilde{\mathcal{A}}(u,v)]\, 2\cos^2\!\left(ku\,\tfrac{x_o}{2}\right) \qquad (12)$$

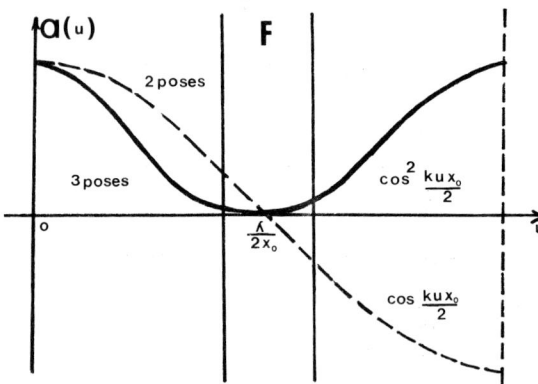

Fig. 6. Amplitude diffused by the identical regions of the two images

In this case, the light proceeding from the identical parts is modulated by the function $2\cos^2 kux_o/2$, while for two exposures it was modulated by the function $2\cos(ku\,x_o/2)$ (figure 6). The envelope presented by the curve representing the function $2\cos^2(kux_o/2)$ in the vicinity of a minimum is very advantageous by comparison to the profile of the curve representing the function $2\cos(kux_o/2)$ in the same region. For a slit F of a given width, centered on a minimum, the noise constituted by the small contribution of the identical parts on this slit is much weaker in the case of three exposures than in the case of two. The noise is still weaker in the case of five exposures, since the envelope is still larger (2,3).

Nevertheless, multiplying the number of exposures lengthens the treatment time. Furthermore, if we directly study a system in evolution, only two successive exposures can be made in real time. Treatment in 3 or 5 exposures is possible only if we have slides made in advance, which we can place successively before the recording plate P. For an evolving system, we will be obliged to form the image of the object studied directly on the plate P, in order to compare its successive states. It is thus indispensable to be able to operate in only two exposures.

Use of Birefringent Systems

We describe here systems permitting us to record three laterally displaced speckles in only two exposures. More complex combinations can also be considered to record a greater number of displaced speckles.

Fixed Division System

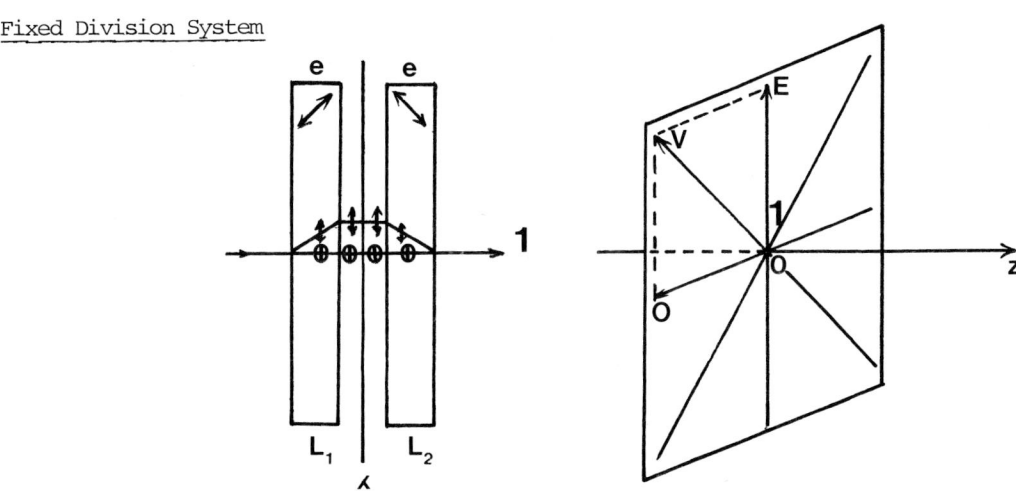

Fig. 7. First exposure: the neutral lines of the half-wave plate are parallel to the extraordinary vibrations E and ordinary vibrations O respectively.

A first system (4) (figure 7) is inspired by the setup proposed by Steel in 1964; it is formed by two parallel-face crystalline plates L_1 and L_2, of the same thickness e, cut at 45° from the axis, between which is a half-wave plate. During the first exposure, the neutral lines of the half-wave plate are parallel to the directions of the extraordinary vibrations E and ordinary vibrations O transmitted by each plate, the incident beam being parallel to the normal to the plates and linearly polarized at 45° to the privilieged directions of the plates L_1 and L_2. In this manner, the birefringent system produces no lateral displacement between the ordinary and extraordinary emerging beams. The intensities of these beams are equal; their vibrations are perpendicular and thus their intensities are added together; we take their sum as unity.

Before the second exposure, the neutral lines of the half-wave plate are turned 45° with regard to their preceeding position.

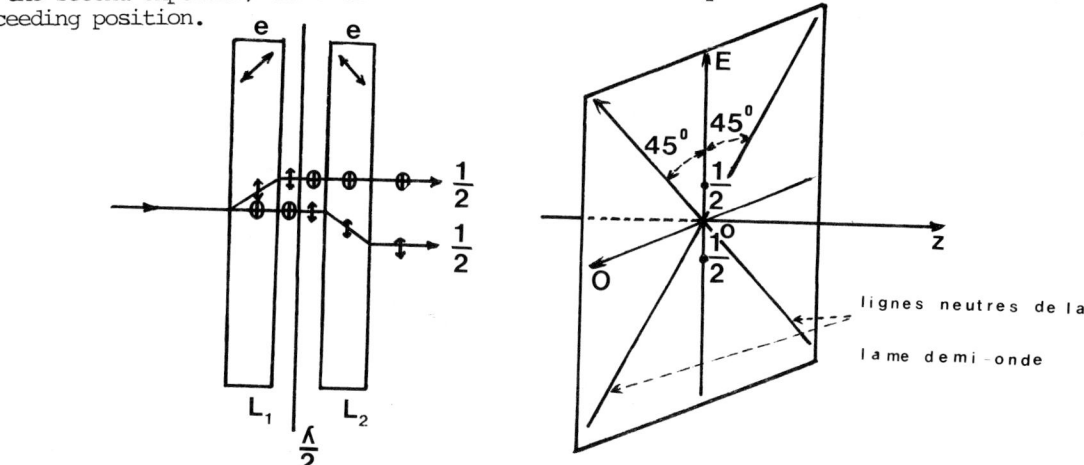

Fig. 8. Second exposure: the neutral lines of the half-wave plate make an angle of 45° with the extraordinary vibrations E and ordinary vibrations O.

The incident beam, polarized as before, is then divided into two beams by the birefringent system. The respective intensities are then 1/2 and 1/2 with regard to the unit intensity of the emerging beam of the first exposure. If this birefringent system is interposed between the unpolished glass D and the photographic plate P (figure 9) in making only two exposures, one with image A, the other with image B, each speckle is exposed three times, with the intensity ratios 1/2, 1, 1/2. In the spectral plane, the noise will thus have the advantageous profile in $\cos^2(ku\,x_0/2)$ instead of $\cos(ku\,x_0/2)$.

Variable Division System

A second system is still more interesting since it provides a variable division, speckles, while in

the preceeding system it is imposed by the thickness e of the plates.
Here we use the association of two Wollaston prisms proposed by Drougard and Wilczynski in 1965.

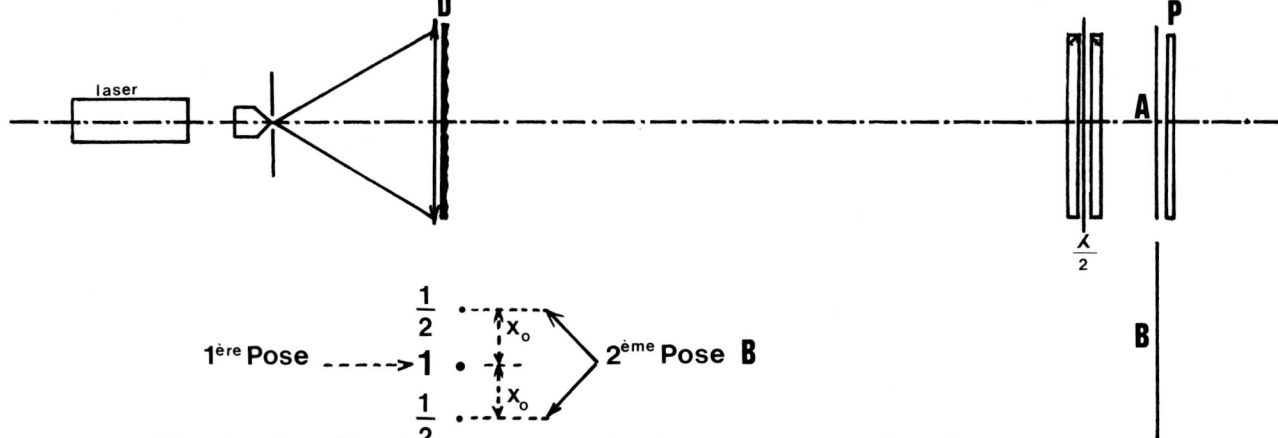

Fig. 9. Recording in two exposures in the presence of a birefringent system

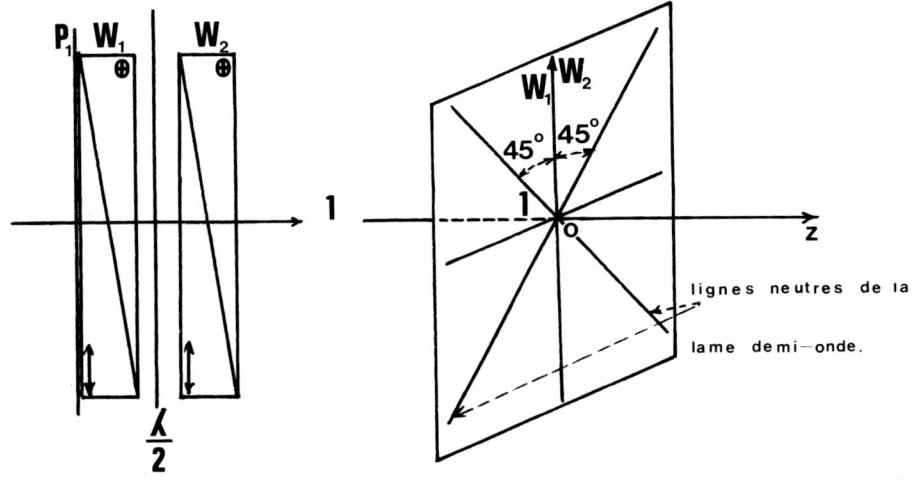

Fig. 10. First exposure: the principal sections of W_1 and W_2 are parallel.

The birefringent system is thus formed by two identical Wollaston prisms W_1 and W_2 (figure 10). A polarizer P_1 is fixed on the entrance face of the prism W_1. The direction of transmission of P_1 makes an angle of 45° with the privileged directions of the prism W_1. No polarizer is placed after the prism W_2. A half-wave plate is interposed between the two Wollaston prisms W_1 and W_2. In the course of the first exposure, the principal sections of W_1 and W_2 are parallel to each other, and at 45° to the principal section of the half-wave plate. The angular division of the extraordinary and ordinary rays is null. Their vibrations are perpendicular; thus their intensities are added together. Their sum is taken as unity.

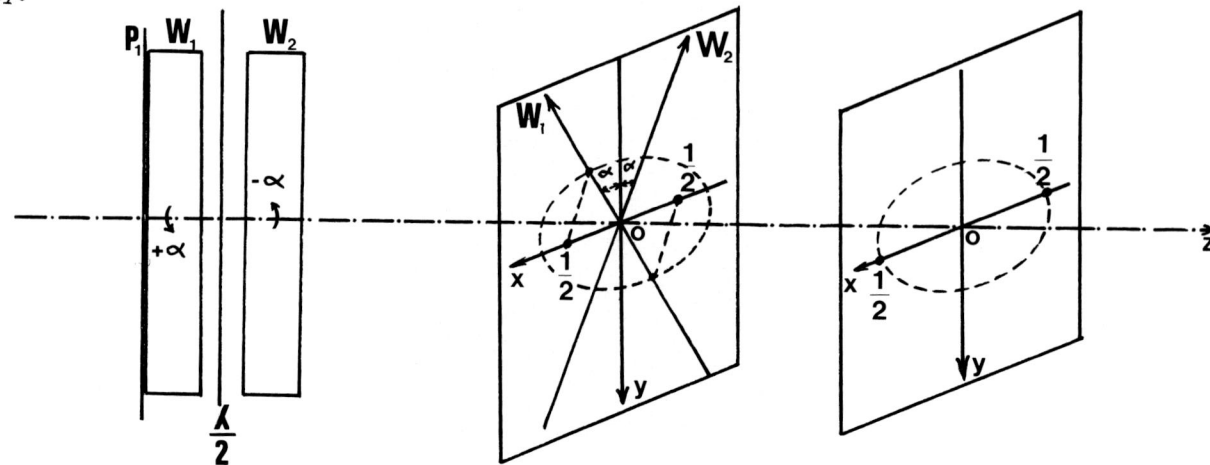

Fig. 11. Second exposure: the prisms W_1 and W_2 are turned through an angle α in opposite directions.

IMPROVEMENT OF THE SIGNAL/NOISE RATIO IN THE METHOD OF SUBTRACTION OF IMAGES BY SPECKLE INTERFEROMETRY

Before the second exposure, the two Wollaston prisms W_1 and W_2 are turned around the axis of the system $\underline{oz}$ through an angle α in opposite directions (figure 11). The angular division of the emerging rays increases with the angle of the two prisms; it is maximal when the two prisms have been turned 90° in opposite directions. The linear division corresponds on the plate P to the same direction $\underline{ox}$, whatever the angle α of the two prisms. Its value depends on the angle α. The intensities of the two types of emerging rays are equal to half of the intensity of the emerging ray in the first exposure. In two exposures, we have thus recorded three speckles displaced laterally in the intensity ratios 1/2, 1, 1/2, as in the case of the first birefringent system. The choice of the displacement x_o can be useful as a function of the size of the small details of the images studied and of the resolution required.

Experimental Results and Conclusion

In these methods of image subtraction by speckle interferometry, the resolution is on the order of 20 microns. When this value is sufficient, the method gives satisfactory results with a good signal/noise ratio. By means of the birefringent systems described in this article, we obtain a signal/noise ratio ten times superior to what it was in the initial case of two exposures. The filtering is easily effected, as we have used a slit 2 cm wide with a focal distance of 120 cm for the optical system at whose focus we effect the filtering.

The photographs in figure 12 show the quality of the images we can obtain. Image A is an intact Kodak sight; image B is the sight from which several details have disappeared. They appear perfectly on the difference.

Image A

Image B

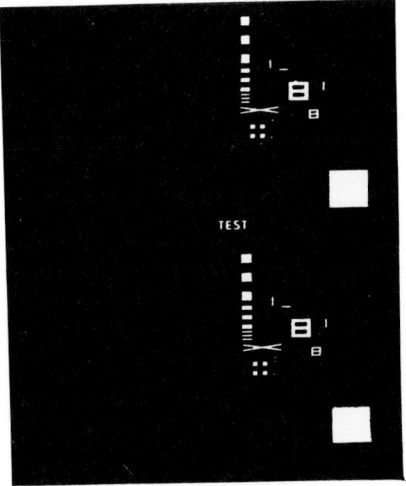

A-B Difference
Fig. 12.

References

1. Debrus, S., Françon, M. and Grover, C.P., "Detection of two differences between two images", Opt. Commun., vol. 4, pp. 4-6, 1971.
2. Debrus, S., Françon, M. and Grover, C.P., "Detection of differences between two images, an improved method", Opt. Commun. vol. 6, pp. 15-17, 1972.
3. Debrus, S., Françon, M. and Koulev, P., "Extraction de la difference entre deux images", Nouv. Rev. Optique, t. 5, pp. 153-168, 1974.
4. Françon, M., Koulev, P. and May, M., "Speckle photography by birefringent plates", Opt. Commun, vol. 17, pp. 163-165, 1976.

DETECTION OF AXIAL DISPLACEMENTS OF A DIFFUSING OBJECT

Yves Dzialowski and Marie May
Institut d'optique et Université Pierre et Marie Curie
Tour 13 – 3ième étage – 4 Place Jussieu
75230 Paris Cedex 05 France

Abstract

The speckles of the image plane of an object are radially shifted and decorrelated when the object is axially translated. It will be shown that the radial shift of the corresponding speckles is removed when the pupil of the optical system is placed in the back focal plane of the imaging lens. A photographic plate twice exposed to the irradiance of image plane and laterally shifted between the exposures, exhibits, after processing, a system of Young's fringes in its Fourier plane. The contrast of these fringes, which represents the correlation degree of the recorded speckle patterns, is only a function of the relative value of the axial shift of the object in comparison with the depth of focus of the imaging lens. If the pupil of the optical system is placed in the plane of the imaging lens, the contrast of the fringes generated in the same way as above, depends on the maximum value of the radial shift suffered by the speckles with respect to their mean size in image plane.

Introduction

It has been shown by different authors [1,2] that, when a diffusing object is axially shifted through a small distance, the speckles of image plane suffer both a radial shift and a decorrelation. In fact, an elementary area of the object, centered at a point M, generates in image space an ellipsoïd [3], the axis of which is the line joining the point M and the center of the pupil of the lens. The speckle of image plane corresponding to this elementary area, is the intersection of the plane and the ellipsoïd. The axial shift of the object produces an angular change of the axis of the ellipsoïds; the corresponding speckles are thus radially shifted. The decorrelation of the speckles is due to the defocusing and depends on the effective numerical aperture of the imaging system. We demonstrate here that the radial shift of the corresponding speckles is only function of the position of the pupil of the imaging lens. Moreover, we calculate the contrast of the Young's fringes generated in its Fourier plane by a photographic plate twice exposed to the irradiance of the image plane (before and after the axial shift of the object) and laterally translated between the exposures.

Pupil in back focal plane of the lens

Let us consider the geometrical scheme of Figure 1 in which the object O, illuminated by a parallel beam of laser light, is imaged in plane I by the lens L, the aperture of which is assumed to be infinite. More-

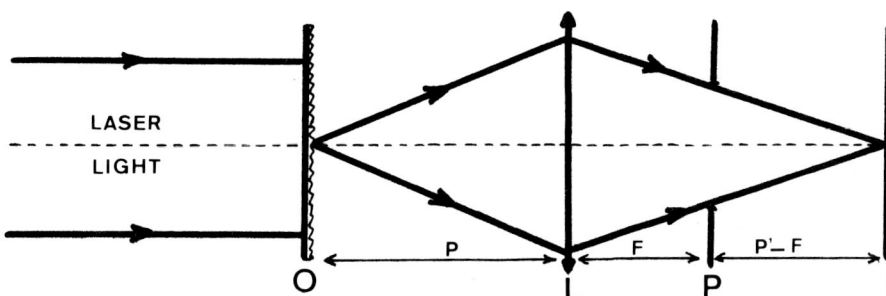

Fig. 1. The object O is imaged in plane I by lens L, through the pupil P placed in the back focal plane of L.

over, let us suppose that the pupil of the imaging system is placed in the back focal plane of L. The amplitude distribution of image plane I is proportional to

$$U_1(x') = \iint O(x)P(X)\exp(-j\frac{2\pi}{\lambda f}xX)\exp(-j\frac{2\pi}{\lambda(p'-f)}xx')\,dx\,dX \qquad (1)$$

where $O(x)$ is the random amplitude distribution of the object O, $P(x)$ is the amplitude transparency of the pupil of the lens and where $1/p + 1/p' = 1/f$. After an axial shift of the object through z, the amplitude distribution in plane I is given by

$$U_2(x') = \iint O(x)P(X)\exp(-j\frac{2\pi}{\lambda f}xX)\exp(-j\frac{\pi z}{\lambda f^2}X^2)\exp(-j\frac{2\pi}{\lambda(p'-f)}xx')\,dx\,dX \qquad (2)$$

The comparison of formulae (1) and (2) shows that the speckles generated by the same elementary area of O before and after its axial shift, are centered at the same point of image plane. However, their correlation degree is not maximum due to the term $\exp(-j\pi zX^2/\lambda f^2)$ of formula (2).

In fact, the axis of the beam scattered by the elementary area M_1 (Figure 2) of the object O_1 is the ray which passes by the geometrical centre of the pupil of the optical system, i.e., the ray parallel to the

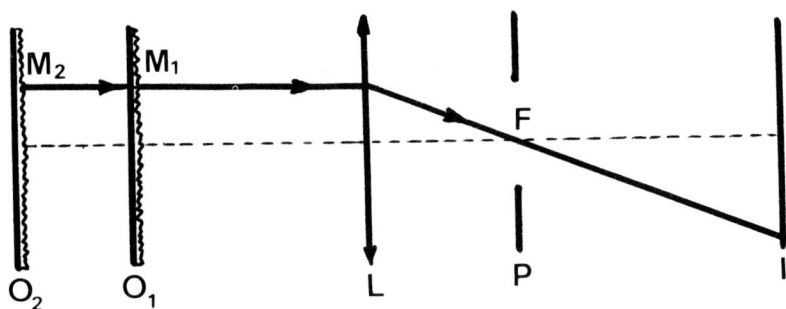

Fig. 2. When the pupil is in the back focal plane of the lens L, an axial shift of the object O involves only a decorrelation of the speckles lying in plane I.

axis of the lens which is issued from M_1 and which passes by the focus F after its refraction by L. After the axial shift of O, the axis of the beam scattered by M_2 is superposed to the earlier and consequently, the corresponding speckles are not radially shifted.

Let us suppose now that the intensity distributions $I_1(x') = U_1(x')U_1^*(x')$ and $I_2(x') = U_2(x')U_2^*(x')$ are successively recorded by a photographic plate H which is laterally translated through x_0' between the exposures. After processing H, illuminated by a parallel beam of monochromatic light, generates a system of Young's fringes [4] in its Fourier plane, the contrast of which represents the correlation degree of the recorded speckle patterns. The amplitude distribution lying in the Fourier plane of H, characterized by its angular coordinate u, is proportional to

$$\tilde{I}(u) = \tilde{I}_1(u) + \exp(j\frac{2\pi}{\lambda}ux_0')\tilde{I}_2(u) \qquad (3)$$

where $\tilde{I}_1(u)$ and $\tilde{I}_2(u)$ are the respective Fourier transforms of $I_1(x')$ and $I_2(x')$. The mean contrast of the fringes is defined by

$$\gamma(u) = \frac{<\tilde{I}_1(u)\tilde{I}_2^*(u)> + <\tilde{I}_1^*(u)\tilde{I}_2(u)>}{<\tilde{I}_1(u)\tilde{I}_1^*(u)> + <\tilde{I}_2(u)\tilde{I}_2^*(u)>} \qquad (4)$$

The photographic plate is assumed to be infinite. It can be shown that

DETECTION OF AXIAL DISPLACEMENTS OF A DIFFUSING OBJECT

$$\langle \tilde{I}_1(u)\tilde{I}_2^*(u)\rangle = \exp(-j\frac{\pi z}{\lambda f^2}(p'-f)u^2) \int\cdots\int \langle O(x_1)O^*(x_2)O^*(x_3)O(x_4)\rangle\, P(X_1)P(X_1-(p'-f)u)P(X_3)P(X_3-(p'-f)u)$$

$$\exp(-j\frac{2\pi}{\lambda f}(X_1(x_1-x_2)-X_3(x_3-x_4)))\,\exp(-j\frac{2\pi}{\lambda f}(p'-f)(x_2-x_4)u)\,\exp(j\frac{2\pi z}{\lambda f^2}X_3(p'-f)u)\,dx_1\,dx_2\,dx_3\,dx_4\,dX_1\,dX_3 \quad (5)$$

If we assume moreover that the size of the speckles incident on the pupil is extremely small compared with the diameter of that pupil and that the statistics of the object is random gaussian circular[5], we may write

$$\langle O(x_1)O^*(x_2)O^*(x_3)O(x_4)\rangle = I_o^2\,(\,\delta(x_1-x_2)\,\delta(x_3-x_4) + \delta(x_1-x_3)\,\delta(x_2-x_4)) \quad (6)$$

where I_o is the mean irradiance of the object.
The object O is limited by a slit of width 2a and the pupil of the optical system is a slit of width 2b. Under these conditions

$$\langle \tilde{I}_1(u)\tilde{I}_2^*(u)\rangle = 4a^2 I_o^2 \exp(-j\frac{\pi z}{\lambda f^2}(p'-f)u^2)\,\{\,(2b-(p'-f)u)\,\mathrm{sinc}(\frac{2\pi}{\lambda f}(p'-f)ua)\,\int \exp(j\frac{2\pi z}{\lambda f^2}(p'-f)uX_3)\,dX_3$$

$$+ \int (\mathrm{sinc}\,\frac{2\pi a}{\lambda f}(X_1-X_3))^2\,\exp(j\frac{2\pi z}{\lambda f^2}(p'-f)uX_3)\,dX_1\,dX_3\,\} \quad (7)$$

The first term of formula (7) is proportional to the Fraunhofer diffraction pattern of the slit limiting O. It decreases very rapidly when u is non zero and can be likened with a delta function. On the same way, the sinc function of the second term can be assumed to be non zero only for $X_1 = X_3$. Consequently

$$\langle \tilde{I}_1(u)\tilde{I}_2^*(u)\rangle = \langle \tilde{I}_1^*(u)\tilde{I}_2(u)\rangle = I_o^2\,(2b-(p'-f)u)\,\mathrm{sinc}(\frac{\pi z}{\lambda f^2}(2b-(p'-f)u)(p'-f)u) \quad (8)$$

and $$\langle \tilde{I}_1(u)\tilde{I}_1^*(u)\rangle = \langle \tilde{I}_2(u)\tilde{I}_2^*(u)\rangle = I_o^2\,(2b-(p'-f)u) \quad (9)$$

Finally

$$\gamma(u) = \mathrm{sinc}\,(\frac{\pi z}{\lambda f^2}(p'-f)u(2b-(p'-f)u)) \quad (10)$$

$\gamma(u)$ is maximum for $u = 0$ and $u = 2b/p'-f$. If z is such that the argument of the sinc function is less than π for all values of u lying between 0 and $2b/p'-f$, the variations of $\gamma(u)$ versus the values of u are represented on Figure 3. The minimum value of $\gamma(u)$ occurs for $u = b/p'-f$ and is given by

$$\gamma_{min} = \mathrm{sinc}\,(\frac{\pi z}{\lambda f^2}b^2) \quad (11)$$

γ_{min} is equal to zero for $z = \lambda f^2/b^2$ which represents the depth of focus of the optical system. When z is more than $\lambda f^2/b^2$, $\gamma(u)$ cancels out for the values of u symmetrical with respect to $b/p'-f$ and given by

$$u = \frac{b}{p'-f} \pm \frac{b}{p'-f}\,(1 - K\,\frac{\lambda f^2}{z b^2})^{1/2} \quad (12)$$

where K is an integer. These values of z correspond to displacements of the object which are larger than the

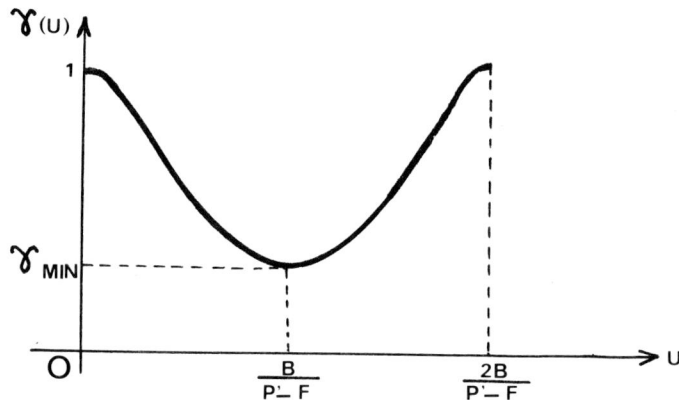

Fig. 3. Variations of $\gamma(u)$ when the pupil is placed in the back focal plane of L.
depth of focus of the optical system.

Pupil in plane of the lens L

We shall assume now, that the object O, limited by a slit of width 2a, is imaged in plane I by the lens L limited by a slit of width 2b (Figure 4).

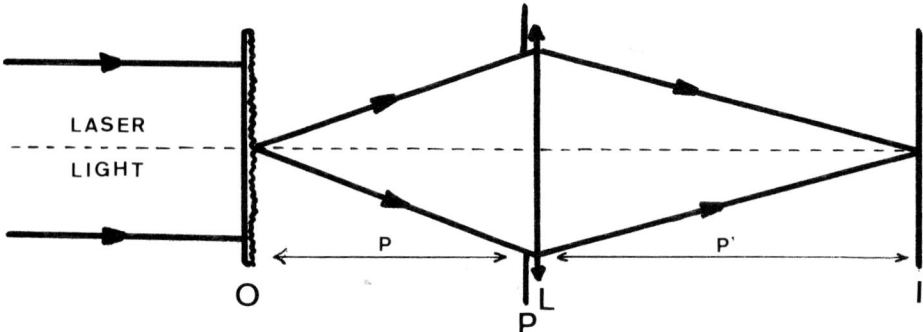

Fig. 4. The object O is imaged by lens L limited by the pupil P

The amplitude distribution in image plane is proportional to

$$U_1(x') = \int O(x) P(X) \exp(j \frac{\pi x^2}{\lambda p}) \exp(-j \frac{2\pi}{\lambda p} xX) \exp(-j \frac{2\pi}{\lambda p'} x'X) \, dx \, dX \qquad (13)$$

After an axial shift of the object through z, Formula (13) may be written as

$$U_2(x') = \int O(x)P(X)\exp(j\frac{\pi x^2}{\lambda p})\exp(j\frac{\pi zx^2}{\lambda p^2})\exp(-j\frac{2\pi}{\lambda p}(1+\frac{z}{p})Xx)\exp(-j\frac{2\pi}{\lambda p'}x'X)\,dx\,dX \quad (14)$$

It can thus be seen that the corresponding speckle patterns are now decorrelated and radially shifted. A photographic plate H records the irradiance of plane I before and after the axial shift of the object in the same way as above. After processing H generates a system of Young's fringes in its Fourier plane, the angular spacing of which is λ/x_0' and the mean contrast of which is now given, with the same statistical assumptions as in formula (10), by

$$\gamma(u) = \operatorname{sinc}(\frac{\pi z}{\lambda p^2}p'u(2b-p'u))\operatorname{sinc}(\frac{2\pi z}{\lambda p^2}p'ua) \quad (15)$$

The first term of $\gamma(u)$ characterizes only the decorrelation of the speckles due to the defocusing. The second term displays the radial shift of the corresponding speckles involved by the axial translation of the object. As a matter of fact, the amount of shift suffered by a speckle of image plane is proportional to its distance from the centre of the image. The maximum value of radial shift d occurs for the speckles which are located at $(p'/p)a$ of the centre of the image and is equal to

$$d = z\frac{p'}{p^2}a \quad (16)$$

Let us suppose that the respective values of b and z are such that z is very less than $\lambda p^2/b^2$. Under these conditions, the first term of formula (15) is practically equal to 1 and

$$\gamma(u) = \operatorname{sinc}(\frac{2\pi}{\lambda}du) = \operatorname{sinc}(\frac{\pi d}{b\sigma}p'u) \quad (17)$$

where $\sigma = \lambda p'/2b$ is the mean size of the speckles of image plane. The Fourier spectrum of the photographic plate is limited by the autocorrelation function of the pupil used during the recording process. Under these conditions, the value of the angle u cannot be more than $2b/p'$. If the maximum radial shift d is less than $\sigma/2$, the contrast $\gamma(u)$ decreases continuously from the center of the field where it is equal to 1 (Figure 5a) to the edge of the field where it is equal to $\operatorname{sinc}(2\pi d/\sigma)$. If $d = \sigma/2$, (i.e. if $z = \lambda p^2/4ab$), the contrast of the fringes decreases from 1 to 0. If d is lying between $\sigma/2$ and σ (i.e. if z is comprised between $\lambda p^2/4ab$ and $\lambda p^2/2ab$) the contrast of the fringes cancels out for a value of u

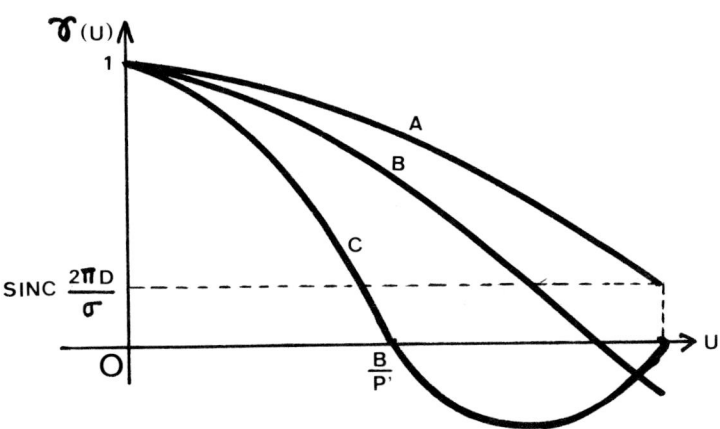

Fig. 5. Variations of $\gamma(u)$ when the pupil is in the plane of L.
(a): $z < \lambda p^2/4ab$ (b) $\lambda p^2/4ab < z < \lambda p^2/2ab$ (c) $z = \lambda p^2/2ab$

lying between b/p' and 2b/p' and is then inversed (fig. 5b). When z is equal to $\lambda p^2/2ab$, the contrast of the fringes (fig. 5c) is maximum for u = 0, cancels out for u = b/p', is inversed and cancels out at the edge of the field.

Figures 6a and 6b represent respectively the Fourier spectra of two photographic plates twice exposed to the irradiance of image plane of a diffusing object. In each case the image of the object is formed with a lens limited by a circular aperture the radius of which is 1 cm. Between the exposures, the object is axially shifted through 135 μ. Figure 6 a corresponds to an object limited by a circular aperture, the diameter

a

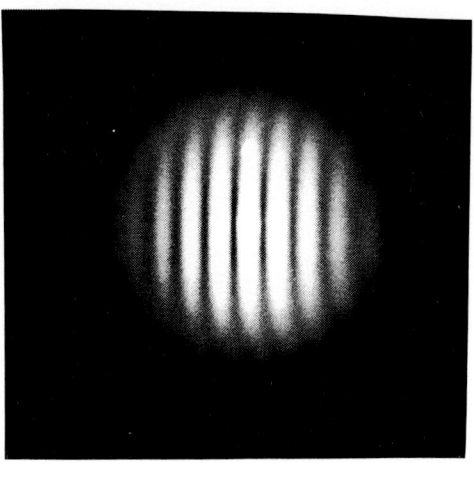
b

Fig. 6 . Fringes lying in the Fourier plane of the processed photographic plate corresponding to z = 135μ , with 2a = 1 cm (6a) , with 2a = 4 cm (6b).

of which is 1 cm. It can be seen that the contrast of the fringes is practically maximum over all the diffused field. In this case z is very inferior to $\lambda p^2/4ab$. The fringes of figure 6b have been obtained with an object limited by a circular aperture, diameter of which is 4 cm (z is approximately equal to $\lambda p^2/2ab$). They disappear at the middle of the field and reappear with an inversed contrast.

Conclusions

We have shown in this paper that the radial shift of the speckles in the image plane, due to an axial shift of the object, is removed if the pupil of the optical system is placed in the back focal plane of the imaging lens. A photographic plate twice exposed to the irradiance of the image plane and laterally shifted between the exposures, exhibits after processing, a system of Young's fringes in its Fourier plane, the contrast of which is only a function of the defocusing. The measure of its minimum value gives the value of z. If the pupil of the optical system is placed in the plane of the imaging lens, the corresponding speckles are radially shifted by the axial translation of the object and the contrast of the fringes generated by a photographic plate, in the same way than above, depends only, when the defect of focus is negligible, on the value of the maximum radial shift suffered by the speckles i.e. on the size of the object.

References

1. Archbold E. and Ennos A.E. , "Displacement measurement from double exposure laser photographs" , Opt. Acta , vol. 19 , pp. 253-271 , 1972.
2. Mendez J.A. et Roblin M.L. , "Relation entre les intensités lumineuses produites par un diffuseur dans deux plans parallèles , Opt. Commun. , vol. 13 , pp. 142-147 , 1974
3. Eliasson B. and Mottier F.M. "Determination of the granular radiance distribution of a diffuser and its use for vibration analysis". J. Opt. Soc. Am. , vol. 61 , pp. 559-565 , 1971.
4. Burch J.M. and Tokarski J.M.J."Production of multiple beam fringes from photographic scatterers" , Opt. Acta , vol. 15 , pp. 101-111 , 1968.
5. Goodman J.W. "Statistical Properties of Laser Speckle Patterns" , Laser Speckle and Related Phenomena , Topics in Applied Physics , vol. 9 , Edited by J.C. Dainty , Springer-Verlag , pp. 9-74 , 1975.

THE DIFFRACTING GAUGE IN EXTENSOMETRY

Jean P. L. Ebbeni
Laboratoire d'Analyse des Contraintes
Université Libre de Bruxelles
87, Avenue Ad. Buyl, 1050 Bruxelles, Belgium

Abstract

The principe of the diffracting gauge is based on the farfield diffraction spectrum of a slit. In extensometry this gauge permits to measure easely and cheaply strain components with great sensibility and fidelity in time.

Introduction

The diffracting gauge has two independant parts : (i) the supports and (ii) thin plates (cfr figure 1). The supports are cimented on a model, and the cylinders are centred on points O_1 and O_2; $O_1O_2 = \ell$ is the basis of the diffracting gauge of direction Ox. The arms are oriented so that the thin plates are coplanair and adjusting screws permits the parellelism of the extremities $A_1 B_1$ and $A_2 B_2$ along the direction Oy perpendicular to Ox. The wave planes of a laser beam are sended normally on the slit and the diffracted waves are received on a screen parallel to the thin plates at a distance L. The theory of diffraction shows that it appears on the screen a family of equidistant rectilinear fringes with an interfringe

$$d_1 = \frac{\lambda L}{a} \tag{1}$$

where λ is the wavelength and a the slitwidth. When the model is deformed, the basis ℓ becomes $\ell' = \ell + \Delta\ell$ and the new interfringe is given by

$$d_2 = \frac{\lambda L}{a + \Delta\ell} \tag{2}$$

From the measured values d_1 and d_2 it is possible to calculate the strain component ε_x according to

$$\varepsilon_x = \frac{a}{\ell} \left(\frac{d_1}{d_2} - 1 \right) = - \frac{a}{\ell} \frac{\Delta d}{d_2} \tag{3}$$

where $\Delta d = d_2 - d_1$.

Sensibility of the method

The accuracy on ε_x depends essentially of the accuracy on the measure of Δd. For $\lambda = 0,6328\ \mu$, $1\ m < L < 25\ m$ and $3\,10^{-4}\ m < a < 5\,10^{-4}\ m$ one obtains easely an accuracy of $10^{-4}\ m$ on Δd which gives the minimum measurable value of ε_x.

$$\varepsilon_{x\ min} = \pm 10^{-4} \frac{a}{\ell d_2} = \pm 10^{-4} \frac{(a+\Delta\ell)}{\lambda L \ell} \simeq \pm \frac{10^{-4} a^2}{\lambda L \ell} \tag{4}$$

where a, ℓ, L and d_2 values are expressed in meters. The different values of $\varepsilon_{x\ min}$ are given by figure 2.

Parasite effects

It can happen that the slit width is not constant due to a relative rotation of the two arms in their plane. By measuring the slit width in two different zones, e.a. at A_1A_2 and B_1B_2 it is easy to show that

$$\Delta\ell = \Delta a = \frac{\delta\ \Delta a\ (A) - (\delta+h)\ \Delta a\ (B)}{h} \tag{5}$$

where $\Delta a(A)$ is the variation of the slit width in A
$\Delta a(B)$ is the variation of the slit width in B
δ, $\delta+h$ are the Oy coordinates of B and A.

In practice :
$$\Delta a(A) = \Delta a(B) + \eta \text{ where } \eta \text{ is of the second order and then } \Delta \ell = \Delta a(B) + m \frac{\delta}{h} \quad (6)$$

If δ is small enough (thin plates near the surface of the model) this effect of rotation is negligeable. A classical computation shows that the out of plan motion of the thin plates is without effect if the distance between the two plates is lower than 3.10^{-4} m for 3.10^{-4} m $< a < 5.10 m^4$. Figure 3 shows that the curves ε in function of the temperature are linear : it is thus easy to calculate the temperature compensation and an error of 1°C gives a fictive value of 4 μ S if the model material is concrete.

Measure by reflection

By cimenting a small miror (thin aluminised piece of glass) on one of the thin plates (cfr figure 1) it is possible to work by reflection. By special set-ups of mirors it is possible to measure the strain components in any direction at places inaccessible by other extensometers.

Applications

Figure 4 shows the good agreement between the results obtained with the diffracting gauge and on electrical gauge : small differences are due probably to the creeping effect of the ciment of the electrical gauge. An application of the method to a problem in civil engineering was presented during the Congress : it consisted to measure the variations of thickness of a crack in a model in concrete where the crack was primarily filled with araldite. The loading system is illustrated by figure 5.

Conclusions

Extensometry by diffracting gauge is interesting by its great sensibility, its low cost, its accuracy and its possibility of applications in zones of the model where it is difficult to have access.

References

1. Born α Wolf, Principles of Optics, Pergamon Press
2. Beurms D., Une nouvelle technique extensométrique : la jauge diffractante - travail de fin d'études 1976 - Service d'Analyse des Contraintes - Université Libre de Bruxelles
3. Bonnelance P., Mise au point de la jauge diffractante, travail de fin d'études 1977, Service d'Analyse des Contraintes, Université Libre de Bruxelles.

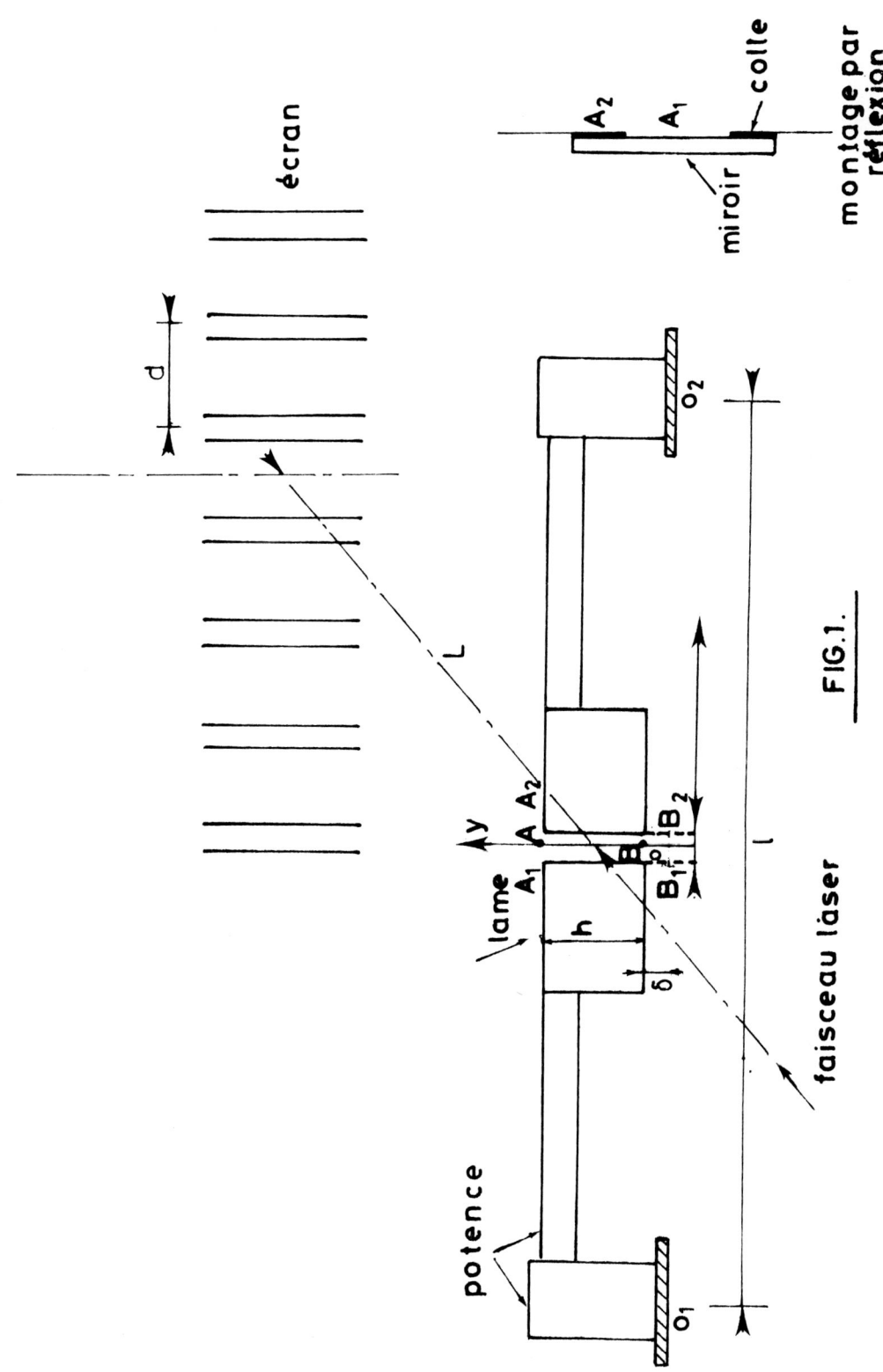

FIG.1.

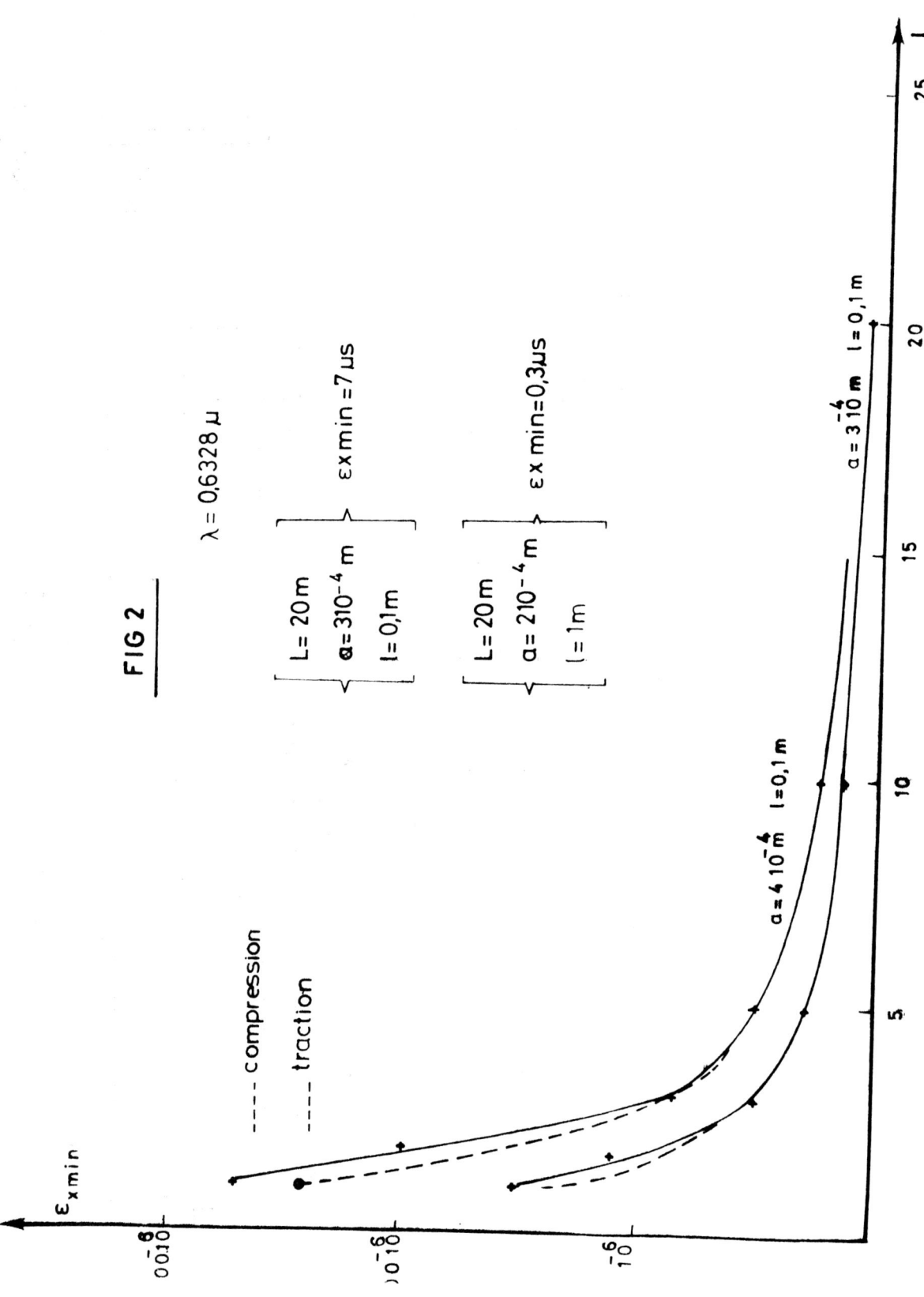

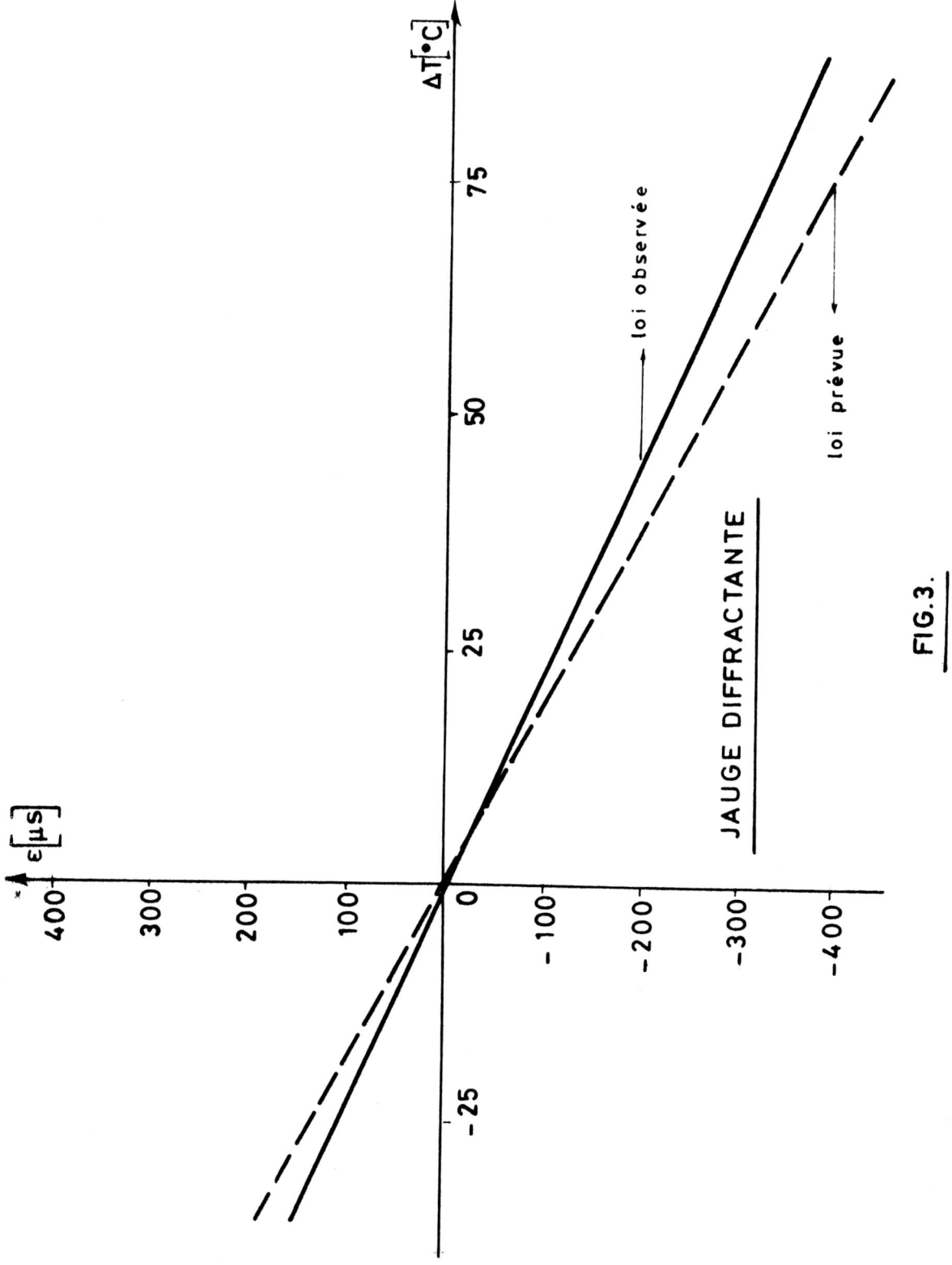

FIG. 3.

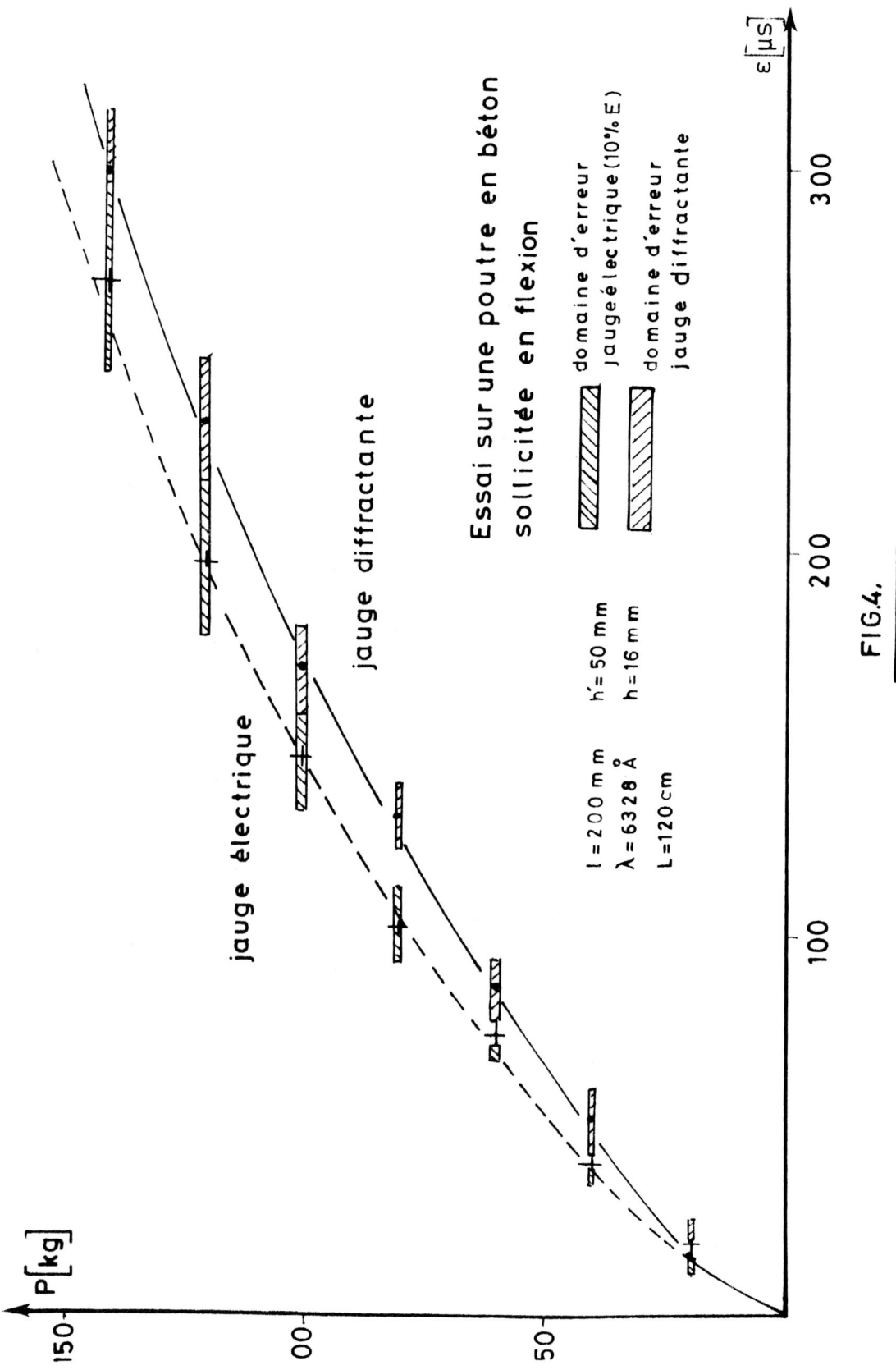

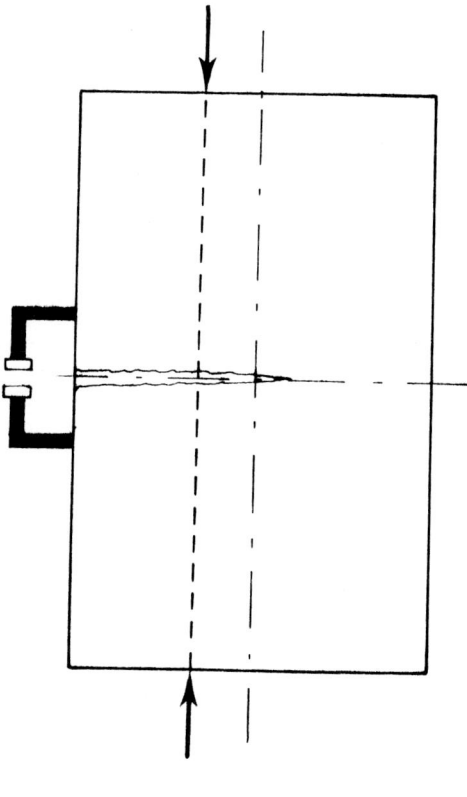

FIG. 5

SPECKLE PHOTOGRAPHY FOR STRAIN MEASUREMENT — A CRITICAL ASSESSMENT

E. Archbold, A. E. Ennos and M. S. Virdee

Division of Mechanical and Optical Metrology, National Physical Laboratory
Queens Road, Teddington, Middlesex, TW11 OLW, UK

Abstract

Speckle photography is a simple technique for measuring displacement in the plane of a surface, so allowing the strain field to be evaluated in two dimensions. A direct experimental comparison of speckle photography with photoelasticity and with finite element analysis on a notched tensile test specimen showed that serious errors could, however, occur due to local surface tilting and to aberrations of the imaging lens. Methods of minimising these effects and those of the speckle de-correlation that can also take place, are discussed.

Introduction

Speckle Photography is a simple non-contacting method for analysing surface displacement. In its basic form, the object to be studied is illuminated with a beam of laser light from any convenient direction, and a double exposure photograph of the surface recorded, on fine-grain film, the object undergoing the displacement to be measured between exposures. The magnitude and direction of the displacement, normal to the light of sight, can then be obtained by forming the diffraction spectrum of the developed negative, which consists of a pattern of parallel equi-spaced dark bands, similar to Young's fringes. Since the object surface is optically imaged on to the photographic plate, a displacement vector that varies from point to point can be analysed by using a sufficiently narrow light probe to form the diffraction fringes.

In a previous paper[1] some assessment was made of the range over which one might use speckle photography for displacement measurement. For very small displacements, it was evident that the greatest sensitivity should be obtained when using a camera lens of large aperture, operating at object-image conjugates near 1:1. Under these conditions an uncertainty of as low as ~ 0.1 μm might be expected in the displacement values, judged from the accuracy of measuring the fringe pattern. However, a number of assumptions were made in respect of the transfer characteristics of the imaging lens, and it was felt that a more critical assessment of the method should be made if absolute values of displacement were to be obtained. This is particularly important if the values of vector displacement are to be used for the measurement of strain, since a very small error in measured displacement can give rise to a large strain error relative to the values encountered in practical cases. Experiments were thus carried out using a model whose surface could be strained in a controlled manner, and whose strain field could be calculated theoretically, and also measured by an independent technique. A laminar tensile test bar made of Araldite epoxy resin was chosen for this purpose, since the stress field within it can be measured photoelastically, and the stresses then converted algebraically into strains using the elastic constants of the material. The theoretical strain distribution over the surface was calculated by finite element analysis.

This paper deals principally with the problems encountered in obtaining speckle photographic data of comparable accuracy to those obtainable by the other two methods. Only the results of the latter will be quoted; a fuller account will appear elsewhere.

Experimental Arrangement

The specimen to be studied was a strip of Araldite C200 plastic, 102 mm wide and 1.56 mm thick, with two V-notches having rounded ends machined into it, as shown in Fig. 1. (This was made in the Department of Mechanical Engineering, University of Surrey). After annealing the plastic, and coating the area to be examined with matt white paint, the strip was mounted in a tensile testing frame, with whipple-tree attachments at either end to distribute the load uniformly. The load was applied by screw action, via a spring and a calibrated load cell for measuring the force. The test frame was mounted vertically and the whitened surface imaged by means of a camera lens on to a photographic plate mounted parallel to the specimen. Illumination was provided by an argon laser, the green light from which was directed obliquely on to the specimen surface. Double exposure photographs were recorded on Agfa-Gevaert 10E56 photographic plates.

Since the strip was held at one end and loaded at the other, the central area containing the notches suffered a large axial displacement as well as becoming deformed. To minimise this movement, which would result in a too-closely spaced Young's fringe pattern, the camera lens and plate holder were mounted as one unit upon an adjustable stage which could be raised by a pre-determined amount to compensate for the displacement at the centre. Initial experiments also showed that it was necessary to control the movement of the thin plate specimen both sideways and in the line of sight direction. Sideways movement was restricted by two rods locating against the edge of the specimen, and two guide-ways coated with PTFE were used to keep the face approximately in the same plane when it stretched.

For the initial tests an 80 mm focal length anastigmat lens, of f/4 aperture, was used to image the surface at 1:1 magnification. A field of view of about 50 mm was covered.

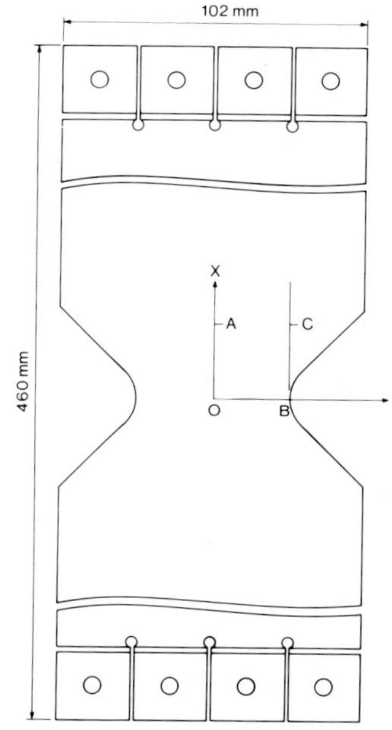

Determination of Strain Field

A double exposure photograph was first recorded, applying the full load of 800 N between exposures. This imposes 1.5 millistrain on ends of the specimen. The processed plate was then mounted on a coordinate slide so that the area around the notch could be analysed point-by-point by directing a laser beam ($\sim \frac{1}{2}$ mm diameter) through it at normal incidence. The Young's fringe pattern generated by the doubled speckle was projected on to a screen covered in graph paper, and the orientation of the fringes relative to the longitudinal axis of the specimen was measured (as angle θ) by rotating the screen about its centre until the fringes and coordinate lines were parallel. The spacing of the fringes was obtained by measuring the distance D between the two extreme bright fringes with a pair of dividers, and noting the number n of fringes between them. The speckle displacement d' on the photographic plate is then given by

$$d' = n\lambda L/D \qquad (1)$$

where λ is the wavelength of the analysing laser light, and L the distance between photographic plate and screen. The lateral displacement d of the object surface is given by

$$d' = d'/M \qquad (2)$$

Fig. 1. Notched tensile test bar (Araldite).

where M is the magnification of the photographic system.

To evaluate the strain field, values of d and θ were measured for points on a square mesh lattice of 4 mm spacing. The components $d_x = d\cos\theta$ (across the specimen) and $d_y = d\sin\theta$ (along it) were determined, and from each of these were subtracted the component values at the centre of the specimen; this eliminated rigid body translation, but not rotation. A graphical illustration of the distortion of the specimen around the notch tip is shown in Fig. 2, where the movement of the mesh points has been magnified 130 times. The strain at any one point (m,n) of the mesh is obtained from the displacement components at surrounding points. If the linear strain in the coordinate directions are ε_x and ε_y, and the shear strain γ_{xy}, then

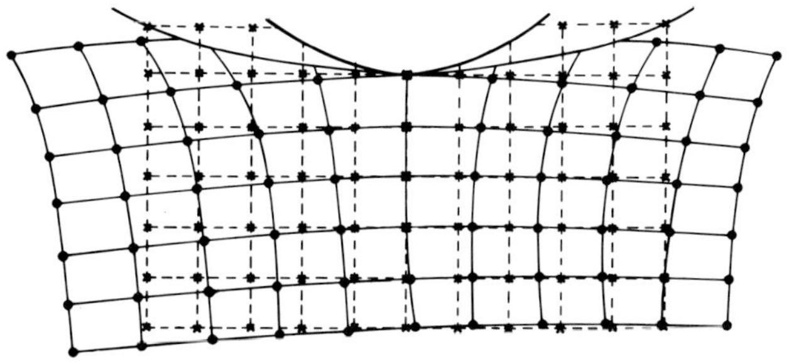

$$(\varepsilon_x)_{m,n} = \frac{(d_x)_{m+1,n} - (d_x)_{m-1,n}}{2\Delta},$$

$$(\varepsilon_y)_{m,n} = \frac{(d_y)_{m,n+1} - (d_y)_{m,n-1}}{2\Delta},$$

$$(\gamma_{xy})_{m,n} = \frac{(d_x)_{m,n+1} - (d_x)_{m,n-1} + (d_y)_{m+1,n} - (d_y)_{m-1,n}}{2\Delta} \qquad (5)$$

Fig. 2. Distortion of bar around the notch, measured by speckle photography.

where Δ is the mesh spacing.

A computer programme to evaluate the strain from these equations was written, using D, n and θ as input. However, for greater accuracy, especially in regions of rapid change of strain, it was found better to plot d_x and d_y graphically against x and y, and then measure its slope in order to obtain the strain.

Experimental Results and Investigation of Errors

In the initial experiments aimed at comparing strain measurements by the three different techniques, the linear strains ε_x and ε_y only were computed along the directions OX, OY and BC (Fig. 1), and the results were presented in terms of strain ratio $\varepsilon/\varepsilon_o$ i.e. the ratio of the strain ε at the measured point to the uniform strain ε_o occurring well away from the notch region. A typical result, shown in Fig. 3,

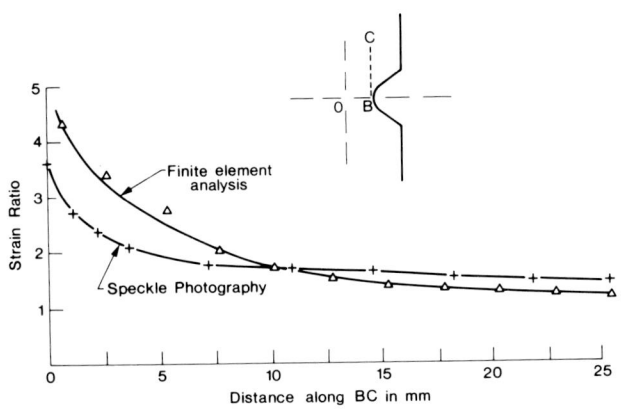

Fig. 3 Comparison of strain distribution, measured by speckle photography, with calculated values.

gives the variation of $\varepsilon_x/\varepsilon_o$ along BC, compared with the variation predicted by finite element analysis and measured photoelastically. Considerable differences in the curves are apparent, particularly at the two extremities of the field. These could not be accounted for by a systematic error in measuring the fringe patterns from the speckle photographs, and it was suspected that they might be due to uncontrolled movements of the strained plastic, such as residual motion in the line of sight direction, and warping (tilting) of its surface. Consideration was therefore given to the effects that these motions might have when superimposed upon a translational displacement in the plane of the object. Geometrical considerations ignoring lens aberrations) predict the following:-

i) <u>Effect of Line-of Sight Motion</u>

If the object moves bodily towards the camera lens by an amount δZ (Fig. 4(a)), the position of a point distant X from the centre of the field will appear to move outwards by $\delta Z \tan \phi = \frac{\delta Z}{Z} \cdot X$, where Z is the object distance. If a double exposure speckle photograph is recorded when this motion occurs, the analysis of it will predict that the surface has suffered a radial strain of value $\frac{\delta Z}{Z}$. In terms of the imaging system used in our experiments, an out-of-plane motion of 50 μm (which is within the allowable depth of focus will give rise to an apparent 300 microstrain.

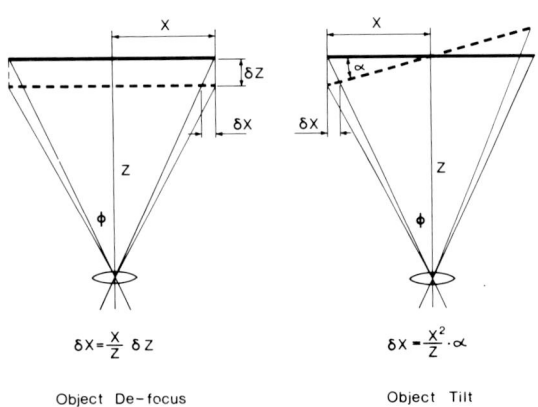
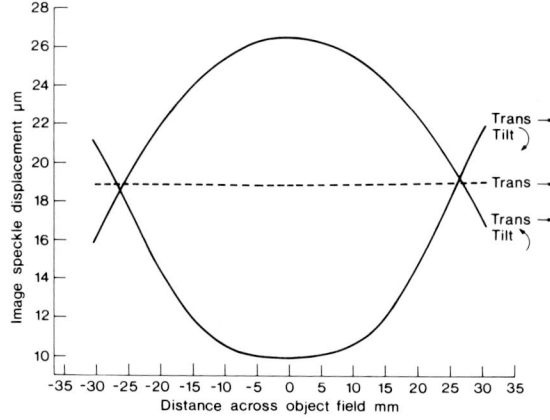

Fig. 4. Effect of object de-focus and tilt on apparent lateral displacement.

Fig. 5. Measured image displacement caused by object translation and tilt.

ii) <u>Effect of Object Tilt</u>

If the surface tilts through a small angle α (Fig. 4(b)), points on the surface will appear to move outwards on one side of the field of view by a distance $\frac{\alpha}{Z} X^2$, and inward by the same amount on the other. This will give rise to an apparent strain that increases linearly with X from the centre. For a 1 mr rotation, for example, its value will be 300 microstrain at the edge of the field.

The above prediction was verified experimentally by imaging a flat white-painted plate that could be translated and tilted in a controlled manner. Fig. 5 shows the apparent image movement measured by speckle photography when a tilt of 2 mr was applied in addition to a 20 μm translation.

Large errors in strain measurement can thus occur when a wide angular field of view is employed, unless the object moves only in its own plane. In principle these can be eliminated by using a telecentric imaging system, in which only those cones of rays scattered from the object that have their axes normal to its surface are selected by the imaging lens. The surface will then be viewed effectively at normal incidence over its entire area, and the geometrical effects of a diverging view eliminated.

Telecentric Imaging System

The telecentric system set up to test the above prediction is shown in Fig. 6. The main imaging lens was a Dallmeyer Octac f/1.5 lens, 80 mm focal length, designed for 1:1 conjugates and stopped down to f/4. A plano-convex lens, focal length, 170 mm was mounted at its focal distance in front of the camera lens, to act

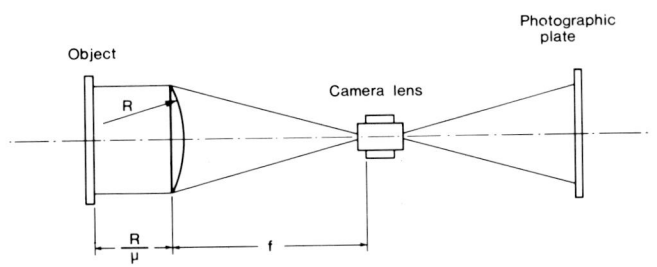
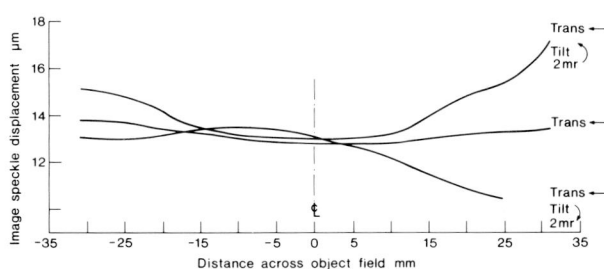

Fig. 6. Telecentric imaging system.

Fig. 7. Apparent image displacement due to object tilt, using telecentric system.

as the telecentric element. The object was mounted at a distance R/μ in front of this lens, in the position that should give minimum image distortion. Double-exposure photographs of the metal plate were recorded with lateral translation of the object and with combined translation and tilt. Analysis of the photographs (Fig. 7) shows that little variation of image displacement occurs across the field for pure translational motion, but that there is a large variation when object tilt is present. The direction of the error (whether larger or smaller than the mean value) depends upon the relative directions of tilt and translation. The 3 μm variation between centre and edge of field for a 2 mr tilt, shown in Fig. 7, represents an apparent strain of 120 microstrain.

The large effect that object tilt has upon measurement of lateral translation can be ascribed qualitatively to aberrations of the imaging system. Although the lens may form an image with low field distortion (giving a linear transference of in-plane translational motion) the field curvature and astigmatism will cause the image speckle pattern to move when the object is tilted. Figure 8 shows how this happens. The surface imaged by the aberrated lens on to a flat photographic plate will not be the flat object surface, but a curved one lying in front of it, shown by the broken line. At the edges of the field, which are thus out of focus by δZ, a tilting of the object through an angle α will cause the speckles in the imaged plane to move laterally by a distance $2\alpha\delta Z$. In terms of the observed variations of translation across the field shown in Fig. 7, it is necessary for the best focus at the edges of the field to be only 0.75 mm different from that at the centre to account for the discrepancy.

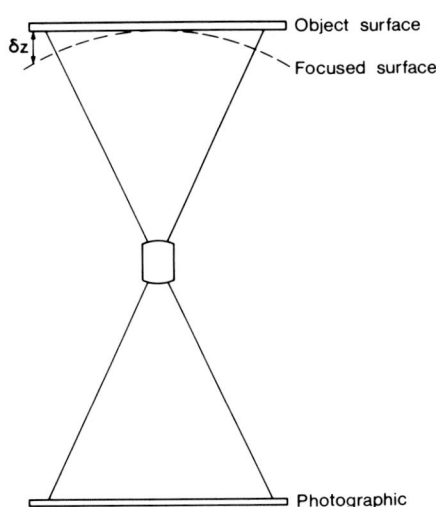

A number of telecentric systems using different powers of planoconvex lens were tested, all giving considerable distortion of the lateral translational motion when surface tilt was present. It is most probable that the single plano-convex telecentric element is responsible for much of the effect. In the absence of a better way of providing a wide-field telecentric system, for example by a novel lens design, it was decided to minimise the off-axis field effects by employing a long focal length imaging lens, without the additional telecentric element, so as to utilise only a narrow angular field.

Narrow-field System

Two aerial survey camera lenses of focal length 750 mm and 500 mm were mounted coaxially, face to face, to give minimum aberrations when object and image were situated in the two focal planes. The angular field of view of this system is less than one quarter that of the telecentric systems previously employed. Speckle photographs recorded with this system showed that the transfer of in-plane motion in the presence of a tilt is much more linear, as shown by Fig. 9. A variation of only 0.5 μm occurs across the field for a 2 mr tilt.

Fig. 8. Effect of lens aberrations (simplified).

Using this narrow field system, a double exposure photograph of the notched Araldite specimen was recorded and analysed. From measurements of displacement, the strain ratios in the two coordinate directions were calculated for points along the lines OX, OY and BC. Two of these results are shown in Fig. 10, together with corresponding curves obtained by photoelastic measurements of the stress field made at the University of Surrey. (The stresses were converted to strains using well-established values of Young's modulus and Poisson's ratio for the material). Finite element analysis curves (provided by the National Engineering

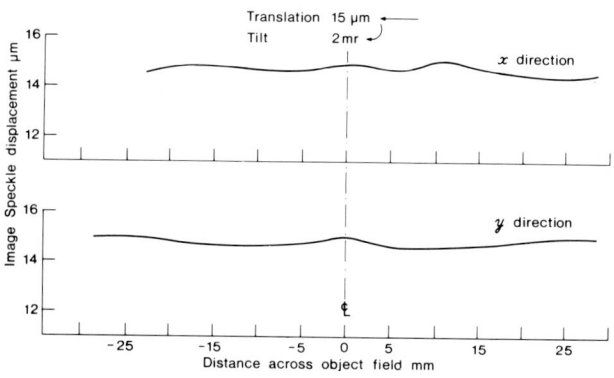

Laboratory) are also plotted on the graphs.

For the direction OY, shown in the curves, the photoelastic and finite element analysis measurements agree well, while speckle photography gives reduced values of strain, especially at the notch tip where the strain is high and changing rapidly. This can partially be accounted for by the necessity for examining a larger area of the specimen when using speckle photography (0.5 mm) than when a photoelastic measurement is made (0.1 mm). For the other directions, discrepancies between the curves of up to 300 microstrain occurred in places, and this can only be ascribed to unknown warping of the surface.

Fig. 9. Transfer characteristics of long focal length, narrow field imaging system.

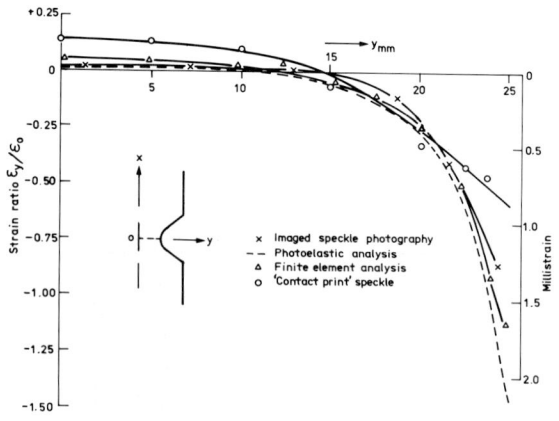
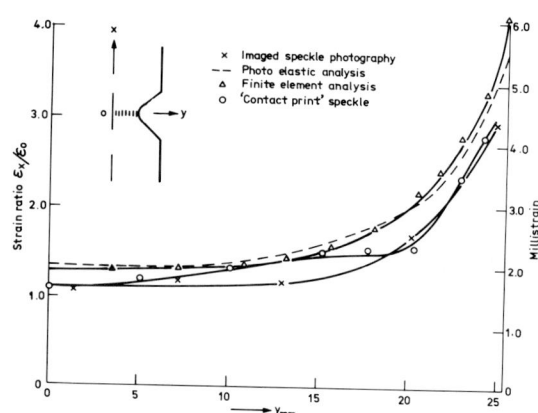

Fig. 10. Strain distribution in notched Araldite bar, measured by speckle photography, photoelastic analysis, and 'contact-print' speckle, compared with calculated distribution. (a) lateral strain along OY, (b) longitudinal strain along OY.

'Contact Print' Speckle Photography

Errors in displacement measurement using imaging systems have been shown to relate to unknown tilts of the surface. For a transparent material such as Araldite epoxy resin, these may be overcome by making a 'contact-print' speckle photograph. For this purpose a photographic plate is mounted parallel to the transparent specimen spaced by 1-2 mm, and a collimated laser beam is directed through the plastic on to the plate, (Fig. 11). The light scattered internally is sufficient to generate a speckle pattern, and although this is superimposed on the directly transmitted light, the 'contact-print' recording will, when analysed, generate Young's diffraction fringes of adequate visibility for measurement purposes. The speckle pattern is formed by scattering centres within the plastic, so some loss in the spatial resolution occurs over the specimen. This will only be serious in regions where the strain is changing rapidly. The effect of specimen tilt will, however, be much reduced, since lens imaging is not used, and forward scattered light is used. A lensless technique of similar nature, described by Boone[2] used the light scattered back from the specimen surface. When this was tried on the Araldite specimen it gave serious speckle de-correlation effects due to the surface tilt.

Results of measurements taken from double exposure contact print recordings of the strained Araldite specimen are also plotted on the graphs of Fig. 10. These diverge from the measurements obtained using the other methods, to different extents depending upon the direction in which the strain is measured. However, in regions of high strain the 'contact print' method generally gives lowered values, presumably due to the reduced spatial resolution caused by imperfect 'contact'.

De-correlation Effects

The ability to analyse surface deformation by speckle photography can be severely limited by speckle de-correlation effects. If the detailed configuration of the pattern in any part of the image changes between the two exposures, the Young's fringe pattern corresponding to that area will be reduced in contrast, and in certain cases the fringes may even vanish. The principal cause of speckle de-correlation is excessive movement of the surface in the line-of-sight direction. Theory predicts that, for a unit magnification system, this movement must be less than $\Delta z \sim 4F^2\lambda$ (the Rayleigh criterion), where F is the effective aperture ratio of the lens. For the systems studied in this investigation, $\Delta z \sim 150$ μm so it is not surprising that great care was required in keeping the specimen movement to one plane, and in preventing surface tilt when it was strained.

De-correlation effects can also manifest themselves as a more selective loss of contrast of the fringe pattern, affecting only part of the diffraction halo field. An example of this is shown in Fig. 12(a). To interpret this effect one must remember that the outer parts of the diffraction halo are generated by components of speckle pattern that are formed by peripheral zones of the imaging lens. For a lens suffering from oblique astigmatism, the focal plane for off-axis sagittal rays will be different from that of the tangential rays, so that, in the diffraction halo, the Young's fringe pattern will be affected by surface tilt in one azimuth, and not in the other. De-correlation will occur when the difference in focal planes is sufficiently large, and surface tilt is present.

A further effect due to lens aberration is shown in Fig. 12(b). The S-shaped fringe pattern is due to spherical aberration, coupled with tilting of the surface. This may be explained as follows:- Since the peripheral regions of the lens aperture focus the light more strongly than the central region, the photographic plate will be imaged back into object space by the outer parts of the lens on to a plane

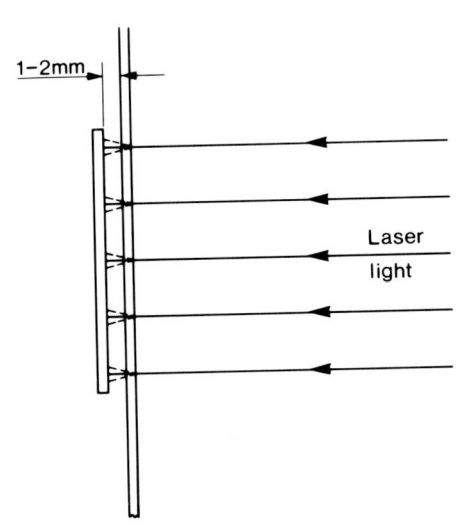

Fig. 11. 'Contact print' speckle photography.

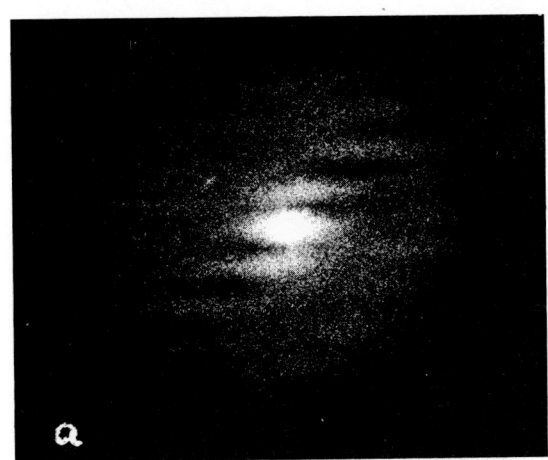
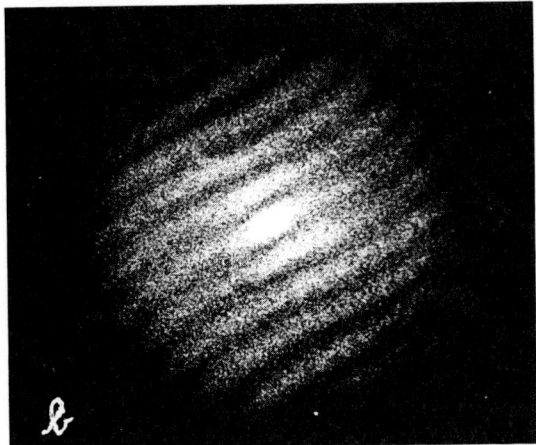

Fig. 12. Effect of lens aberrations combined with object tilt on Young's fringe patterns, (a) astigmatism, (b) spherical aberration.

lying in front of the object. Speckles in this plane are sensitive to object tilt, so that a modification to the apparent displacement vector will be caused. This manifests itself in the diffraction halo as a small change in direction of the fringes.

Both of the above types of decorrelation effect will reduce the accuracy of in-plane displacement measurement, since they give rise to an uncertainty in the true spacing of the Young's fringes, or limit the angular field within which they can be measured.

Conclusions

Speckle photography has many advantages when used for measuring in-plane displacement, especially when a reduced sensitivity (compared with interferometric methods) is required. However, if applied at the highest sensitivity, for example to measure strain fields, considerable errors may arise which are shown to be due to lens aberration effects. Even under the best conditions of a flat object being imaged by a well-corrected lens of limited field angle, uncertainties of ~ 100 microstrain may still occur. The situation is worsened if the object is not flat, when a tilting of the surface will give rise to a large apparent strain.

Acknowledgements

The authors are indebted to Prof. I M Allison, University of Surrey, Guildford, for provision of the

Araldite tensile test specimen and for carrying out the photoelastic stress measurements on it. Also to Mr M H Hodges of the National Engineering Laboratory, East Kilbride for performing a finite element strain analysis of the structure.

References

1. Archbold, E. and Ennos, A.E., 1972, Optica Acta 19, 253.
2. Boone, P.M., 'The Engineering Uses of Coherent Optics', Cambridge Univ. Press, London, p. 81.

1st EUROPEAN CONGRESS ON OPTICS APPLIED TO METROLOGY

Volume 136

SESSION 6.2

SPECKLE

Session Chairmen
Fossati
Veret

STUDY OF THE DISTRIBUTION OF VELOCITIES IN A FLUID BY SPECKLE PHOTOGRAPHY

R. Grousson and S. Mallick
Laboratory of Optics, Curie University
T 13, 4 place Jussieu, 75230 Paris cedex 05, France

Abstract

We present an experiment based on speckle photography, to study the distribution of velocities in a fluid at a given instant. This technique is complementary to laser velocimetry (LDV), in which the velocity is measured at a certain point of the fluid, as function of time. A section of the fluid is lit by a laser beam widened in one direction by a cylindrical lens and an image of this section is formed on a photographic plate. When polystyrene balls are added to the fluid, the image of the section lit is speckled. The laser beam is modulated so that the section of the fluid being studied is lit by two short pulses separated by a known interval of time. The photographic plate thus records two speckles displaced from each other; this displacement is variable in the plane of the photographic plate, because of the distribution of velocities in the lighted plane. An analysis of this photographic plate permits us to construct the map of velocities of the illuminated section of the fluid.

* * *

The study of the measurement of velocities in fluids by diffusion of laser light has been greatly developed in the last few years[1]. This light is diffused by particles of microscopic dimensions, present in or artificially added to the fluid. The majority of these methods are based on the Doppler shift of the frequency of the diffused light; this displacement is related to the velocity of the particles, thus to the velocity of the fluid. An important limitation of these different methods is that they provide the instantaneous velocity at only one point. It will thus be necessary to make a great number of experiments to obtain the distribution of velocities throughout the fluid. On the contrary, the method based on speckle photography which we describe permits us to attain simultaneously the velocity at all the points of a determined section of the fluid. This experiment is based on the following well-known phenomenon: two identical speckles displaced from each other produce, in the Fourier plane of the photographic plate on which they are recorded, a system of parallel fringes whose spacing and orientation depend on the relative displacement of these two speckles.[2]

The experimental setup is sketched in Figure 1. A flow of liquid is produced in a parallelepipedal tank. Polystyrene balls are added to the liquid to increase the diffused light; the diameter of these balls is chosen such that they follow, without disturbing, the macroscopic movement of the fluid. A laser beam, enlarged by a cylindrical lens, lights a section of the liquid. A modulator placed in the path of the incident laser beam allows us to light the section of the liquid by two successive pulses. The duration τ of each pulse and the interval T between two pulses can easily vary, τ being at any rate chosen to be small compred to T. We can also use a series of pulses of equal duration τ and constant interval T. In all cases the total time of the experiment must be inferior to the correlation time of the fluid.

A lens L forms an image of the section of the liquid on a fine-grained film (Kodak microfilm or Agfa coexpan). The optical axis of the system forming the image is perpendicular to the plane lit. In this experiment we measure in effect the projection of the velocity of the fluid in this plane. The image of this plane is a speckle; the dimension of the grain of the speckle is a function of the angle from which we see the lens L from the plane of the film. This speckle will be of high contrast if the pulse time τ is such that the displacement of the polystyrene balls during this interval can be considered as negligeable. This section of liquid being lit by two successive pulses, we thus record on the photographic plate two speckles displaced from each other. This displacement is variable in the plane of the film because of the distribution of velocities in the illuminated plane. To extract from this speckle photograph the maximum of information about the distribution of velocities, it is necessary that the two speckles recorded at two different instants be correctly correlated. Since the Brownian movement tends to decorrelate these two speckles, it will thus be necessary, in a preliminary experiment, to determine the time at the end of which the two speckles are totally decorrelated.

The exposed film is developed and analysed to construct the distribution of velocities. Two methods of analysis are possible[3]; one of them consists in measuring, point by point, the velocity; the other gives a map of velocities over the whole of the image.

In the first method, a small region of the film, in which we can suppose the velocity to be constant, is isolated with the aid of a diaphragm and lit by a parallel beam of light. The observation is made in the focal plane of a convergent lens placed after the film. A system of fringes represented by the following equation is observed:

$$I(\mu_1) = \cos^2 \left[\frac{\pi}{\lambda f} (VT) \mu_1 \right] \qquad (1)$$

where VT is the relative displacement of the two speckles in the region of the film considered. The period of these fringes is $P = \frac{\lambda f}{VT}$ and their orientation is perpendicular to the velocity vector. A point by point analysis of the film will thus provide us with the map of the velocities in the lighted section of the fluid.

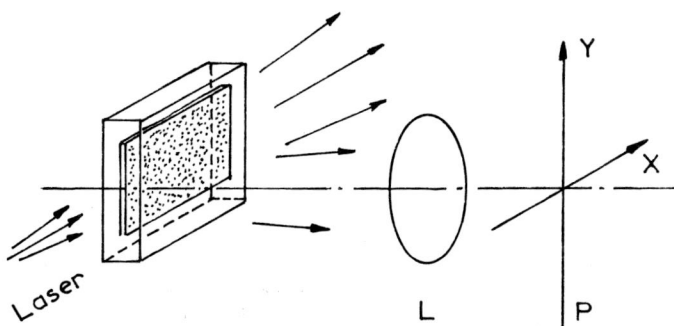

Fig. 1. Recording setup of the speckle produced by the polystyrene balls present in the fluid. The photographic plate is placed in the xy plane.

To obtain a complete image of the spatial distribution of velocities, the photographic film is observed in a filtering setup (figure 2). A plane wave of monochromatic light illuminates the film situated in the (x,y) plane; a lens L forms the image in the (x'y') plane. A diaphragm constituted of a small hole h $(R, \emptyset)$ is placed in the (x_1, y_1) plane, the focal plane of the lens L. The intensity in the image point of the point M having the velocity V in the direction μ can be calculated in the following manner. The region around the point M of the photographic film produces, in the plane of the frequencies, a system of fringes given by the following equation:

$$A(\mu_1) = \cos\left[\frac{\pi}{\lambda f}(VT)\mu_1\right] \tag{2}$$

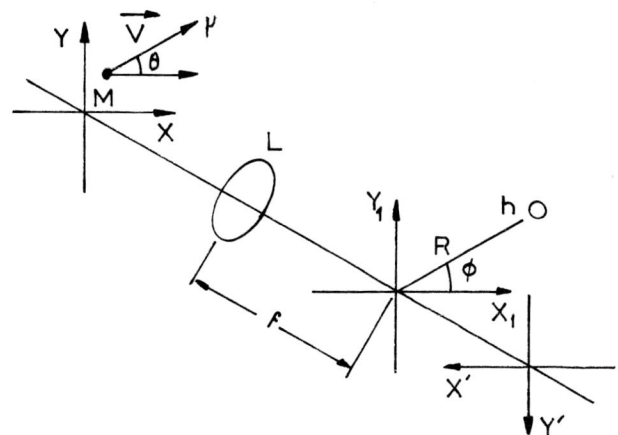

Fig. 2. Setup for observation of the photographic plate by filtering

The amplitude $A(R, \emptyset)$ at the point h of coordinates $(R, \emptyset)$ can be determined by substituting for μ_1 the coordinates of h, or

$$\mu_1 = \cos(\emptyset - \theta) \tag{3}$$

in equation (2).

$$A(R, \phi) = \cos\left(\frac{\pi RT}{f} [V \cos(\phi - \theta)]\right) \qquad (4)$$

The dimensions of the filtering hole h **are** chosen small compared to the smallest period of the fringe systems present in the Fourier plane; the distribution of amplitude in the aperture plane can be considered as uniform. The resolution of the optical system forming the image thus becomes, in this case, smaller than the displacement undergone by the object. In these conditions, the intensity in the image plane is linearly related to the square of the module of $A(R, \phi)$, or:

$$I(x', y') = \cos^2\left(\frac{\pi RT}{f} [V \cos(\phi - \theta)]\right) \qquad (5)$$

Since V and θ vary in the (x, y) plane we obtain a system of fringes in the image plane. These fringes represent the contour of $V \cos(\phi - \theta)$, that is, the component of the velocity following the azimuth of the filtering hole. By changing the distance R, the spacing between two fringes is modified.

We tested our method in a simple case, the study of a lamellar movement. This movement is created by the fall of the liquid subjected to its own weight present in the tank, the flow of the liquid being controlled by a faucet. The liquid used is glycerine, and the polystyrene balls have a diameter of 0.5 micron.

The velocity vectors (figure 3) are everywhere parallel to the y axis. The velocity is maximal at the center of the tank, and zero at the walls. Theoretically, the distribution of velocities can be represented by the expression:

$$V(x) = V_{max}\left(1 - \frac{x^2}{r^2}\right) \qquad (6)$$

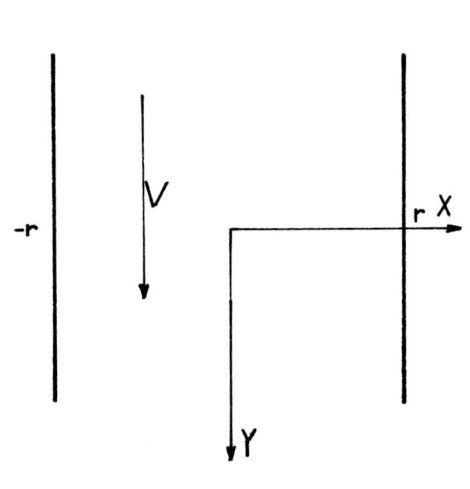
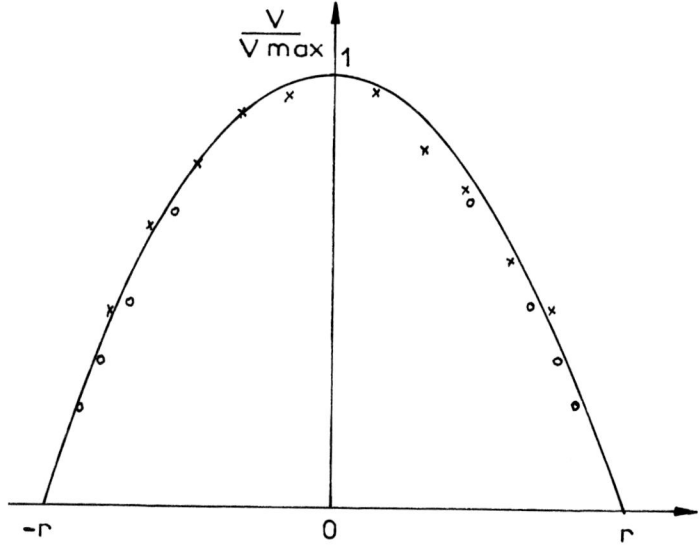

Fig. 3. Representation of the lighted section of the fluid. The walls of the tank have as abscissa $x = \pm r$.

Fig. 4. The continuous curve represents function (6); the x's, the experimental values obtained by the point by point analysis, the dots those obtained by the filtering method.

We have traced this curve (figure 4) as a function of x, and we have noted the experimental values found on the same figure. The points represented by an x **indicate** the values determined by a point by point analysis; and the points represented by a dot, those determined by the filtering method. The concordance between the experimental and theoretical results is very satisfying.

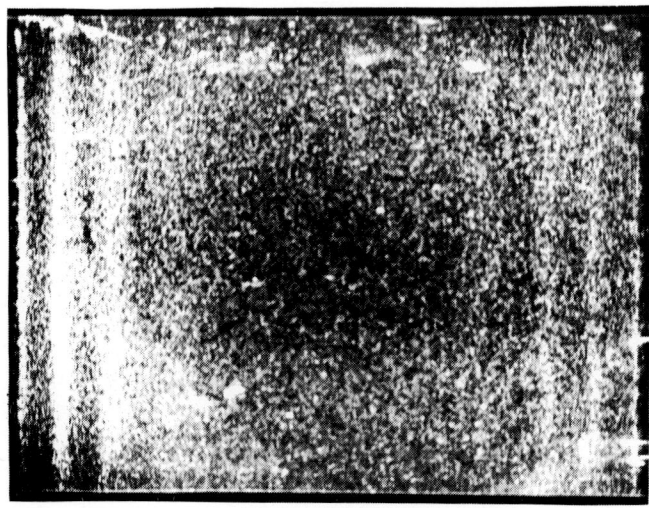

Fig. 5. Photograph of the illuminated section of the fluid, after filtering. The fringes near the wall of the tank through which the light enters show better contrast than those near the exit face. This is due to the fact that light being propagated in a diffusing fluid progressively loses its spatial coherence.

Figure 5 shows a photograph of the section of the liquid, obtained after filtering. The velocity vector is parallel to the y axis (figure 3). In the image plane the fringes obtained are parallel to the walls of the tank (that is, to the y axis).

The fringes are closer together in the region of the walls, since the velocity varies rapidly. In the center, on the other hand, the variations of velocity, and thus intensity, are small. For the bright fringes the argument of the cosine is a whole multiple of π, thus:

$$\frac{nRT}{\lambda f} V = n\pi \qquad (n = 0, 1, 2...) \qquad (7)$$

n is zero at the walls, and equal to 1 for the first fringe, and so on until the center of the tank. For the dark fringes

$$\frac{nRT}{\lambda f} V = (n + 1/2)\pi \qquad (n = 0, 1, 2...) \qquad (8)$$

The maximal velocity which we measured in this experiment is on the order of 1 mm/sec. The values of τ (duration of each pulse) and T (interval between two pulses) are respectively 5 and 25 milliseconds. Higher values of velocity can be measured by using much smaller times τ and T. To study turbulent movements, for example, it seems necessary to use a pulsed laser. Ideally, two lasers would be necessary to be able to vary T from zero to infinity.

In our experiment, the visibility of the fringes in the Fourier plane does not play a fundamental role. But if we study the visibility of these fringes as a function of T (time between two poses) in the absence of macroscopic movements, we can deduce information about the microscopic movement of the polystryrene balls, and thus about the properties of the fluid.

References

1. See for example: B.M. Watrasiewicz and M.J. Rudd, Laser Doppler Measurements, Butterworths, London 1976.
   F. Durst, A. Melling and J.H. Whitelaw, Principles and Practice of Laser Doppler Anemonetry, Academic Press, London, 1976.
2. J.M. Burch and J.M.J. Tokarski, Opt. Acta 15, 101 (1968).
3. See for example: L. Delaya, J.M. Jonathan and S. Mallick, Optics Comm. 18, 496 (1976).
   R. Grousson and S. Mallick, Appl. Opt. 16, 2334 (1977).

DEFORMATION MEASUREMENTS ON CONNECTING RODS BY SPECKLE PHOTOGRAPHY

Alfons Happe
Volkswagenwerk AG, Division of Research
and Development — Messtechnik/Optik
D-3180 Wolfsburg 1, West Germany

Abstract

The connecting rod, which transmits the forces due to the pressure of the gas from the piston to the crankshaft, is also subjected to great traction forces due to the inertia of the piston. Deformations caused by these forces should not be too great, in particular not in the zone of contact with the crankshaft; the lubricant layer must remain continuous to avoid excessive wear.

Such deformations are created, using static simulations of the traction forces, and were measured by using double exposure speckle photography.

Introduction

To illustrate the problem of deformations of connecting rods let us remember the principle of the internal combustion engine as it is used in the most cases for automobile propulsion.

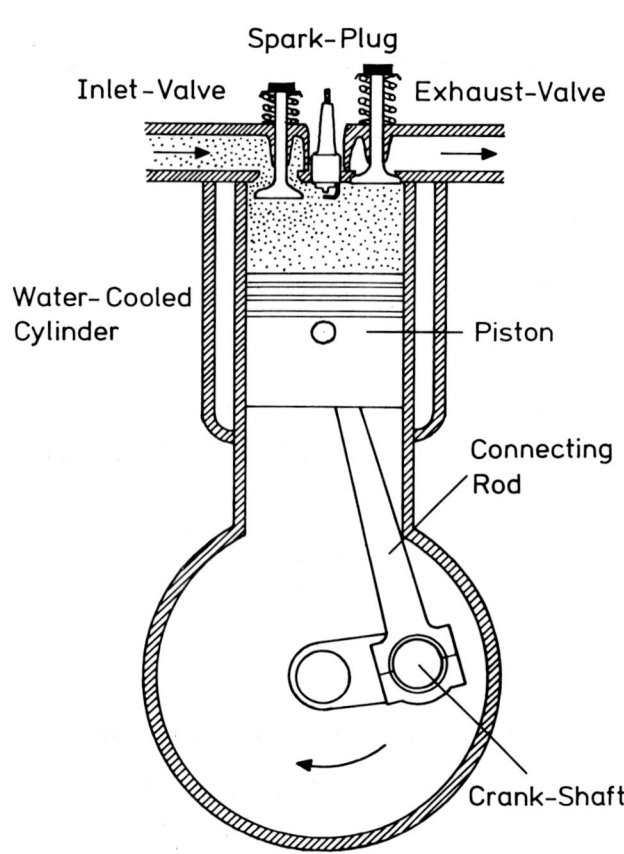

Fig. 1. Principle of a four-stroke internal combustion engine. First stroke.

The water-cooled cylinder, fitted with mechanically operated inlet- and exhaustvalves and a spark plug, encloses a piston which is connected by a connecting rod to the crankshaft.

The complete cycle of operation consists of four strokes. The figure shows the piston starting downward on the first stroke. During this stroke the piston has to be accelerated by the connecting rod. Therefore this is submitted to a traction force. From the second to the fourth stroke the connecting rod is loaded by a compressive force.

That means, within each cycle of operation the load of the connecting rod is changed from tensile force to compressive force.

The most critical load is the tensile force at maximum speed of rotation. It reaches values of nearly 15 K Newton.

The degree of elastic deformation of the bearing shells of the connecting rod, caused by this load, must not reach such values that the lubricant film can be destroyed.

Stiffness calculations of connecting rods are done by the finite element method. A point-by-point method to measure the in-plane displacement, as it has been described by Archbold, Burch, Ennos et.al. (1, 2, 3, 4, 5), is therefore very useful for an experimental control of the calculated displacement values of single meshes of the finite element system. In the case of disagreement between calculated and measured values the results of speckle photography offer the possibility of a realistic change of the boundary conditions for the computations.

Experimental Arrangement

Figure 2 shows a photograph of a connecting rod. The big end bearing, by means of which the rod is connected to the crankshaft, consists of two halves, which are pressed together by the connecting rod bolts. What had to be measured is firstly the oval deformation of the bearing and secondly the relative movement of the halves within the encircled area of separation, if the rod is submitted to a tensile force.

DEFORMATION MEASUREMENTS ON CONNECTING RODS BY SPECKLE PHOTOGRAPHY

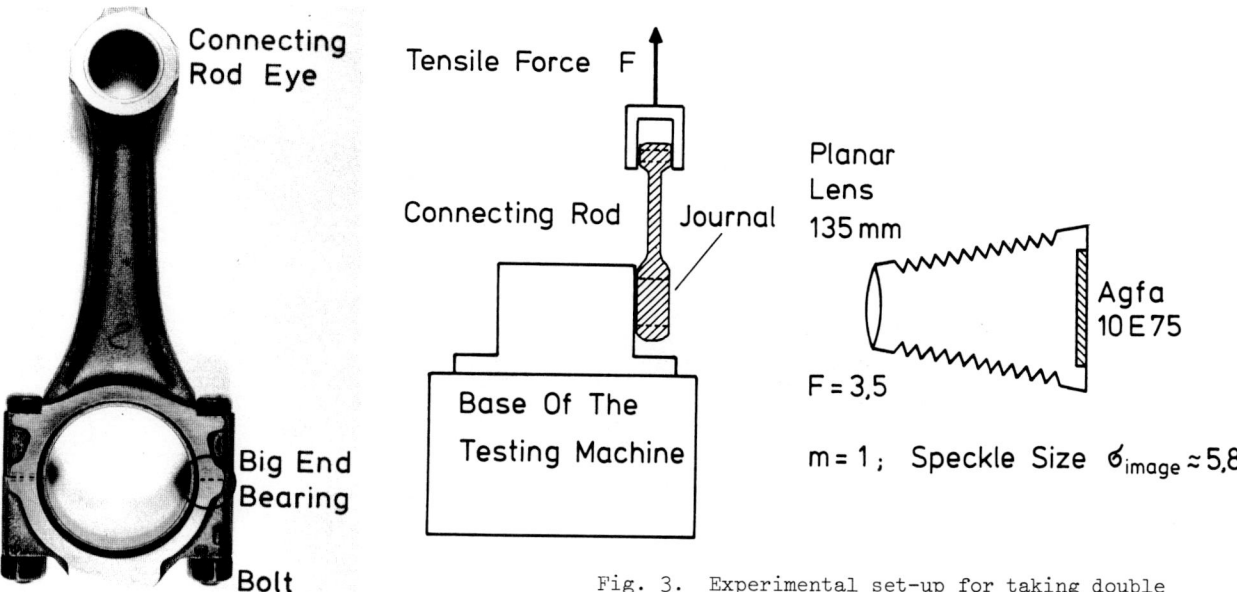

Fig. 2. Photograph of a connecting rod. The separation of the big end bearing is marked by a dotted line.

Fig. 3. Experimental set-up for taking double exposure speckle photographs of the total big end bearing and the journal.

The experimental set-up to measure the relative movement between the crankshaft and the connecting rod's big end bearing or with other words its oval deformation is shown by figure 3. In order to get also the deviation of the "crankshaft" the big end bearing of the connecting rod has been fitted to a journal on a heavy steel-block. The front of this journal is in-plane with the side-surface of the connecting rod's big end. Both are focussed on a photographic plate 9 x 12 cm with unit magnification (m = 1). The steel-block was clamped to the base of a testing machine, by which the connecting rod could be submitted preselected tensile forces. The camera was mounted to the steel-block.

Measurements

Double exposure photographs of the laser illuminated connecting rod bearing were recorded on Agfa Scientia 10 E 75 plates for incremental loads from 0,5 to 24 KN. The purpose of the 0,5 KN preload was to overbridge the bearing play. For each exposure, the camera shutter was opened and the focussed surface illuminated by a ruby-laser-flash. The coherent laser light generates a fine structure in the image, called speckle pattern. The size of the speckles depends on the wavelength of the laser-light λ, the F-number of the lens and the demagnification factor m. It is given by the formula

$$\sigma_{image} \approx 1,2 \cdot (1 + m) \cdot \lambda \cdot F. \tag{1}$$

In our special case we have

$$\lambda = 0,694 \, \mu m, \; F = 3,5 \text{ and } m = 1 \longrightarrow \sigma_{image} \approx 5,8 \, \mu m.$$

The sensitivity of this simple method of speckle photography is given by the speckle size σ_{image}. If the image movement between the two exposures is greater than one speckle diameter σ_{image}, individual areas of the processed plate will scatter a laser beam into a diffraction halo with Young's fringes. The angular spacing α of the fringes depends on the lateral motion D of the object. The dependence is given by the formula

$$\sin \alpha = \frac{\lambda \cdot m}{D} \approx \frac{a}{b} \longrightarrow D \approx \frac{\lambda \cdot m \cdot b}{a} \tag{2}$$

a is the distance between two neighbouring Young's fringes on the screen and b the distance between the specklegram and the screen. The direction of the fringes lies at right angles to the direction of displacement.

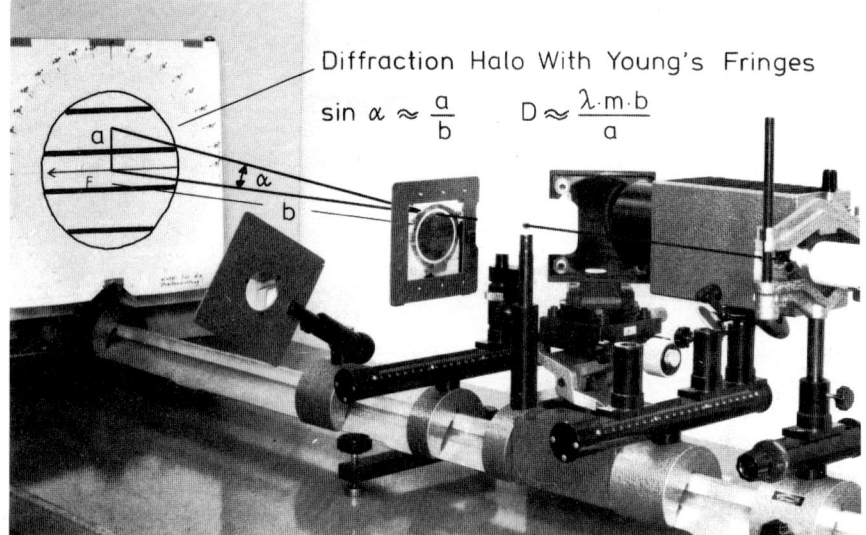

Fig. 4. Specklegram evaluation arrangement
  a. The specklegram, illuminated simultaneously by the white light system and a laser beam, is focussed on the screen by the imaging lens. The exact location and the diameter of the area being illuminated by the laser can be seen on the screen.
  b. The white light system has been switched off and the imaging lens has been swung out. A diffraction halo with Young's fringes appears on the screen.

The evaluation apparatus, shown in figure 4a + b, is very similar to that, which Archbold and Ennos used to measure the deformation along a weld crack.

A narrow laser beam which was brought to focus on the screen passes the specklegram. An auxiliary white light system was provided to image the specklegram on to the screen. This consists of a lamp with condenser lens, a 45°-mirror with a small hole in its centre through which the laser beam passes, and an imaging lens that can be swung in and out of position.

With both, laser and white light systems operating, the exakt location and diameter of the area being illuminated by the laser can be identified (figure 4a). The position of the specklegram can be changed in x- and y-directions, if the laser beam is in z-direction.

After switching off the white light system and swinging out the imaging lens we see on the screen a diffraction halo with Young's fringes (figure 4b).

Figure 5 shows a specklegram recorded with the set-up shown in figure 3 and the diffraction halos with Young's fringes belonging to the points marked (1), (2), (3) and (4).

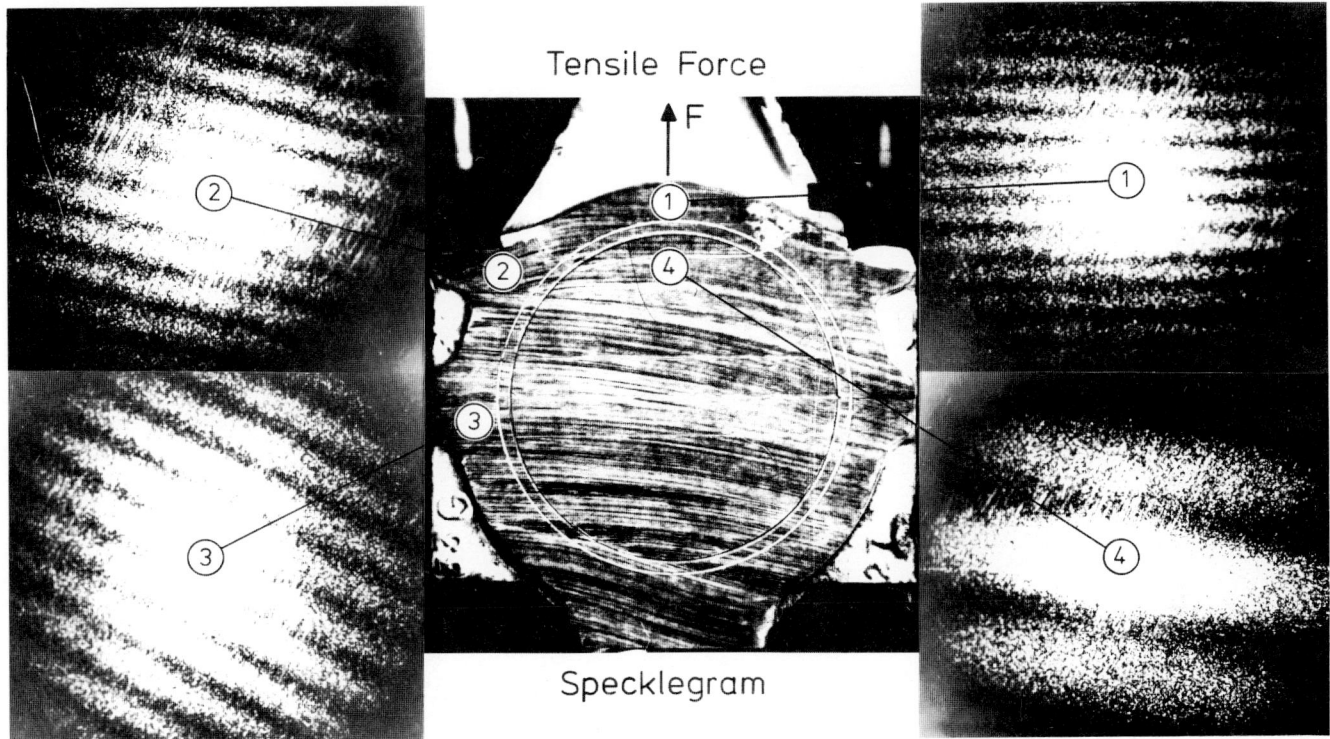

Fig. 5. Double exposed specklegram of a connecting rod's big end bearing. Right and left are the diffraction haloes with Young's fringes which appear if the laser beam passes the points (1), (2), (3) or (4).

The angular spacing of the fringes belonging to (4) is relative big, that means, the displacement of the journal is relative small. The direction of the displacement is vertical to the direction of the fringes and nearly the same as that of the tensile force F. The points (1), (2) and (3) were meshes of the finite-element-grid for rigidity calculations. From the fringe systems we got the polar coordinates of the displacement vectors. The following Table 1 shows these values transformed to rectangular coordinates X_M and Y_M. These coordinates include the elastic displacement of the journal. On the other hand the displacement of the different points of the bearing relative to the journal is of greater interest. Therefore the table 1 gives these values in the columns $X_{corr.}$ and $Y_{corr.}$ for two or three increments of the tensile force. The amount of the components of the total displacement relative to the journal is equal to the sum of the partial displacement values belonging to the successive increments of the tensile force F.

To control if there is a relative movement of the two halves of the big end bearing within the encircled area of separation in Fig. 2, a measurementsystem with higher sensitivity than 5,8 μm is necessary.

Figure 6 shows a photograph of the modified experimental arrangement which has been used for this measurement.

A 35 mm camera was fixed to the base of the testing machine and arranged to view the encircled field with the separation with normal incidence. The camera, focussed on the frontside surface of the big end bearing, recorded an image of the surface with demagnification factor m = 0,5. The F-number of the lens was 2,0. The object has been illuminated again by ruby laser light. After equation (1) this leads to an image speckle size $\sigma_{image} \approx$ 2,5 μm.

Table 1: Displacement values of a big end bearing relative to the crankshaft under different tensile forces F.

Point of Measurement	I. Standard Bolts				II. Improved Bolts			
	X_M	Y_M	$X_{corr.}$	$Y_{corr.}$	X_M	Y_M	$X_{corr.}$	$Y_{corr.}$
	F = 0,5 — 12 KN				F = 0,5 — 10 KN			
①	53,96	3,77	44,98	2,19	36,68	2,56	28,20	1,67
②	39,20	1,58	30,20	0	27,59	4,86	19,11	3,97
③	28,27	8,71	19,29	7,13	21,36	10,42	12,80	9,53
④	8,98	1,58			8,48	0,89		
	F = 12 — 24 KN				F = 10 — 20 KN			
①	36,77	0,34	26,98	0	31,78	1,11	23,61	0,82
②	25,75	4,54	15,96	4,2	24,75	5,26	16,58	4,97
③	21,27	7,74	11,48	7,4	19,14	7,73	10,97	7,44
④	9,79	0,34			8,17	0,29		
					F = 20 — 25 KN			
①					16,34	0	12,21	0
②					11,70	2,06	7,57	2,06
③					11,52	4,19	7,39	4,19
④					4,13	0		
	F = 0,5 — 24 KN				F = 0,5 — 25 KN			
①			71,96	2,19			64,02	2,49
②			46,18	4,20			43,26	11,00
③			30,77	14,53			31,24	21,16

All values are given in μm!

Fig. 6. Photograph of the experimental arrangement for controlling the relative motion of the two halves of the big end bearing by double exposed speckle photography. The area under test is encircled. The laser is a ruby laser.

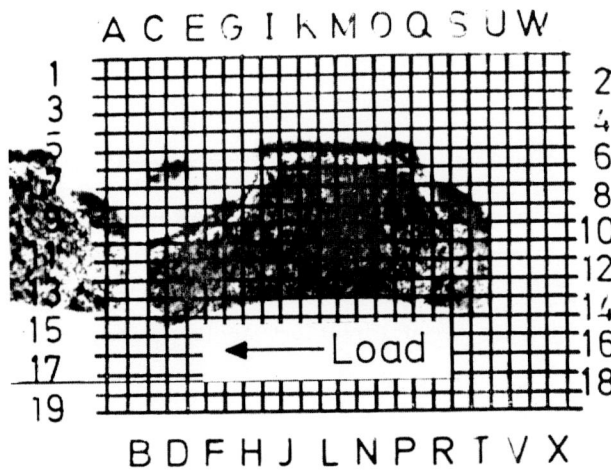

Fig. 7. Specklegram of the encircled area of Fig. 6 with superimposed square-grid.

Figure 7 shows on its right side a double exposed speckle photograph taken with the arrangement of Fig. 6 with a superimposed square grid. The relative position of the specklegram and the grid has been adjusted in that way, that the separation of the crankshaft bearing is between the columns L and M, the outer edge between the lines 5 and 6. Within each mesh of the grid which is covered by the specklegram a reproducible measurement of the lateral surface motion is possible.

Figure 8 shows a point-by-point specklegram evaluation along 3 lines. The values are polar-coordinates r, φ of the local in-plane displacement vectors. The unit for r is μm. This method of measurement delivers displacement values relative to the camera with the relative high sensitivity of 2,5 μm.

The displacement values shown in figure 8 belong to an increment of the tensile force from 0,5 to 4 KN. On the same way it is possible to measure the displacement vectors belonging to further increments of the tensile force up to its highest possible value. The total of all successively measured components of the displacement vector delivers the components of the total displacement at the highest possible tensile force.

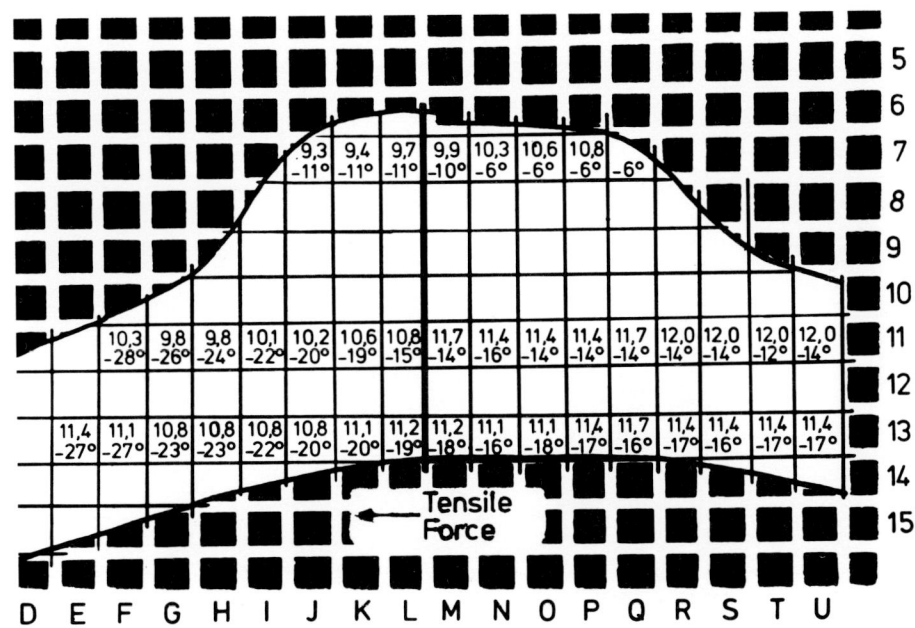

Fig. 8. Evaluation of a specklegram. The values of the local displacement vectors are given by polar-coordinates r, φ. The upper numbers within the evaluated meshes are r-values in μm, the lower numbers are φ-values in degrees.

Conclusions

Testing machines, by which one can subject an object to tensile or compressive forces, and photographic cameras are normally a part of the standard equipment of a testing laboratory.

The additional providing of a coherent light source, e. g. a pulsed ruby laser or a CW-laser of at least 15 mW power, enables in such cases in-plane deformation measurements by double exposure speckle photography without any mechanical contact to the object under test. The sensitivity of this inexpensive method is with 2,5 μm relative high.

Acknowledgments

The author thanks Dr. H. Klingenberg and Mr. H. Hartwig for supporting this work. Furthermore he is very grateful for the skillful help of Mr. G. Hoppe in performing the experimental work.

References

1. Archbold, E., Ennos, A.E., "Displacement measurements from double exposure laser photographs", Optica Acta 19 (1972), 253-271.
2. Archbold, E., Ennos, A.E., "Laser photography to measure the deformation of weld cracks under load", Non-Destructive-Testing, 1975, 181-184.
3. Archbold, E., Burch, J.M., Ennos, A.E., 1970, Optica Acta, 17, 883.
4. Luxmoore, A.R., Amin, F.A.A., Evans, W.T., "In-plane strain measurement by speckle photography; a practical assessment of the use of Young's fringes", J. Strain Analysis 9 (1974) 26.
5. Cloud, G., "Practical speckle interferometry for measuring in-plane deformation", Appl. Optics, Vol. 14 (1975) 878-884.

ANALYSIS OF THE DIFFRACTION SPECTRUM OF A POPULATION OF PARTICLES

J. Fleuret, H. Maitre, J. F. Thery
Ecole Nationale Supérieure des Télécommunications, Image Laboratory
46, rue Barrault Paris 13ème, France

Abstract

We study the diffraction spectrum of a population of circular objects randomly distributed. We present a rigorous study of the deconvolution of the intensity of a diffracted spectrum, permitting us to obtain the complete histogram of the distribution of sizes. (The method is applied notably to multi-modal histograms).

The granularity term corresponding to the random positionings of diffracting objects was studied by means of a statistical model based on the independence of the positionings of the objects. We find that the size of the speckle ω_o is proportional to the inverse of the width of the object field studied. In the field of frequencies greater than ω_o we find the conventional properties of the speckle, which permits us to verify the approximations made in the complete expression of the spectral intensity recorded.

Introduction

We describe a method permitting the complete determination of histograms of the sizes of circular objects randomly distributed in a plane. This is a global method based on the analysis by deconvolution of the intensity of the Frauenhoffer spectrum. Contrary to other existing methods, we obtain, after treatment, the complete histogram of the distribution of sizes.

Reference #1 describes a first version of the method, with the corresponding results, in the case of narrow histograms. The very simple measurement consists in recording, in the Fourier plane, a radial section of the diffraction spectrum (obtained in coherent light) by means of any photodetection device (photomultiplier, mobile photodiodes, matrix of photo-detectors). The treatment described[1] used the amplitude of the diffraction spectrum, which entailed notably the necessity of restoring, heuristically, the sign of this amplitude. We propose here a more general treatment bearing on the intensity of the diffraction spectrum, which validates the method in the case of any histogram whatsoever (multi-modal, for example).

Moreover, the granularity in the spectrum is a phenomenon which we must try to avoid, as far as possible. Because of this we were led to move the object at the moment of recording the spectrum, which leads, according to reference #2, to an improvement of the signal/noise ratio and, in our case, makes possible the use of the diffracted spectrum for the treatment considered.

We present in this article a theoretical study of the granularity term corresponding to the random positions of the diffracting objects. We will describe first of all a simple model of random process, then we will describe the more adequate study of a Poisson process. We will thus establish the principal statistical results, in the second order, bearing on the intensity of the granularity term.

Method of Treatment

Recording the Spectrum

Let us consider first of all, in the object plane N_a, circular disks, of radius a, and whose centers x_i, y_i are "randomly" distributed. The complex amplitude diffracted in the Fourier plane is expressed as a function of the angular spectral coordinates (u,v) by:

$$f_{N_a}(u,v) = S(u,v) \sum_{i=1}^{N_a} e^{jk(x_i u + y_i v)} \quad (1)$$

$$(k = \frac{2\pi}{\lambda})$$

where $S(u,v)$ represents the amplitude of the diffraction spectrum of a centered disk:

$$S(u,v) = 2\pi a^2 \frac{J_1(ka\omega)}{ka\omega} \quad (1a)$$

$$(\omega = \sqrt{u^2 + v^2})$$

The exponential sum in (1) represents a granularity term, which expresses the rapid fluctuations of interferences, characteristic of the speckle[3].

The intensity observed thus depends on the positions, supposed randomly distributed, of the totality of the particles. It is thus a realization of a random number, for which we can obtain conventionally an average value by varying the positions of the particles according to a certain law of probability. Another means of obtaining this average consists, in practice, in causing the entire object lamella to vibrate and integrating the luminous intensity over a certain time interval. We thus effect, in the laboratory, a simulation of the recording of an average for the total. This will be designated by < > and will mean, in what follows, either the result of the calculation of probabilities described above, or that of the laboratory experiment such as we have described it. We will also use the condensed notation: $\hat{x} = \langle x \rangle$

In the case of a population of objects of different radii $a_m (m = 1, M)$, the recorded intensity will be:

$$\hat{I}(u,v) = \left\langle \left| \sum_{m=1}^{M} S_m(u,v) \, g_m(u,v) \right|^2 \right\rangle \tag{2}$$

with-- $S_m(u,v)$ the amplitude of the diffraction spectrum of a disk of radius a_m

- $g_m(u,v) = \sum_{p=1}^{N_m} e^{jk(x_{p,m} u + y_{p,m} v)}$ granularity term
- N_m number of objects of the class m (of radii a_m)
- $(x_{p,m}; y_{p,m})$ coordinates of the $\underline{p^{th}}$ object of the class m

The development of (2) permits us to distinguish the squared terms from the crossed terms, or:

$$\hat{I}(u,v) = \left\langle \sum_{m=1}^{M} S_m^2 |g_m|^2 \right\rangle + \left\langle \sum_{\substack{m; m' \\ m \neq m'}} S_m S_{m'} \, g_m g_{m'}^* \right\rangle \tag{3}$$

Next we will admit the statistical independence between different classes, the second term of (3) becoming:

$$\sum_{\substack{m; m' \\ m \neq m'}} S_m S_{m'} \, \hat{g}_m \hat{g}_{m'}^*$$

Given the rapid fluctuations of the phase of the granularity terms, we will admit from now on the following estimations:

$$\begin{cases} \hat{g}_m \neq 0 & (4) \\ \widehat{|g_m|^2} \neq N_m & (4a) \end{cases}$$

These two relations will be justified in the following paragraph, for the domain of frequencies defined:

$$\omega = \sqrt{u^2 + v^2} \gg \omega_o = \frac{\lambda}{L} \qquad (L = \text{width of object field observed})$$

The final expression obtained for the intensity is thus:

$$\hat{I}(u,v) \neq \sum_{m=1}^{M} S_m^2(u,v) \, N_m \tag{5}$$

which incidentally justifies, in the framework of our hypotheses, the intuitive notion that "it is as if" the light used were spatially incoherent.

In the case of circular objects, (5) is written, after (1a):

$$\hat{I}(u,v) = I(\omega) \neq 4\pi^2 \sum_{m=1}^{M} a_m^4 \left\{ \frac{J_1(ka_m \omega)}{ka_m \omega} \right\}^2 N_m \tag{5a}$$

Treatment of the Intensity $I(\omega)$

In the case of a continuous histogram ($N_m \to N(a) \, da$) we will write (5a) in the form:

$$I(\omega) \propto \int_0^\infty a^4 \left\{ \frac{J_1(ka\omega)}{ka\omega} \right\}^2 N(a) \, da \tag{5b}$$

The obtaining of $N(a)$ from (5b) results from a treatment analogous to that described in reference #1 and which is based on the use of a Hankel transform. In fact, the transform of (5b) is:

$$\begin{aligned} F(t) &= \int_0^\infty I(\omega) \, J_o(k\omega t) \, \omega \, d\omega \\ &= \int_0^\infty a^4 N(a) \, \frac{1}{a^2} \, C\left(\frac{t}{2a}\right) da \end{aligned} \tag{6}$$

where C designates the autocorrelation function of a circular disk [4]:

$$C\left(\frac{t}{2a}\right) = \begin{cases} \text{Arc}\cos\left(\frac{t}{2a}\right) - \frac{t}{2a}\sqrt{1-\frac{t^2}{4a^2}} & t < 2a \\ 0 & t \geq 2a \end{cases}$$

Let us derive (6) twice under the sum sign. All calculations made, we obtain:

$$\frac{d^2F}{dt^2} = t \cdot \int_{\frac{t}{2}}^{\infty} \frac{N(a)\,da}{\sqrt{a^2 - \frac{t^2}{4}}} \qquad (7)$$

We see that (7) is easily reduced to an equation of convolution by means of a simple change of variables ($a^2 = A$; $t^2/4 = T$); we can thus obtain the distribution $N(a)$ by the conventional techniques of deconvolution applied to (7).

In summary, the stages of treatment are the following:
- recording the light intensity $I(\omega)$
- Hankel transform
- double derivation
- deconvolution

Study of the Granularity Term

The granularity term corresponding to a class of objects of the same radius is written (considering a mono-dimensional model to simplify):

$$g(u) = \sum_{p=1}^{N} e^{j\varphi_p(u)} = \rho(u)\,e^{j\Phi(u)} \quad \text{with } \varphi_p = kux_p$$

In a first approach, the terms $\widetilde{\varphi}_p = \varphi_p$ modulo 2π can be considered as random equi-distributed over $[0, 2\pi[$; we are then brought back to models of "random process" from which it results that $\rho = |g(u)|$ follows Pearson's law, and the global phase Φ is equi-distributed over $[0, 2\pi[$. Since this simple model is acceptable only at a great distance from the origin of the frequencies, we are led to choose better ones, notably the Poissonian model. Let us suppose then that the centers x_p form a Poisson process. We will call the average number of objects per unit of length ($N \# \rho L$).

Estimation of $G(u) = |g(u)|^2$

This calculation involves the law of probability of the intervals between any two objects: $x = x_p - x_q$ ($p > q$). In the case of a Poissonian process, this law is given by:

$$P(x)\,dx = \frac{\rho^n}{(n-1)!} e^{-\rho x} x^{n-1}\,dx \qquad (8)$$

where $n = p-q$
$\rho = N/L$

From its definition, $\hat{G}(u)$ is written:

$$\hat{G}(u) = N + 2 \sum_{q=1}^{N-1} \sum_{p=q+1}^{N} \langle \cos ku(x_p - x_q) \rangle$$

$$\hat{G}(u) = N + 2\,\text{Re} \sum\sum \int_0^{\infty} e^{jkux} P(x)\,dx$$

which involves the characteristic function corresponding to Poisson's law. Or in setting $v = ku/\rho$:

$$\hat{G}(u) = N + 2\,\text{Re} \sum\sum \frac{1}{(1-jv)^{p-q}}$$

and after summation of geometrical series:

$$\hat{G}(u) = N + \frac{2}{v^2}\left\{1 - \text{Re}\left(\frac{1}{1-jv}\right)^{N-1}\right\} \qquad (9)$$

This function, which is reflected by the frequential variations of the diffracted intensity, is traced in figure 1.

fig. 1

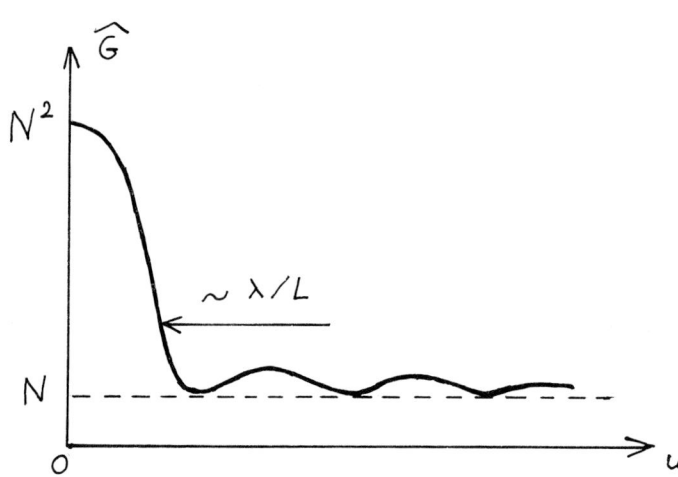

$\hat{G}$ is presented in the form of a central peak of width proportional to $u_o = \lambda/L$. For frequencies $u \gg u_o$, we can legitimately assimilate $\hat{G}(u)$ to N, which justifies (4a). The justification of relation (4) is made by a very similar calculation.

We thus find results analogous to those of Stark[5], who used for his study a model of non-overlapping objects generated by a jitter phenomenon.

Finally, it must be noted that the central peak is not particularly bothersome for our study; in all cases its width was inferior to that of the direct beam and a simple occultation sufficed to get rid of it. Nonetheless, the preceeding study gives the possibility of taking account of it with all due rigor.

Properties of the Second Order

Strictly speaking, $G(u)$ is not stationary in the first order, and its autocorrelation function does not exist. Nonetheless, let us place ourselves in the frequency domain $u \gg u_o$, for which we can admit

$$\begin{cases} \langle \cos kux \rangle = 0 \\ \hat{G}(u) = N \end{cases}$$

with, however, $\langle \cos ku_o x \rangle \neq 0$, and let us calculate the function $\Gamma(u_o) = \langle G(u) G(u+u_o) \rangle - \langle G(u) \rangle \langle G(u+u_o) \rangle$
We have:

$$\langle G(u) G(u+u_o) \rangle = \langle \{N + 2 \sum_{\substack{p,q \\ p>q}} \cos ku(x_p - x_q)\} \{N + 2 \sum_{\substack{p',q' \\ p'>q'}} \cos k(u+u_o)(x_{p'} - x_{q'})\} \rangle$$

given the independence of the intervals $(x_p - x_q)/(x_{p'} - x_{q'})$ except if $p = p'$ and $q = q'$.

$$\langle G(u) G(u+u_o) \rangle = N^2 + 4 \sum \langle \cos(kux) \cos(k(u+u_o)x) \rangle$$

We are brought back to a calculation identical to that effected above for $G(u)$. Or, all calculations made:

$$\Gamma(u_o) = 2 \, \text{Re} \sum_{p>q} \sum \left(\frac{1}{1-jv_o}\right)^{p-q} \qquad v = ku/a \qquad (10)$$

From which we deduce the following consequences:

- for the hypotheses considered (notably $u \gg u_o$) we can consider the intensity $G(u)$ as stationary in the second order;
- the corresponding correlation radius (that is, the average size of the speckle studied) thus results and we verify that it is of the order of magnitude of $u = \lambda/L$;
- $\sigma^2 = \Gamma(0) = N(N-1)$
  and the ratio S/B has the value $\langle C \rangle / \sigma = \sqrt{N/(N-1)} \quad \# 1$

The last two results are perfectly well verified experimentally, from the examination of the diffracted spectrum.

Let us note finally that the results obtained are comparable to those of Goldfischer[6], which concern the granularity due to the illumination of a diffuse surface.

Conclusion

The proposed method of determination of histograms is global, fairly convenient, and applicable in a fairly simple manner. It was developed on the occasion of the study of populations of blood cells, but could be applied perfectly well to analogous problems.

Furthermore, the study of the granularity term permitted us to set off the essential properties of the speckle and to specify the conditions of validity of frequently used hypotheses.

References

1. J.F. Thery "Estimation des tailles de particules circulaires", National Colloquium on Signal Treatment and its Applications (April 1977).
2. S. Lowenthal, D. Joyeux, H. Arsenault, "Déplacement fini d'un diffuseur mobile", Optics Communications, Sept. 1970.
3. M. Françon, Optique. Formation et Traitement des Images Masson (1972).
4. J. Goodman, Introduction to Fourier Optics McGraw Hill (1968).
5. H. Stark "Diffraction patterns of non-overlapping circular grains" J.O.S.A., May 1977.
6. L.I. Goldfischer, "Autocorrelation Function and Power Spectral Density of Laser-Produced Speckle Patterns" J.O.S.A., March 1965.

DECORRELATION PRODUCED IN THE IMAGE OF A DIFFUSING OBJECT BY A ROTATION OF THE INCIDENT WAVE—APPLICATION TO THE STUDY OF SURFACE STATES

Michel Menu and Marie-Louise Roblin
Laboratory of Optics—Institute of Optics
Université Pierre et Marie Curie T. 13, 3eme etage
4. Place Jussieu — 75230 Paris Cedex 05

Introduction

The goal of this experiment is to demonstrate local differences in surface state on a diffusing object by double speckle photography. We record on a single holographic plate <u>two</u> images of the diffusing object lit with coherent light; between the two exposures we slightly modify the angle of incidence of the light beam. We then use the interference fringes produced in the Fourier plane of this double photograph; their visibility characterizes the correlation between the two speckle figures, related to the local surface state. By an appropriate filtering setup it is then possible to visually localize on the image zones of different rugosity.

The formula relating the visibility of these fringes to the rugosity RMS σ of the surface has previously been established in the case of an "objective speckle pattern" (recording at a finite distance from the diffuser, without intermediary optics). It is written:

$$V = \exp - \left(\frac{2\pi}{\lambda} \sigma \sin\theta \, \delta\theta \right)^2 \qquad \text{for incidence } \theta$$

$$V = \exp - \left[\frac{2\pi}{\lambda} \sigma \frac{(\delta\theta)^2}{2} \right]^2 \qquad \text{for incidence } 0 \tag{1}$$

λ is the illumination wavelength and $\delta\theta$ is the rotation introduced between the two exposures.

In the case of a "subjective speckle" (optical image of the diffuser) this expression is appropriate given two conditions:
- a perfect optical system
- a large diffraction spot compared to the correlation length of the profile.

These two conditions then determine the practical realization of the experiment. The first determines the choice between the two possible setups. The second will impose certain conditions and will permit discussion of the results.

1. Choice of Setup

To achieve a variation of the angle of incidence on the object, there are in effect two possibilities:
1. Turn the incident wave through the angle $\delta\theta$ (figure 1a); the object, the optical system and the photographic plate M **remain fixed**.
2. Turn the object itself (figure 1b) while conserving the position of the wave and the optical system; it is then necessary to turn the photographic plate symmetrically to compensate the defect in focus.

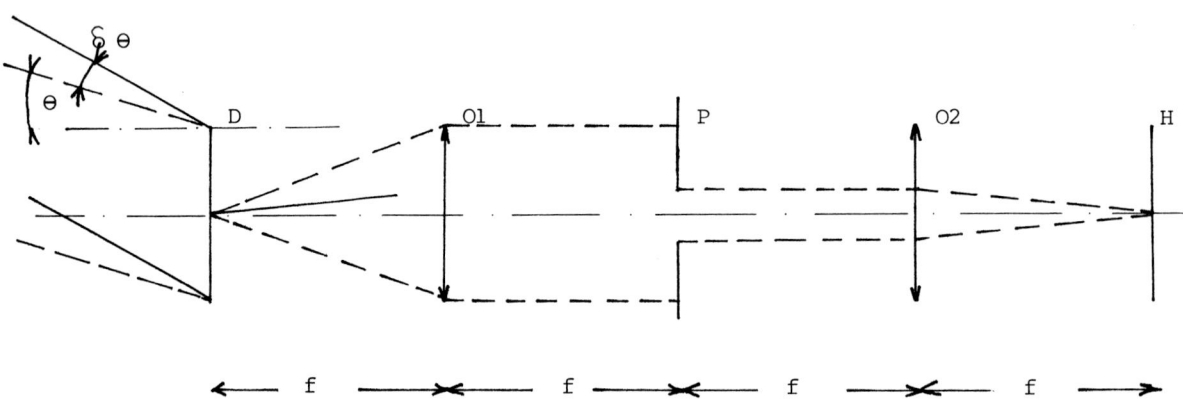

Fig. 1a.

DECORRELATION PRODUCED IN THE IMAGE OF A DIFFUSING OBJECT BY A ROTATION OF THE INCIDENT WAVE—APPLICATION TO THE STUDY OF SURFACE STATES

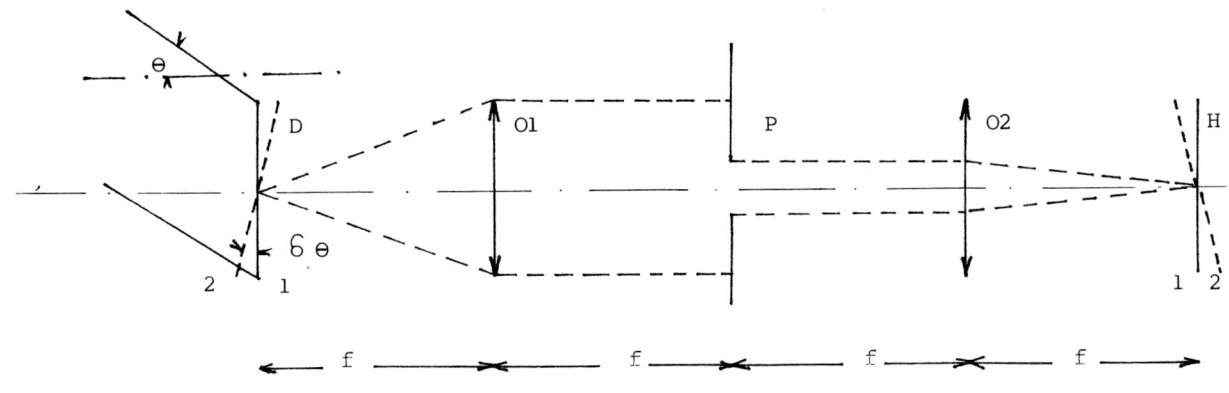

Fig. 1b.

To study the influence of the aberrations it is necessary to consider the phenomena entailed by the rotation $\delta\theta$ at the level of the pupil of the optical system; the diffusing object produces a speckle figure in this plane. This figure is diaphragmed by the contour of the pupil. Furthermore, the speckle function is multiplied by a variable factor of phase on the pupil, characterizing the aberrations of the system. The image is then obtained by optical Fourier transform from the speckle function thus modified. For the two setups, we show that the rotation $\delta\theta$ produces a lateral displacement of the speckle function in the plane of the pupil. There results, in the first place, a different diaphragmation of the speckle figure in the second exposure, which can be compensated simply by an identical translation of the pupil. But the relative displacement of the speckle function with regard to the aberration function cannot be compensated: it produces at each point of the pupil a phase variation proportional to the differential of the aberration function, which entails effects of decorrelation in the image (if the system presents aspherical aberration and coma) and variable displacements (if the system presents astigmatism or curvature). The effects of decorrelation due to the aberrations produce a decrease in visibility of the fringes and thus disturb the study of the rugosities. They are quantitatively related to the dimensions of the pupil, which we must reduce in consequence. For a given aperture the tolerances on the aberrations are inversely proportional to the value of the displacement of the speckle on the pupil, a value which varies considerably between the two methods proposed.

For the setup in figure 1b the displacement in the pupil is equal to $\cos\theta\,\delta\theta$ where θ is the incidence; for setup 1b: $(1 - \cos\theta)\,\delta\theta$. For the incidence zero this gives respectively $\delta\theta$ and 0.

Setup 1b, for incidence zero, is practically insensitive to aberrations. For an incidence of 30° (resp. 0.87$\delta\theta$ and 0.13$\delta\theta$) a ratio of 6 is conserved. We thus chose setup 1b for the study of rugosities; setup 1a with rotation of the wave is developed as a method of checking optical systems.

The choice of the normal incidence seems completely favorable from the point of view of the aberrations. Nevertheless, reasons of sensitivity and domain of use lead us to use in general an oblique incidence. For a rotation between the two exposures of $\delta\theta = 6°$ we have

for $\theta = 30°$ $\qquad\qquad \sin\theta\,\delta\theta \sim 5 \times 10^{-2}$

for $\theta = 0°$ $\qquad\qquad \dfrac{\delta\theta^2}{2} \sim 5 \times 10^{-3}$

which in the expression of the visibility[1] multiplies the coefficient of σ by 10, then increasing the sensitivity in oblique incidence.

In normal incidence, the range of detectable rugosities extends from 10 microns to several tens of microns. The use of oblique incidence permits us to displace the domain of use towards the small rugosities. A limit towards the large rugosities will be imposed by the correlation length, as will be specified below.

2. Experimental Results

Figure 2 gives the experimental results obtained in the case of the setup using the rotation of the object. The latter is formed by two unpolished glasses of different rugosities (1μ and 8μ). The angle of incidence θ is around 20°. The rotations are respectively equal to 0, 3° and 5°. The filtering setup sketched in figure 2 includes a diaphragm at a finite distance on the axis of the system: it permits us to observe fringes over the entire surface of the object. The deformations and variations of spacing of the fringes related to the astigmatism of the optical system are visible but do not detract from observation. The variation of visibility is considerable between the two halves of the field corresponding to the two different rugosities.

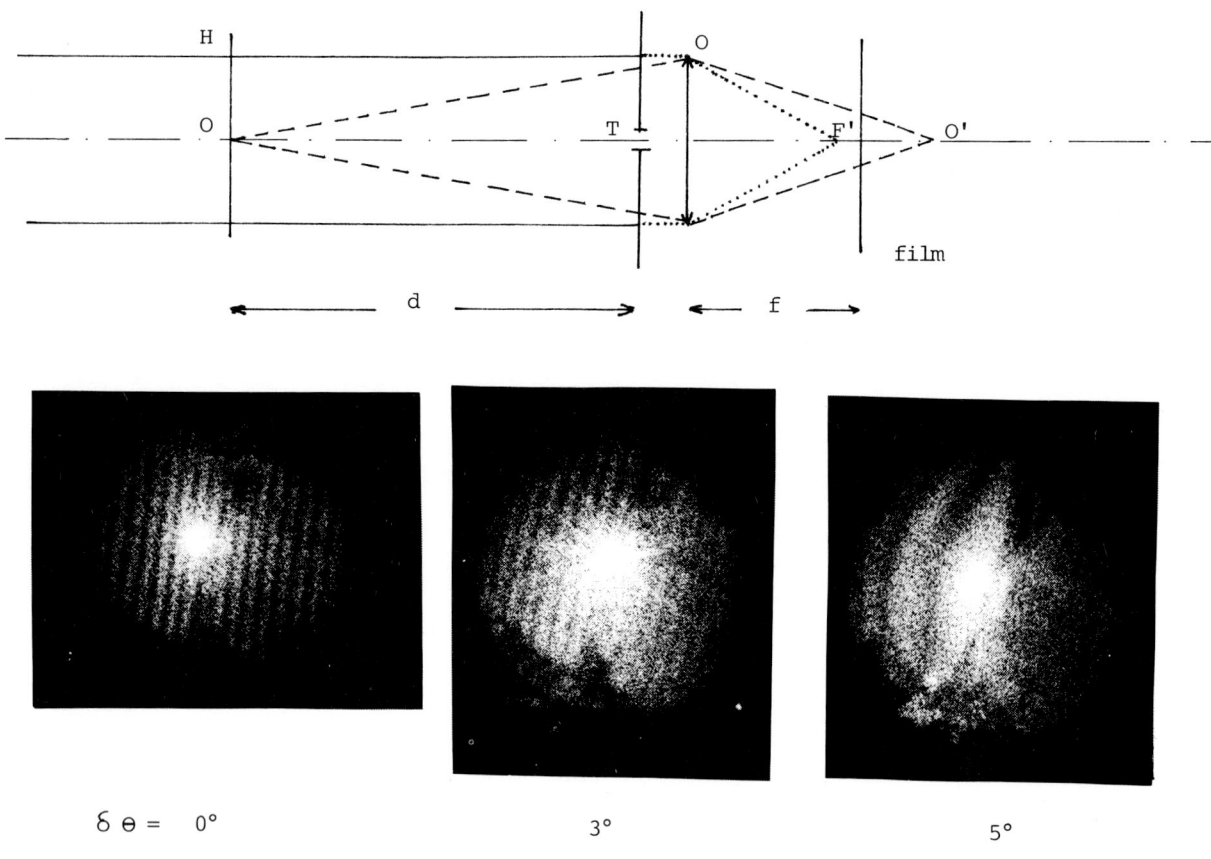

δθ = 0°   3°   5°

Fig. 2.

3. Discussion of Results

We will now study the second condition: the existence of a diffraction spot which is large by comparison to the correlation length.

Without going into the details of the theoretical calculations, we can remark that only the points of the object contributing at a point of the image to the formation of the speckle participate in the decorrelation effect; it is thus necessary that on the average in a zone equal to the diffraction spot we find all the possible values of the profile function of the object so that the real rugosity can play a role.

To achieve a sufficiently large spot the aperture of the system can be diminished: nonetheless the aperture also conditions the dimensions of the diffused halo, which must be sufficiently extended to permit the observation of the fringes. In doing the experiment with a focus defect, the diameter of the diffusion spot can be increased without modifying the dimensions of the halo.

Figure 3 illustrates well the increase of the influence of the rugosity as we move away from the focus plane; the wave rotation experiment was effected with a diffusing dihedron with angle parallel to the recording plane. The filtering effected in the Fourier plane permits us to obtain fringes drawing the contour lines of the object; we observe a decreasing contrast of the fringes beginning with the section throught the focus plane.

The systematic experimental study of the visibility as a function of the focus defect has been made. This study has given the following results.

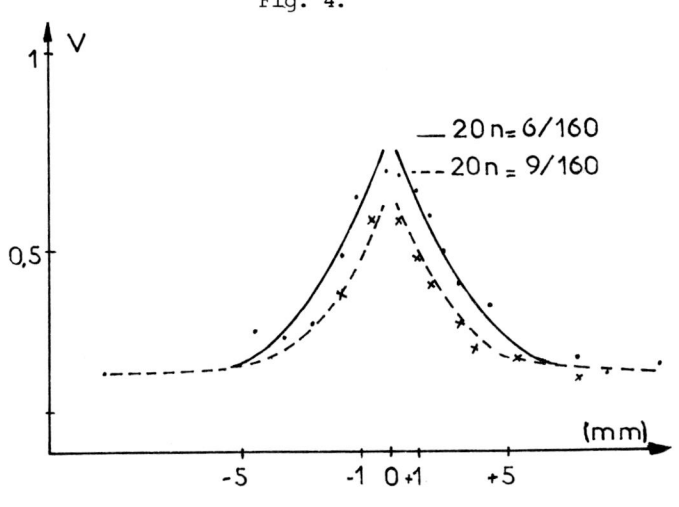

Fig. 4.

Figure 4 represents the variations of the visibility as a function of the focus defect for a given object and two values of the aperture. The visibility decreases as we move away from the focus plane ($\varepsilon = 0$), then becomes constant. We see that the plateau is reached more or less quickly depending on the aperture of the system, but that its height (and thus the value of the corresponding visibility) is independent of the aperture. We can thus admit that the theoretical visibility characterizing the rugosity has been reached.

The curves on the left in figure 5 correspond to two objects of different rugosity and to a single numerical aperture. The height and the position of the plateau vary, which corresponds well to different rugosities and correlation lengths.

The two curves on the right correspond to a single object and two values of $\delta\theta$. The height of the plateau varies as foreseen, but the abscissa of the beginning of the plateau is noticeably the same, which also conforms to what was expected.

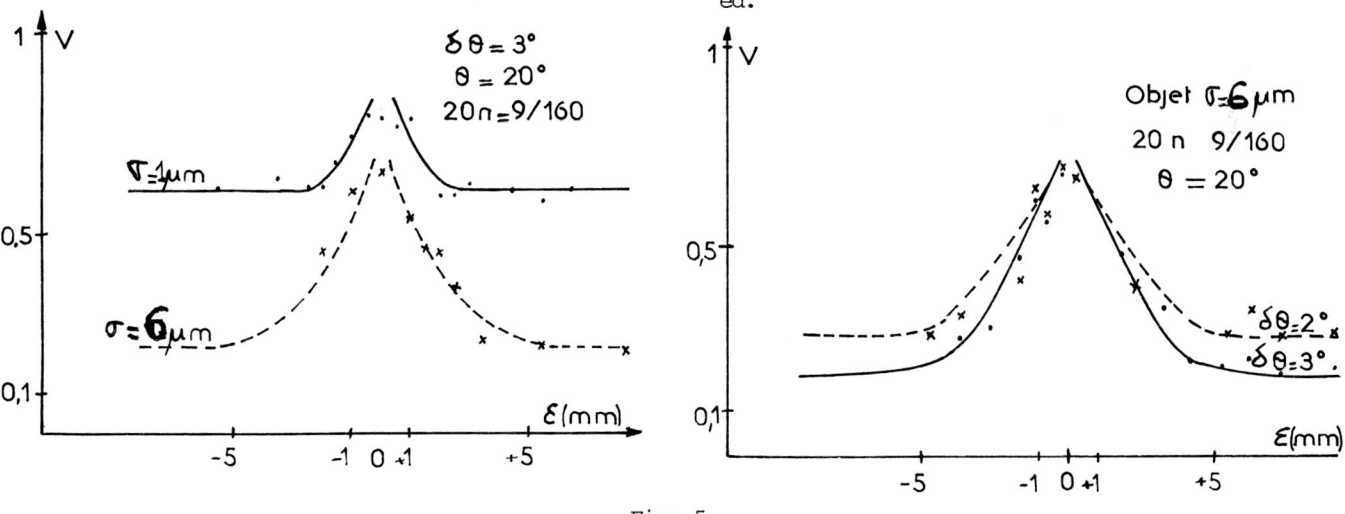

Fig. 5.

Conclusion

On a practical level this study has permitted us to define the optimal experimental conditions: it is necessary to operate with a focus defect of several mm for a focal length of 120 mm and an aperture on the order of F/10. The results also permit us to envisage determining by this experiment both the rugosity and the correlation length of the profile, by using the curve relating the visibility to the focus defect.

References

1. M.L. Roblin, M. Shalow, B. Chourabi, "Interférometrie différentielle des aberrations d'un système optique par photographie de speckles", J. Optics (Paris) 1977, 8, n° 3, pp. 149-158.
2. M. Menu and M.L. Roblin, "Détermination de rugosité par corrélation des speckles dans l'image de la surface diffusante", Opt. Comm., 1977, 21, n° 3, pp. 355-360.
3. T. El Dessouki, M. Menu and M.L. Roblin, "Détermination des lignes de niveaux par double photographie en lumière cohérente", Opt. Comm., 1977, 22, n° 3, pp. 307 311.

TWO WAVELENGTH SPECKLE IMAGES APPLIED TO CONTOURING*

Gilbert Tribillon

Laboratoire de Physique Générale et Optique, Faculté des
Sciences et des Techniques, 25030 Besancon Cedex, France

Abstract

A method is proposed to use the speckles created in front of a curved surface to approach the slope contours of this surface. By focusing the camera on a plane in front of the surface and photographing the speckles contained in this plane before and after changing of the laser light wavelength, a speckle interferogram is obtained.

Various methods of analysing speckle photographs are presented to obtain some informations related to the form of the objects. Two methods use the diffraction properties of the speckle patterns in coherent optics. A third method of viewing speckle photographs is to use a small light source which may have a limited spatial coherence and is often polychromatic. A fourth method is suggested to observe in real time the contour lines of the object.

Introduction

Multiple wavelength laser beam scattered by a rough surface produces various speckle patterns. In a far field plane we can define a correlation degree depending of the microstructure of the surface of the shape of the object and of the spectral difference $\Delta\lambda$. For a plane object, theoretical approaches [1][2][3] and practical applications [4][5] have been developped. The amplitude of the spectral difference $\Delta\lambda$ is very important. If we choose, $\Delta\lambda$, very small, the correlation degree is just represented by an homotecy [6][7]. Using an optical system we can localize the evolution factor with the slope of the object [8]. Recording system and analysis methods are described in this paper.

Optical Recording System

A rough surface is illuminated by means of a dye laser beam, the reflected wavelets coming from different parts of the surface will interfere to create speckles. If a defocused image is recorded as it is represented on the figure 1 we obtain subjective speckles controlled by the aperture size of the lens L.

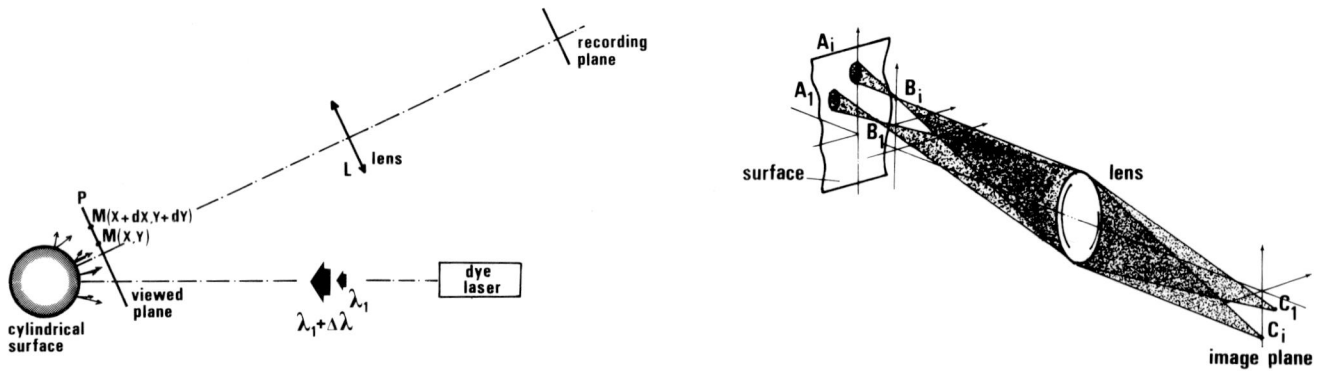

Fig. 1. Recording arrangement.   Fig. 2. Optical system of a defocused image formation.

The speckle pattern contained in plane P is photographed before and after the change of wavelength. The result is a double exposure speckle photography. Before to describe the analysis methods it is necessary to understand the evolution of speckle patterns in plane, P, when we change the wavelength of the laser light.

Let us consider figure 2. Point B_1 is a typical speckle dot in the object plane of the lens. The lens images this dot at C_1 in its image plane. The finite aperture of the lens defines a circular region A_1 on the surface. The speckle formed at B_1 is the contribution of the region A_1. We can consider the surface as the succession of elementary areas. At different points B_i correspond different regions A_i.

* This research work has been performed under DRME Contract N° 76.36.177.

TWO WAVELENGTH SPECKLE IMAGES APPLIED TO CONTOURING

The incidence angle, i, varies with the different regions A_i. Angles, r, of the scattered light in the direction B_i vary in the same order.

The figure 3 represents the wavelength dependence of the speckle patterns when a small spectral shift $\Delta\lambda$ is applied. We known that a small variation of the wavelength of the laser light introduces an homotecy between the speckle patterns. The homotecy factor is directly depending of the spectral shift $\Delta\lambda$.

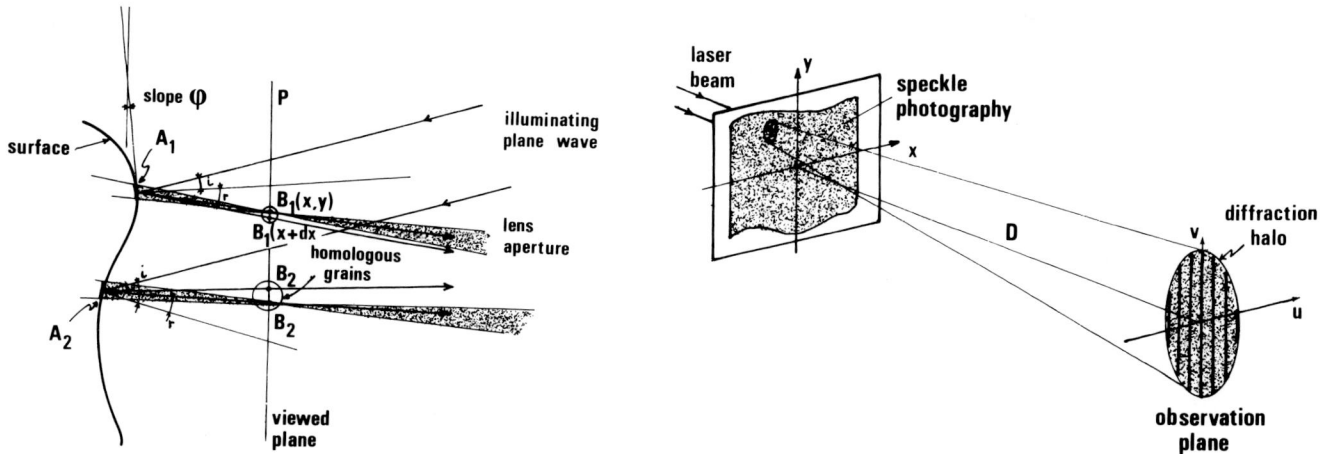

Fig. 3. Wavelength dependance of individual speckle grains with the slope variation.

Fig. 4. Pointwise filtering.

In the viewed plane, P, a speckle grain at the position $B_1(x,y)$ has a homologous grain at the position $B_1(x + dx ; y + dy)$. The displacement (dx, dy) depends of the angle of incidence on the surface and consequently depends of the surface slope ϕ. For an other point, B_2, the displacement is different.

Analysis methods

Pointwise filtering

In a first step, the analysis method of photographs by pointwise [9] filtering has been developped. The optical arrangement is as shown on the figure 4. Illuminating the speckle photography with a narrow beam of laser light, diffracted light is observed on a ground glass at some distance D. We observe a well known phenomena i.e. a diffraction halo modulated by a series of cosine squared fringes. The beam size diameter is about 1mm - 2mm in diameter. The fringe spacing and orientation are according to the local displacement and consequently to the surface slope ϕ. The surface slope information is accessible point by point.

Figure 5 represents the results for two cylindrical surfaces. The upper lines evidence the influency of curvature. The third line notes the influency of the spectral shift $\Delta\lambda$.

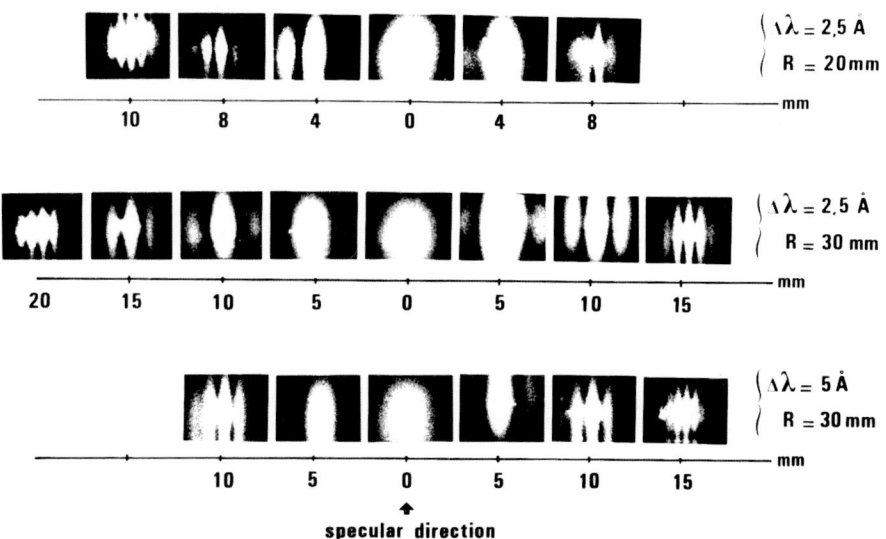

Fig. 5. Fringe patterns observed by pointwise filtering of a two wavelength speckle photography for two values of the curvature.

Filtering in Double Diffraction System

The isothetics have been determined on the speckle photographies by optical spatial filtering [10]. An optical Fourier transform is performed on the amplitude transmission function of the recorded images. We insert the photographic plate containing the speckle patterns in front of an optical correlator as represented on the figure 6. The light intensity distribution spectrum can be described as the product of the diffraction halo of the speckle pattern modulated by an infinity of interference systems. One place a mask with a small aperture in this filtering plane. Bright fringes satisfying at $d.u. = n\lambda f$ can be observed in the image plane.

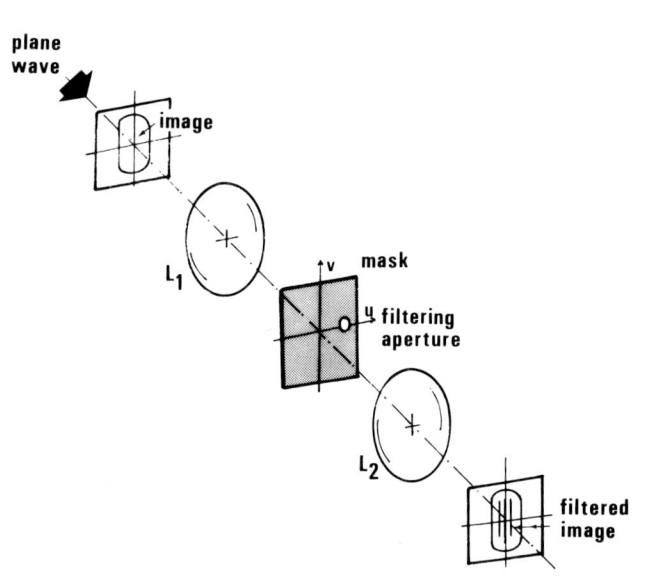

Fig. 6. Double diffraction system of analysis of a two wavelength speckle photography.

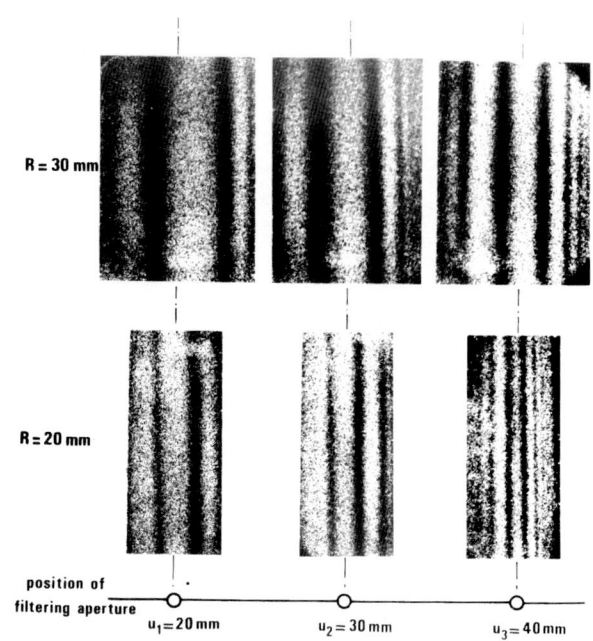

Fig. 7. Fringe patterns corresponding to isothetics observed by filtering of a two wavelength speckle photography.

Fringe pattern on two cylindrical object with different sensitivity are presented on the figure 7. The sensitivity depends of the position of the small aperture placed in the filtering plane.

Observation in Partially Coherent light

An other analysis method of speckle photographs is to use a small polychromatic light source with a limited spatial coherence to observe the locus of virtual fringes seen through the photograph as shown on the figure 8. Interpretation has been given by D. A. GREGORY [11]. When we observe a defocused speckle photograph, the eye acts as a selective aperture in the far field projection plane and the eye selects particular portions and orientations of the real Young's fringes coming from the different parts of the photograph. The selected real fringes enter on the eye before to be combined accross the eye lens. The finite aperture of the eye limits the contrast and resolution ; contrast and resolution are optimum when a fringe just covers the eye aperture. Another problem is that the eye does not focus on the emulsion surface of the defocused speckle photograph. For visual observation this is not a problem (important depth of field). The finite eye aperture defines the size of area which is creating a virtual image. We must use a compromise between the size of the photographic plate, the eye aperture, the fringe resolution.

A mercury arc source with a green narrow band interference filter gives a minimum noise and eliminates the speckle noise. If we use a polychromatic light we observe white light fringes located in the zero order any where on the speckle photograph. The only lens used is the eye or the camera imaging lens. To observe fringes with a good contrast it is useful to place the eye just outside the direct line of projector source.

The interpretation of the virtual fringes in white light is very complex, due to the large number of variables.

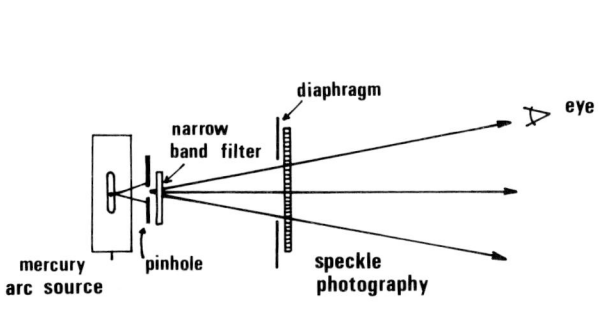

Fig. 8. Optical arrangement to analyze a two wavelength speckle photography under a partially coherent light.

Fig. 9. Virtual fringes observed on a small area.

Let us summarize the advantages of this technique

- we can observe large parts of a speckle photograph with rapid appropriate choice of the distance and direction.
- Using a white light source, the fringes are coloured and are located in the zero order anywhere on the surface.
- Using light with low temporal coherence, noise due to speckle phenomena is minimum.
- The sensitivity can be as good as that for contouring holography.

In contrary the main disavantage of this technique is virtual fringes the measurements require the projection on a monitor or some other devices.

Let us consider on the figure 9, the virtual fringes from the point of view of understanding the topological variations on defocused speckle photographs.

The speckle pairs in the small area defined around of C have a constant magnitude and direction. The small area, illuminated by a small collimated beam will project Young's fringes into the far field observation plane. The observation plane is parallel to the projection plane. It is a direct mapping between the both planes. The eye position can adjust the fringes spacing and orientation. When the eye position changes, the locus fringes is modified. When a part of the photograph is projecting the nth real fringes accross the eye aperture, the eye sees one point of the nth locus virtual fringe.

This analysis has a great advantage on the conventionnal methods because the eye and brain are able to explore the speckle photograph with high speed and to give the suitable sensitivity for the variables as the distance L between the speckle photograph and the observation plane.

Experiments have been performed on a metallic plate where the surface slope had approximately a sinusoïdal profile with an amplitude of 10 mm. Observations are shown on the figure 10. Partially coherent light observations are presented on (a) and white light observations are presented in b. Variations od sensitivity are represented as a function of the eye position.

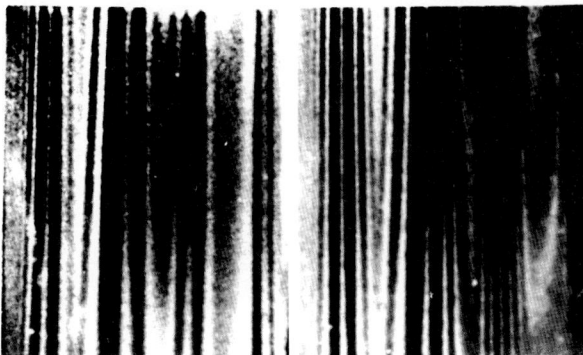

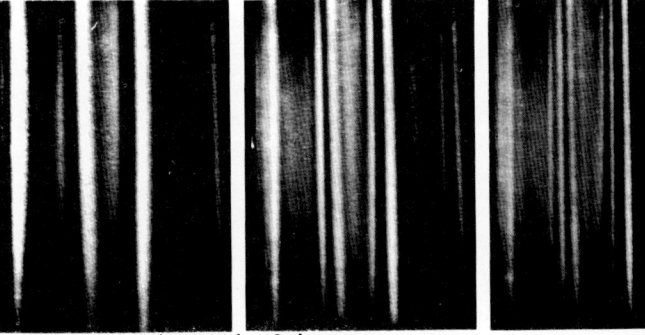

achromatic fringes in the specular direction

Fig. 10.a. Observations with mercury arc source - low temporal source - for two values of $\Delta\lambda$: $\Delta\lambda = 2,5$ Å, $\Delta\lambda = 5$ Å.

Fig. 10.b. Observations in white light source. Sensitivity is represented as a function of the eye position.

Real time Observation Method

We would like to suggest an observation method of the slope of a complex surface by the observation of the speckle patterns directly. Let us recall the Massey's method (12)(13) applied to the observation of the mode shape of a vibrating panel.

When the panel is vibrated, certain regions of the object will perform oscillatory tilting about fixed axes. The related speckle dots for these regions will be moving. For vibrations with period shorter than the persistance vision about 0.1 s the moving dots will appear as stationnary streaks.
Non tilting regions of the panel will leave stationary discrete speckle in the object plane. Monochromatic coherent light and vibrating object are used in this case.

If the wavelength illuminating a non moving object changes continuously on a narrow spectral domain -0 - 10 A for example - speckle patterns appear in the observation plane in according to the surface slope and the spectral shift $\Delta\lambda$. Wavelength variations are performed with an electronically tunable system. Continuous scanning is selectable and the sweep frequency can be adjusted to obtain a scan about 50 ms. Figure 11 gives the experimental arrangement. As an illustration is presented the case of a surface characterized by a sinusoïdal profile.

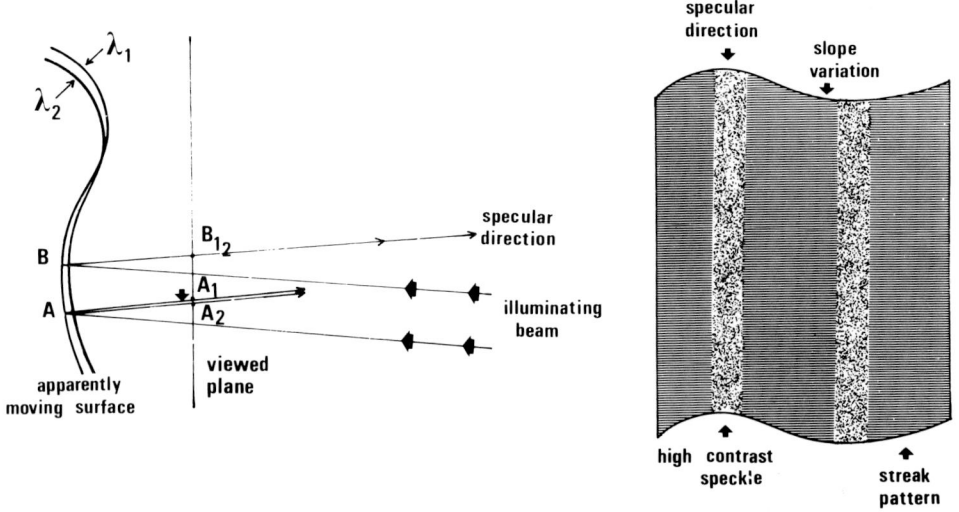

Fig. 11. Real time observation method suggested to visualize the slope variation by contrast variation.

Conclusion

Speckle photography methods using multiple wavelengths produced by a dye laser demonstrate which they are an intersting approach as contouring methods. Real time observation has been suggested to developpe a practical tool for some industrial applications.

References

1. Goodman, J.W., Stanford Electronics Lab., Tech. Rept., n° 2303-1, 1963.
2. Parry, G., Opt. Acta, vol. 21, p. 763, 1974.
3. Pedersen H.M., Opt. Acta, vol. 22, p. 15, 1975.
4. George, N., Jain, A., Appl. Phys., vol. 4, p. 201, 1974.
5. George, N., The wavelength Sensitivity of Backscatting, ICO. X, Prague 1975.
6. Mendez, J.A., Roblin, M.L., Opt. Comm., vol. 13, p. 142, 1975.
7. Tribillon, G., Bobillier, J.F., Wavelength Dependence of correlation between two amplitudes or intensity speckle patterns, ICO. X, Prague 1975.
8. Tribillon, G., Garcia-Garcia, M., Opt. Comm., vol. 20, p. 229, 1977.
9. Khetan, R.P., Chiang, F.P., Appl. Opt., vol. 15, p. 2205, 1976.
10. Archbold, E., Ennos, A.E., Opt. Acta, vol. 19, p. 253, 1972.
11. Gregory, D.A., Opt. Laser Tech., vol. 9, p. 17, 1977.
12. Massey, G.A., Nasa Rept., n° 68-14070, Springfield Va. 22151, 1968.
13. Eliasson, B., Mottier, F.M., J.O.S.A., vol. 61, p. 559, 1971.

APPLICATION IN CIVIL ENGINEERING OF A SANDWICH SPECKLE METHOD

Jean P. L. Ebbeni
Laboratoire d'Analyse des Contraintes
Université Libre de Bruxelles
87, Avenue Ad. Buyl, 1050 Bruxelles, Belgium

Abstract

The sandwich speckle technique described in the paper permits to cancel the influence of uniform displacements field due to the deformation of the supports of the loaded model. On the other hand it is possible to follow the evolution in time of deformation by a step by step recording.

Introduction

Double exposure speckle techniques record the "total" displacements fields resulting of the deformations of the object but also of those of the supports.
These last displacements are of no interest and limit the field of applications. If they are too important, the superposed zones of the neutral and deformed object are no more correlated and fringes of interference disappear. By the "sandwich" technique, the images of the neutral and deformed object are recorded on separate plates. After developing the two plates are superposed and by adequate translations and rotation it is possible to eliminate the displacements field resulting of the deformation of the supports. Speckle grains can be obtained by scattering of laser light on the model, or more simply by the direct recording of the microstructure due to the roughness of the model illuminated by a white light source. This microstructure can be obtain by special techniques, e.g. with retroreflective paint.

Sandwich plates

The plane object is lighted by plane waves and is imaged by the lens L on the photoplates A_1 and B_1, superposed gelatin to gelatin (cfr figure 1a). After the deformation of the object a similar recording is made with the photoplates A_2 and B_2 (cfr figure 1b). After developing the plates A_2 and B_1 (or A_1 and B_2) are placed gelatin to gelatin in a beam of plane waves in normal incidence including a zone of the object where the displacements field is considered as uniform. We obtain better results by using only two plates, B_1 for the neutral object and A_2 for the deformed object. The support of A_2 is before recording translated longitudinaly by the value $\frac{e}{n}$, where e is the thickness of the photoplate and n its index (cfr figure 1 c and d). The technique used for the reposition of the plates A_2 and B_1 in superposition for canceling the uniform displacement field is the following

a. The plate B_1 is fixed and A_2 is on a (XYZ, O_x, O_y, O_z) table. One adjusts the orientations so that B_1 and A_2 are parallel (e.g. by laser beam reflection).

b. The two plates are then moved close together and by in plane rotations and translations we arrange the system to exhibit YOUNG fringes.

c. On an undeformed part of the object, or on a reference object, one translates A_2 vertically until the fringes are vertical with a direction non sensitive to an horizontal translation.

d. The vertical fringe are eliminated by horizontal translation.

e. Normally, the family of fringes cannot desappear because of the longitudinal magnification and we end up with circular fringes. By adjusting the longitudinal translation we finally cancel this family circular fringes.

In practice we are left in fact with one parasitical circular fringe of about 7 cm of diameter. Then by the classical technique we can analyse each part of the object and obtain the displacement components from the YOUNG fringes (cfr figure 1 e and f). The same technique is used for incoherent speckle with white source of light. The sandwich method permits the determination of displacement fields in a lot of applications e.g. in problems in soil engineering, mechanics, plasticity, bitumineous material in constructions. It is also now possible to study the evolution in time of a phenomena by a step by step method. The sensitivity is 1 µ for the coherent technique and 50 µ for the incoherent technique.

Applications

The first application consists to follow the evolution of a plastic zone in a <u>notched</u> specimen intraction (cfr figure 2). By measuring the contrast of the YOUNG fringes, it is possible to determine with an accuracy of 1/10 mm the boundary of the plastic zone. In the plastic zone the state of surface is completely changed, the correlation is destroyed and no fringe are thus visible. An other application with incoherent lighting by mean of a flash was presented during the conference. It consisted to follow the evolution of a crack by loading a model in concrete.

Acknowledgments

The author washes to thank Mr. J.J. DEVILLERS for the applications realized in the laboratory.

References

1. Ebbeni, "Introduction à l'analyse expérimentale des contraintes", Notes de cours, Service d'analyse des contraintes, Université Libre de Bruxelles
2. Watte, "Speckling et plaques sandwich", Travail de fin d'études 1976, Service d'analyse des contraintes, Université Libre de Bruxelles
3. Forno, "White-light Speckle photography", Optics and Laser Technology, vol. 7, n° 5, p. 217, 1975.
4. Burch et Forno, "Optic Engineering", vol. 14, p. 178 à 185, 1975
5. Devillers, "Etude de déformations par speckling", travail de fin d'études 1977, Service d'Analyse des Contraintes, Université Libre de Bruxelles
6. Ebbeni, La jauge diffractante en extensométrie.

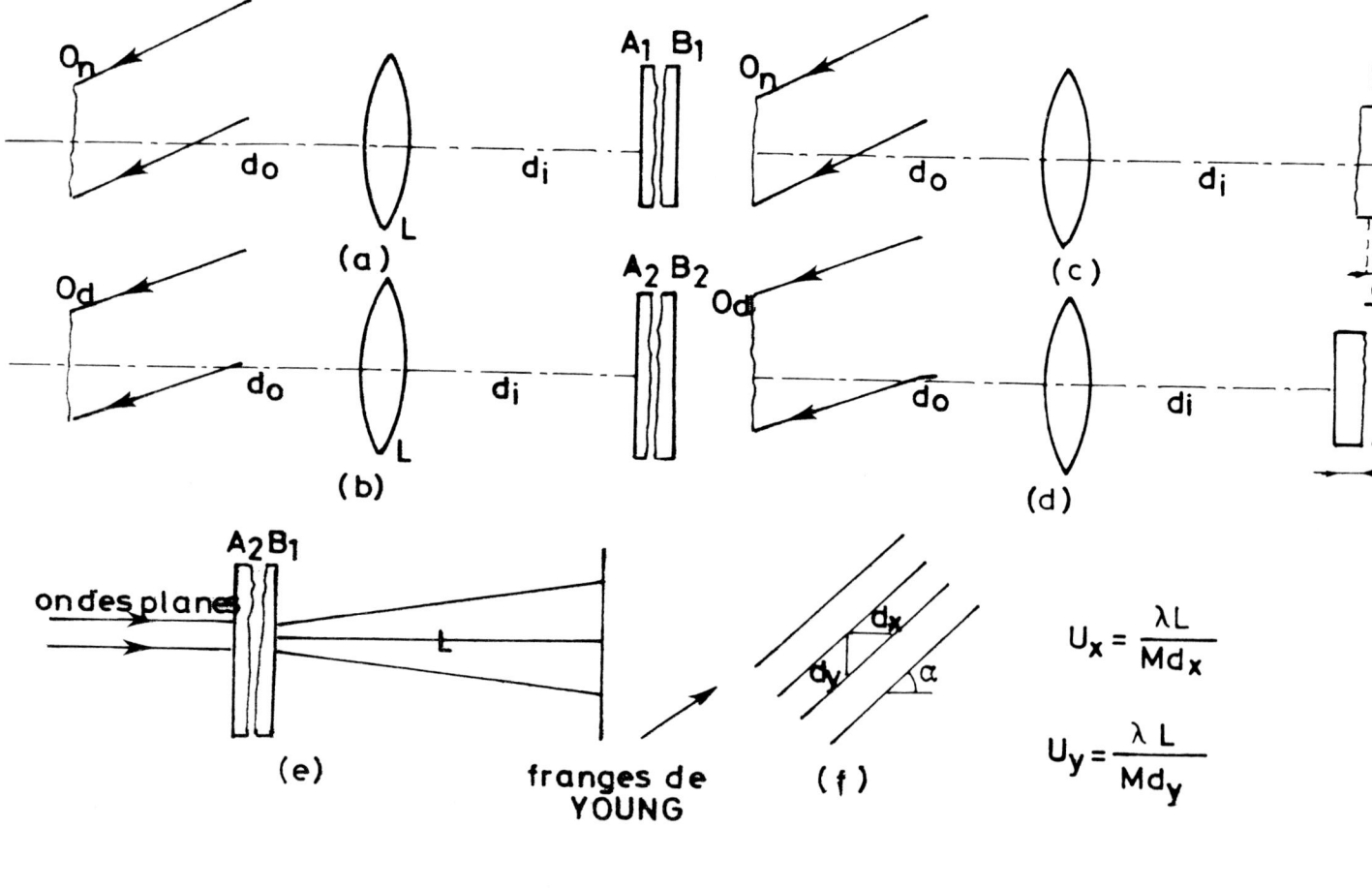

FIG.1.

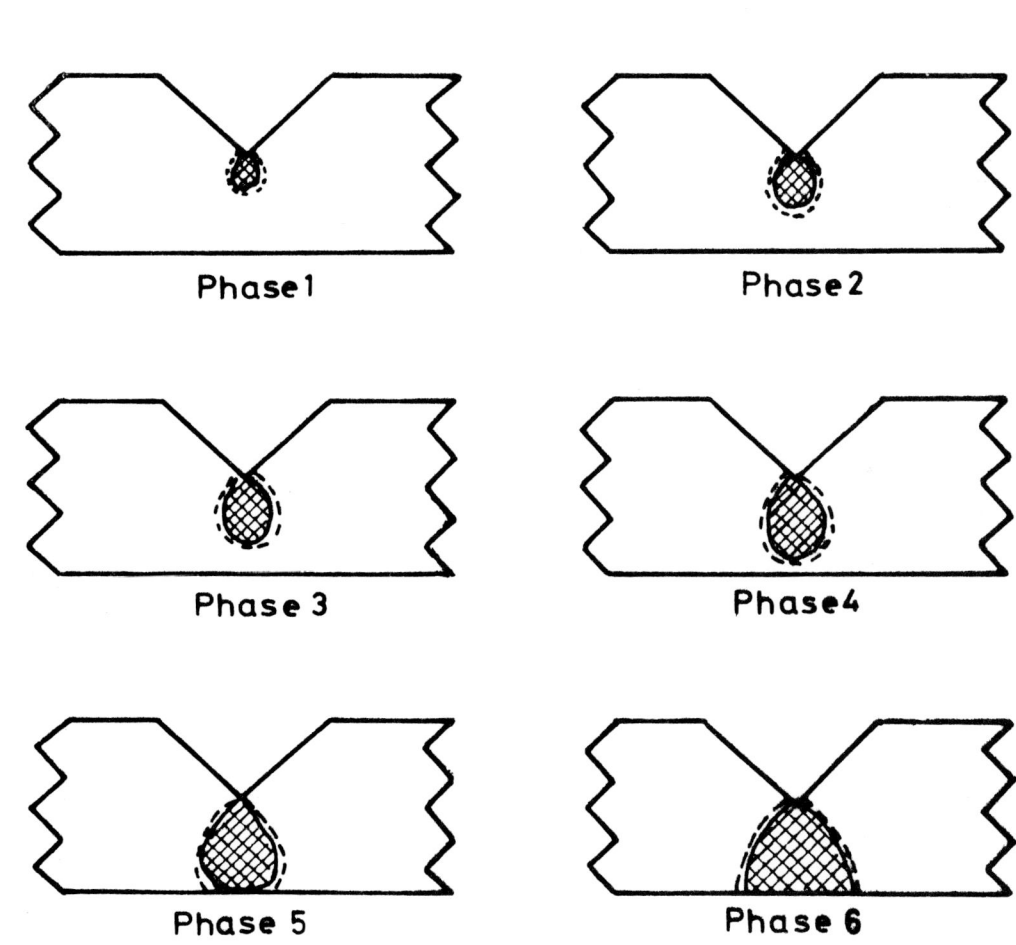

FIG. 2

1st EUROPEAN CONGRESS ON OPTICS APPLIED TO METROLOGY

Volume 136

SESSION 7

MOIRE

Session Chairmen
Bulabois
Denisyuk

OPTICAL DIFFERENTIATION OF MOIRE-HOLOGRAPHIC FRINGES BY WAVEFRONT RECONSTRUCTION WITH WHITE LIGHT SOURCES

Carlo Brutti, Giuseppe Di Chirico, Umberto Pighini
Instituto di Macchine e Tecnologie Meccaniche
Università di Roma

Abstract

Moiré-holograms contain a parallel-line grating recorded during the exposure process of the model grid under three-dimensional deformation.
Utilizing the diffraction of a white light beam striking the holoplate, frozen moiré patterns can be observed, under proper angles, directly by reflection.
With a Fourier transform optical system, the reconstruction is achieved by means of a white source, filtering one among the chromatic wavefronts at the first diffracted order. If the wavefronts of two complementary colors are allowed to pass through the filter, direct derivatives of displacement fringes will appear on the planes where the images are slightly shifted, before or after overlapping. By a grounded glass the image plane is determined where the differentiation fringes exibit the neatest pattern, which may be recorded on a photographic film.

Introduction

Some wavefront reconstruction techniques by white light sources have been studied successfully in the last fifteen years by several researchers, such as Denisyuk[1], Leith and Upatnieks[2], Rosen[3] and others.
All these methods are founded essentially on construction of holograms with particular diffraction properties or on diffraction caused by a grating between the light source and the hologram.
Starting off from this consideration and from the fact that moiré-holograms already contain a parallel line grating, after showing the possibility to reconstruct moiré patterns by reflection of a white light collimated beam, this paper proposes a method to obtain the immediate vision of direct derivatives of displacement u and v.

Observation of fringes by reflection

A white light collimated beam, passing through the grating existing on moiré-hologram surface, is diffracted in various orders and the relative wavefronts spread according to the directions defined by angles ϑ_n:

$$\sin \vartheta_n = n \frac{\lambda_i}{p} \quad (n=0, \pm 1, \pm 2, \ldots)$$

where n is the diffraction order, λ_i the wavelength relative to each color of the spectrum and p the pitch of the grating.
By observing the hologram following the scheme in fig.1, the angular dispersions of the wavefronts, corresponding to colors of the light spectrum, are compensated, thus determining a unique black and white image of the moiré pattern, located on the hologram surface [4].
When the beam is not perfectly collimated or the illumination is not uniform, some chromatic disomogeneities may appear; nevertheless they do not disturb the reconstruction process of the fringes, which appear sharp and clear from various observation points.
The evident advantage of this techinique is that a moiré-hologram, recorded in a laboratory with coherent light, may be observed in any given environment, almost as a common photographic film, with a merely collimated beam, or else diffuse light coming from a sufficient distance.

Wavefront reconstruction by means of a Fourier transform optical system

The direct observation of moiré fringes is made possible also by trasmitted light; however the sight is always incomplete from a single point of observation, due to chromatic dispersions.
On the contrary, an excellent reconstruction of wavefronts is obtained with the optical system in fig.2.
In the plane of trasformed images, F, at each order of diffraction a spectrum appears, formed by monochromatic bundles, ranging from bleu to red.
Each one is an exact reconstruction of the fringe patterns of displacement components. Moiré-holograms of components u and v are actually obtained by double exposure, by interfe-

rence of symmetrical wavefronts, diffracted by the model grid under the effect of plane or threedimensional deformations.

After the development of the plate, the transmission function, related to the x axis, is expressed by [5]

$$T(x) = 2A_n^2 \left[2 + \cos 2\pi \frac{2nx}{p_1} + \cos 2\pi \frac{2nx}{p_2(x)} \right]$$

where ± n is the couple of diffraction orders filtered during the recording process; p_1 is the undeformed pitch, p_2 is the deformed pitch.

In the reconstruction process with a white light collimated beam, a diffraction pattern is obtained, where the intensity of each monochromatic wavefront, relative to orders plus one and minus one, is modulated respectively by one of the following terms:

$$A_n^2 \left[e^{-2\pi i \frac{2nx}{p_1}} + e^{-2\pi i \frac{2nx}{p_2(x)}} \right]$$

$$A_n^2 \left[e^{2\pi i \frac{2nx}{p_1}} + e^{2\pi i \frac{2nx}{p_2(x)}} \right]$$

deriving from the previous equation.

At the diffraction order plus one, the light intensity of each monochromatic radiation is expressed by an equation of the type:

$$I(x) = 2A_n^2 + A_n^2 \cos 2\pi x \left[\frac{2n}{p_2(x)} - \frac{2n}{p_1} \right]$$

for points where $x = x_0 + u$, and similarly for points $y = y_0 + v$:

$$I(y) = 2A_n^2 + A_n^2 \cos 2\pi y \left[\frac{2n}{p_2(y)} - \frac{2n}{p_1} \right]$$

These equations describe the displacement components u and v and show that the number of moiré fringes is increased by a factor 2n. Therefore, since the hologram grating has a high number of lines, the angular dispersion in the focal plane of L_2 is generally sufficient to spread polychromatic light into bundles of discrete wavelenghts, which may be filtered separately.

Optical differentiation of moiré fringes

If two monochromatic images are filtered together, at the same diffraction order, as in fig.3, there will be a plane in which they fully overlap.

Moving the screen back and forth along the optical axis, a lateral shift of the images is obtained. When the number of moiré fringes is high enough, using two complementary colors of the light spectrum, these shifts originate new systems of fringes which are the direct derivatives $\varepsilon_x = \partial u/\partial x$ and $\varepsilon_y = \partial v/\partial y$ of the displacements u and v [6], with a differentiation scale equal to $p/\Delta x$ or $p/\Delta y$.

The orders of fringes are integers. In fact, at the points of zero derivative, namely points which do not shift during the displacement of the images, two complementary colors overlap.

In the same way the next fringes of the orders 1,2,.... are outlined.

This method shows several advantages as compared to other optical differentiation systems. Essentially, a single photographic exposure is sufficient to reconstruct direct derivatives of displacements from the hologram. Moreover the shift of the images is visually controlled, in order to obtain clear and meaningful fringes. Finally, the quality of the reconstructed image is generally better than the equivalent one obtained with coherent light; therefore, also sharp fringes of crossed derivatives $\partial u/\partial y$ and $\partial v/\partial x$ may be simply obtained by double exposure of the photographic film, shifting the back of the camera in the proper direction.

Examples of application

Figure 4,5 and 6 show the pattern of direct derivatives of displacements u of three clamped circular disks: the first one has a circular hole and is loaded on the hole edge, the others have a square hole and are loaded on the four or on the two opposite sides. A grid with 20x20 dots/mm² was printed on the perspex models (Kodak Photoplast), dia.100 mm and thickness 3 mm. By filtering +3 and -3 horizontal orders, the pitch of the equivalent grating was of 120 lines/mm. Derivative fringes were obtained filtering two images relative to green and orange (λ=490 and 605 nm), with the screen located along the optical axis in order to obtain a shift of 4 or 5 mm on the x-axis. The reconstructed patterns were recorded on Tri-X-Panchromatic film.

The differentiation scale is of 2 mm/m for the models in fig.4 and 5, and of 1.5 mm/m for the model in fig.6.

The accuracy of derivative values, obtained with the above mentioned method, were tested on a model of clamped centrally loaded disk (P=40 kg), having the same sizes of the previous ones.

Tables 1 and 2 show the agreement between theoretical and experimental values of ε_r and ε_t at the points where fringes intersect the disk radius

Table 1

r (mm)	ε_t theoretical	ε_t experimental
14	8.03×10^{-3}	8×10^{-3}
19	5.96×10^{-3}	6×10^{-3}
26	4.12×10^{-3}	4×10^{-3}
36	1.92×10^{-3}	2×10^{-3}

Table 2

r (mm)	ε_r theoretical	ε_r experimental
10	3.84×10^{-3}	4×10^{-3}
13	2.19×10^{-3}	2×10^{-3}
18	0.13×10^{-3}	—
26	-2.18×10^{-3}	-2×10^{-3}

Conclusions

The performed experimental work outlines the feasibility and the advantages of the method proposed to optically differentiate moiré-holographic fringes by wavefront reconstruction with white light sources.

In fact, derivative patterns are recorded by a single exposure of the film; besides, it is easy to select the shift of two complementarily colored images, in order to visually control the appearance of clear and meaningful fringes. It should be remarked also that erroneous evaluations of fringe orders are avoided by following the fringe growth on the screen, while the lateral shift is increasing.

Although the described technique permits to observe only direct derivatives of the displacement components, it does not seem impossible to obtain directly also the crossed derivatives by a somehow more sophisticated optical arrangement. At any rate, since the quality of the reconstructed images is excellent, crossed derivatives may be always obtained by a simple shift of the back of the camera, in a double exposure process.

References

[1] Y.N. DENISYUK, "Photographic Reconstruction of the Optical Properties of an Object in its own Scattered Radiation Field", Sov. Phis. Dok. 7,543, 1962.

[2] E. LEITH, J. UPATNIEKS, "Photography by Laser", Scient. Amer. 212,6,1965.

[3] L. ROSEN, "Focused-Image Holography with Extended Sources", Appl. Phys. Lett. 9,337, 1966.

[4] G. DI CHIRICO, C. BRUTTI, "Ricostruzione di ologrqmmi moiré in luce bianca", Atti del V° Convegno A.I.A.S., Bari, Sett. 1977.

[5] C.A. SCIAMMARELLA, G. DI CHIRICO, F. CHIANG, "Moiré-Holographic Technique for Three-Dimensional Stress Analysis", J. of Appl. Mech., 3, 1970.

[6] A. DURELLI, V. PARKS, "Moiré-Patterns of Partial Derivatives of Displacement Components", J. of Appl. Mech., XII, 1966.

OPTICAL DIFFERENTIATION OF MOIRE-HOLOGRAPHIC FRINGES BY WAVEFRONT RECONSTRUCTION WITH WHITE LIGHT SOURCES

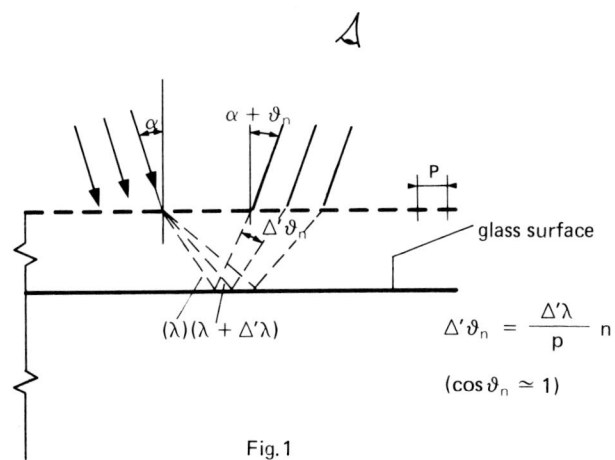

Fig. 1

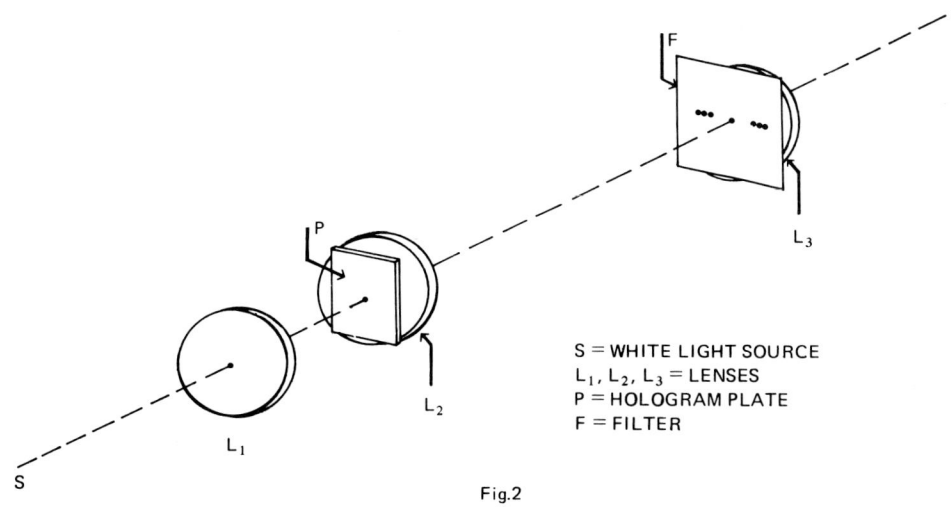

S = WHITE LIGHT SOURCE
L_1, L_2, L_3 = LENSES
P = HOLOGRAM PLATE
F = FILTER

Fig. 2

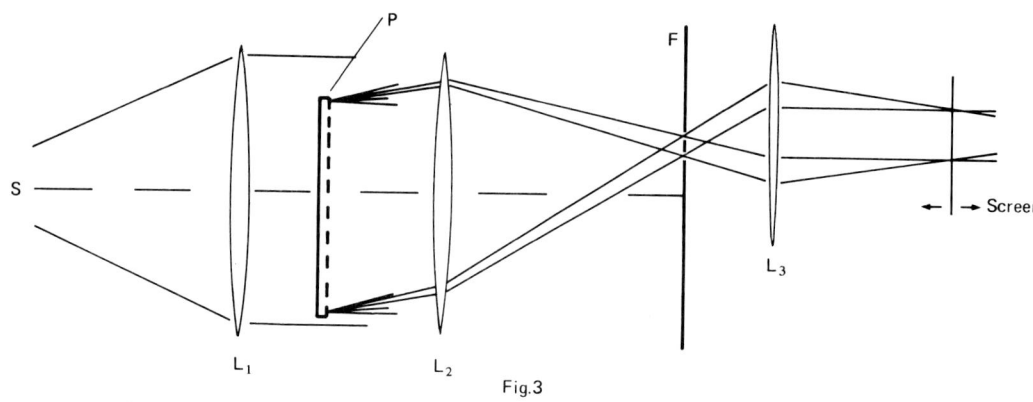

Fig. 3

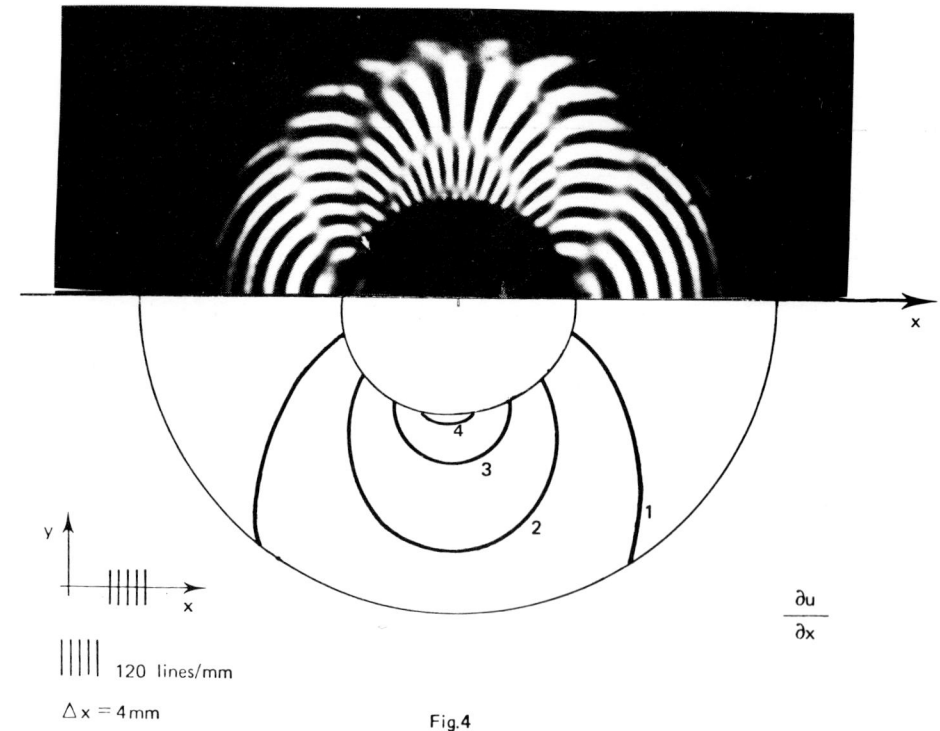

120 lines/mm

$\Delta x = 4$ mm

$\frac{\partial u}{\partial x}$

Fig. 4

OPTICAL DIFFERENTIATION OF MOIRE-HOLOGRAPHIC FRINGES BY WAVEFRONT RECONSTRUCTION WITH WHITE LIGHT SOURCES

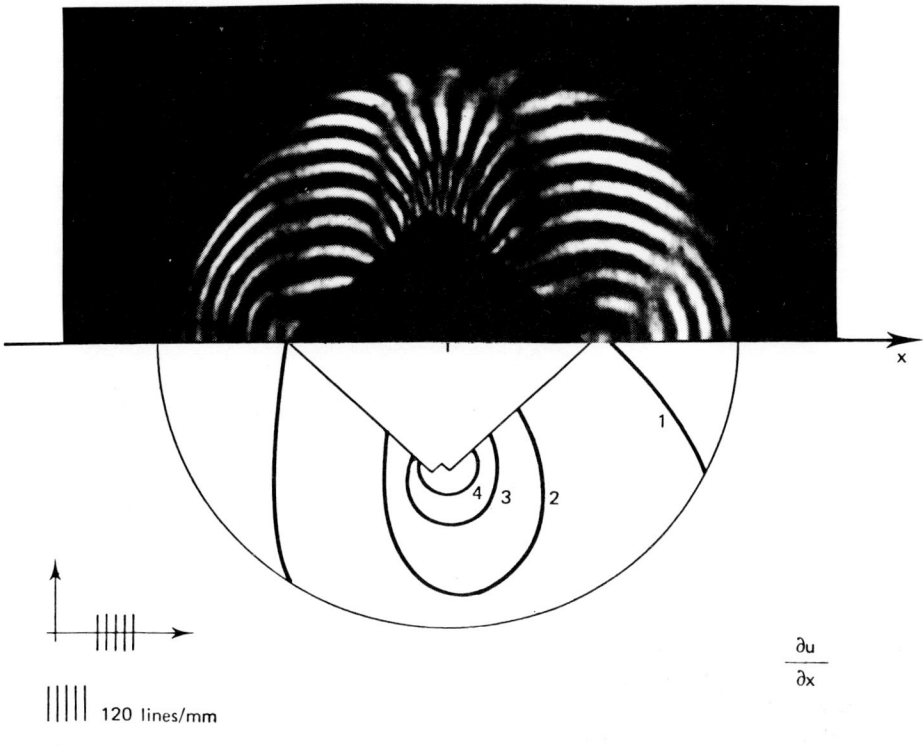

Fig.5

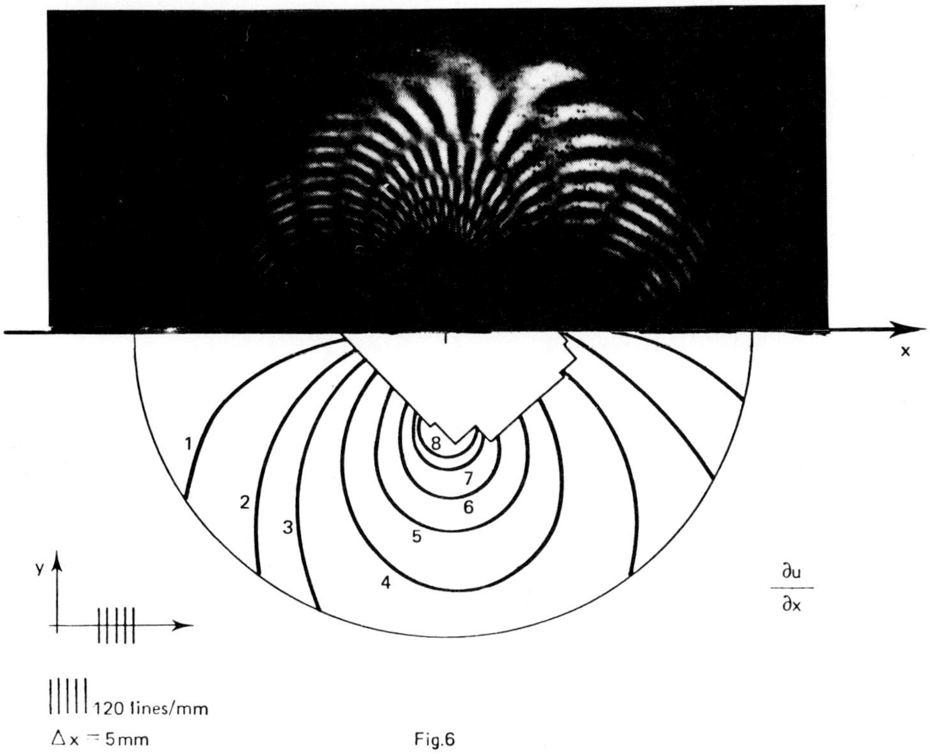

Fig.6

THE MEASUREMENT OF RESIDUAL STRESS BY A MOIRE FRINGE METHOD

C. A. Walker, J. McKelvie
Centre for Industrial Innovation, University of Strathclyde,
100, Montrose Street, Glasgow, G4.0LZ. Scotland, U.K.

Abstract

An assessment of the residual stress pattern around a butt-weld has been made using a hole-drilling technique along with a sensitive moire fringe method. The method of applying the specimen grid is described, and also the procedures for extracting stress data from the moire fringe patterns. This technique may be useful in situations where the residual stress varies rapidly, or where it is inconvenient to apply numbers of strain gauges.

Introduction

The analysis of residual stresses in engineering materials is a matter of continuing interest due to the effect that such stresses can have upon the performance of engineering structures. Our understanding of residual stress has been fostered by the measurements made, first of all by dilatometric techniques, and more recently by X-ray, strain gauge and Barkhausen noise methods. The fact remains, however, that these methods do exhibit drawbacks when applied to real engineering structures[1]. It is for these reasons that a sensitive moire method of in-plane strain measurement has been applied to the measurement of residual stress distributions.

The Moire Method

Since the sensitivity of measurement in any moire method depends directly upon the density of the grating, efforts have been made[2],[3], to increase the sensitivity either by increasing the grating line density[1],[2], or by using a Fourier-filtering interferometer to extract fringe information from the higher diffracted orders[4],[5]. The optical system used in this study is essentially a development of that proposed by Post[4] Figure 1.

In general, the specimen grids used for this technique are photographic emulsions[6], or photoresist layers[2],[3]. Since these types of grid suffer from low diffraction efficiency in the first instance, and fragility in the second, the authors have developed a method of casting a grid in epoxy resin, with the grid pattern impressed in the surface[11]. These grids have excellent diffraction efficiency, and will withstand environmental degradation. The problems involved in aligning a laser interferometer with a massive specimen are avoided by taking a cast replica from the surface of the specimen grating after it has been deformed, Figure 2. As a result of these two features, a routine sensitivity of .001 mm/fringe can be obtained and has been used for this study.

Residual Stress Measurement by the Hole-Drilling Method

The measurement of residual stress by interpretation of the surface deformation when a hole is drilled in it remained only a theoretical possibility until the development of sensitive strain gauges. The method has since been developed to a high level of accuracy[7],[8], backed up by a fund of theoretical work to establish the relationship between the residual stresses existing in a material and the elastic/plastic recovery which takes place when a hole is drilled in the surface. In essence, an array of strain gauges is used to measure the elastic recovery which takes place upon hole drilling. The strain gauges must be positioned accurately in relation to the centre, and edge, of the hole. The effects both of hole shape, and of the stresses induced by the machining process have been analysed theoretically and experimentally[9]. The result of this work is a rather widely used technique in which a 120° rosette of three radially disposed strain gauges is used to measure the relaxation strains from a centrally drilled hole. This allows of the accurate measurement of residual stresses which are spatially invariate over the area delineated by the strain gauge array. Cordiano and Salerno have shown that a residual stress pattern which has a linear biaxial variation can be analysed by the hole-drilling technique in conjunction with an array of eight gauges spaced at 45° intervals around the hole. In general, the stress system around a weld is complex, and certainly shows spatial variations of considerable magnitude[9].

The advantage to be gained from a moire system arises where a stress distribution, rather than a spot measurement, is desired. Once the specimen grid is laid down, holes can be drilled in it at any location, either in conformance with a desired matrix of sampling points, or as indicated by results obtained from the first few holes. In addition, the precision with which the hole must be positioned is not critical, since the hole itself is used as the origin of the co-ordinate system.   Further, the fringe pattern will demonstrate a fine structure of the residual stress distribution which cannot easily be visualised using strain gauges.

Finally, as will be evident from the above discussion, the amount of data extracted from the moire fringes around each individual hole can be varied dependent upon whether the spatial variation of the stress system is simple or complex.   This option does not operate with strain gauges, where the worst case must be assumed at the time of laying down the strain gauge rosette.

Materials

The work reported in this paper relates to the residual stresses along a butt-weld between two mild steel plates 25 mm thick, and each 300 mm long and 150 mm wide.   The plates were given a standard weld preparation, Figure 3, resulting in a weld approximately 12 mm wide.   After welding by the manual metal arc method, using a balanced welding technique to minimise distortion, the surface was dressed flat on both sides. A consideration of the welding process demonstrates the origin of the residual stresses, for as the molten weld pool cools, and solidifies, a state of residual stress will result from the shrinkage of the weld material, and its restraint by the parent material, Figure 4.

Experimental Procedure

A specimen grid (orthogonal 40 line/mm) was laid down over a 100 mm x 100 mm area covering the weld. This grid itself is epoxy resin, and the grid structure is formed by a casting process, using a silicone rubber master grating[10].   An array of holes was drilled through the grid into the metal, each hole being 3 mm in diameter and 3 mm deep.   The holes are offset from each other so that the deformation patterns do not interfere with one another.   After drilling, the surface deformations were recorded by making a cast of the surface in transparent silicone rubber.   This case is then analysed in a fringe-multiplying interferometer, Figure 1, in which the basic moire sensitivity of 40 line/mm is increased to the equivalent of 1000 line/mm.   Each moire fringe therefore denotes a relative displacement of 1 micron, relative to adjacent fringes.   The resulting deformation patterns around each hole, caused by the relief of the residual stress show how residual stress varies both across and along the weld.

Quantification of Residual Stress

Since the effect of machining-induced stresses falls off rapidly away from the edge of the hole, a convenient rule of thumb is to measure the relaxation strain at a distance of 1 hole diameter from the edge of the hole[7]. By this means, the effects of machining stresses are avoided, and if no relaxation strain is observed at this point, then the residual stress is below the limit of measurement.

Following the results of Nawwar et.al.[7], the simplest method of measuring the residual stress is to identify the fringe spacing at a distance of one hole diameter from the hole edge.   If this measurement is made in the direction of the principal stress, then the stress at the hole edge is a factor of three more than the figure measured one diameter away, Figure 5.

Alternatively, when the deformation component distributions are being quantified to produce figures for maximum stress, a useful technique is the introduction of a degree of rotational mismatch, Figure 8.   By this means, the measurement of fringe separation at a point can be made by drawing a tangent to the fringe and obtaining the fringe spacing from the relationship between the angle of the tangent, and the amount of rotational mismatch introduced.   The use of this technique avoids the problems of gauge length in a rapidly varying strain field, and also of the lack of whole fringes when the residual stress is low.   Residual stresses as low as 2.5 ton in^{-2} can be detected.

Results

The patterns of relaxation occurring after the hole-drilling operation are shown in Figures 6, 7 and 8.   It will be seen that, as one might intuitively anticipate, the residual stress is high within the weld itself and has a lower value away from the weld.   The actual values of the residual stress are plotted in Figure 9.

The fact that the principal strain is aligned along the axis of the weld is in agreement with the results of Cordiano and Salerno.   The magnitude of the residual stress will be seen to be a large fraction of the yield stress in this material; in addition the minor lobes displayed by the deformation patterns point out the complexity of the residual stress distribution.

Conclusion

A sensitive moire technique has been used to visualise and quantify the residual stress distribution around a butt weld between two steel plates. Maximum stress levels of 432 MNm^{-2} have been measured in the region of the weld, falling to 170 MNm^{-2} in the parent material away from the weld. The technique shows promise for use in complex engineering situations where a multiplicity of strain gauge rosettes would be required; in addition to the example described here it has been applied successfully to blast furnace shrouds, steel mill rolls, ball bearing races and heavy gauge plate T-butt welds.

References

1. Gardner, C.G., Proceedings of a Workshop on NDE of Residual Stress, San Antonio, Texas, Southwest Research Institute, 1975, pp.1-8.
2. Boone, P., "Laser Produced Moire Gratings", Strain, April 1969. p.43.
3. Marchant, M., & Bishop, S., "Interference technique for the measurement of in-plane displacement of opaque surfaces". Jnl. Strain Analysis, 1974, 9 p.36-43.
4. Post, D., "Moire Fringe Multiplication with a Nonsymmetrical Doubly Blazed Reference Grating". Applied Optics, Vol.10, No.4, 1971, p.901.
5. Sciamarella, C.A., "Moire Fringe Multiplication by means of a filtering and a wavefront reconstruction Process". Experimental Mechanics, April 1969, p.179.
6. de Caluwe' M., "Production of amptitude gratings with preset black to white line width ratio". Optics and Laser Technology, 1970, 2, p.189.
7. Cordiano, H.V., & Salerno, V.L., "A Study of Residual Stresses in linearly biaxial stress fields". Experimental Mechanics, January, 1969, p.17.
8. Nawwar, A.K., McLachlan, K. & Shewchuk, J., "A Modified hole-drilling technique for determining residual stress in thin plates". Paper given at SESA Spring Conference, Chicago 1975.
9) Beaney, E.M. & Proctor, E., "A critical evaluation of the centre hole technique for the measurement of residual stress". Strain, January 1974, 10, 1, p.7-14.
10) Bjorhovde, R, et.al., "Residual Stresses in Thick Welded Plates". Welding Research Supplement, August 1972, p.292-3.
11) McKelvie, J., Walker, C.A., "A practical multiplied moire fringe system". Experimental Mechanics, To be published.

Appendix

Measurement of fringe spacing using a known rotation mismatch.

If it is assumed that we have a fringe density of p, fringes per cm at a point, at an angle to the axis (see Figure). Suppose, then, a rotational mismatch corresponding to Q fringes per cm parallel to the component direction is introduced. The angle of the deformation fringes will change to, say but the spacing along the component direction will remain fixed.

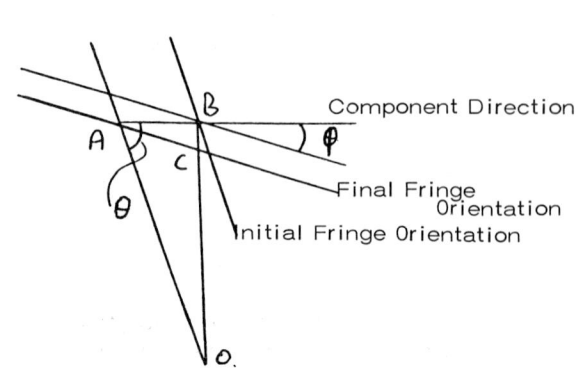

Then,
$$AB = \frac{1}{p}$$
$$BC = \frac{1}{r} \text{ where } r = \text{vertical fringe density after rotation.}$$
$$BO = \frac{1}{q} \text{ where } q = \text{vertical fringe density before rotation}$$
$$\frac{AB}{BC} = \cot\phi \quad (1), \qquad \frac{AB}{BO} = \cot\theta \quad (2)$$

Thus,
$$\frac{\frac{1}{p}}{\frac{1}{r}} = \cot\phi$$

$$p = r \tan\phi$$

Now $r = q + Q$
and from (2) $q = p \cot\theta$

and Q can be measured in a stress-free region, or by measuring the change in vertical fringe density induced by the rotation.

$$p = (p \cot \phi + Q) \tan \phi$$
$$p (1 - \cot \theta \tan \phi) = Q \tan \phi$$
$$p = \frac{Q \tan \phi}{1 - \cot \theta \tan \phi}$$

Hence, knowing Q, θ and ϕ, we can measure p in situations where we only have one fringe, and hence where the measurement of fringe spacing is difficult.

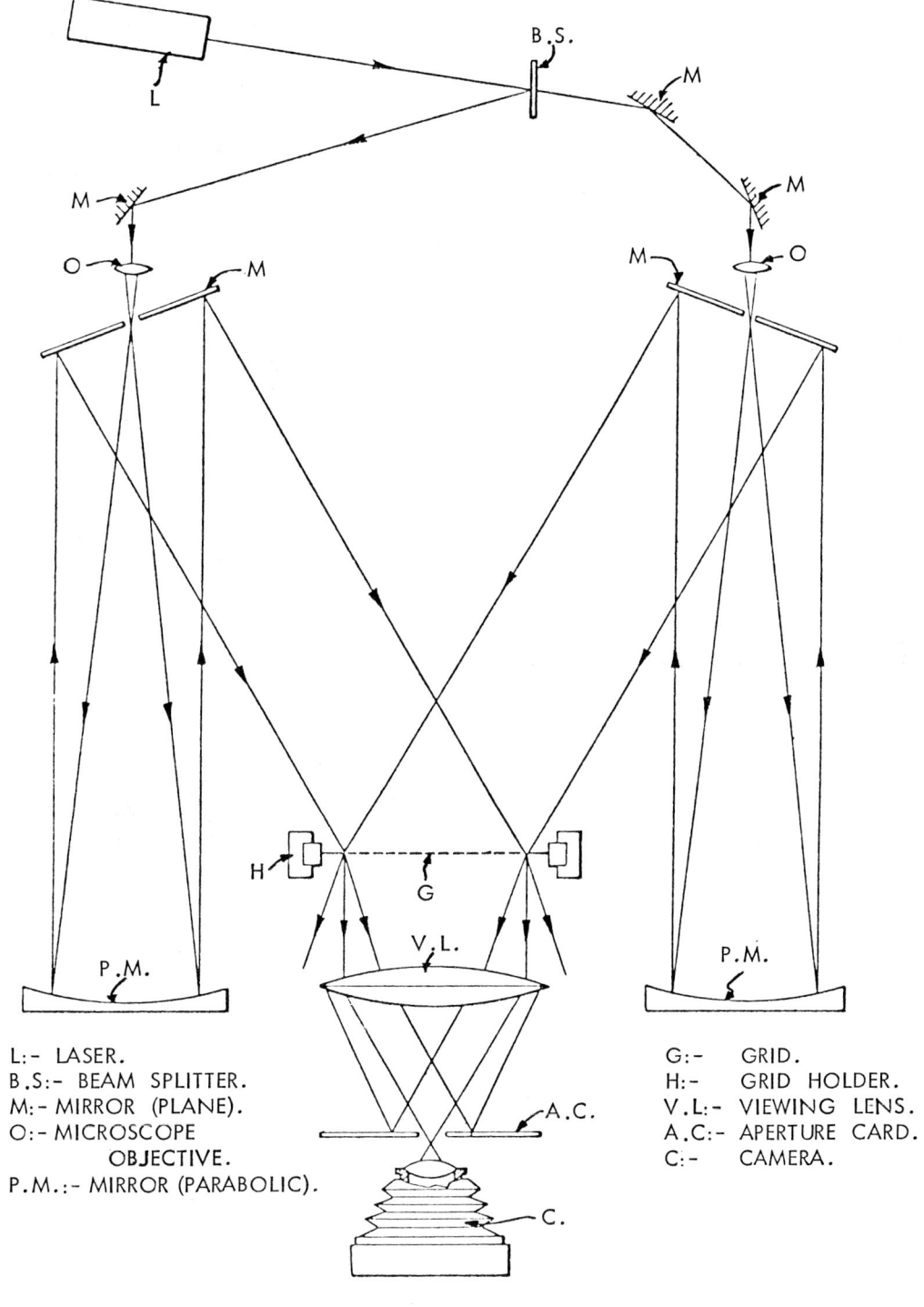

L:- LASER.
B.S:- BEAM SPLITTER.
M:- MIRROR (PLANE).
O:- MICROSCOPE OBJECTIVE.
P.M.:- MIRROR (PARABOLIC).

G:- GRID.
H:- GRID HOLDER.
V.L:- VIEWING LENS.
A.C:- APERTURE CARD.
C:- CAMERA.

FIG. 1.   Fourier Filtering Interferometer.

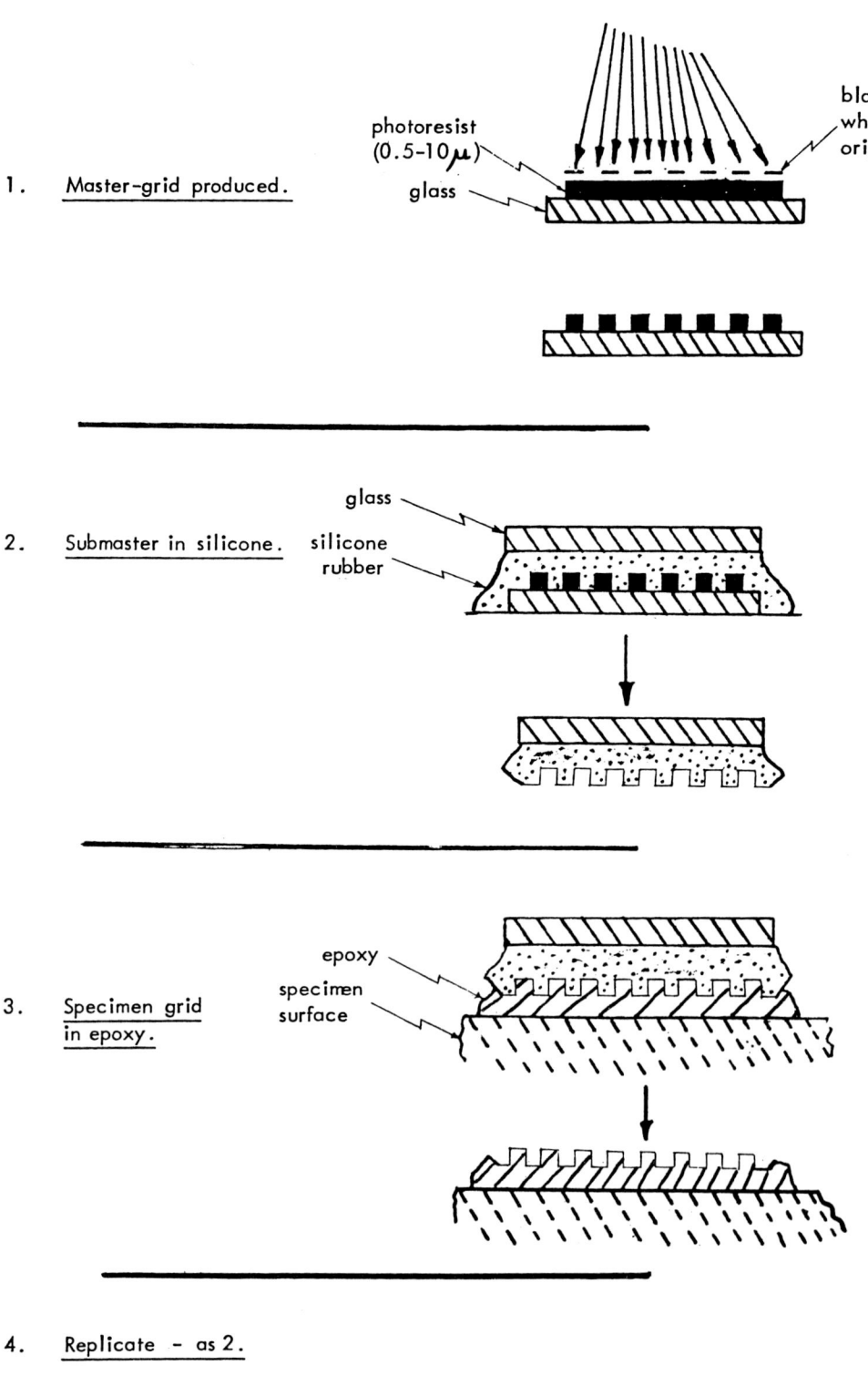

FIG.2. Moire Grid Application Process.

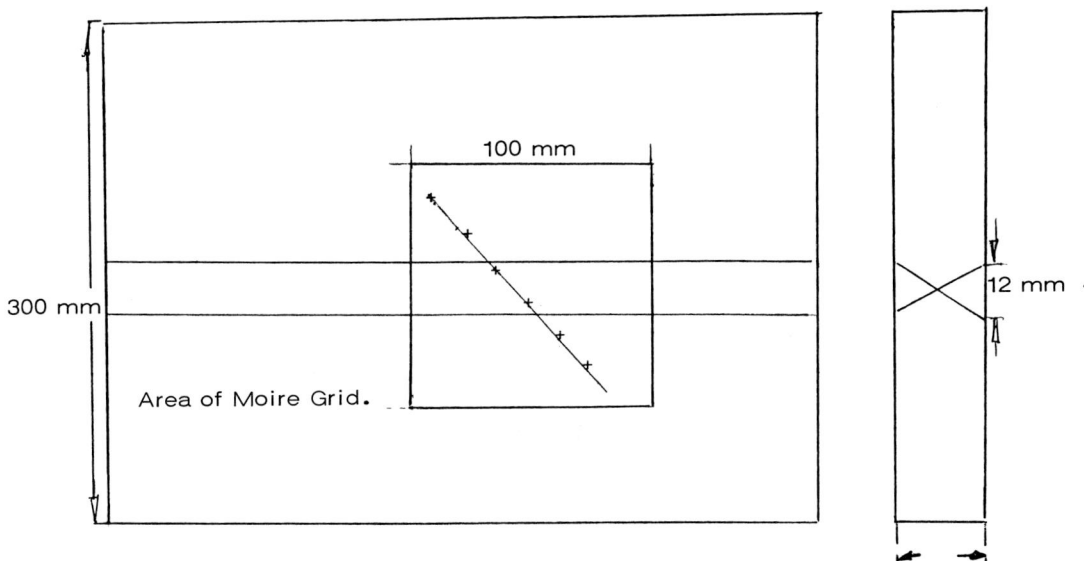

FIG. 3.

Welded Specimen & Layout of Holes.

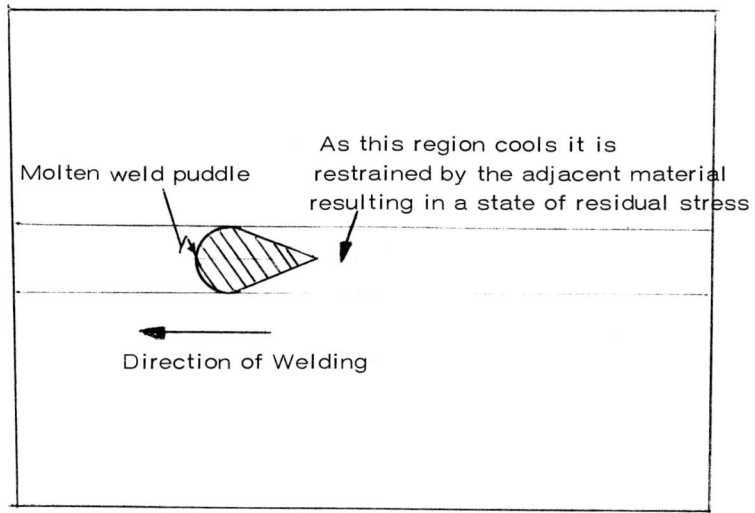

FIG. 4.

Origin of Residual Stresses due to Welding.

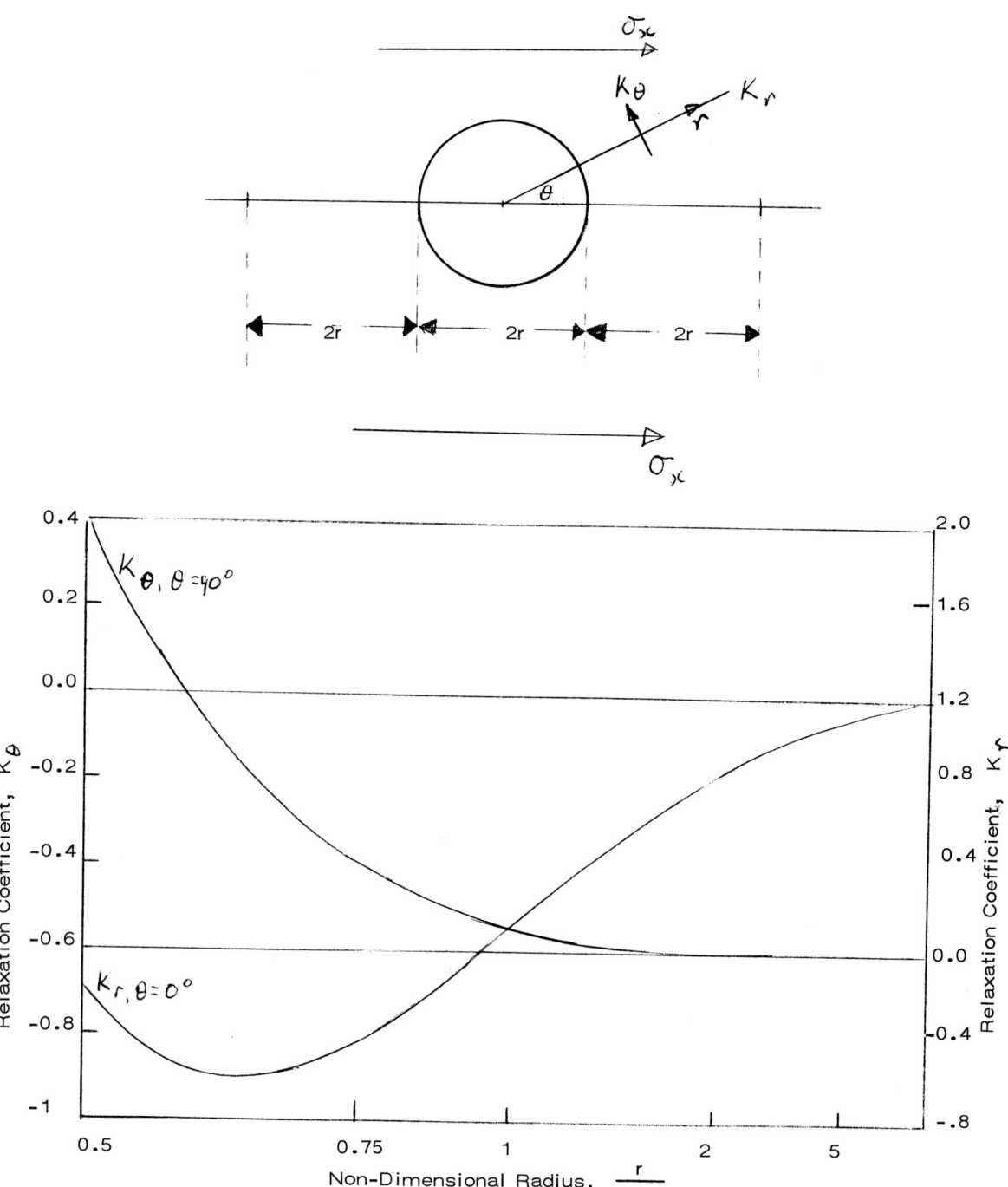

Residual Stress Quantification. (after Nawwar et.al.)

$$\sigma_x = K_r \frac{\epsilon_r}{E} = K_\theta \frac{\epsilon_\theta}{E}$$

FIG. 5.

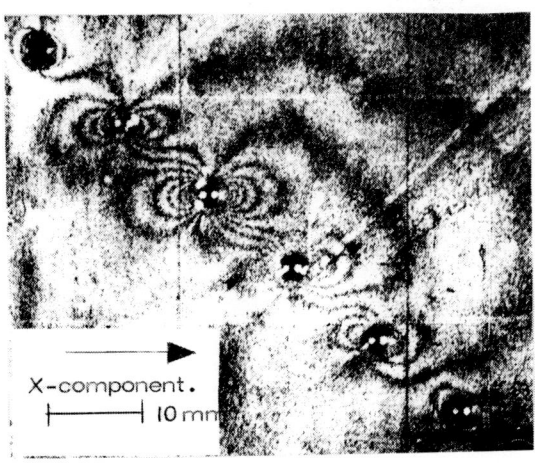

FIG. 6.

X-component of relaxation
after hole drilling.

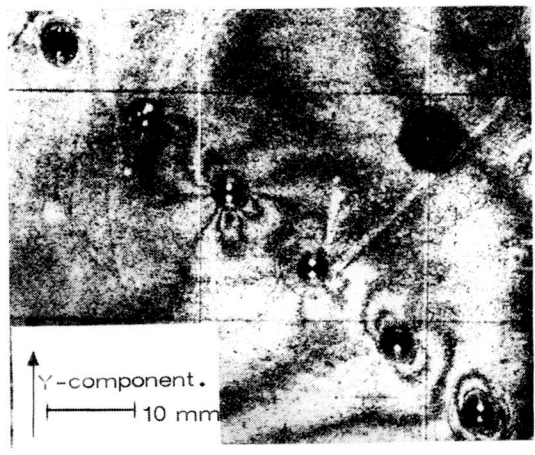

FIG. 7.

Y-component of relaxation
after hole drilling.

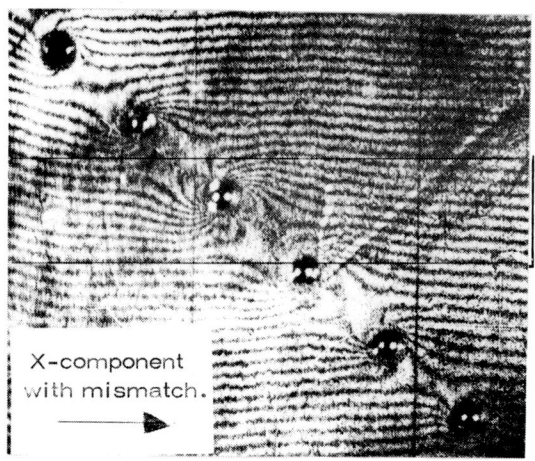

FIG. 8.

X-component of relaxation after hole drilling.

A degree of rotational mismatch has been introduced.

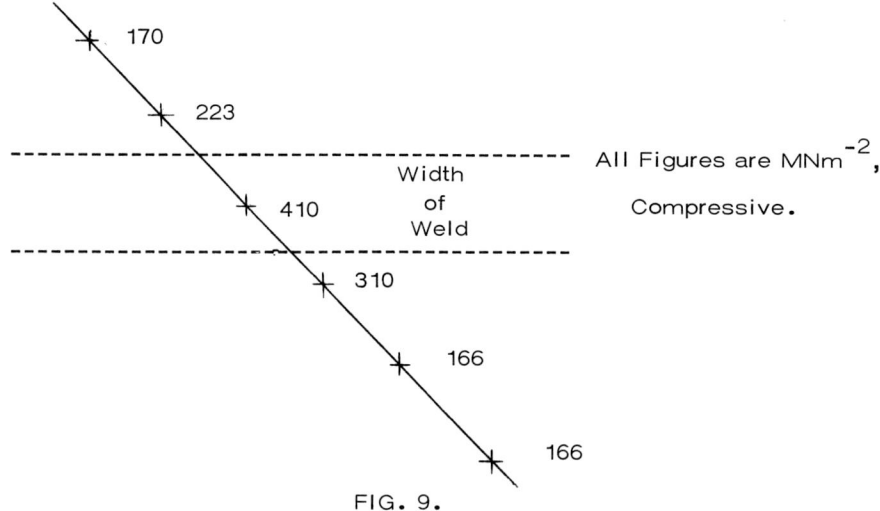

FIG. 9.

Variation of Residual Stress across weld.

ELECTRONIC PROCESSING OF MOIRE FRINGES.
APPLICATION TO MOIRE TOPOGRAPHY AND ELEMENTS OF COMPARISON WITH PHOTOGRAMMETRY

J. C. Perrin and A. Thomas

Centre d'Etudes et de Recherches de la Compagnie Electro-Mecanique
49, rue du Commandant Rolland — 93350 Le Bourget, France

Abstract

Some elements of comparison are given between the projection-type Moiré topography and photogrammetry. The basic mathematics of photogrammetry is applied to reconstruct the 3.D. Shape of the object from the Moiré pattern. Some particular aspects of the Moiré method are discussed and a general methodology is proposed. In connection with this, an opto-electronic technique is described, which measures the Moiré phase with high resolution and sign determination. The experimental results show that this technique is especially suitable for high accuracy automatic reconstruction of 3.D. Shapes.

Introduction

Several authors have shown the possibility to achieve 3.D. shape measurements using projection-type Moiré topography. A first difficulty encountered in this technique is to made a correct mathematical analysis of the entire process. Some authors [3], [4] have developped some intrinsic mathematics whose application to computerization is not always clear. The corresponding methodology to define what must be measured, how and with which accuracy is not put in evidence. In the first part of this paper, we have developped an analogy between this method and the photogrammetry. It is shown that the Moiré method is an unconventional way to measure a basic quantity : the parallax. On this basis, it is possible to define a general methodology. More, once the elements of comparison with photogrammetry are given, it is clear that the basic relationships of photogrammetry can be used to reconstruct the relief from the Moiré data. A second difficulty encountered in this technique, is the problem of the poor accuracy given by conventional densitometric processing of the Moiré pattern. This is related to the problem of fringe interpolation, but by no way it is intrinsic to the Moiré method. For this reason, we have developped in the second part an opto-electronic technique which measures the Moiré phase with high resolution and sign determination. This technique is based on heterodyne double beam holographic interferometry which was developed by R. Dändliker, B. Ineichen and F.M. Mottier [6] to solve the experimental problem of fringe interpolation in processing holographic interferograms for mechanical deformation measurements. The adaptation of this technique to process Moiré fringes is described with particular emphasis on experimental results.

Basic Relationships and Elements of Comparison with Photogrammetry

We are concerned with projection-type Moiré topography (fig. 1). The basic components are the projector and the camera. O_1 (respectively O_2) are the nodal points and (ω_1, m_1, n_1) (respectively (ω_2, m_2, n_2)) are the focus planes. The projector illuminates the object with a reference grating of period p. The resulting negative taken by the camera is then superimposed on the reference grating, creating Moiré fringes which, under certain conditions, are the contour lines of the object.

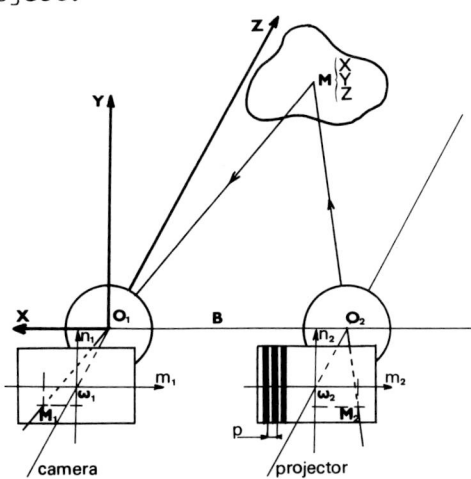

Fig. 1. Set up for projection-type Moiré topography.

The elements of resemblance with photogrammetry are straightforward : both methods are visibly based on the stereoscopic effect given by the pair of perspective centers O_1 and O_2. $B = O_1 O_2$ is the corresponding basis. A first important difference is the fact that the object is illuminated. This restricts our comparison to close range photogrammetry and the corresponding applications. Another difference is the way the result is obtained. In photogrammetry two cameras (or the same translated) are used. Any point M of the object has therefore two photographic images M_1 and M_2. These two homologous images are then identified and their photographic coordinates m_1, n_1 (respectively m_2, n_2) are measured by using a stereo-analyser. It is worth noting that such apparatus is essentially manual. Some efforts have been made in the past to automatize this step, but the resulting equipment is highly sophisticated and costly. Basically, the data collection is always long and tedious. Hundred points per minute can be considered has a good performance. In any case, the points M must be well defined by marking the object. In this way, one get four measurements (m_1, n_1) and (m_2, n_2) from which the three coordinates of M can be deduced. This is made by applying the general relationships of the phototriangulation. Such formulas are extensively developed in the specialized litterature [5] and we do not intend here to get involved in the corresponding mathematics. We shall only consider the idealized arrangement of fig. 1 in which the two optical axis are parallel and oriented perpendicularly to the basis. We suppose equally that $\omega_1 m_1$ and $\omega_2 m_2$ are parallel to the basis. The corresponding equations are particularly simple. First is measured the stereoscopic parallax :

$$a = m_1 - m_2 \qquad (1)$$

The three coordinates X, Y and Z of M are then derived using the coordinate system $(O_1\ X\ Y\ Z)$ so that $O_1 Z$ is the optical axis and $O_1 X$ is the basis :

$$Z = -\frac{Bf}{a}$$
$$X = -\frac{m_1 Z}{f} \qquad (2)$$
$$Y = -\frac{n_1 Z}{f}$$

f is the principal distance $O_1 \omega_1 = O_2 \omega_2$
It appears on equations (2) that the contour lines of the object defined by it's intersections with planes perpendicular to Z axis are equiparallax lines. We shall see that under the same conditions the Moiré method gives directly these contour lines in the analogic form of Moiré fringes.

Moiré and stereoscopic parallax
─────────────────────────

We suppose that the lines of the reference grating are parallel to $\omega_2 n_2$. In a first step, we calculate the intensity $E(m_1, n_1)$ in the image plane ($\omega_1 m_1 n_1$). We shall only consider the case of uniformly diffusing objects, so that the intensity at M_1 is proportionnal to the intensity $T(M_2)$ at his homologous point M_2 where $T(M)$ is the transparency of the reference grating. We suppose that $T(M)$ is given by :

$$T(M_2) = a + b \cos 2\pi \nu m_2 \qquad (3)$$

with $\nu = \frac{1}{P}$ = spatial frequency. a and b are constants. Consequently, it follows from equations (3) and (1) :

$$E(m_1, n_1) = a' + b' \cos 2\pi \nu (m_1 + \frac{Bf}{Z}) \qquad (4)$$

This relation is easy to interpret in the case where the object is a plane perpendicular to the optical axis : $Z = Z_0$ = cte. The image is identical with the reference grating translated along m_1 from the parallax $a = \frac{Bf}{Z_0}$. In the general case, the structure of the image is much more complex because m_1 and n_1 are function of Z. But in any case, equation (4) shows that the parallax a modulates in the plane ($\omega_1\ m_1\ n_1$) the phase of the reference grating. In the following, equation (4) will be written in the form :

$$E(m_1\ n_1) = a' + b' \cos\left[2\pi \nu m_1 + \varphi(m_1, n_1)\right] \qquad (5)$$

with
$$\varphi(m_1, n_1) = 2\pi \nu \frac{Bf}{Z} = 2\pi \nu a \qquad (6)$$

A negative is then taken by the camera. After developpment, the final transparency of the negative is given by :

$$T(m_1, n_1) = T_0 \left[1 + C \cos\left(2\pi \nu m_1 + \varphi(m_1, n_1)\right)\right] \qquad (7)$$

The Moiré fringes are then formed by superimposing the final transparency on the reference grating. This can be done in different way but the general mechanism can be interpreted as an heterodyne demodulation in which the carrier frequency ν is filtered and the phase recovered. The resulting intensity in the detector plane is then given in the general form :

$$E(m_1, n_1) = E_0 \left[1 + C' \cos \varphi (m_1, n_1) \right] \quad (8)$$

The electronic processing of the Moiré pattern leads to the determination of the Moiré phase φ, with sign determination, at any point in the image field and even for non-uniform transparencies. But in any case, this measurement is made with a $2k\pi$ ambiguity (k integer). Apart from this ambiguity on k, and according to equation (8), it appears that this Moiré method leads to the determination of $\frac{a}{p}$:

$$\frac{a}{P} = \frac{\varphi}{2\pi} \quad (9)$$

p can be measured independently with high accuracy. We then deduce : $a = P \frac{\varphi}{2\pi}$. We now see that the Moiré method is an unconventionnal way to measure a basic quantity : the stereoscopic parallax a.

This leads to a much more simple interpretation of the Moiré pattern. For exemple, we can derive from equations (2) and (7) that the dark Moiré fringes $\varphi = \pm (2n + 1)\pi$ are non-equidistant contourlines. Their distribution along Z axis is hyperbolic, following the law :

$$z_n = \frac{2\nu fB}{2n + 1} \quad (10)$$

This result was derived precedently using a more complicated procedure [2], [4].

Methodology for Moiré data processing

Apart from the ambiguity on k, we see that the procedure to process the data and recover the 3.D. shape can be considered as basic. Once the parallax a is measured, X, Y, and Z are determined by applying the general formulation of the photogrammetry. Here in the situation of Fig. 1 the equations are particularly simple because the Moiré fringes are the equiparallax lines of the object, that is it's contour lines. We can therefore speak of analogical Moiré method. But for a more general set-up, for exemple with convergent axis, the mathematics becomes much more complicated, involving analytical calculations [3], [4]. But even in such situations, the general mathematics of photogrammetry can be applied with much profit. On this way, the methodology for Moiré fringes processing appears quite clear. First the indetermination on k must be removed for at least one point M_0 of the object. This can be done by measuring Z_0 (or a_0) at this point by other means. Then we have to check that any point M of the object is monotonously related to M_0 so that the absolute phase difference $\varphi(M) - \varphi(M_0)$ can be measured. For any point M are then measured the two photographic coodinates in the image plane (m_1, n_1) and the absolute phase difference relative to the reference point M_0 : $\varphi(M) - \varphi(M_0)$. The three measurements are then introduced in the computer and the three coordinates X, Y and Z are calculated by applying the basic mathematics of photogrammetry. The entire process can be automatized because the measurements are made in the image plane (ω_1, m_1, n_1) only. In comparison, the automatization of stereoanalysers for photogrammetric compilation is much more hazardous and costly. But for this purpose, a major difficulty which must be preliminarly overcome is the problem of high accuracy and sign sensitive Moiré phase measurement. Let us take a practical example to illustrate this requirement. An indetermination δa on parallax measurement corresponds to an indetermination δZ on Z given by equation (2) : $\delta a = \frac{Bf}{a} \delta z$. In practice, following parameters are fixed : mean magnification $\frac{Z_0}{f} \simeq 10$; Perspectiv $\frac{Z_0}{B} \simeq 3$. We then get $\delta a = \frac{\delta z}{30}$. Now if an accuracy of the order of 0.1 mm is required on Z, which is just sufficient for the control of large mechanical objects, the corresponding accuracy on the parallax a is given by $\delta a \simeq 3 \mu m$. In practice, we use a grating's period of $p = 120 \mu m$. This is a good compromise because a to small period gives difficulties to get highly contrasted and uniform negatives mainly due to the surface roughness of the object. Therefore, the corresponding accuracy which is necessary on the Moiré phase measurement is of the order of $\delta \varphi 10°$ (see eq. 9). Such an accuracy on Moiré fringes interpolation can not be obtained by densitometric measurement. More, densitometry is not capable to give the sign of the Moiré phase without sophisticated arrangements. This experimental problem can nevertheless be solved by an electronic phase measurement technique.

Electronic processing of Moiré fringes

Principle

The electronic processing of the negatives is performed on a separate bench. It's principle is based on two reference beam holographic interferometry. This method was developped by authors in reference (6) to overcome the problem of fringe interpolation accuracy which is a serious limitation in processing double exposure interferograms for the quantitative determination of mechanical deformations. The principle adapted to Moiré fringes processing is shown on figure (2). Two coherent plane waves A_1 and A_2 are creating an holographic grating in the plane (ω, m, n). The period is given by $p = \frac{\lambda}{2 \sin \theta}$. This grating recorded on a photographic plate is used as reference grating in figure (1). The negative given by the camera is then put in the plane (ω, m, n) and oriented so that the coordinate systems (ω, m, n) and (ω_1, m_1, n_1) coincide, creating the Moiré fringe pattern. The fringes are detected in the image plane given by the observation system which achieves low pass optical filtering and eliminates the carrier frequency. Electronic processing is carried out by introducing a small frequency shift $\Omega = 110$ KHz between the two light beams A_1 and A_2, resulting in an intensity modulation at the beat frequency Ω for any point of the field, given by :

$$E(m, n, t) = E_0 \left[1 + C' \cos\left(\Omega t + \varphi(m, n)\right) \right] \qquad (11)$$

Note that the frequency shift results in a uniform transverse motion of the reference grating. This could be obtained by other means, but much less easily.

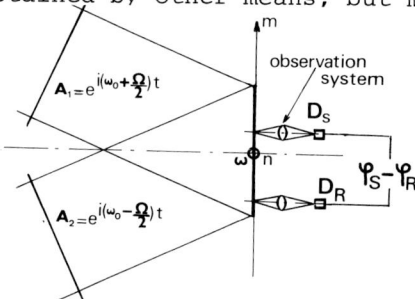

Fig. 2. Principle for opto-electronic Moiré phase measurement.

The intensity is converted into an electrical signal by the detector D_S. The phase difference $\varphi_S(m, n) - \varphi_R(m_0, n_0)$ of the two signals given by this detector and a reference detector D_R fixed at a predetermined point can be measured very accurately by electronic phase measurement. The corresponding accuracy is constant in the field and is independent in a large dynamic range of the varation of E_0 and C'.

This method of Moiré fringes processing differs from double beam holographic interferometry about two important points. First is the fact that we have to deal with the energy transparencies of the negatives. This can be a difficulty, because coherent processing generally works on amplitude transparencies. The second difference is that the negative considered as an hologram, is put here in the image plane whereas in holographic interferometry, it is generally placed near the pupil of the optics forming the image (7). The sources of errors resulting from misalignements and drifts are therefore quite different.

An important point is the way by which the observation system achieves low-pass filtering. We have first tested coherent filtering by limiting the coherent band-pass of the observation system. This was done by placing a pinhole in it's focal plane. With such filtering, the reproducibility of the measurements was found very poor. As a matter of fact, this is due to the fact that the amplitude transparency of the negative depends on complex factors as the emulsion's thickness. This rule out the possibility of using coherent filterings in the Moiré method, because such method is essentially incoherent. So we have finally used incoherent filtering by placing a diffuser in the image plane of the observation system. The low-pass filtering is achieved by spatial integration in this plane.

Experimental Results

According to equation 9. a phase error $\delta\varphi$ corresponds to a parallax error $\delta a = p \frac{\delta\varphi}{2\pi}$. With the grating's period $p = 120 \mu m$, we get $\delta a = 0.3 \mu m$ per degree. In the following, both quantites δa and $\delta\varphi$ will be indicated. But it should be kept in mind that δa only is significant. For given parameters f an B, δa determines the accuracy of the method independently of p which is a secondary parameter.

The integration time of the phasemeter being of 700 ms, the observed phase fluctuations are within $\pm 1°$ ($\pm 0.3 \mu m$). They can be attributed to residual vibrations of the components and to turbulences. They diminue if reference and signal detectors D_S and D_R are close together. For a spacing of 5 mm, the observed fluctuations are within $\pm 0.25°$ ($\pm 0.08 \mu m$).

The photographic reference grating being put back in place, a constant phase is measured in the whole field. But a long term drift appears after some days. After two months the resulting phase defect is of the order of $\pm 5º$ ($\pm 1.5 \mu m$) accross the field. This can be practically a serious difficulty. In any case, a special attention should be devoted to the mechanics.

The reproducibility of the measurement was then tested on a large number of negatives taken from the same object. In this work, we have been using the Agfa-Gevaert 23.D.56 emulsion. This emulsion is mainly devoted to photogrammetric work. It is especially suitable for Moiré methods as well. With this emulsion and the experimental conditions described precedently, the phase reading $\varphi(m, n) - \varphi_R$ was only reproducible to within $\pm 10º$ ($\pm 3 \mu m$) in the whole field. The errors can be attributed to mechanical misalignments of the plates mainly due to the transfer from the camera to the processing set-up. A practical solution, to get a better definition of (ω_1, m_1, n_1) axis should be to use a frame camera and to include facilities in the processing bench to achieve a good alignement. In this experiment we have used standard components and we didn't get those facilities. Nevertheless, in order to evaluate the smallest detectable parallax variation independently from misalignement defects, we have made a differential phase measurement over an object constituting a step of variable height along Z axis. The differential measurement was made with two indentical holes fixed together in the image plane and spaced of 5 mm. The geometry of the object and the conjugated images corresponding to the differential pair of holes is shown on figure (3). The step is made with two parallel and indentical 8 centimeters side cubes, one of which being mounted on a table in translation along Z axis.

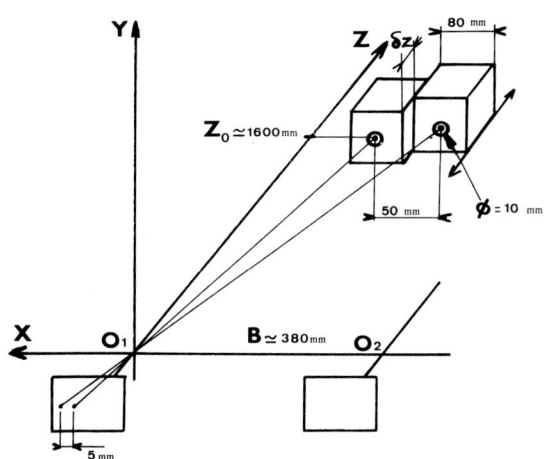

Fig. 3. Set up for testing the accuracy and the reproducibility of the Moiré phase measurement.

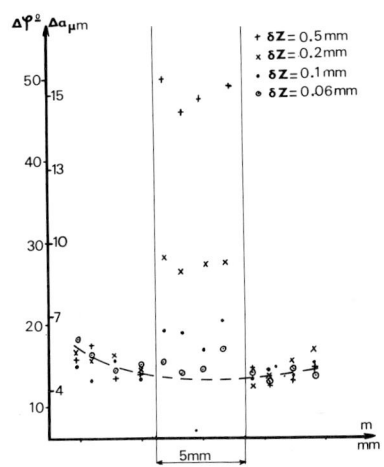

Fig. 4. Experimental result : $\Delta \varphi(m) = f(\delta Z)$.

The phase difference $\Delta \varphi(m)$ measured by translating the differential pair of holes along m axis for four negatives corresponding to four different height δZ are shown on figure 4. The corresponding parallax difference $\delta a(m)$ are also indicated. The dashed curve corresponds to the mean value. The fact that the mean value is near 12º indicates that the cubes are not strictly perpendicular to Z axis. It appears that the phase difference reading at any point is reproducible to within $\pm 3º$. The smallest parallax difference detectable is correspondingly $\delta a = \pm 0.9 \mu m$. For the values of the parameters f,B and Zo as indicated on fig. 3, the corresponding smallest height δZ detectable is given by

$$\delta Z = \frac{\delta a}{fB} Z_o = 0.05 \text{ mm}$$

This is clearly illustrated on fig. 4. It is clear that this method measures parallax differences which could hardly be detected by photogrammetric means, taking account that no mark exists on the object and that the measurement is automatic.

Illustration of Automatic 3.D. Shape Measurement

By way of illustration of the two preceding sections, a 3.D. profil measurement was achieved over a 1.50 m high turbine blade. Due to it's large twist angle, it was only possible to measure half of the blade at once. For this purpose, a reference point determined by φ_R, Z_R was put in the field. It's Z_R coordinate was preliminarily measured by other means. Then the blade was put in place and linked to the reference point with a paper tape. The corresponding Moiré pattern is shown on the fig. 5.

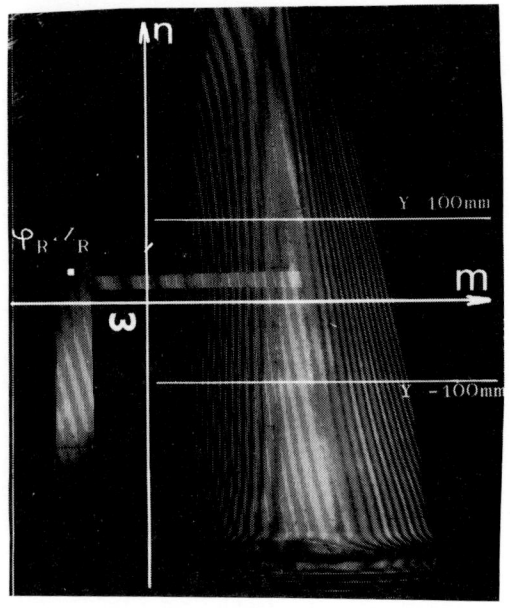

Fig. 5. Moiré pattern observed on a turbine blade.

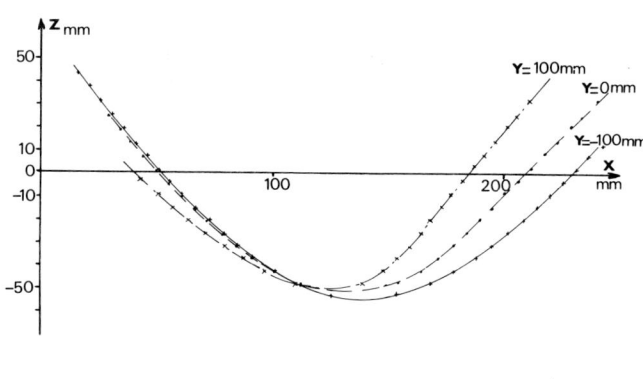

Fig. 6. Plotting of the measured profiles at the three heights. y = 0 and y = ± 100 mm

Each point of the blade is determined by it's two photographic coordinates (m, n) and it's absolute phase difference $\varphi(m, n) - \varphi_R$. Using the general methodology described precedently, we have measured the profil of the blade at the three heights y = 0 and y = ±10cm Fig. 6. The measured profiles were found correct to ± 0.5 mm. Another experiment made with a well known cylinder confirmed this order of value. At this stage, we point out that this illustration was made with standard laboratory equipements. It's clear that the final accuracy could easily be improved by optimizing the components. In particular the distorsion could be minimized for the mean magnification γ_0. Another important source of error that was mentionned before is the lack of accuracy in the determination of (ω_1, m_1, n_1) and (ω, m, n) axis.

Nevertheless, this experiment illustrates the possibility of the method : the 3.D. shape of the object is entirely reconstructed from the negative. The collection of the Moiré data and the associated calculations to deduce X, Y and Z can be automatized. The rythm of the entire process can be very fast. The limit results from the integration time of the phase meter. According to the high frequency of work (Ω = 110 kHz), the limit of our phase-meter is of 10 ms. This possibility made this method very attractive.

Conclusions

Both the mathematical formalism and the automatical Moiré fringes readout made the method especially suitable for 3.D. measurements. In many applications it can compete with conventionnal photogrammetric equipments. Following advantages are :
- Automatic readout and 3.D. reconstruction with high speed affordable.
- Relatively simple and low cost reading's equipment.
- No marking points on the object.
- Monochromatic illuminations of the object and corresponding simplifications in the conception of the camera and projector optics.

Main drawbacks are :
- Angular limitations in the relief of the objects.
- Limitations concerning the surface roughness of the objects.
- Relatively moderate spatial resolution.

This work was supported by "Comité de la Mécanique de la D.G.R.S.T.".

Acknowledgments

The authors wish to thank R. Dändliker for many helpful discussions and for the lending of the radial grating.

References

1. P. Benoit, E. Mathieu, J. Hormière, and A. Thomas, Nouvelle Revue d'Optique 1975, T. 6, n° 2, p.p 67-86.
2. B. Dessus, J.P. Gérardin and P. Mousselet, Opt. Quantum Elec. 7,15 (1975).
3. Masanori Idesawa, Toyohiko Yatagaï and Takashi Soma, Applied Optics, Vol. 16, n° 8 Aug. 1977.
4. J. Hormière, Thesis - Université Paris XI 1977.
5. Manual of Photogrammetry, Third Edition - Falls Church, Va. American Society of Photogrammetry.
6. R. Dändliker, E. Marom, F.M. Mottier, Opt. Communications, 9, 412-416 (1973).
7. R. Dändliker, E. Marom, F.M. Mottier, J. opt. Society of America 66, 23-30 (1976).

DIMENSIONAL METROLOGY OF LARGE OBJECTS BY PROJECTION MOIRE TECHNIQUES

Luciano Pirodda
Professor of Engineering Mechanics
Politecnico di Milano, Istituto di Meccanica
20133 Milano, Italy

Abstract

The paper deals with the problem of dimensional testing of the limiting surfaces of objects having comparatively large frontal dimensions and transversal dimensions in the same order of magnitude. The solution is sought by an extension and generalization of the projection moiré technique. A geometry is considered where two sets of curves are obtained by projecting a grid from a point light source over a master and test object respectively, while an optical system having central point at finite distance records them. The conditions are analyzed under which the moiré fringes produced by the superposition of the two sets are the contours of the differences between the objects.

Introduction

The content of this paper is connected with the problem of mechanical quality control, and in particular with the problem of finding a simple, practical and real-time technique for the on-line dimensional testing of the external surfaces of machine parts. In the limiting cases said surfaces are supposed to be rather large (larger frontal dimension in the order of magnitude of 1 m) and their depth to be in the same order of magnitude. Absolute or difference contouring them with reference to a master surface is the object of the control. Due to their simplicity and ease of employment plus their ability to assure the required accuracy in this type of measurement, moiré methods appear to be particularly suited for the above application. A consistent developement of this technique is therefore looked for in the following. The principles of shadow and projection moiré shall be of course the basis for said developement (see [1],[2],[3],[4],[5] and [6],[7],[8],[9]). The distinction between the two techniques is in fact somewhat artificial, as the unifying approach adopted in the following analysis will point out.

The main scope of this paper is a generalization of the geometric theory of projection moiré, resulting in measurement techniques and apparatus suitable for application on objects the size formerly described and requiring a reasonable space amount. Diverging projection and observation is a particular aspect of this generalization, since it dispenses from the use of large collimating lenses or of long projection distances (in the case of quasi-parallel projection).

A General Projection Moiré System

A projection moiré system, assumed to be in a fairly general form, is presented in fig. 1. The system is composed of an optical device projecting a plane grid over the object whose surface is to be mapped and of a lens acting as recording and observing mean, both at finite distance from the surface.

Whatever the constructive details of these two optical components, it is assumed that

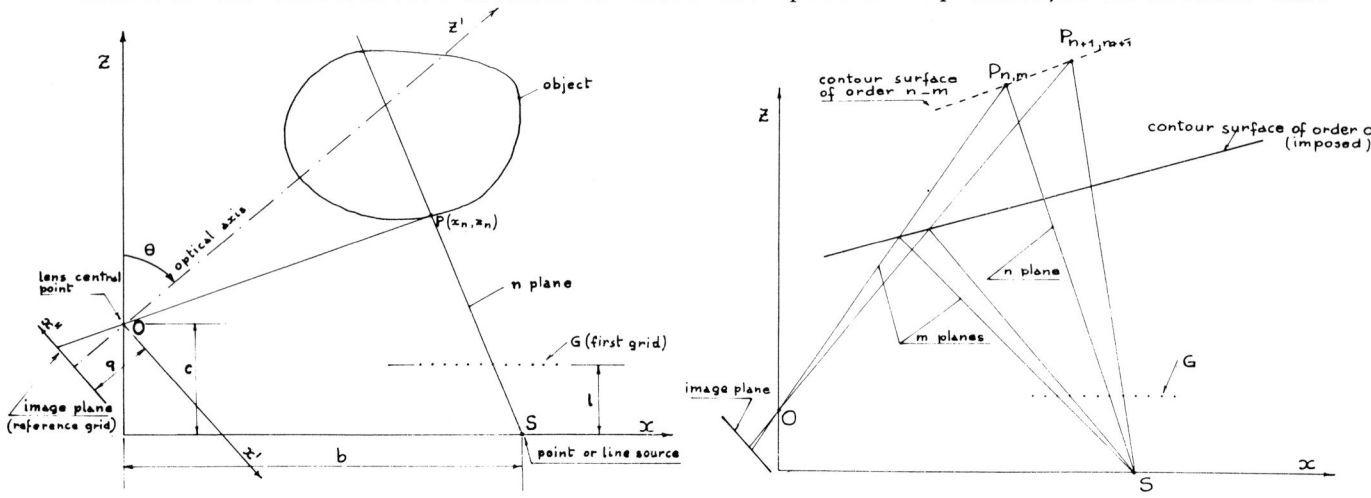

Fig. 1. Schematical representation of a general projection moiré system.

Fig. 2. General geometry for determining the contour surfaces.

they virtually and effectively operate as in their schematical representation of fig. 1, which may be therefore taken as a basis for quantitative analysis. In other words, it is assumed that the projecting process is equivalent to shadowing a grid G over the object by means of a point source S (in principle the source could also have the form of a segment of a straight line, parallel to the lines of the grid), and that the observing or recording process is equivalent to projecting the points of the object over the image plane through a central point O.

With no lack in generality, a reference frame x,y,z can be chosen so as to introduce in the analysis only the independent geometrical parameters, i.e. on the following assumptions.

The xy plane is parallel to the plane of the grid.
The y axis is parallel to the lines of the grid.
The source S lies on the x axis, and the central point O on the z axis.

On the other side we have to assume, for generality, that the z coordinate of the point O has a value $c \neq 0$ (the straight line connecting O and S is not in general parallel to the plane of the grid, or, in case of a linear source, the plane defined by O and S is not in general parallel to the plane of the grid).

Finally, generality requires that the optical axis can have any direction relative to the reference frame. We somewhat relax generality, without any practical consequence, by assuming that this axis is contained by the xz plane, while we retain, as the independent variable defining it, the angle θ it forms with the z axis.

Summing up, the independent geometrical parameters of the system are: b, l, c and θ, as represented in fig. 1, plus the pitch p of the grid G.

The projection of a line of the grid from the source S is effected through a plane whose equation, for a line of order n (starting to count from the yz plane), is:

$$x = b - \frac{b - np}{l} z \quad \text{(n an integer)} \tag{1}$$

This plane intersects the limiting surface of the object according to a space curve whose points have x and z coordinates satisfying the same equation:

$$x_n = b - \frac{b - np}{l} z_n \quad \text{(n an integer)} \tag{1'}$$

Consequently we shall regard equation (1)' as a relation between the x and z coordinates of points belonging to the surface of the object and to one curve of the set defined by the same equation.

Let us now consider a second reference frame x',y',z', whose origin is at point O, while z' coincides with the optical axis of the observing lens and y' is parallel to y. The following equations connecting the x',z' to the x'z coordinates apply:

$$\begin{aligned} x' &= x\cos\theta - (z - c)\sin\theta \\ z' &= x\sin\theta + (z - c)\cos\theta \end{aligned} \tag{2}$$

Finally, we set a reference x",y" in the image plane, x" being parallel to x' and y" to y and y'. For any point P(x,y,z), and in particular for the points of the surface of the object, we have:

$$x" = \frac{x'}{z'} q \tag{3}$$

where q is a constant factor.

<u>Absolute contouring</u>

Contour lines of the limiting surface of the object are obtained by the superposition of the image of the set (1)' of space curves with a reference plane grid placed on the same image plane.

We assume that the reference grid is always composed of straight lines parallel to the y (and y', y") axis, while we separately perform the analysis of two different cases.

<u>Case a): The Pitch of the Reference Grid is Constant</u>. Let the equation of the reference grid on the image plane be:

$$x"_m = m p_r \quad \text{(m an integer)} \tag{4}$$

p_r being the pitch of the reference grid.

The x'' coordinates of the points of a space curve $(1)'$ are obtained through the equations (2) and (3):

$$x''_n = \frac{x_n \cos\theta - (z_n - c)\sin\theta}{x_n \sin\theta + (z_n - c)\cos\theta} q \tag{5}$$

For the points of a clear subtractive moiré fringe resulting from the superposition of the two plane patterns (4) and (5) the condition: $x''_m = x''_n$ must be fulfilled, namely:

$$mp_r = \frac{x_n \cos\theta - (z_n - c)\sin\theta}{x_n \sin\theta + (z_n - c)\cos\theta} q \tag{6}$$

We first set:

$$p_r(1 + \gamma)\frac{1}{q} = p \tag{7}$$

where γ is a constant to be determined later, then we make some manipulation on (6), by selectively introducing the value of x_n coming from the $(1)'$, by taking into account the (7) and successively the same $(1)'$ again. We obtain at last, having dropped the n index for x and z, which means that we now run over the whole field of n and m values, provided $(n - m)$ is constant:

$$\frac{1}{z}\left[b + c\,\mathrm{tg}\theta + (c - x\,\mathrm{tg}\theta + \gamma z)\frac{x - (z - c)\mathrm{tg}\theta}{x\,\mathrm{tg}\theta + z - c}\right] = b + l\,\mathrm{tg}\theta - (n - m)p = \text{Constant} \tag{8}$$

The moiré fringes are therefore images of curves, lying on the surface of the object, which are loci of constant value for the first member of (8), a complicated function of the x,z coordinates of the points of the locus and of θ.

Equation (8) can also be interpreted as the equation of a directrix of a set of cylindrical surfaces whose generatrices are parallel to the y axis. Said surfaces intersect the surface of the object according to the described loci. We name them accordingly "contour surfaces", while the loci are contour lines.

Upon discussion of equation (8), it can be seen that the only condition assuring the contour surfaces to be planes is:

$$c = \theta = 0$$

In this case equation (8) reduces to:

$$z = l\frac{b + \gamma x}{b - kp} \qquad (k = 0, 1, 2, \ldots) \tag{9}$$

When the reference grid is the image of the first grid, supposed to extend to cover the field of the lens, we have to set: $\gamma = 0$ (consider equation (7) and the present assumptions for c and θ) and equation (9) reduces to the classical equation of generalized shadow moiré (see (1) to (5)). The contour surfaces are planes parallel to the xy plane.

When γ is different from 0, i.e. when the pitch of the reference grid is different from the image of p, the contour surfaces are planes having positive slopes towards the xy plane if the reference grid is denser than the image ($\gamma > 0$), negative slopes in the opposite case. The contour planes are not parallel, since their slope is a function of their order k, which of course makes this sort of contouring less practical than the classical one.

Case b): The Pitch of the Reference Grid is Variable. To analyze this point we make use of the concept of contour surface, as already introduced. It is not difficult to realize, on the basis of the preceding discussion, that a contour surface of order $(n - m)$ is generated by all the straight lines (parallel to the y axis) which are intersection of a plane n, projecting the nth line of the first grid through S, with a plane m, projecting the mth line of the reference grid through O, and for which $(n - m)$ is constant (see fig.2). Reversing the above reasoning, we can make any surface (cylindrical, having generatrices parallel to the y axis) to be a contour surface simply by materializing it and by photographing on it (according to any angle θ) the projected first grid. If the negative thus obtained is taken as a reference grid, said surface is a contour surface (actually is the 0th order one) for the projection moiré system having the same geometry.

Starting from the above ideas, the general problem arises of finding the equations of the complete set of contour surfaces when the "imposed" surface is given. In the following said analysis is confined to the case where the "imposed" surface is a plane, in order to investigate whether a set of parallel planes can be obtained as contour surfaces under less stringent geometrical conditions than in case a). In fact the results shall be only partly positive in this insight, since said geometrical conditions are going to be relaxed only as regards the angle θ, which is now free. On the other side the above result is almost self-evident: as fig.2 points out, the contour surfaces are defined by a number of geometrical conditions (like the position of O and S, etc.) but they are independent from the position of the image plane, and in particular from the angle θ. Changing this plane only changes the form of the reference grating, as recorded by photographing the projectd first grid on the "imposed" surface.

Referring to fig.2, suppose that the first grid is projected over a material plane surface having the equation:

$$z = a + \alpha x$$

Upon photographing the projected first grid and taking the negative as a reference, we have a moiré system with said plane as contour surface of order 0. The directrices of the contour surfaces of order $k = n - m$ are defined by points like $P_{n,m}$, $P_{n+1,m+1}$, etc. in fig.2 (in the represented case $n - m = 4$).

Finding the equation of such a directrix is a simple problem of analytic geometry. We first write equation (1) for the projecting plane $\overline{SP}_{n,m}$ in the form:

$$np = b + l \frac{x - b}{z} \tag{10}$$

while the equation of the projecting plane $\overline{OP}_{n,m}$ is found (after some computation we omit here) as:

$$mp = b + l \frac{b(z - c) - x(a + \alpha b - c)}{\alpha c x - a(z - c)} \tag{11}$$

The equation of the surface generated by all the intersections of the planes (10) and (11) for which $n - m$ is a constant (contour surface of order $k = n - m$) is obtained by equating the differences of both members of (10) and (11). After some algebraic computation one obtains:

$$l(z - a - \alpha x) \frac{c(x - b) + bz}{z(az - ac - \alpha c x)} = (n - m)p = \text{Constant} \tag{12}$$

Equation (12) is of course satisfied by the coordinates of the "imposed" plane contour surface for $k = n - m = 0$, but is no equation of a plane for $k \neq 0$, if $c \neq 0$. Setting $c = 0$ we have:

$$z = lb \frac{a + \alpha x}{lb - (n - m)ap} \tag{13}$$

$n - m = k$ being the order of the contour surface, taking positive integer values for those lying above the "imposed" 0th order one and negative integer for those below the 0th order one.

Thus it comes out again that the only condition for the contour surfaces to be planes is that the line connecting O and S is parallel to the plane of the first grid. When this condition is satisfied the contour surfaces are a set of parallel planes only if $\alpha = 0$, i.e. if the "imposed" contour plane is parallel to the plane of the first grid.

When the "imposed" contour plane coincides with the plane of the first grid ($\alpha = 0$, $a = l$), equation (13) reduces again to the classical equation of shadow moiré.

It is worthwhile to point out, as a result of the present approach, that the set of contour planes in shadow moiré is independent of the angle θ of the optical axis.

Difference Contouring

Let us suppose that the limiting surface of the object undergoes a changement, and that the same general projection moiré system as represented in fig.1 is used in order to produce on the image plane moiré fringes mapping the differences between the original and the changed surface.

Said moiré fringes are caused by the superposition of the images of the two sets of curves (1)' belonging to the original and to the changed surface respectively. Let us clarify their meaning.

The relation between the x and z coordinates of the points of one curve in the set coming from the intersection of the projecting planes through S with the changed surface of the object, is represented by the equation, similar to (1)':

$$x^{o*}_m = b - \frac{b - mp}{l} z^{o*}_m \quad (m \text{ an integer}) \tag{1}''$$

where the asterisk has been used to distinguish the x,z coordinates of the two surfaces.

The x" coordinates of the points of the two sets (1)' and (1)" can be obtained by the (2) and (3):

$$x''_n = \frac{x_n \cos\theta - (z_n - c)\sin\theta}{x_n \sin\theta + (z_n - c)\cos\theta} q \tag{14}$$

$$x^{o*}_m{''} = \frac{x^o_m \cos\theta - (z^o_m - c)\sin\theta}{x^o_m \sin\theta + (z^o_m - c)\cos\theta} q \tag{15}$$

The moiré condition is:

$$x''_n = x^{o*}_m{''}$$

Equating the second members of (14) and (15) and performing some algebraic computation, one obtains:

$$x_n(z^o_m - c) = x^o_m(z_n - c) \tag{16}$$

The angle θ has disappeared, which means that the moiré equation is independent of this angle. This result could have been anticipated by a less formal reasoning, as already done for the case of absolute contouring.

We now substitute in (16) the values given by (1)' and (1)" for x_n and x^o_m. After some algebraic computation, including a second use of (1)', the following equation can be obtained:

$$\frac{z^o_m - z_n}{z^o_m z_n} \frac{bz_n - cb + cx_n}{z_n - c} = (m - n)\frac{p}{l} = \text{Constant} \tag{17}$$

Interpreting the (17) requires some reflexion. Note at first that (fig.3) the same straight line trough O connects the points $P^o(x^o_m, z^o_m)$ and $P(x_n, z_n)$ to the corresponding point on the moiré curve of order (m - n) in the image plane. $(z^o_m - z_n)$ is therefore the component along the z axis of the vector $\overline{PP^o}$ measuring the local distance between the two surfaces along the OP line. We name said component $\Delta \bar{z}$, and we remark that in general $\Delta \bar{z}$ is different from Δz, i.e. from the local distance between the two surfaces measured along a parallel to the z axis through P. Generally speaking, $\Delta \bar{z}$ and Δz coincide only when the tangent plane to the original surface in point P is parallel to the xy plane. If the surface slopes against said plane are small the two components may be considered approximately equal.

We can now write equation (17) in the form:

$$\frac{\Delta \bar{z}}{(z + \Delta \bar{z})z} \frac{bz - cb + cx}{z - c} = k\frac{p}{l} \tag{18}$$

having dropped the now unnecessary index n.

A moiré fringe is therefore the image of a line, lying on the surface of the object, which is a locus of constant value for the first member of (18), a function of $\Delta \bar{z}$ and of the coordinates of the points of the locus. In (18) k is the order of the difference contour fringe, taking positive integer values for the fringes mapping positive values of $\Delta \bar{z}$ and negative ones for negative values of $\Delta \bar{z}$.

The result just found is more suitable for applications if c = 0. In such a case the moiré fringes are images of loci of constant value for:

$$\frac{\Delta \bar{z}}{(z + \Delta \bar{z})z} = k \frac{p}{bl} \qquad (19)$$

Equation (19) can be in principle used to measure $\Delta \bar{z}$, provided the surface (original) has been previously absolute-contoured, i.e. the function $z = z(x,y)$ is known. The same equation is an exact result for any $\Delta \bar{z}$, since no assumption has been made about the magnitude of this component. However in many cases $\Delta \bar{z}$ is inferior to z by at least one order of magnitude(in fact it is several orders of magnitude for the metrologic problems originating this paper). In this case it is legitimate to rewrite equation (19) as follows:

$$\frac{\Delta \bar{z}}{z^2} = k \frac{p}{bl} \qquad (20)$$

To complete the interpretation of the results so far obtained, we remark that the original and the changed surfaces may be referred to a polar coordinates system having point O as center (fig.4). Given a point $P(x,y,z)$ of the original surface and its polar vector $\overrightarrow{OP}$, the corresponding point in the changed surface is defined by the collinear polar vector $\overrightarrow{OP^o}$. The difference in magnitude of the two vectors, $|\overrightarrow{PP^o}|$, is the local distance between the two surfaces according same coordinate system. It can be easily shown that the following equation applies:

$$|\overrightarrow{PP^o}| = |\overrightarrow{OP}| \frac{\Delta \bar{z}}{z} = \sqrt{x^2 + y^2 + z^2} \frac{\Delta \bar{z}}{z} \qquad (21)$$

If the function $z = z(x,y)$ is known, equation (21) can be used, in conjunction with (19) or (20), to evaluate the difference $|\overrightarrow{PP^o}|$ at any point.

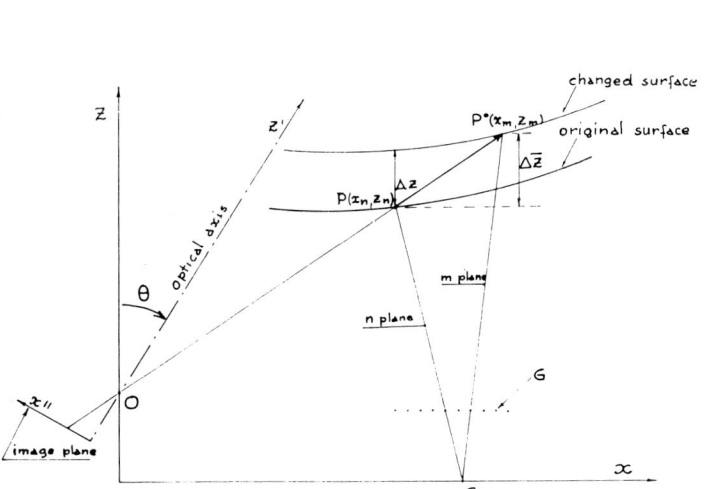

Fig. 3. General geometry for difference contouring analysis.

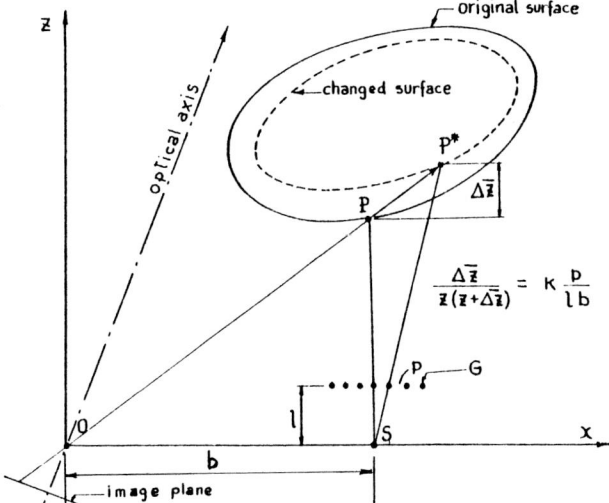

Fig. 4. Actual geometry for difference contouring.

Fig. 5. Contouring of the change of shape of an object.

It may be worthile to emphasize the three-dimensional character of the difference-contouring law expressed by the (19), (20) and (21). The third variable y has never appeared in the course of the analysis simply becouse the planes projecting the grid G from S are parallel to the y axis, and therefore their equation does not contain y.

As an example of application of the difference-contouring technique just exposed and represented in fig.4, the contouring of the change of shape of a gasoline tank, due to a small variation of the internal pressure, is presented in fig.5. Here the average sensitivity factor $z^2 p/bl$ is about 0.5 mm.

References

1. Pirodda,L.,<u>Principi e applicazioni di un metodo fotogrammetrico basato sull'impiego del moiré</u>,Rivista di Ingegneria,n.12,December 1969.

2. Dykes,B.C.,<u>Analysis of displacements in large plates by the grid-shadow moiré technique</u>, Fourth International Conference on Experimental Stress Analysis,Cambridge (U.K.),6th,10th April 1970.

3. Meadows,D.M.,Johnson,W.O.,Allen,J.B.,<u>Generation of surface contours by moiré patterns</u>, Applied Optics,Vol.9,n.4,1970.

4. Takasaki,H.,<u>Moiré topography</u>,Applied Optics,Vol.9,n.6,1970.

5. Collet,J.P.,Marasco,J.,Pflug,L.,<u>Le moiré d'ombre:une méthode expérimentale et ses possibilités</u>,Bulletin Technique de la Suisse Romande,n.9,April 1974.

6. Abramson,N.,<u>Interferometric holography without holograms</u>, Laser Focus,December 1968.

7. Der Hovanesian,J.,Hung,Y.Y.,<u>Moiré contour-sum,contour-difference and vibration analysis of arbitrary objects</u>,Applied Optics,Vol.10,n.12,1971.

8. Benoit,P.,Mathieu,E.,Hormière,J.,Thomas,A.,<u>Characterization and control of three-dimensional objects using fringe projection techniques</u>,Nouvelle Revue d'Optique,Vol.6, n.2,1975.

9. Chiang,F.P.,Khetan,R.P., <u>A general analysis of projection moiré methods</u>, Proceedings of the 15th Midwestern Mechanics Conference,Chicago,Ill.,23th-25th March 1977.

ANALYSIS OF GRATING IMAGING AND ITS APPLICATION TO DISPLACEMENT METROLOGY

R. M. Pettigrew

National Engineering Laboratory, East Kilbride, Glasgow, United Kingdom

Abstract

The analysis of shadow casting by fine grids is extended to include the effects of diffraction. The transfer function of an incoherently illuminated 'pupil' grid is derived and shown to equal the autocorrelation of a modified pupil. Achromatic image casting properties having no geometric analogue are predicted and their application to a displacement transducer is described.

Introduction

In the past there has been considerable attention given to the imaging of periodic objects by imaging systems capable of imaging the general object[1]. There has also been interest in the self-images of periodic objects formed in coherent light[2]. This article considers a restricted class of image-forming systems which embodies aspects of both the above phenomena. It is shown that a system, comprising as its single element a pupil of periodic transmission, is capable of forming images of periodic objects.

The shadow images of objects cast through periodic grids have been used previously for applications such as incoherent character recognition[3], surface topography and strain analysis[4]. The description of such systems based on geometric optics is quite straight forward[5], while the effects of diffraction have also been included by one author[6] and used to obtain magnified displacements of fine grids. In this paper the above results are generalized to describe the response of a periodic system illuminated by a periodic object and it is shown that the transfer function is given by the autocorrelation of a modified pupil function. The results derived show that images are formed under conditions not predicted by geometric optics.

The main application of such a system considered here is in the design of a transducer for use in metrology. It is well established that the linear displacement or rotation of a finely divided grating scale can be observed by the movement of moiré fringes formed on a second similar grid called the index[7]. Non-contacting operation is normally achieved by forming an image of the scale in the plane of the index by means of a lens system or by self-imaging of the scale. These methods have disadvantages that are most apparent when using fine gratings where limited image depth of focus leads to exacting tolerances in the alignment and location of components. Obtaining sufficient source collimation or suitable lens quality can also present problems. Those difficulties encountered have led to the wide spread application of coarse gratings in measuring systems and to the use of large amounts of moiré fringe analogue interpolation in order to achieve the desired resolution. Direct digital measurement from an accurate primary scale is undoubtedly more attractive, thus effort has been applied to the design of simple transducers, for use with fine gratings that incorporate less critical operating restrictions. One such transducer has evolved from the study of the more general image-forming properties of gratings. Before proceeding to practical implementation a closer examination will be made of the types of system shown above.

Analysis of Grating Imaging

Consider the optical system described by Figure 1. An object in plane $\underline{x}_2$ illuminates a pupil of transmission $g(x)$ in plane $\underline{x}$. The irradiance in an output plane represented by $\underline{x}_o$ can be fully described if the mutual coherence of the object irradiance is known, namely:

$$I_o(\underline{x}_o) = \int_{object} \Gamma_2(\underline{x}_2, \underline{x}_2') h(\underline{x}_o, \underline{x}_2) h^*(\underline{x}_o, \underline{x}_2') \, d\underline{x}_2 \, d\underline{x}_2' \tag{1}$$

where $\underline{x}_2$ and $\underline{x}_2'$ are locations in the $\underline{x}_2$ plane, Γ_2 is the mutual coherence of the source and $h(\underline{x}_o, \underline{x}_2)$ is the system impulse response. Since the application concerns incoherently illuminated one-dimensional grids, the analysis is restricted to this case and equation (1) reduces to

$$I_o(x_o) = \int_{object} \Gamma_2(x_2) |h(x_o, x_2)|^2 \, dx_2. \tag{2}$$

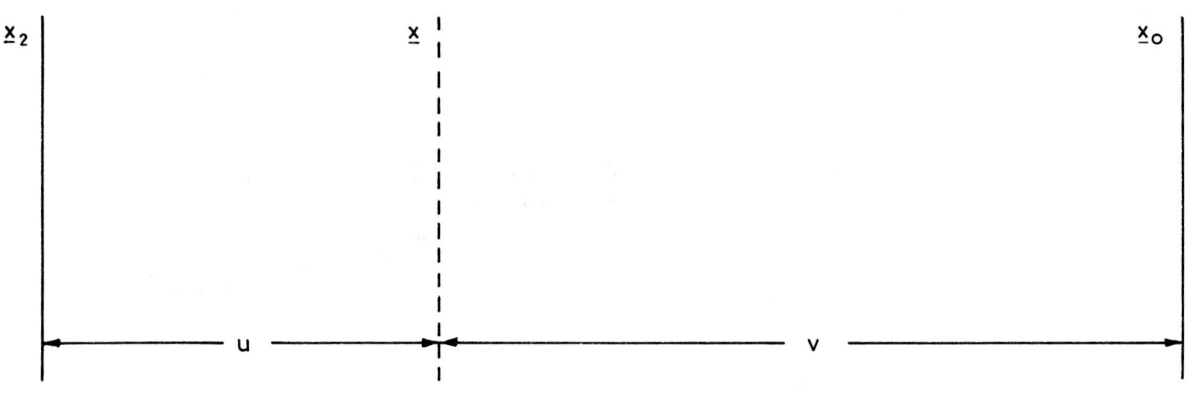

Fig. 1. Optical system.

If the pupil function is periodic then the impulse function may take the form of a self-image of the pupil. From the Fresnel-Kirchoff formula it can be shown that a unit point source at x_2 illuminating the pupil $g(x)$ will give rise to a field $h(x_o, x_2)$ in the image plane where

$$h(x_o, x_2) = A \exp\{\phi(x_o, x_2)\} F_T\{g(x) \exp(ikx^2/2F)\}. \tag{3}$$

F_T is the Fourier transform operation evaluated at co-ordinates

$$f = (1/\lambda v)\{x_o + (v/u)x_2\} \tag{4}$$

and

$$1/F = 1/v + 1/u \tag{5}$$

A is a constant and the phase $\phi(x_o, x_2)$ can be neglected since the image irradiance is the quantity of interest.

If the pupil is a one-dimensional grating of infinite extent and of spatial frequency N with unit cell $g_c(x)$, it can be expressed as

$$g(x) = \sum_{p=-\infty}^{+\infty} g_c(x) \circledast \delta(x - p/N). \tag{6}$$

The image plane irradiance can then be shown to be proportional to

$$|h(x_o - \tilde{x}_2)|^2 = \sum_{p}^{+\infty} \sum_{q}_{-\infty} A_p A_q \left[\exp\{-i\alpha(p^2 - q^2)\}\right]\left[\exp\{2\pi i\beta(p-q)(x_o - \tilde{x}_2)\}\right] \tag{7}$$

where A_p and A_q are Fourier coefficients equal to the transform of $g_c(x)$ evaluated at co-ordinates p/N, q/N respectively, p and q are integers and

$$\alpha = \pi\lambda FN^2$$
$$\beta = \{u/(u+v)\}N$$
$$\tilde{x}_2 = -(u/v)x_2. \tag{8}$$

Thus magnified reconstructions, or Fourier images, of the grid will be obtained for

$$1/u + 1/v = (1/n)N^2\lambda \tag{9}$$

n being an integer. An odd value of n gives rise to an image displaced by half a pitch from the equivalent shadow image of the grid. Non-integer values give rise to harmonic images in intermediary planes which are called Fresnel images[2]. Their properties have been fully discussed by a number of authors[2,8,9].

The spectrum of the image can contain all harmonics of β, their magnitude being modulated by the quadratic phase factor in equation (7). The modulation is not applied to those components in the spectrum satisfying the condition $p = -q$, ie to these interference fringes formed between symmetrically disposed order of diffraction of the pupil grid. The effects can best be seen in the case of a sinusoidal pupil grid.

$$g(x) = a_0 + a_1 \cos 2\pi N x. \tag{10}$$

Solving equation (3) leads to

$$|h(x_0)|^2 = a_0^2 + \tfrac{1}{2}a_1^2 + 2a_0 a_1 (\cos \alpha)(\cos 2\pi\beta x_0) + \tfrac{1}{2}a_1^2 \cos 2\pi 2\beta x_0. \tag{11}$$

It is evident that the second harmonic components in the image spectrum is always in focus independent of the distances separating object image and pupil and is achromatic. The fundamental component in the spectrum corresponds to the geometric shadow image though it exists separately for each source wavelength satisfying condition (9).

In order to determine the general response to a periodic source it is convenient to define a modified pupil by

$$p(x) = g(x) \exp\{(ikx^2)/2F\}. \tag{12}$$

The impulse response can then be expressed, apart from constant terms, as

$$h(x_0 - \tilde{x}_2) = P(x_0 - \tilde{x}_2) \tag{13}$$

where
$$P = F_T(p)$$

and the image plane irradiance is then

$$I(x_0) = \int I_2(-u/v, \tilde{x}_2) |P(x_0 - \tilde{x}_2)|^2 d\tilde{x}_2$$

$$\equiv I_{2P}(x_0) \circledast |P(x_0)|^2 \tag{14}$$

where
$$I_{2P}(x_0) = I_2(-u/v, x_0).$$

$I_2(x_2)$ being the object irradiance. $I_{2P}(x_0)$ represents the geometrical 'pinhole' image of the object.

The image frequency spectrum is

$$G_0(f) = G_{2P}(f) H(f) \tag{16}$$

where
$$G = F_T(I) \text{ and}$$

$$H(f) = p(f) * p(f) \tag{17}$$

is the autocorrelation of the modified pupil evaluated at the appropriate co-ordinates. It is the optical transfer function of the system and is an array of delta functions in frequency space; thus the system will only image certain spatial frequency components in the object, whose values are determined by the spectrum of the pupil and by the object-to-image separations.

The significance of the above relations can be appreciated by considering again the sinusoidal pupil, equation (10). The frequency response of this system is calculated from equation (17) and is given by

$$H(f) = (a_0^2 + \tfrac{1}{4} a_1^2)\delta(f) + (a_0 a_1 \cos \alpha)\{\delta(f + \beta) + \delta(f - \beta)\} +$$

$$+ \tfrac{1}{4} a_1^2 \{\delta(f + 2\beta) + \delta(f - 2\beta)\} \tag{18}$$

-f is the spatial frequency of image space, (u/v × object space).

From this equation it is evident that two object spatial frequencies will be imaged by the sinusoidal grid. The first of these represents an object grid of spatial frequency

$$N_2 = \{u/(u + v)\}N. \tag{19}$$

This image is similar to the shadow image predicted by geometric optics and is accordingly called the 'geometric image'. Its existence is restricted by the conditions imposed by condition (9) which can be rewritten using equation (19) to yield

$$u = n\{1/(\lambda NN_2)\} \qquad (20)$$

thus the object distance is confined to discrete values.

The sinusoidal pupil will also image an object of spatial frequency

$$N_2 = 2\{v/(u + v)\}N. \qquad (21)$$

In this case the object image distances are not restricted to discrete values and the image is achromatic. Moreover this image does not have a geometrical equivalent since it arises from the mutual interference of the first orders of diffraction of the pupil grid and thus it is called the 'diffraction image'. It has low contrast, however, unless the zero order a_o of the pupil spectrum can be suppressed.

The relations derived above are summarized in Figure 2, where the 'optical lever' relation connecting displacement of the image to displacements of the pupil grid is noted.

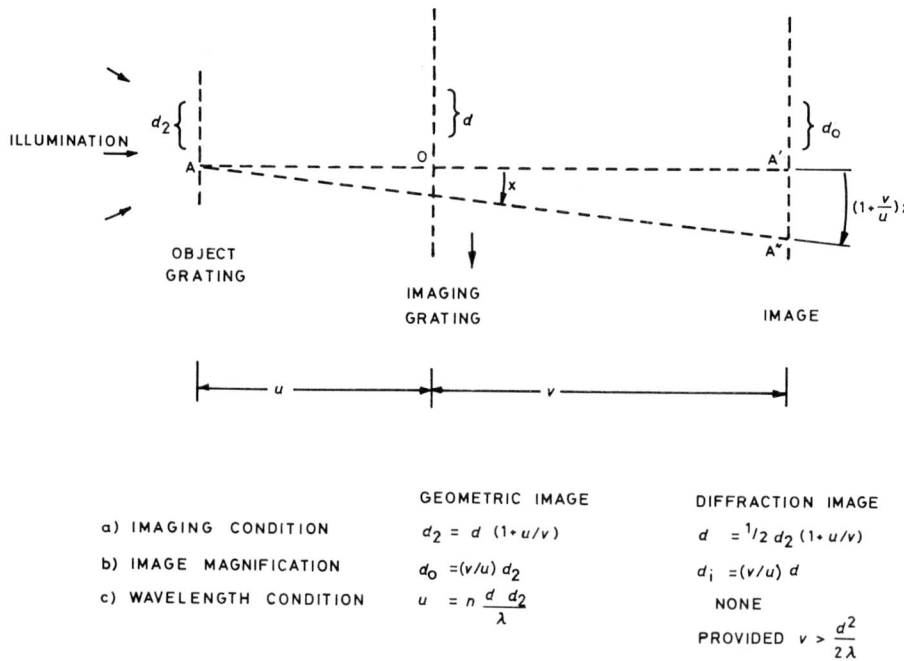

Fig. 2. Grating image forming conditions.

Application of Grating Imaging to the Measurement of Displacement

It is apparent that moiré fringes will be obtained on a grid of the appropriate pitch placed in the image plane of the systems shown in Figure 2 and that movement of the pupil grid will cause a magnified movement of the image and hence of the moiré fringes. An arrangement such as this is shown in Figure 3. Three photographic 500 l/cm gratings are separated by equal distances. The first, the object grating, is diffusely illuminated with white light. The second grating is the pupil and the third grating is used to detect the image as evidenced by the formation of moiré fringes. The relations between the grating separations (equal in this case) and line densities are that the 'diffraction image' is formed. When using photographic bar and space grids as shown, a moiré fringe contrast of about 25 per cent is obtained, while the use of a bleached 'phase' grating, as object, reduces the contrast to less than 5 per cent. However a pupil grating bleached to suppress the zero diffraction order can significantly improve the contrast and, as will be shown, values of 40 per cent are obtainable. If the pupil is replaced by a 100 l/cm grating then coloured moiré fringes are observed, the preferred wavelength satisfying condition (20).

Fig. 3. Arrangement of three gratings to produce moiré fringes.

A displacement transducer for automatic measuring machines would be unduly complex if it comprised three gratings positioned as shown in Figure 3, the depth of focus being no better than that obtained in conventional transducers. It is possible to reduce the system to two gratings by using a reflecting pupil[10,11] as shown in Figure 4. By this means the location tolerances are relaxed and the depth of focus limitation is removed. By use of an object grating ('the index') and a reflecting pupil grating ('the scale') of the same line density a diffraction image is formed by reflection in the plane of the index grating. Moiré fringes are observed which record the displacements of the scale grating; the displacement of the image being twice that of the scale by virtue of the relation in Figure 2.

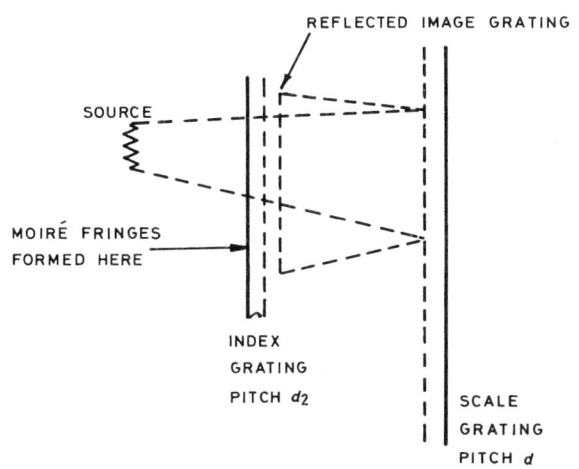

FOR GEOMETRIC TYPE TRANSDUCER $d_2 = 2d$

FOR DIFFRACTION TYPE TRANSDUCER $d_2 = d$

Fig. 4. Reflection system.

The image remains exactly superimposed on the index grating despite changes in gap, or of misalignments occurring between the grating planes. The effects of gap changes are shown in Figure 5. The vertical axis represents the image contrast as measured from the moiré fringes and horizontal axis is the separation between the index and scale grating planes. A high average contrast of approximately 45 per cent is obtained by using a reflecting grating etched on a quarter-wave thick deposit of chrome on glass with a final thin deposit of chrome to enhance the reflectivity and to improve the suppression of the zero diffraction order. Favourable results can also be obtained from etched polished steel. Table 1 records results obtained from a variety of scale gratings using a photographic index grating. The results given in Figures 5 and 8, and Tables 1 and 2 were obtained using a retro-reflector

Table 1. Three Grating System Diffraction Imaging

25 lines/mm Scale   25 lines/mm Index
Source Sensor Unit - Spectronics
Separation Between Index and Scale - 10 mm

Gratings	Depth of Modulation		
	a.c. signal peak-peak µA	Mean d.c. signal µA	Modulation per cent
GC25	16	17.75	45
GCC25	30.5	29.25	52
GCC20	30	25	60
GC20	13.25	15.875	43
MCA	25	33.5	37.3
MCB	28.5	35.25	40.4
MEB2	11	14.5	38

Code G - Glass grating
     M - Metal grating

Table 2. Three-grating Reading Head Alignment Tolerances

Retro-Reflector Unit	Orientation of Reading Head W.R.T. Grating Lines	Depth of Focus ±3 dB Signal Points mm	Grating line Density lines/mm	Peak Fringe Photo-current µA	Mechanical Alignment Tolerances ±3 dB Signal Points		
					Yaw deg	Tilt deg	Roll deg
Texas TIL139	Parallel	2-6	100	40	±3.0	±0.2	±1.5
	Perpendicular	2-6	100	50	±0.5	±0.2	±1.5
	Parallel		25	36	±3.0	±1	±1.5
Hafo 69X72	Perpendicular	25-75	100	240	±4.0	±0.2	±2.5
	Perpendicular	25-75	25	600	±4.0	±0.8	±2.5

Yaw - out of plane parallel to lines
Tilt - in the plane of the grating lines
Roll - out of plane perpendicular to lines

unit (infra-red LED and photo-transistor detector) attached directly to the index face (Figure 6). Table 2 list the misalignments between index and scale that can be tolerated; movement of the scale corresponding to yaw will alter the moiré fringe phase and hence cause measurement inaccuracy. Figure 7 shows a displacement transducer on a 100 l/cm scale grating. It illustrates the compact nature of the transducer and the insensitivity to the gap separating the transducer and scale.

Much higher moiré fringe contrast can be obtained using the geometric image which is formed if the index grating pitch is twice that of the scale grating. The superposition of this image on the index is also exact; however, it has finite depth of focus as predicted by equation (20). The contrast of the moiré fringe will thus depend on the gap separating the two gratings. The result of this can be seen in Figure 8, where using a relatively narrow band source of illumination a large number of image orders are available before the image contrast peak falls below 50 per cent. The depth of focus limitation is similar to that of the normal incidence transducer which uses a self-image of the scale in collimated light. The geometric image transducer does have the advantages of not requiring coherent collimated illumination and of having twice the depth of focus of the normal incidence transducer for the same scale grating. This is illustrated in Figure 9 where the performance of the three transducers is compared for the same pitch of scale grating.

The construction of a practical transducer is not considered further here, suffice to say that it remains a complex problem of optics, electronics and engineering to design a transducer to withstand the rigours of the hostile environment often associated with automatic machine tools. The interested reader is referred primarily to Guild[12].

Conclusions

Image casting by grids has been analysed taking the effects of diffraction into account. It has been shown that in coherent illumination, the image casting grid may perform as a linear imaging system for certain spatial frequencies and form images to pinhole images of geometric theory. Other achromatic harmonic images may also be formed which have no geometric analogy. The application to the measurement of grating displacement was considered and a novel transducer described which is particularly effective in measuring the displacements of fine grids.

Acknowledgement

This paper is presented by permission of the Director, National Engineering Laboratory, Department of Industry. It is British Crown copyright.

References

1. Hopkins, H. H., 'On the Diffraction Theory of Optical Images', Proc. Roy. Soc. A., Vol. 217, p 408. 1953
2. Cowley, J. M. and Moodie, A. F., 'Fourier Images', Proc. Roy. Soc. B., Vol. 70, p 486. 1957.
3. Leifer, I. and Rogers, G. L., Incoherent Fourier Transformation: a New Approach to Character Recognition, Optica Acta, Vol. 16, p 535. 1969.
4. Theocaris, P. S. and Koutsabessis, A., Surface Topography by Multisource Moiré Patterns, Exp. Mech., Vol. 8, p 82. 1968.
5. McCurry, R. E., Multiple Source Moiré Patterns, Appl. Phys., Vol. 37, p 407. 1966.
6. MacGovern, A. J., Encoder Readout System, United States Patent No 3 812 352. 1974.
7. Burch, J. M., The Metrological Applications of Diffraction Gratings, Progress in Optics, Vol. 2, p 75. 1963.
8. Rogers, G. L., A Diffraction Theory of Insect Vision, Proc. Roy. Soc. B., Vol. 157, p 83. 1962.
9. Winthrop, J. T. and Worthington, C. R., Theory of Fresnel Images, J. Opt. Soc. Amer., Vol. 55, p 373. 1965.
10. Russell, A. and Pettigrew, R. M., Recent Trends in Angular Measurement Using the Optical Grating, Proc. of Nelex 76 Metrology Conf., Paper 14, East Kilbride, Glasgow, National Engineering Laboratory, 1976.
11. Pettigrew, R. M., U.K. Patent Application No 44 522/74.
12. Guild, J., 'Diffraction Gratings as Measuring Scales', Oxford University Press. 1960.

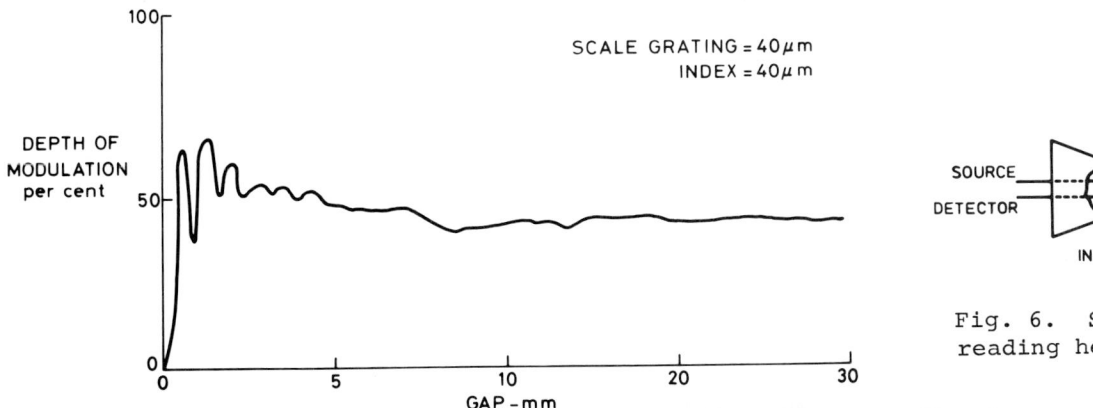

Fig. 5. Diffraction image - effect of change in gap.

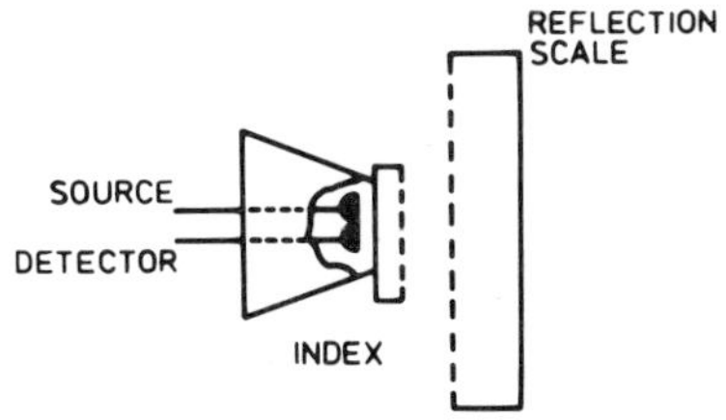

Fig. 6. Solid-state reading head.

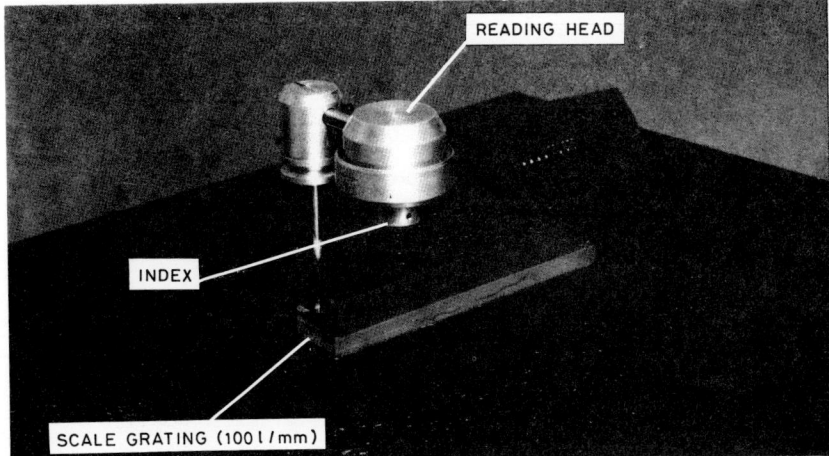

Fig. 7. Displacement transducer

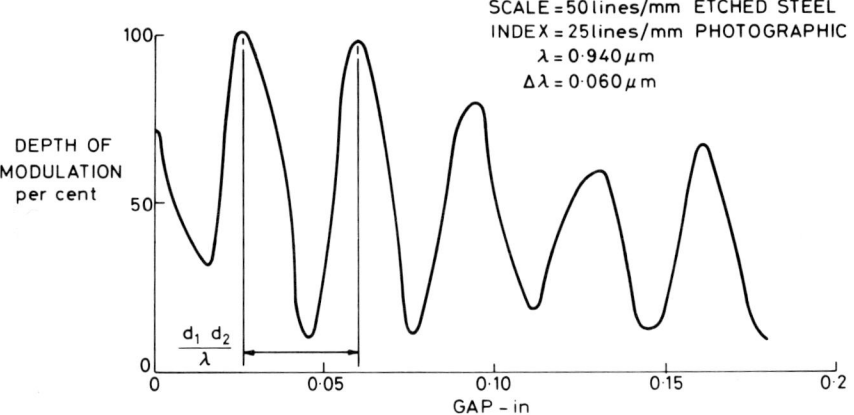

Fig. 8. Geometric image - effect of change in gap

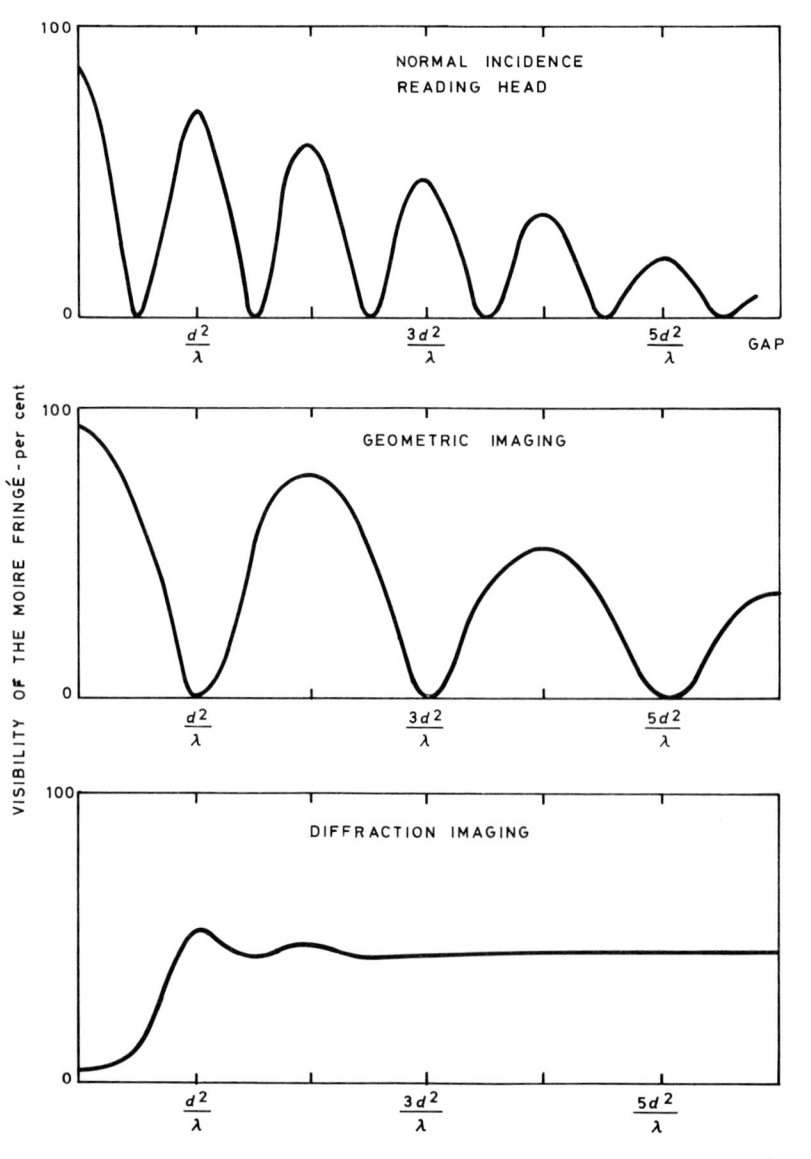

Fig. 9. Reading head sensitivity to gap changes

MEASUREMENT IN REAL TIME OF TRANSVERSAL MICRO-VIBRATIONS (DOWN TO 1 Å) ON DIFFUSING OBJECTS: RANDOM MOIRE CAPTORS

D. Joyeux
Institut d'Optique, Université de Paris-Sud
B.P. 43, 91406 Orsay-Cedex, France

Abstract

We studied an optical method of diffuse wave interferometry based on a random moiré principle, permitting, in real time, the measurement of very small displacements or vibrations under experimental conditions which are unusual by reason of their weakly constraining character. In fact, - the field of application is that of objects with unpolished surfaces, observed perpendicularly to this surface, for displacement vectors situated in the plane of this surface; - the setup demands only ordinary precision ; it is composed of a lighting system with two coherent (laser) waves and a low-resolution imagery system followed by a detector ; - the mode of operation consists in the detection of variations of lighting in a laser image granularity spot ; under certain conditions, these variations are proportional to the displacement of the point corresponding to the object ; - the performances : the sensitivity is interferometric ; the theoretical limit is on the order of 0.1 Å$\sqrt{Hz}$ under standard experimental conditions. Values close to this limit have been attained experimentally. The precision of the measurements is on the order of 1%.

Introduction

The problem of investigating the vibration state of a mechanical object has received a variety of optical solutions,[1] which generally perform one only of the following tasks :
1) *the spatial analysis* : one obtains the spatial repartition of simple parameters such as amplitude or phase of the vibration at the surface of the vibrating object ; 2) *the temporal analysis* : one obtains the displacement vs time function at a selected point of the vibrating surface.

In addition to these major orientations, the different experimental situations yield the different aspects of the various methods. The following parameters are of particular interest : the amplitude range to be measured, the surface state of the vibrating object and the direction of the vibration with respect to the surface and to the direction of observation. The method we describe below is a "temporal" solution to a seldom investigated experimental situation: 1) unprepared objects are used (i.e. objects can have any surface state except specular); 2) the vibrations are in-plane, perpendicular to the direction of observation ; 3) an interferometric sensitivity is wanted, i.e. amplitudes from a few thousand angström to less than one angstrom are to be measured with typically 1% accuracy; 4) as implied by "temporal", the method has to realize the transduction of the instantaneous displacement into variations of a photometric quantity.

Although the method can be studied in the framework of real time speckle interferometry, we found preferable to describe it as a generalization of a classical real time moiré transducer. This will become clear after some simple considerations ; such an approach allows to derive the principle characters of the method with very few mathematics.

In order to demonstrate the proposed generalization, let us first briefly recall the principle and properties of a classical moiré displacement transducer.

The classical moiré transducer

Let us consider the simple idealized moiré experiment shown in Fig. 1. The (X,Y) plane contains a grating whose amplitude transmittance is $\cos 2\pi\nu X$. The grating is then displaced by x along X : its amplitude transmittance becomes $\cos 2\pi\nu(X-x)$. Let us illuminate the shifted grating by a fringe pattern with an amplitude of $\cos 2\pi\nu_o X$ (e.g. produced by the interference of two plane waves).

The object plane is imaged through a linear shift-invariant optical system with a cut-off frequency ν_c. The amplitude A in the image plane is very easy to compute ; one obtains, dropping constant multiplicative factors :

$$A = 0 \qquad \text{if } \nu_c < |\nu_o - \nu| \qquad (1\text{-a})$$

$$A = \cos 2\pi\{(\nu-\nu_o)X - \nu x\} \quad \text{if } \nu_c > |\nu_o - \nu| \qquad (1\text{-b})$$

The image illuminance I is therefore :

$$I = 0 \qquad \text{if } \nu_c < |\nu_o - \nu| \qquad (2\text{-a})$$

$$I = 1 + \cos 2\pi\{2\nu x + 2(\nu_o - \nu)X\} \quad \text{if } \nu_c > |\nu_o - \nu| \qquad (2\text{-b})$$

Eqs. (1-b) and (2-b) exhibit the twofold aspect of moiré phenomenon : 1) the spatial variation of A and I, i.e. the variation along the image coordinate X, when the object position x is constant. In this case, the difference $|\nu-\nu_o|$ determines their variation speed ; 2) the variations of A and I with the object position x, X being given. The speed of these variations is determined by the object frequency ν itself.

This second aspect is indeed the interesting one : it is responsible for the use of this kind of system as displacement transducer. The characteristic function I vs x is a sinusoid. In particular, one can fix an operating point and transduce linearly very small displacements of the object grating. Such a transducer is well-known ; it satisfies all the experimental constraints listed in the introduction except one : it does not work with unprepared objects, since a grating is needed as object.

We shall now examine how the same setup can work as a transducer with a rough object placed in the object plane.

Extending the moiré transducer to rough surface objects

The moiré phenomenon is not entirely described by Eq. (1-b) (or 2-b), but by the set of Eqs.(1) (or 2) including the associated inequations. Eq. (1-a) (or 2-a) in particular tells that if the frequencies of the grating and of the fringe pattern are too different, then no signal (illuminance) is available in the image plane. Therefore, if we have as object a grating with a spatial frequency ν satisfying $\nu_c < |\nu-\nu_o|$, it is equivalent, from the point of view of image phenomena, to having no grating in the object plane.

Since the optical imaging system of Fig. 1 is linear in amplitude, the above remark can be applied to the whole spatial spectrum of any object. In the case of a random, wide-spectrum object (i.e. a diffusing object), this is equivalent to say that the phenomena in the image plane are not dependent on the real object spectrum (Fig. 2-a) but on a truncated version of it (Fig. 2-b). This virtual spectrum is equal to the real one inside two bands determined by $|\nu-\nu_o| < \nu_c$, and is zero outside these bands. It is therefore a pass-band random spectrum; its central frequency is the frequency of the fringe pattern ; its bandwidth is twice that of the imaging system. Since these two parameters are setup parameters, they can be adjusted at will. As a consequence, the ratio (bandwidth/central frequency) can be made as small as desired. From the theory of narrow-band random processes, this means that the virtual object can be approximated by a sinusoïdal one (in amplitude-transmittance) with an accuracy as high as desired.

In other words, the variations of illuminance at any fixed point of the image plane have the same aspect as if a sinusoidal grating was placed in the object plane. Consequently, the principle of use of the system for measuring very small vibrations is the same as for a real-time moiré system. We do not develop this point here ; more details are given in Ref. 2.

Discussion

The result we have obtained could seem paradoxical. As a matter of fact, it must be emphasized that the quasi-equivalence between the classical moiré transducer and a random moiré transducer has been obtained at a certain cost, namely : 1) the accuracy of the measurement cannot be refined beyond a certain limit, which depends on the actual values of two setup parameters. In the practise, the accuracy limit can be in the range 1% - 0,1% quite easily ; 2) the elimination of all the spatial spectrum of the actual object, except for two narrow bands, leads to a waste of the energy incident on the object plane ; thus, one obtains a much lower signal to noise ratio for the illuminance measurement than can be obtained by using the equivalent classical moiré system. In other words, the random-moiré system is certainly useful when nothing (grating) can or must be placed on the tested surface (e.g. for experimental reasons). But, in all other cases, the classical moiré system must be preferred.

We shall not discuss here the practical realization of the random moiré system. A detailed analysis is given in Ref. 2. However, the experimental results given below will illustrate its actual possibilities.

Experimental results

In order to evaluate correctly the presented results, we state again the experimental conditions of the measurements. 1) All the following measurements were performed with unprepared diffusing objects such as the surface of a metallic, not polished cylinder, or the electrode of a piezoelectric ceramic. 2) The observed displacements are in-plane and the object surface is observed perpendicularly. 3) At least for small amplitudes (i.e. < 1000 Å), the system works as a true displacement transducer, i.e. it yields as output (illuminance variations) a replica of the displacement vs time function.

Moreover, it should be mentioned that an electronic processing was needed to increase the generally very low signal/noise ratio of the detector output (this low signal/noise was due to the very low level of the image illuminance).

MEASUREMENT IN REAL TIME ON TRANSVERSAL MICRO-VIBRATIONS (DOWN TO 1 Å) ON DIFFUSING OBJECTS: RANDOM MOIRE CAPTORS

1. Displacement vs time display by using a sampler averager

Fig. 3 shows the displacement induced at the interface of an ultrasonic vibrator by repetitive (3 kHz) trains of periodic (34 kHz resonance) voltage pulses. The settings have been adjusted to display the time interval between two successive pulse trains. Two records were performed : 1) without anything touching the vibrator (Fig. 3-a) ; 2) with a piezoelectric transducer coupled with it (Fig. 3-b). In case 1), the permanent regime amplitude is 635 Å p-p and the decay time constant is 0,32 ms. In case 2), these values become 470 Å and 0,17 ms respectively.

The response of a mechanical system driven by a piezoelectric ceramic is shown in Fig. 4. The excitation voltage consists in isolated voltage pulses. The amplitude resolution is about 2 Å.

2. Amplitude and phase of a sinusoidal vibration

Fig. 5 shows the resonance curve of a piezoelectric device, i.e. amplitude and phase vs frequency. The electronic processing consists of a synchronous detection with a time constant 1 s. The noise level is better than 3 Å p-p.

Conclusion

We have presented the theoretical basis of the random moiré displacement transducer. Like classical moiré displacements transducers, it works in real time and presents a very high sensitivity (down to Angström or less) which is basically an interferometric sensitivity.

The particular features of the random system is to be able to work with diffuse objects; the accuracy of measurements is in the percent range.

The performances and possible applications have been illustrated by some examples of real measurements.

References

1. Cloud, G., Appl. Opt., vol. 14, pp. 878. 1975.
2. Joyeux, D., Appl. Opt., vol. 15, pp. 1241 and pp. 1248. 1976.

Figures

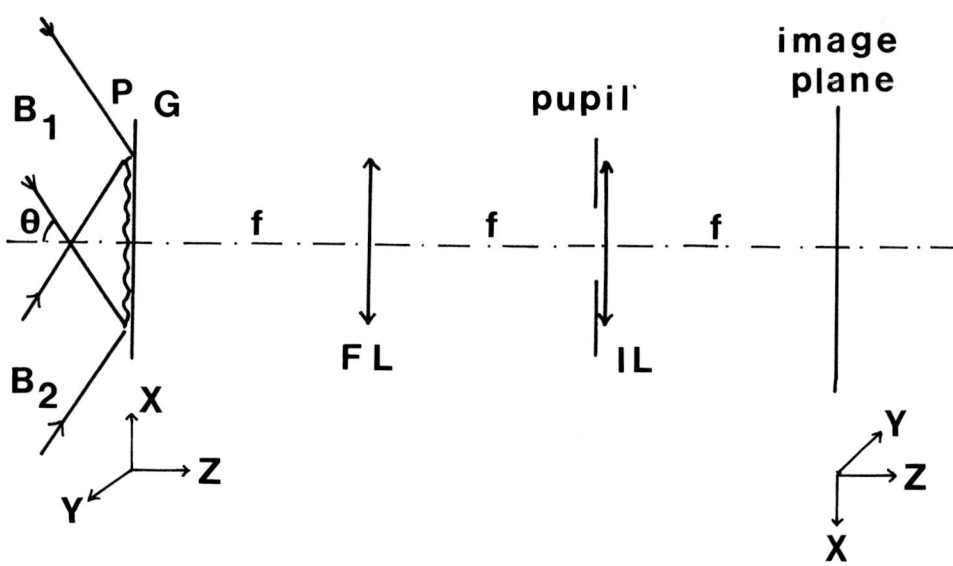

Fig. 1. The object plane (X,Y) contains either a grating G or a random diffuser. It is illuminated by a fringe pattern P, produced by the interference of two coherent plane waves B_1 and B_2. A shift-invariant system yields the image of the object plane.

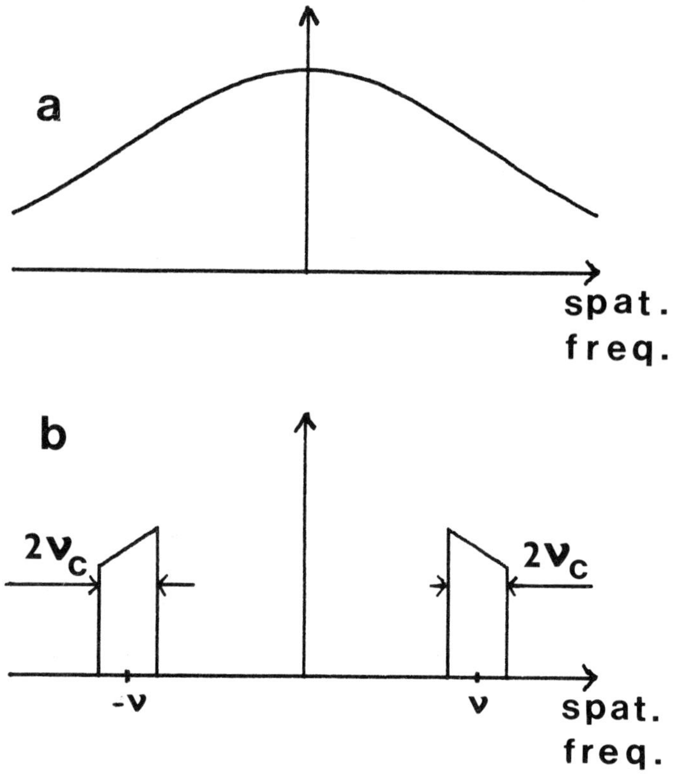

Fig. 2. Assuming 1) the object spectrum plotted in Fig. 2-a, 2) an imaging system with a rectangular transfer function centered on the null frequency, and a cut-off frequency ν_c, then the moiré depends only on the virtual spectrum plotted on Fig. 2-b. ν is here the fringe pattern frequency.

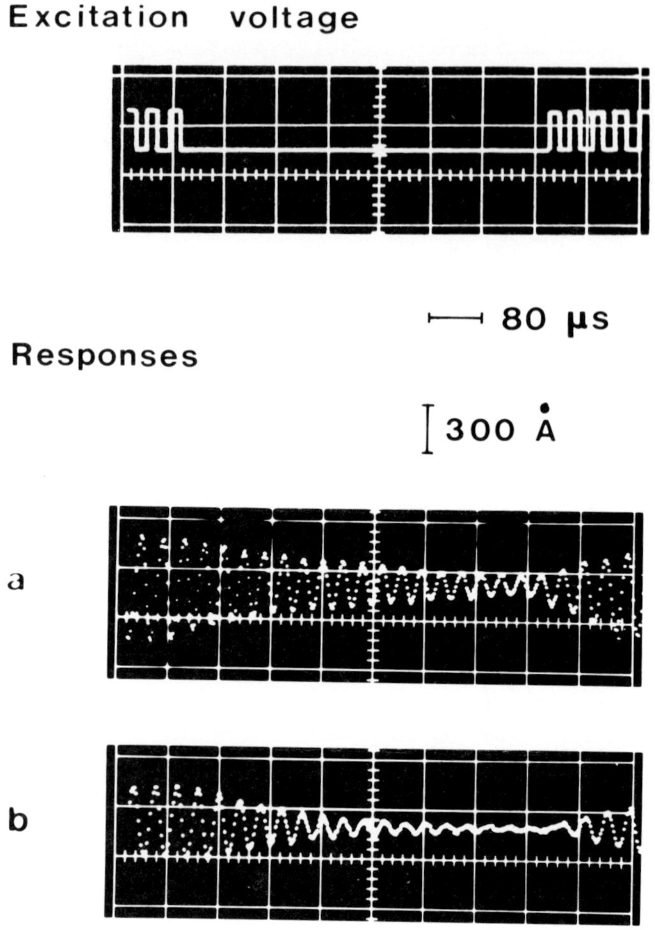

Fig. 3. Waveform restoration by digital sampling averaging. An ultrasonic vibrator is excited by repetitive trains of voltage pulses. The excitation voltage is shown at the top, using the same time scale as for the response (see text). a) Response of the vibrator when excited without mechanical load. b) Same, when a piezoelectric transducer is used as a mechanical load.

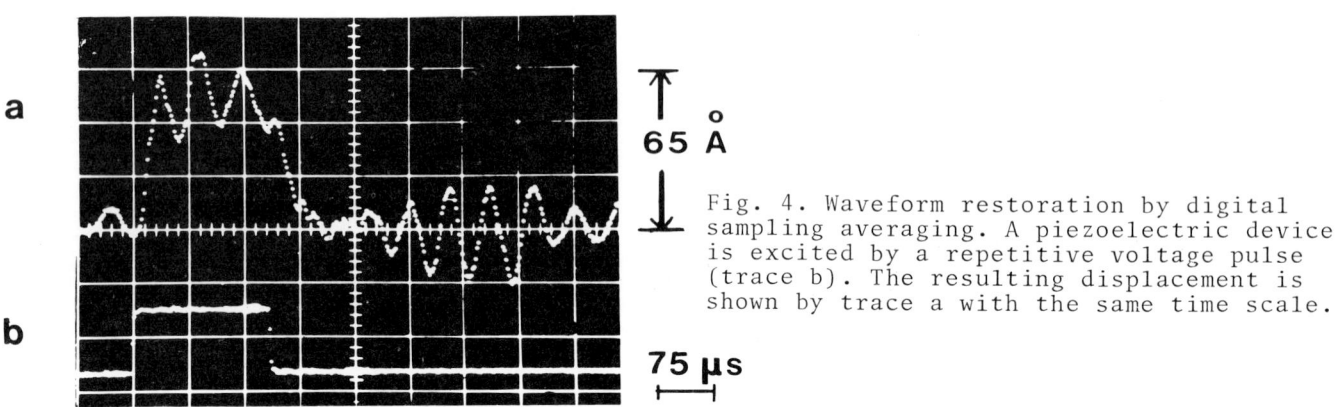

Fig. 4. Waveform restoration by digital sampling averaging. A piezoelectric device is excited by a repetitive voltage pulse (trace b). The resulting displacement is shown by trace a with the same time scale.

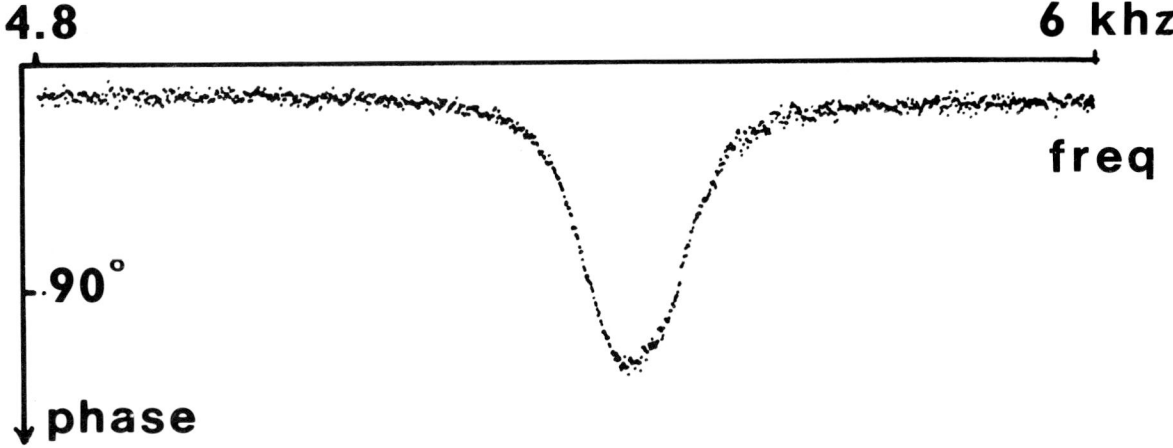

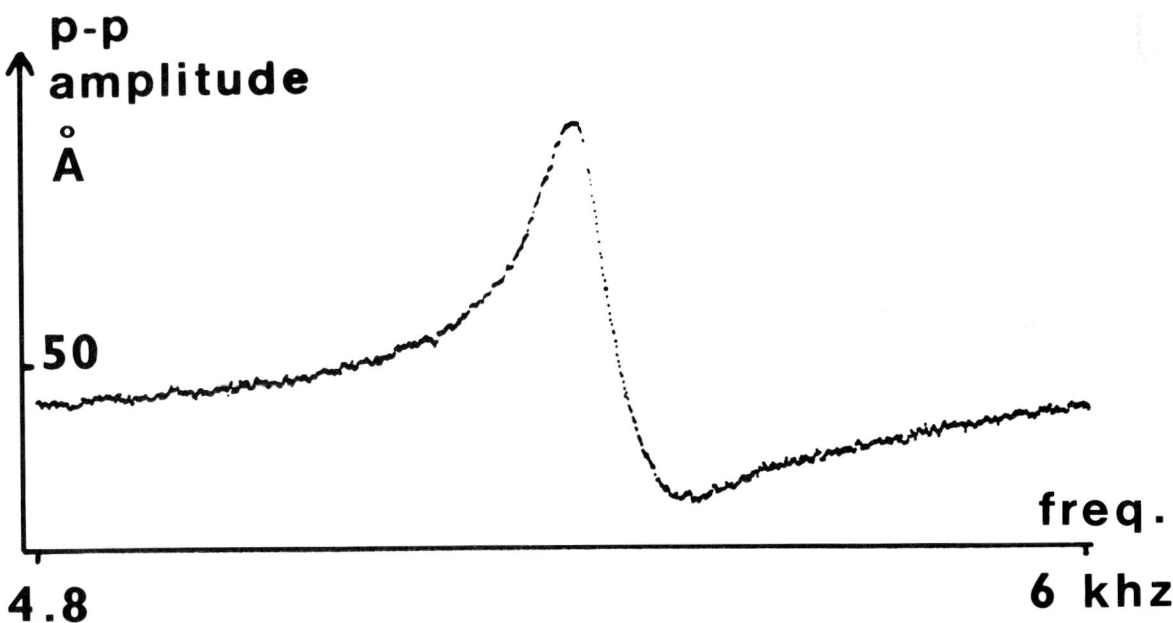

Fig. 5. Resonance curves of a piezoelectric device : amplitude and phase vs frequency plot. The piezoelectric device is excited by a constant amplitude sinusoidal voltage with slowly variable frequency. Detection by a lock-in with time constant 1 second.

1st EUROPEAN CONGRESS ON OPTICS APPLIED TO METROLOGY

Volume 136

SESSION 8

ACOUSTO-OPTICS

Session Chairmen
Meyrueis
Grosmann

CONTRIBUTIONS OF ACOUSTICAL HOLOGRAPHY TO MECHANICAL METROLOGY

J. Pasteur and Y. Seyzeriat
Laboratory of General Physics and Optics, C.N.R.S. N° 214
"Holography and Optical Treatment of Signals"
University of Franche-Comté, 25030 Besancon Cedex, France

Abstract

Holographic transposition of ultrasonic waves into light waves can be exploited not only qualitatively, as a technique for non-destructive testing of opaque materials, but also quantitatively. In fact, as a method of visualization, acoustical holography translates the variations of acoustical transparence of the object inspected into the form of variations of lighting. Generally, the non-linearities of this process limit the accessible information solely to the contour of the variations in transparence as well as their localization, which leads to a simple dimensional metrology. It is nonetheless possible to exploit the conservation of the phase to effect, by means of optical techniques, measurements on the parameters characterizing the inhomogeneous zones of a material. The conditions of linearity of the transfer of the information by holography are examined in the case of the surface relief method and some examples of the optical treatment of the acoustical information are presented.

Introduction

Non-destructive testing of material by ultrasound has been considerably developed in the last decade, impelled by technological advances. In particular, the association of scanning techniques and powerful methods of detection by echo-sounding (reflection of impulses) has permitted achievement of a two-dimensional representation of a section of the material being tested, with resolution on the order of a millimeter for frequencies of several MHz.

The transposition into acoustics of the techniques of optical holography, demonstrated by Mueller and Sheridan[1] in 1966, has further enlarged the field of applications of ultrasound. Acoustical holography has appeared as a new means of visualization of ultrasound. The goal of our paper is to show that it can also serve as a base for the development of methods of measurement of certain parameters of mechanical pieces.

Ultrasonic Imagery by Holography: Examples

Diverse methods of recording ultrasonic holograms have been proposed[2]. We used the method of surface relief[3], in which the photosensitive detector is a liquid-gas interface which is deformed by the action of the ultrasonic radiation pressure and which is simultaneously lit by a monochromatic wave: the deformed interface constitutes a phase hologram which, by diffraction, reconstructs a luminous image of the ultrasonic field. Figures 1 and 2 provide some examples of images obtained by this process and permit us to compare these images to the photographs of the objects examined. In the thick plates (fig. 2), the artificial internal defects appear clearly.

These images translate the transparence of the object into ultrasound; this transparence is a function of the absorption of the material and, principally, of the variations in acoustical impedance. They reveal the internal structure of the material examined and permit us to localize details and to determine from them the form and the dimensions. Ultrasonic holography can thus be considered as a means of dimensional metrology. The precision of the recordings, which is related to the resolution of the system of imagery, is limited in fact by the wavelength of the ultrasound in the medium examined: in the range of frequencies commonly used (between 2 and 10 MHz), the absolute uncertainty is, at the minimum, some tenths of a millimeter.

The metrological potential of acoustical holography is not limited solely to the aspect which has just been described. In fact, the transposition of the ultrasonic wave into a light wave facilitates the exploitation of the information contained in the phase term in the sense that proven optical techniques can be used. Thus we can envisage treatment by Foucaultage, striscopy, interferometry or filtering, not only to improve the vision of the defects of the material examined, but also to determine its characteristics. The validity of the results of the measurements furnished by the optical techniques remains nonetheless related to the hypotheses of the existence of a resemblance between the complex amplitude of the acoustical wave and that of the light wave. We will also examine the conditions in which the transfer of the wave is effected in the case of the surface relief method.

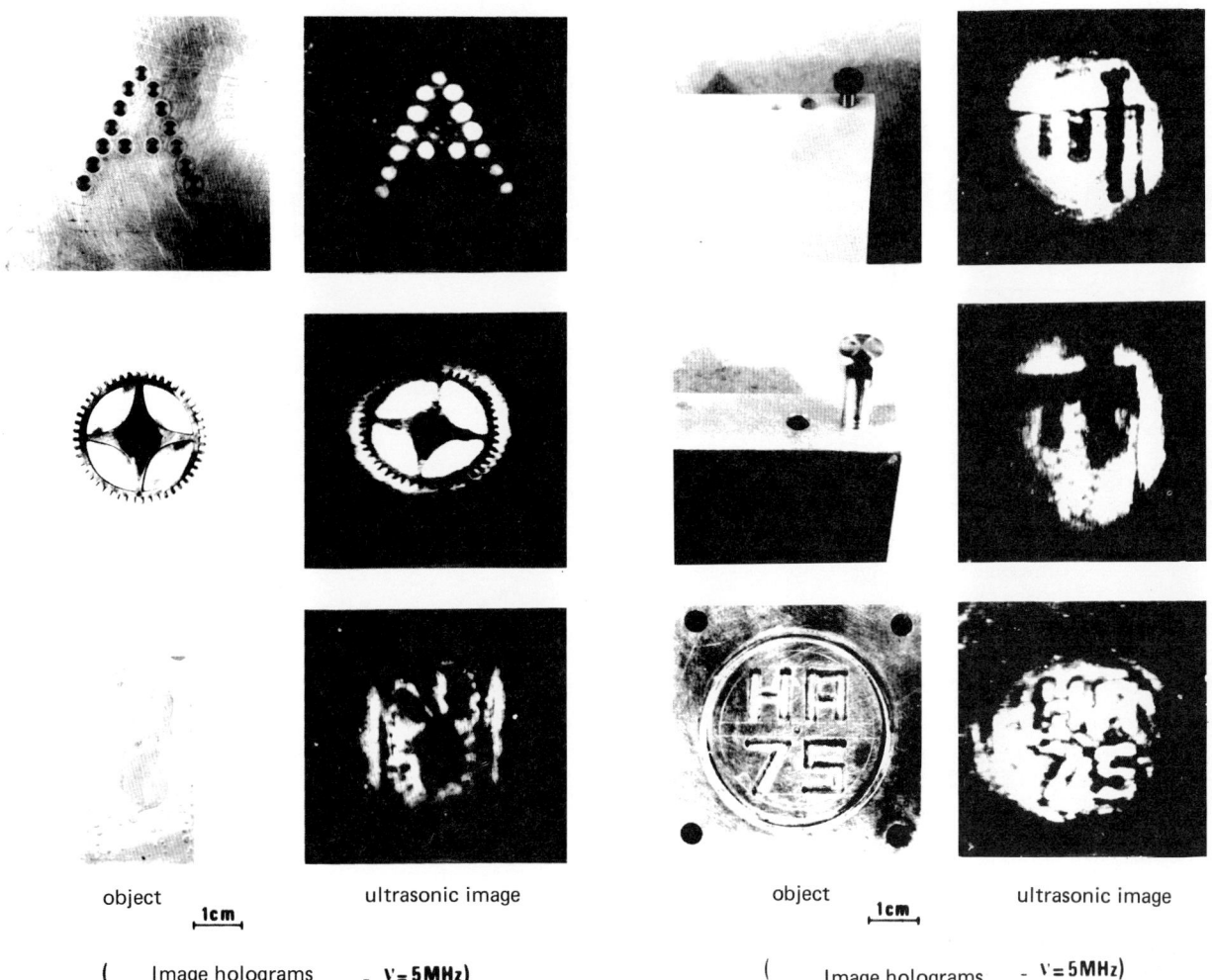

Fig. 1. Ultrasonic images of pierced metallic pieces, cut out or mounted.

Fig. 2. The ultrasonic images display the position and the depth of the holes pierced in the thickness of the plates.

Transfer of Information in the Surface Relief Method: Conditions of Linearity

The principle of the surface relief method is illustrated in figure 3. The acoustical hologram is inscribed in the form of ripples, of depth 2B, which are superposed at an average elevation A from the surface of the liquid. This reflecting surface, lit by the wave Σ_{or}, constitutes a phase hologram capable of reconstructing a light wave Σ_{∞} similar to the acoustical wave Σ_{ao}.

To treat the problem of the resemblance between the object wave and the reconstructed wave, let us examine the case of Fourier holography sketched in figure 4: the object is situated in the $x_1 O_1 y_1$ plane in the path of the plane wave Σ_a and the spectrum of its acoustical transparence is formed in the plane of the holographic receptor xOy. S_o represents the punctual reconstruction light source, Σ_{∞} designates the reconstructed direct wave and Σ^*_{∞} the conjugated wave.

The calculation of the deformation $z(x,y)$ of the liquid-gas interface translates the "recording" stage—more precisely, the "formation" stage of the hologram.

We designate:

P_o and P_r the constant amplitudes of the "object" and "reference" beams
$\delta(x_1, y_1)$ the Dirac measurement
$T(x_1, y_1)$ the acoustical transparence of the object
$\otimes$ the symbol of the product of convolution

The amplitudes of the ultrasonic waves in the $x_1 O_1 y_1$ plane are written:

(1)

(2)

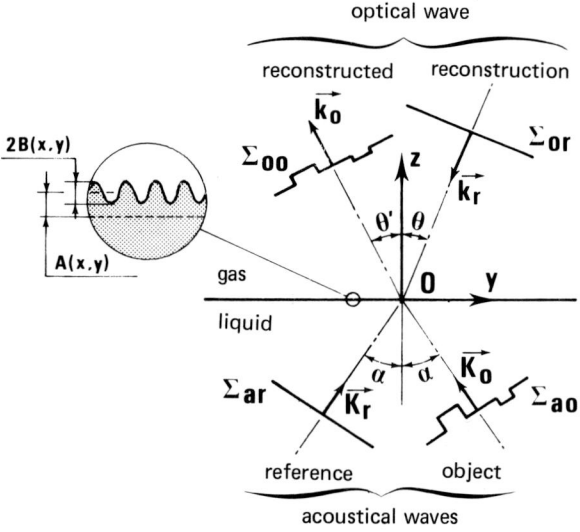
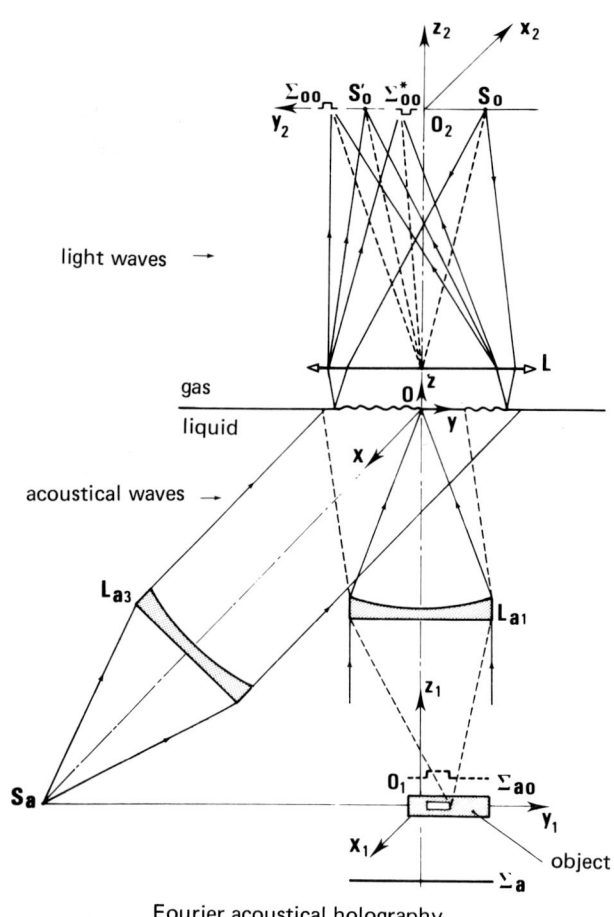

Fig. 3. The holographic detector is the liquid-gas interface; the fringes are materialized as ripples.

Fig. 4. Principle of Fourier holography in the surface relief method.

Fourier acoustical holography

In the xOy plane, the distributions of amplitude are proportional to the Fourier transforms (TF) of U_o and U_r:

(3)

where: f_a is the focal length of the acoustical lenses L_{a1} and L_{a3},
$K = 2\pi/\lambda$ is the module of the acoustical wave vector, which gives:

(4)

(5)

The acoustical radiation pressure on the interface, in the xOy plane, is expressed:

(6)

at each point, it is compensated by the gravitation (g) and the superficial tension (γ), so that the equation of equilibrium of the interface is written(1,2):

(7)

The solution of equation (7) provides a solution which, under conditions of impulsion, can be put in the form[4]:

(8)

with

(9)

C_1 and C_2 being constants linked to the experimental parameters and the time (5).
The reading of the hologram (reconstruction stage) is effected by lighting the interface with a light wave of wavelength λ, issued from a point source S_0 situated in the focal plane $x_2 O_2 y_2$ of the objective L. The amplitude of this wave can be written in the form:

$$ \tag{10} $$

The distribution of amplitude of the wave reflected by the interface $z(x,y)$ with the reflection coefficient R, is:

$$ \tag{11} $$

or keeping in mind (8) and for $z = 0$:

$$ \tag{12} $$

with

$$ \tag{13} $$

designating by J_n the Bessel function of the 1st type of order whole $\underline{n}$.
In the $x_2 O_2 y_2$ plane, the distribution of diffracted amplitude is proportional to the Fourier transform of $a'(x,y,0)$, which is written, taking into account the orientation of the axes,

$$ \tag{14} $$

It is clear that the calculation of the integral (14) cannot be made simply, unless we pose:

$$ \tag{15} $$

If these last approximations are satisfied and if we designate by Γ the transversal magnification of the system such that:

$$ \tag{16} $$

we show that the wave Σ_{∞} reconstructed in the **order** $n = 1$ has the amplitude

$$ \tag{17} $$

in which T' is the Fourier transform (14) of

$$ \tag{18} $$

that is, taking account of the properties of the Fourier transformation:

$$ \tag{19} $$

This demonstrates the existence of a resemblance between the light wave reconstructed in order 1 and the initial acoustical wave. This resemblance, of ratio Γ, exists only if the approximations (15) are verified; these last thus represent the conditions of linearity of the transposition of an acoustical wave into a light wave. On the experimental level, the validity of the relations (15) is expressed by two imperatives (5): 1) the ultrasound must be emitted by limited wave trains and 2) the acoustical intensity must be below a certain threshold. Figure 5 clearly illustrates the degradation of the reconstructed image when this threshold is surpassed.

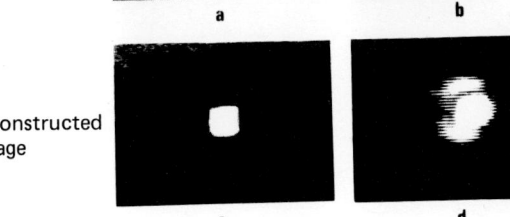

Fig. 5. Illustration of the respecting (a and c) or failure to respect (b and d) of the conditions of linearity in Fourier holography.

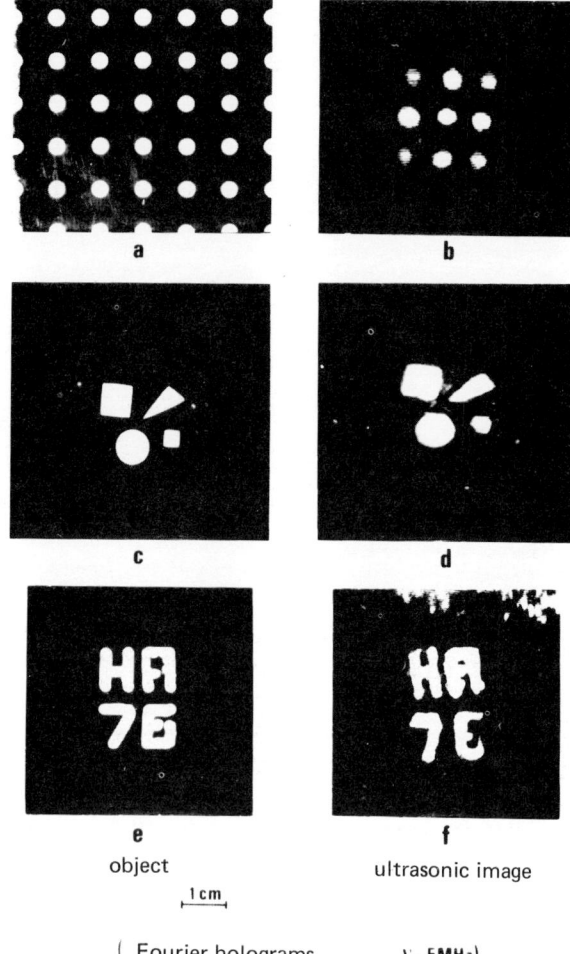

Fig. 6. Photographs (a,c,e) of cutout metallic plates and reconstructed images (b,d,f) in Fourier acoustical holography.

Fourier holography could be used to visualize acoustical fields for the same reason as image holography. The results which we can expect from this technique (figure 6) should be, in principle, comparable to those furnished by image holography in that which concerns the resolution. Nevertheless, the latter is smaller in Fourier holography, due to the restricted dynamics of the liquid-gas interface used as holographic receptor.

object   ultrasonic image

(Fourier holograms ν = 5MHz)

Optical Treatment of Ultrasonic Information

Whatever its limitations, Fourier acoustical holography offers the advantages of providing a transcription of the ultrasonic spectrum of an object in the form of light; it thus permits us to exploit, for metrological puposes, the acoustical data by means of diverse optical techniques.

We have previously cited the case of strioscopy: it is known that such a high-pass filtering, by eliminating the coherent background, facilitates the examination of the details and, in so doing, improves the precision of the dimensional measurements. Moreover, the phase contrast and the interferential contrast permit the observation of inhomogeneous zones not delimited by sudden variations of acoustical impedance. Let us also cite interferometry, which can lead to the measurement of acoustical phase differences, with a precision superior to that of electronic methods. We can finally apply (5,6) more recent techniques of optical treatment of information, for example that of correlation-filtering. The setup used is sketched in figure 7: an optical correlator (L_2-L_3) is placed after the holographic device; it is aligned in the direction of propagation of the wave Σ_{oo} and positioned longitudinally so that the reconstructed image and the image of S_o are formed in its entrance plane. In the first stage, we proceed to the recording, in the plane (F), of the hologram-filter of a particular object t chosen as reference. After development, the filter is replaced in the plane (F), the image of S_o is obscured, and we can proceed to the replacement of t by an object t_i. In the exit plane of the device, the observation of the correlation term contributes a response as to the existence and position of the object t memorized in the filter among the collection of objects t_i analyzed. In the example presented in figure 7, the object t is the number 7. Another case of recognition of form is illustrated in figure 8: the correlation term indicates the position of the object "A" and points out the existence of an object of similar form.

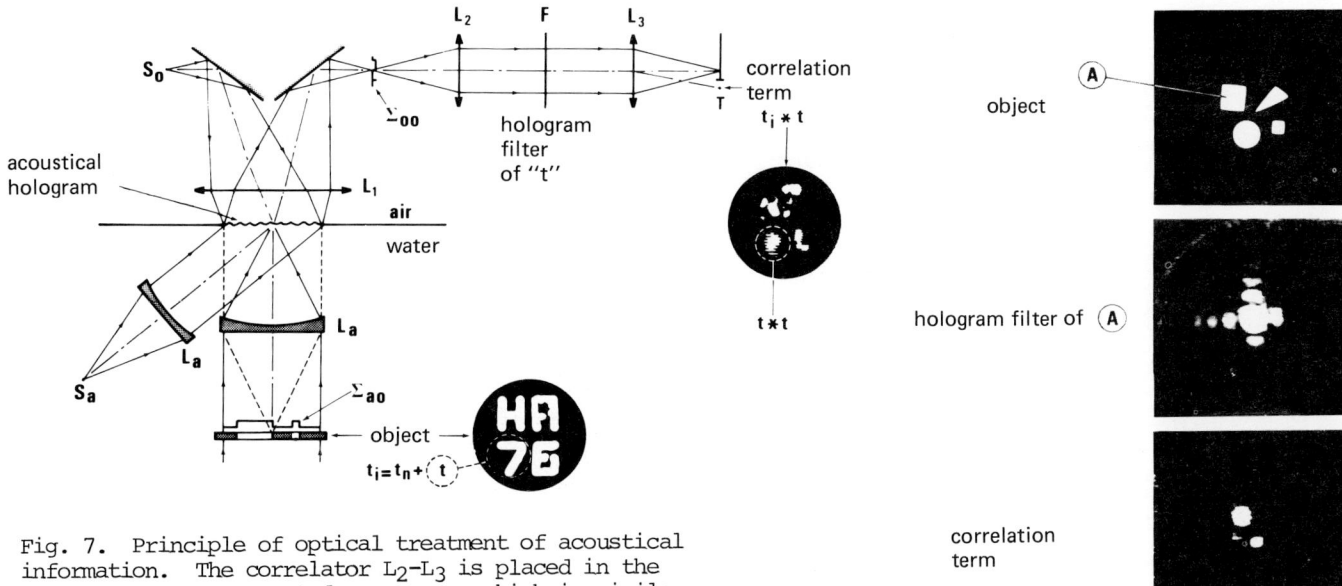

Fig. 7. Principle of optical treatment of acoustical information. The correlator L_2-L_3 is placed in the path of the reconstructed wave Σ_{oo}, which is similar to the ultrasonic wave Σ_{ao}.

Fig. 8. Recongnition of the square A among four geometrical figures.

These two examples of optical treatment of acoustical information represent, from our point of view, only a first step in the proceeding undertaken. They prove the validity of the process, even if the form of the objects is particularly simple (on this subject, we should for that matter remark that a more marked differentiation of the forms could only reinforce the acuity of the response).

Conclusion

The association of optical techniques and ultrasonic holography constitutes a new direction for mechanical metrology. Despite the drawbacks which we have indicated, it offers the major advantage of compensating, by the great precision of the optical techniques, the weak resolution due to the value of the wavelength of the radiation which captures the information.

References

1. Mueller, R.K., and Sheridan, N.K., "Sound Holograms and Optical Reconstruction", Appl. Phys. Letters, vol. 9, pp. 328-329, 1966.
2. Hildebrand, B.P., and Brenden, B.B., An Introduction to Acoustical Holography, Plenum Press, New York, 1972.
3. Ferriere, R., Application de l'holographie et de la diffraction de la lumière par les ondes sonores à la visualisation d'objets placés dans un champ acoustique, Thesis, Univ. of Besançon, 1975.
4. Pille, P., and Hildebrand, B.P., "Rigorous Analysis of the Liquid-Surface Acoustical Holography System", pp. 335-371, in Acoustical Holography, vol. 5, Plenum Press, New York, 1974.
5. Seyzeriat, Y., Holographie acoustique: visualisation en temps réel et traitement par voie optique d'informations ultrasonores, Thesis, Univ. of Besançon, 1977.
6. Pasteur, J., and Seyzeriat, Y., "Holographie acoustique; application au traitement optique de l'information acoustique", Optica Acta, vol. 24, n° 8, pp. 859-875, 1977.

1st EUROPEAN CONGRESS ON OPTICS APPLIED TO METROLOGY

Volume 136

POSTER SESSION

President: Mr. J. P. Christy, Director of the Department for the Study of Characterization of Photosensible Emulsions (SECEP)

Experts: Dr. Gugliemette, Director of the Organic Synthesis Laboratory at the University of Brest

Dr. Huignard, Principal Laboratory for Research of Thomson-CSF

Dr. Robillard, Director of the Firm ISSEC, SA

Mr. Sagaut, SECEP

Mr. Weber, Director of Research for the Firm La Cellophane

The Firms AGFA-Gevaert and Kodak were officially represented by:
Dr. DeWinne, AGFA-Gevaert, Antwerp
Dr. Joly
Mr. Doreau, AGFA-Gevaert, Rueil-Malmaison
Mr. Lecart, Kodak-Pathe, Paris

INVESTIGATION OF CAVITATION BUBBLE DYNAMICS BY HIGH SPEED RUBY LASER AND ARGON ION LASER HOLOCINEMATOGRAPHY

Karl Joachim Ebeling
Universität Göttingen, Drittes Physikalisches Institut
Bürgerstr. 42-22, D-3400 Göttingen, Fed. Rep. Germany

Abstract

For the investigation of the dynamics of tiny, fast moving cavitation bubbles we developed high speed holocinematographical recording techniques. In most experiments we used a multiply Q-switched ruby laser as illuminating light source. The maximum number of holograms which could be recorded in one sequence was limited to eight in this case. More recently, we examined the cavity-dumped argon ion laser for recording longer hologram series. The experimental devices operate in the range of 10 000 to 20 000 holograms per second with ruby laser illumination and at pulse rates up to two kilohertz with argon ion laser illumination. For separating successively recorded information we employed spatial multiplexing techniques in both cases. Reconstructed hologram series taken of acoustically produced cavitation bubbles demonstrate the performance of the experimental arrangements. From the recordings bubble motion can easily be evaluated.

Introduction

High speed photographic investigations of highly dynamic phenomena necessarily fail when the objects to be studied move out of focus during the experiments are under way. In such cases only high speed holographic techniques can point out a way out of this dilemma. As has been already shown [1-5] we have successfully employed high speed holocinematographical techniques to study the dynamics of laser produced and acoustically produced cavitation bubbles. The bubbles can be quite small, they are distributed over a volume much larger than their size, and they are subject to fast motions. From the knowledge of the dynamics of cavitation bubbles we hope to get a deeper understanding of the origins of cavitation noise and cavitation damage.

In our experiments we mostly used a multiply Q-switched ruby laser as illuminating light source. The maximum number of holograms which could be recorded in one sequence was limited to eight in this case. More recently, we examined the multiply cavity-dumped argon ion laser for recording longer hologram series. For separating successively recorded holograms we applied spatial and spatial frequency multiplexing techniques. In this context we confine ourselves to results obtained by the spatial multiplexing technique. Spatial frequency multiplexing techniques are discussed elsewhere[4].

Hologram recording with the multiply Q-switched ruby laser

The holographic set-up is shown in Fig. 1. The geometry is conventional. It corresponds to a common set-up for single exposure holography. The ruby laser is multiply Q-switched and produces up to eight light pulses for hologram exposure at a maximum repetition rate of 20 kHz. Each pulse has a duration of about 30 ns and an energy of about 2 millijoule. All Q-switch pulses go the same path. The collimated reference beam has a diameter of about 3 cm. The object beam diffusely illuminates a water-filled container through a ground glass plate. Cavitation bubbles to be investigated holographically are produced by focussing a giant pulse from a second ruby laser into the water, or they are acoustically generated inside a hollow cylindrical piezoelectric transducer. The cylindrical transducer is indicated in Fig. 1 by the dashed rectangle inside the cuvette. The bubbles are premagnified by means of two lenses. Diameters and focal lengths of the lenses and the distance between the lenses are chosen that an object volume of about 2 cm diameter and 10 cm depth can be recorded. As the key element of the whole device a rotating disk with apertures is placed directly in front of the holographic plate (Agfa 8 E 75) and selects that small portion which is to be exposed by a single pulse of the sequence.

The disk applied in the xperiments is shown in the left hand side of Fig. 2. The disk is black, the openings are displayed as bright rectangular areas. Because of the symmetric arrangement of the openings it is not necessary to synchronize disk rotation and laser emission. The disk rotates so fast that each pulse illuminates just another portion of the photographic plate. This can be seen from the right hand side of Fig. 2 where a developed holographic plate of size 6 by 9 cm is displayed. The plate is exposed by a sequence of seven pulses. The information of one simple hologram is stored in one small rectangular area. The different densities of the small areas indicate that not all pulses of the series had the same energy and cross-sectional intensity distribution. The speed of rotation necessary to avoid overlapping of areas exposed by different pulses depends upon the arrange-

INVESTIGATION OF CAVITATION BUBBLE DYNAMICS BY HIGH SPEED RUBY LASER AND ARGON ION LASER HOLOCINEMATOGRAPHY

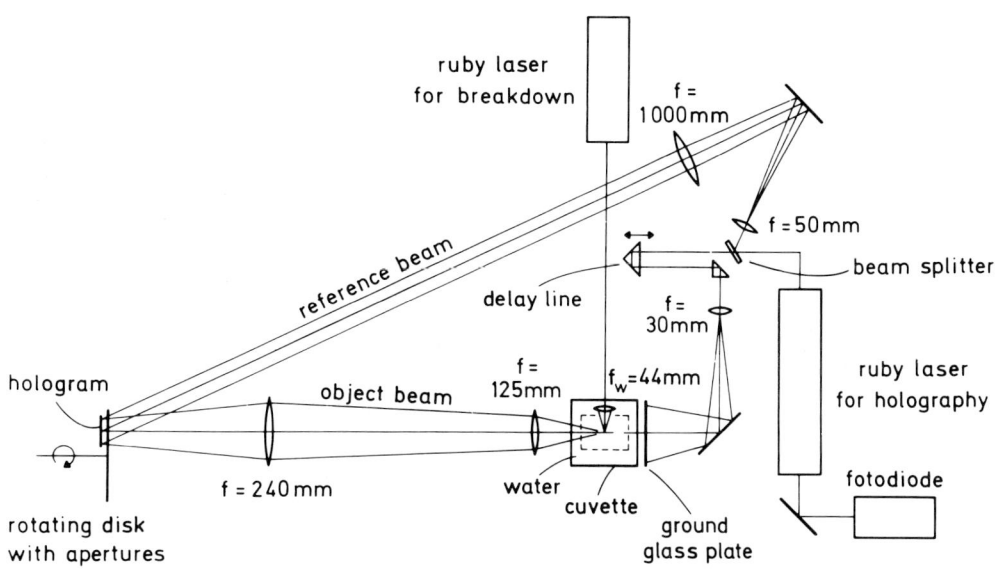

Fig. 1. Set-up for recording a hologram series with the multiply Q-switched ruby laser.

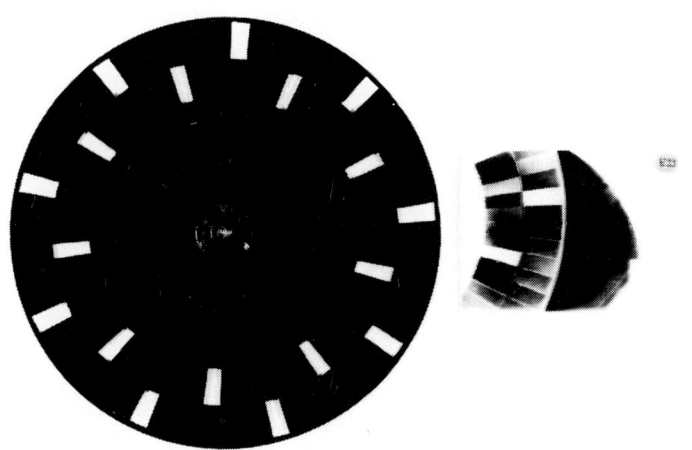

Fig. 2. The disk with apertures and a developed holographic plate (6 x 9 cm).

ment of the openings in the disk and the pulse rate applied. In our case at a pulse repetition rate of 20 kHz the needed speed is about 400 cycles per second. This is furnished by a conventional grinding machine.

Hologram series showing the behaviour of laser produced bubbles are presented elsewhere [1,2,3]. Here, in Fig. 3, we show a typical reconstructed hologram series taken of an acoustically produced cavitation bubble field. The bubbles perform forced oscillations at the 19 kHz resonance frequency of the piezoelectric transducer. The holograms are recorded at times of maximum expansion of the bubbles, the time delay between successive recordings being one or two periods of the driving sound field. Each hologram is reconstructed in another column. To demonstrate the great depth of observation four different planes in depth containing interesting parts of the bubble field are reproduced in the four rows of Fig. 3. The bubbles are displayed as small dark spots in front of a speckling background which is caused by the diffuse illumination. Not all frames have the same quality due to different intensities and coherence properties of the light pulses used for recording the sequence. The cloudy background in some frames could be caused by higher transversal modes in the light pulses. Several bubbles arranged along branches (which often lead to a densely populated centre) are in focus in the two upper rows of Fig. 3. The size of the smaller bubbles, demonstrating the resolution of the method, can be estimated from the size of the frames (about 12 mm) to some ten microns. In the two lower rows of Fig. 3 the interest is focussed onto two heavily oscillating larger bubbles.

The configuration of the branches remains stable for many periods of the driving sound field as earlier investigations have already shown. Single bubbles, however, are in rapid motion driven by the radiation forces in the sound field. In order to evaluate the motion of single bubbles a special reconstruction technique is used. Reconstructed bubble images are viewed by a vidicon camera tube and displayed magnified on a television monitor. When

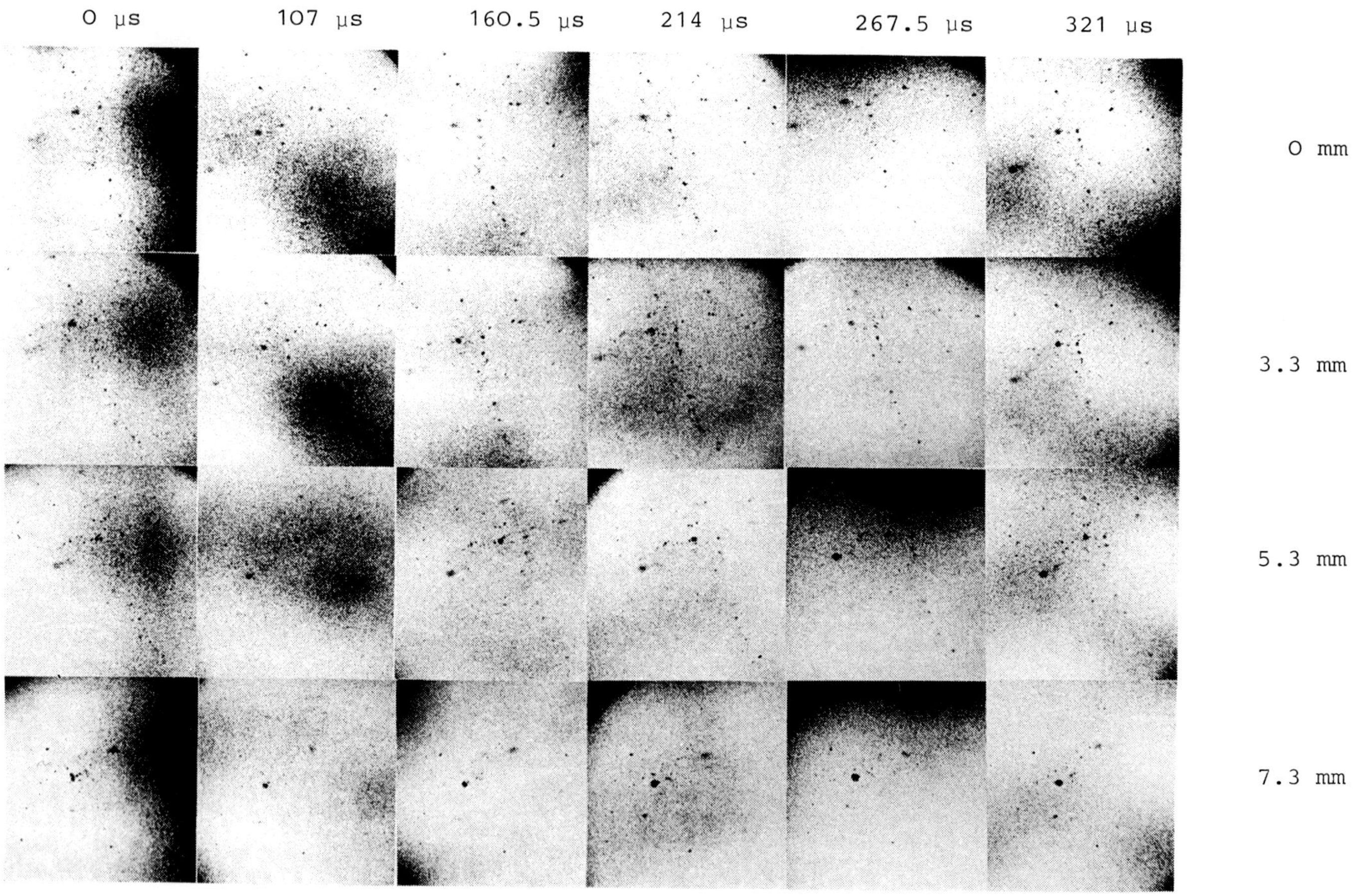

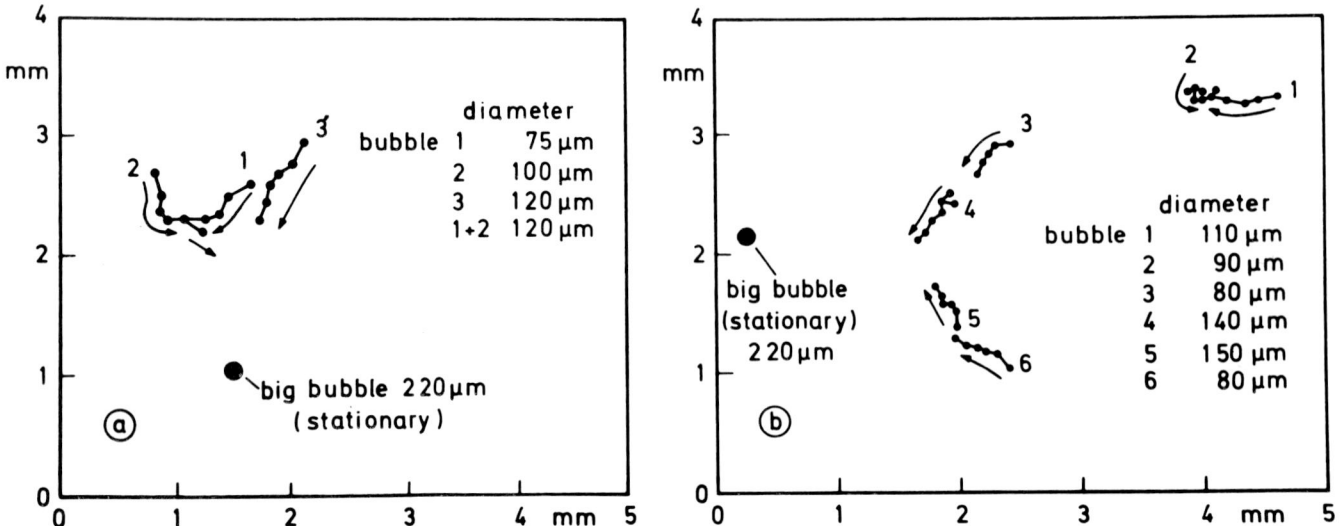

Fig. 3. Reconstruction of a sequence of six holograms showing acoustically generated cavitation bubbles. The columns show different planes at the same time (from the top to the bottom: 0 mm, 3.3 mm, 5.3 mm, 7.3 mm). The rows show the same plane at different times (from the left to the right: 0 µs, 107 µs, 160.5 µs, 214 µs, 267.5 µs, 321 µs). The frame area is 11 mm x 12 mm.

Fig. 4. The paths of single bubbles in the sound field.
a) Coalescence of two bubbles, numbered 1 and 2.
b) Part of Y-shaped path of some bubbles running in the direction of a big central bubble.

the holograms of a sequence are reconstructed shortly one after the other, for instance by rotating the disk with apertures in front of the holographic plate during reconstruction, the motion of distinct bubbles brought into focus can be watched directly. Paths of some larger bubbles taken from the sequence of Fig. 3 are depicted in Fig. 4. Fig. 4a shows the coalescence of two bubbles numbered 1 and 2 on their way in the sound field. The arrows indicate the direction of bubble motion. Fig. 4b shows a section of a typical Y-shaped branch of the field. Bubbles run in the direction of a large central bubble or bubble cluster. Maximum bubble drift velocities of about 2.5 m/sec were measured.

Hologram recording with the multiply cavity-dumped argon ion laser

The holographic set-up for recording hologram series with the multiply cavity-dumped argon ion laser is shown in Fig. 5. In our experiment the argon laser produces about 60 light pulses of 30 ns pulse length at a repetition rate of 1 kHz for recording the hologram series. The pulse energy is about 0.3 µWs at a pumping power which corresponds to a cw-laser power of 1 Watt. In order to prevent a preexposure of the holographic plate by the leakage light of the argon ion laser a mechanical shutter is inserted into the beam. This shutter triggers the emission of the argon laser. The reference and object beams are formed in the usual manner. The cavitation bubbles to be holographed are acoustically generated below a Mason horn oscillating at about 20 kHz. The water filled cuvette into which the Mason horn is inserted is directly illuminated. Because of the relative low energy of the light pulses the portion on the holographic plate (10 E 56) which is exposed by a single pulse is limited to an area of about 3 by 4 mm. For separating successive exposures the holographic plate itself rotates so fast that each pulse illuminates another portion of the plate.

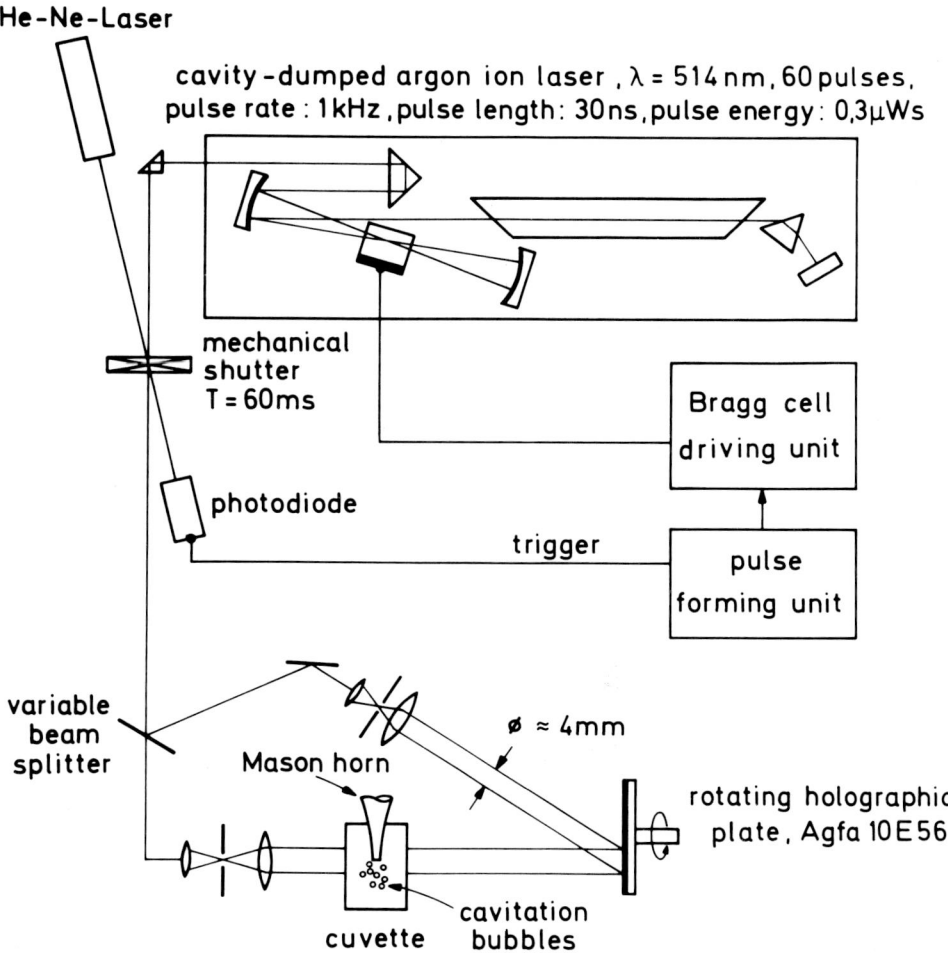

Fig. 5. Set-up for recording a hologram series with the multiply cavity-dumped argon ion laser.

Fig. 6. A developed holographic plate exposed by a sequence of ca. 60 holograms.

A developed holographic plate exposed by a series of about 60 holograms is shown in Fig. 6. The information of each single hologram of the sequence is stored in one of the darkly colored small areas arranged to the large ring by the rotation of the plate. The uniform appearance of the exposed areas is due to the uniform and stable argon laser emission in the lowest transversal mode. It should be mentioned that the pulse rate of 1 kHz requires a rotation speed of only 16 cycles per second. At this low rate the interference pattern on the plate is not blurred during the exposure time of 30 ns. At much higher rates, however, blurring will occur and in addition the photographic emulsion will be deformed by the centrifugal forces.

Fifteen successively recorded holograms out of a whole series of about 60 holograms are reconstructed in Fig. 7. The series shows the development of a cavitation bubble field below the Mason horn in time steps of one millisecond. The bubbles are imaged as dark spots in front of a bright background. In all frames it is focussed on the same plane in depth. One finds great changes in the arrangement of the bubbles from one hologram to the next. Periods in which many small bubbles are present in the field of view follow periods of rather rare bubble density. This is a typical behaviour found in many other series. Single bubbles cannot be identified in successive frames. This is due to the long time interval between successive holograms which corresponds to 20 periods of the oscillating Mason horn. In some frames, for example in the second and in the twelfth, somewhat blurred dark circles appear. As can be concluded from other investigations this is a question of shock waves emitted by collapsing bubbles.

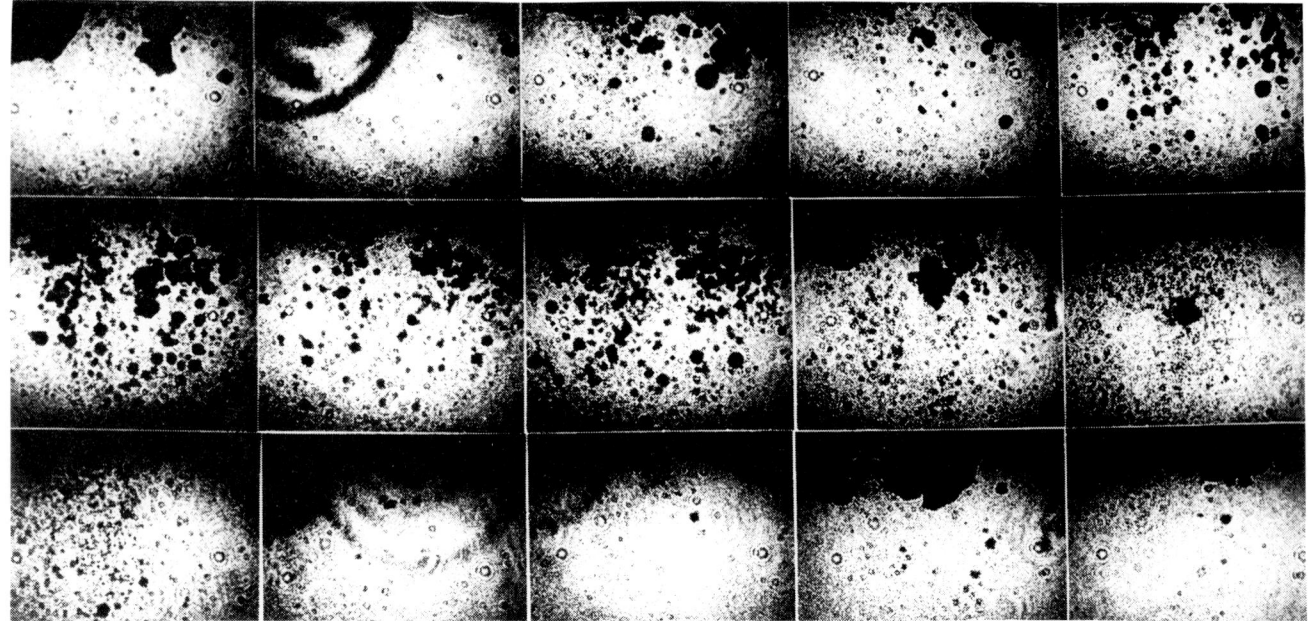

Fig. 7. Reconstruction of a series of 15 holograms showing the cavitation bubble field below an oscillating Mason horn. All frames show the same plane in depth. The time interval between successive frames is 1 ms. The sequence runs from the left to the right and from the top to the bottom. The frame area is 3.0 mm x 2.8 mm.

Finally, in Fig. 8 we have reconstructed six different planes in depth of the eighth frame of Fig. 7. Now, the effects of defocussing the bubbles can be studied. The reconstructed depths lie 400 µm apart. When the image of a small bubble is considered in different frames one can easily determine the location in depth. Bubbles of 10 µm diameter are clearly resolved. For these small bubbles one determined the location in depth to better than 400 µm.

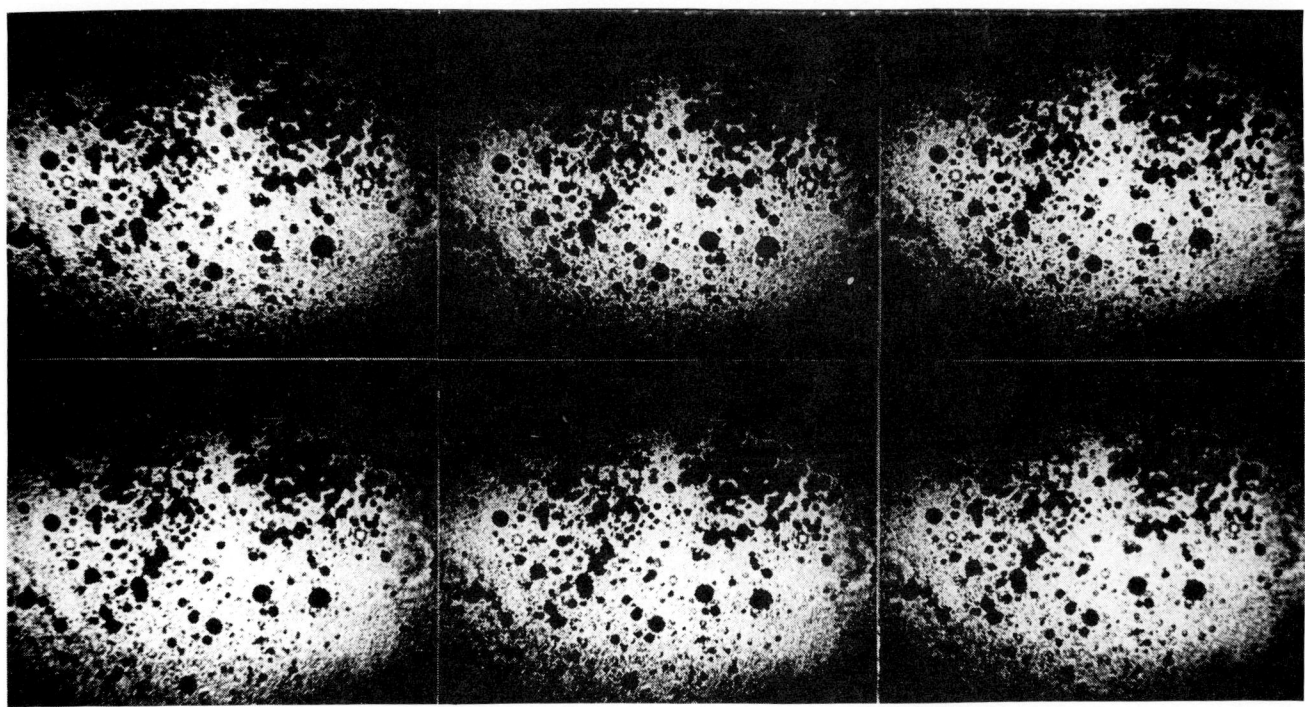

Fig. 8. Different planes in depth of the eighth hologram of the series reconstructed in Fig. 7. Successive planes lie 400 µm apart. The sequence goes from the left to the right and from the top to the bottom.

Comparison of the two methods

The experiments show that hologram series of high quality can be recorded with argon ion laser illumination as well as with ruby laser illumination. Each system has its specific drawbacks and advantages.

With the ruby laser a high pulse energy can be obtained. Thus, relative large scenes and large hologram areas can be illuminated and various (energy consuming) multiplexing techniques can be applied [1,4]. On the other hand, problems arise because of the difficulties in producing long, stable, and coherent pulse series. In our system pulse series of maximal eight pulses at a maximum repetition rate of 20 kHz could be obtained reproducibly by multiply Q-switching the ruby laser during the same pumping period. The duration of the pumping period gives an upper limit for the length of the pulse series.

With the cavity-dumped argon laser transversal coherent pulse series with repetition rates from 1 MHz to 0 Hz can easily be produced. The number of pulses is unlimited. Unfortunately, the pulse energy is relative low (< 1 µWs). With such pulses only small holograms of some square millimeters area taken of transmissive or highly reflective small objects can be recorded on most sensitive highly resolvable photographic material like the Agfa 10 E 56 plate. Furthermore, only light energy saving multiplexing techniques can be applied. However, as demonstrated by the experiments the multiply cavity-dumped argon ion laser seems to be an ideal light source for recording hologram series of small objects at repetition rates up to some kilohertz.

Acknowledgement

The work was supported by the Deutsche Forschungsgemeinschaft. I thank Dr. W. Lauterborn and Dr. K. Hinsch for many discussions on the subject.

References

1. Ebeling, K. J., Lauterborn, W., Optics Communications 21 (1977) 67.
2. Ebeling, K. J., Optik 58 (1977) 383 and 481.
3. Lauterborn, W., Ebeling, K. J., Appl. Phys. Letters 31 (1977) 663.
4. Ebeling, K. J., Lauterborn, W., Appl. Optics, in press.
5. Ebeling, K. J., Acustica, in press.

MINUTES OF THE ROUND TABLE ON PHOTO-SENSITIVE SURFACES

J. P. Christy, J. Sagaut and J. L. Tribillon

President: Mr. Christy, Director of the Department for the Study of Characterization
of Photosensible Emulsions (SECEP)

Experts: Dr. Guglielmetti, Director of the Organic Synthesis Laboratory at the
University of Brest

Dr. Huignard, Principal Laboratory for Research of Thomson-CSF

Dr. Robillard, Director of the Firm ISSEC, SA

Mr. Sagaut, SECEP

Mr. Weber, Director of Research for the Firm La Cellophane

The Firms AGFA-Gevaert and Kodak were officially represented by:
Dr. De Winne, AGFA-Gevaert, Antwerp
Dr. Joly
Mr. Doreau, AGFA-Gevaert, Rueil-Malmaison
Mr. Lecart, Kodak-Pathe, Paris

The round table discussion, composed of two parts, had approximately sixty participants.
The first part, to which the majority of time was devoted, dealt with non-silver materials presently under research and development or having recently appeared on the market.
The second part was concerned with problems of choice, use and supply of silver emulsions. Due to the interest shown by the participants, the discussion of most of these products was primarily oriented around the use of sensitive surfaces in holography.
Consequently certain materials, such as photochromes and europium oxide films, which have different uses, were in fact abandonned during the open discussion.

NON-SILVER MATERIALS

OPENING LECTURES

The complete text of these lectures will be published elsewhere. Five materials were presented:
- Europium oxide films, by Dr. SELEZNEZ (U.S.S.R)
- Photopolymers, by Dr. JEUDY (ISSEC)
- Photochromes, by Dr. GUGLIELMETTI
- Photothermoplastics, by Dr. INEICHEN (BRAUN-BOVERI)
- Electrooptical crystals, by Dr. HUIGNARD

DISCUSSIONS

After taking into account the major concerns of the participants, three materials were the subject of in depth discussion:
- Photopolymers
- Photothermoplastics
- Electrooptical crystals of the BSO type

Photopolymers. It is apparent that important progress has been made in the last two years in the preservation of these products before and after recording. These materials offer the following advantages:
- They are self-processing.
- Amplification can be obtained after recording by exposure to a light beam which can be the reconstruction beam.
- They have a high diffraction efficiency, which nevertheless diminishes in terms of time of reading. Typically, it changes from 80% at the start of the reading to approximately 20% after eight hours for a beam of $.5$ mW/mm^2.
- They have a resolution better than 3000 lines/mm.
- They can be coated on large surfaces and by great thicknesses.

- They can be sensitezed to mosts wavelengths of the visible spectrum.

- Photothermoplastics. In these materials, information is recorded in the form of variations of thickness of the thermoplastic. One therefore directly obtains a relief hologram, which can later be duplicated by embossing techniques. The principal points made were:
- The diffraction efficiency is good, but depends heavily on exposure.
- The reproducibility of the intrinsic characteristics of this type of material from one sample to to the next has been questioned; however, Dr. INECHEN declared that this problem has been resolved for the products used by his firm.
- The spatial bandwidth of the thermoplastic increases when the thickness diminishes, but at the cost of lower diffraction efficiency. This problem of bandwidth was discussed at great length, with the conclusion that the required bandwidth depends strongly on the application envisioned.

- Electrooptical crystals of the type BSO ($Bi_{12}SiO_{20}$). In this type of material, information is recorded as variations of the index of refraction in the bulk of a crystal. BSO crystals exhibit:
- Extremely high resolution when the recording is made in the transverse configuration.
- Excellent sensitivity, around 100 $\mu J/cm^2$ in the blue or green region of the spectrum. This sensitivity is comparable to that of the 649-F plate, and is superior by many order of magnitudes to the usual sensitivity of non-silver materials.
- Almost perfect symmetry of the write-erase cycle, which makes them unsuitable for multiple hologram storage, but well adapted for quality control at the end of a line of products by holographic interferometry, for example.
- Absolutely no fatigue (contrary to the thermoplastics).
They are currently available in sizes of approximately $1cm^2$.

CONCLUSIONS

Thermoplastics are now commercialized. Certain photopolymers are beginning to appear on the market, while others are still undergoing research, development and experimentation. Electrooptical crystals are still under development, and some of the most promising among them are available only in small sizes. Nevertheless, one can foresee the availability of BSO crystals in the neighborhood of $10cm^2$ in a close future.

These three materials were judged to be of promise and meet some of the users needs in holography, mostly because of the absence of wet processing. However, it was stated that certain problems are not entirely resolved, and need still a good deal of work.

- Sensitivity for the whole of these materials remains inferior to that of the silver processes. But thanks to the progresses made in this area, one can imagine that their other characteristics will enable them to be introduced on the market.
- Much research is underway in order to provide those materials with better amplification processes. To illustrate his work, Dr. ROBILLARD presented a process based on variations of the dielectric properties of an organic semi-conductor under the action of light. The amplification is supplied by a high frequency electric field used for development.

SILVER EMULSIONS

DISCUSSION

The questions addressed to Agfa-Gevaert and Kodak concerned holographic plates. They dealt mainly with:
- The possibility of coating special products, very thick emulsions, for example.
- The minimum order possible.
- The maximum size of the plates or film.
- The supply delays.

The answers were:
AGFA-GEVAERT:
- Supply delays are in the area of two to three months.
- Sizes can be obtained on demand, theoretically there are no limitations.
- Orders are accepted even for small quantities.

KODAK:
- Supply delays are about the same, if the laboratory is already qualified to receive products classified as "strategic". This qualification is easy to obtain.
- The sizes of the films or plates standard and cannot exceed 60 x 70 cm (approx.).
- For standard products the minimum order is very small.

These two companies stressed the importance of consulting them for all special problems so that they may examine with the customer the best way to solve the problem. To this end they told how to contact the competent persons in case of need, in France and in Europe.

MINUTES OF THE ROUND TABLE ON PHOTO-SENSITIVE SURFACES

Questions of a more general nature were posed on the manufacturers's interest in the holographic market. While each of the representatives of the two firms confirmed his firm's interest in this market, neither was willing to supply information on the size, relative or absolute, for the business concern which he represented.

CONCLUSIONS

While the disproportion between the small needs of the users of sensitive surfaces for holography, and the production capability of the industrial manufacturing groups creates certain problems, the two companies present asserted their willingness to examine each particular case to best resolve the difficulty.

The round table ended with a brief survey of the new high resolution products which have recently appeared on the market.

PRINCIPLE OF THE HOLOGRAPHIC CINEMATOGRAPHY

Victor G. Komar
Cine & Photo Research Institute (NIKFI)
Moscow, USSR

Abstract

This paper describes the principles of holographic cinematography developed in NIKFI. The following ideas are underlying the principles of the holographic cinematography: 1) using lenses of large aperture, about 200 mm, for shooting and projection; 2) using a point-focusing multiplying holographic screen for projection; 3) using a holographic film with a thick emulsion layer of the order of 10 micrometres; 4) using two methods of shooting. In the first method a pulse coherent light is used for indoor shooting. In the second method a common non-coherent light is used for outdoor shooting. In the latter one a lenticular plate is put in the camera. A space colour image is stored on a colour film stock. This image is converted into a hologram film during printing.

Introduction

For the past ten years some proposals concerning holographic cinematography have been suggested in several countries. We have evaluated these proposals at NIKFI (1). We found that these earlier proposals could not be realized for theatrical cinematography. There were no solutions of the hologram motion picture for a large auditorium.

Principles of holographic cinematography with a colour, three dimensional and large picture, viewed by a large audience, were developed by the author in 1974 (2). They have been verified experimentally in NIKFI in 1976 (3).

Shooting of the hologram film

Fig.1 illustrates the scheme of holographic shooting with pulse lasers of three (or four) wavelengths in the blue, green and red parts of the spectrum.

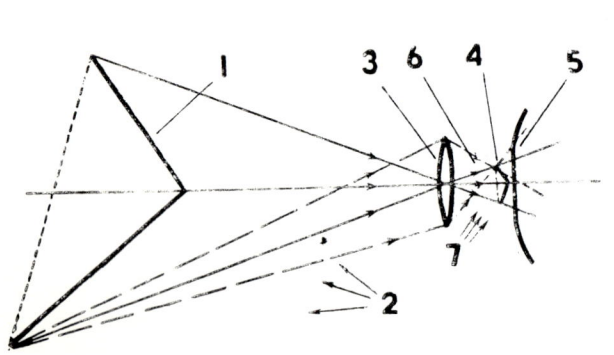

Fig. 1. Hologram film shooting using coherent light.

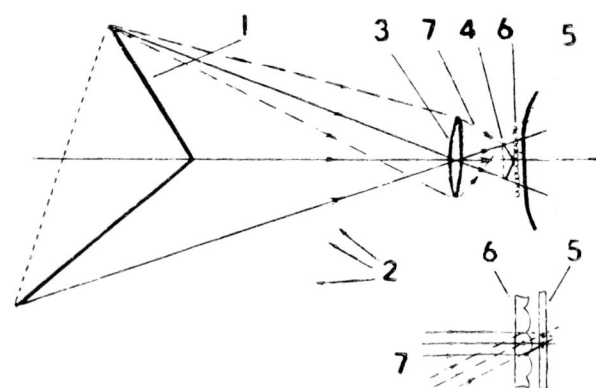

Fig. 2. Raster film shooting in conventional non-coherent light.

Object (I) is illuminated by a coherent light (2). The light reflected from the object (I) passes through a lens (3), having a large aperture, and forms a reduced space image (4). This image (4) is stored on a holographic film stock (5). Beside an object beam (6), the film stock (5) is illuminated by a reference beam (7) from the same laser as the light beam (2). The film motion in the camera is intermittent. A multitude of sinusoidal hologram gratings is recorded on the holographic film (5). Each elementary hologram grating contains information on the brightness of the small object surface element for single wavelength.

Several pulse lasers with an impulse energy from 2 to 10 joules, coherent length from 10 to 20 metres, frequency from 16 to 24 pulses per second are needed to shoot scenes of 10-20 metres. We can achieve the optimum wavelengths providing the best colour rendition, for instance, about 440 nm (blue), 510 nm (green), 570 nm (yellow), 640 nm (red).

An important problem is the creation of film stock for colour pulse laser shooting. Such hologram films must have a sensitivity, corresponding to an operation exposure from 2 to 4 micro-joules per square centimetre. They must also have a high diffraction efficiency and a low level of noise. For the holographic cinematography it is necessary to develop shooting lenses having large aperture 200 mm and focal length of about 150 mm. The lenses must have long enough distances between the lens and the film for the introduction of the reference beam.

Fig.2 shows the scheme of raster shooting in conventional non-coherent light. An object (I) is illuminated by non-coherent light (2). The light reflected from the object (I) passes through the lens (3), having a large aperture and forms a reduced space image (4). The space image (4) is stored on a colour film stock (5). The film (5) is located in the focal planes of small spherical lenses of the raster (6), also called lenticular plates. Therefore, the elements of the image determined by the rays (7), which have different directions, will be stored in various spots under the small spherical lens of the raster. Each direction of rays (7) corresponds to a certain perspectives of a space image.

Printing of holographic film

The primary film frames, shot by the different methods mentioned earlier, can be printed as holograms for a single holographic film. Fig.3 illustrates the scheme of holographic printing by the simple non-optical method. The primary hologram film (I) is located

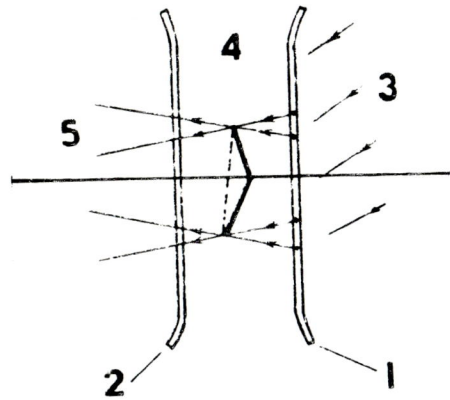

Fig. 3. Holographic film printing by the non-optical method.

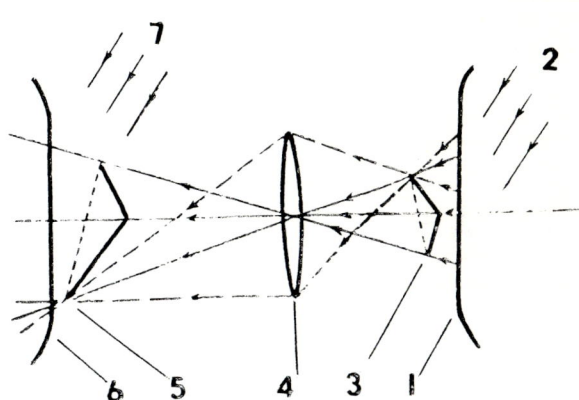

Fig. 4. Holographic film printing by the optical method.

near the second holographic film stock (2). A single reference beam (3) of coherent light reproduces a space image (4). This space image (4) is recorded on the second holographic film (2) with the same reference beam (3). During exposure both films are immovable, if a continuous laser is used.

Fig.4 shows the scheme of holographic printing by the optical method. The primary holographic film (I) is illuminated by coherent light (2). A reproduced space image (3) is transformed, by the intermediate optics (4), into a secondary image (5). The secondary space image (5) is recorded on the secondary film stock (6) with a reference beam (7) from the common laser.

During printing both by the optical and non-optical method, it is possible to correct the colour image by changing the relative intensity of the reference beams having different wavelengths. Moreover, it is possible to slightly shift the space image relative to the film. It may be necessary for improving the configuration of the camera, printer or projector.

A space image shot through the lenticular plate can be printed on a holographic film. Both the optical and the non-optical methods can be in this case used for printing. Fig.5 shows the scheme of optical printing of the raster image into the hologram. The primary film (I) with a stored space raster image is illuminated by coherent light (2) through an opal glass (3). The coherent light passes through a lenticular plate (4) and forms a space image (5). This image (5) is transformed by optics (6) into a space image (7), which is recorded on a holographic film stock (8) with a reference beam (9) of the common laser.

A raster camera and a raster printer shall have a high registration of about 5 micrometers. This value refers to the postion of film with regard to the raster.

The achievement of holographic film stock for holographic printing is an important problem. This film must possess sensitivity in the blue, green and red parts of the spectrum.

It is supposed necessary to have a sensitivity, i.e. an operation exposure of 300-600 micro-joules per square centimetre, a resolution power of 10000 lines per millimetre, a diffraction efficiency of 60-70%, small light scattering.

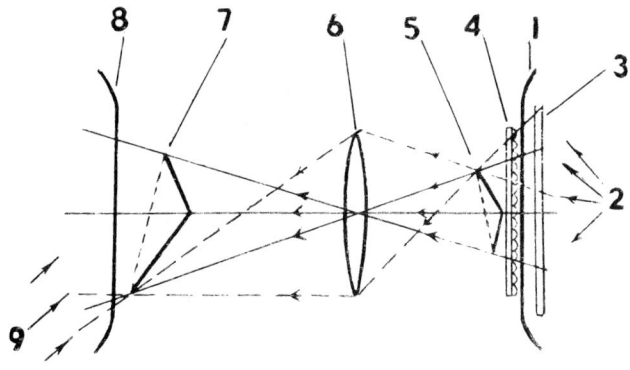
Fig. 5. Optical film printing of the raster image into the hologram.

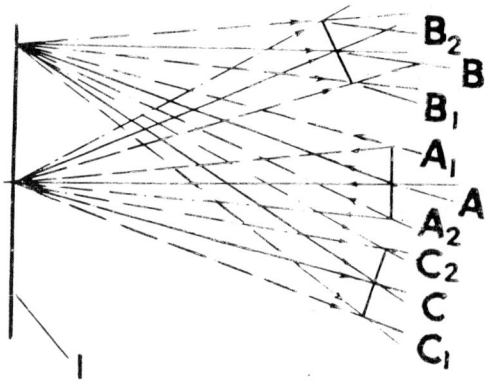
Fig. 6. Holographic screen.

Projection of holographic films

An important part of the holographic projection installation is a holographic screen. It is an optical device having point focusing and multiplying properties.

Fig.6 shows that a holographic screen (I) has a primary focusing centre (A) and several secondary focusing centres (B, C, ...). If a point light source is in the primary centre (A) of the screen, the reflected light is focused in the secondary centres (B, C, ...). If a point light source is displaced from the primary centre (A) to a closely located point (A_1), the reflected light is collected in points (B_1, C_1, ...) which are located near the corresponding secondary centres (B, C, ...). Therefore, a light beam falling to the screen (I) through an aperture (A_1A_2), containing a primary centre (A), forms several zones of reflected light (B_1B_2, C_1C_2, ...) containing secondary centres (B, C, ...).

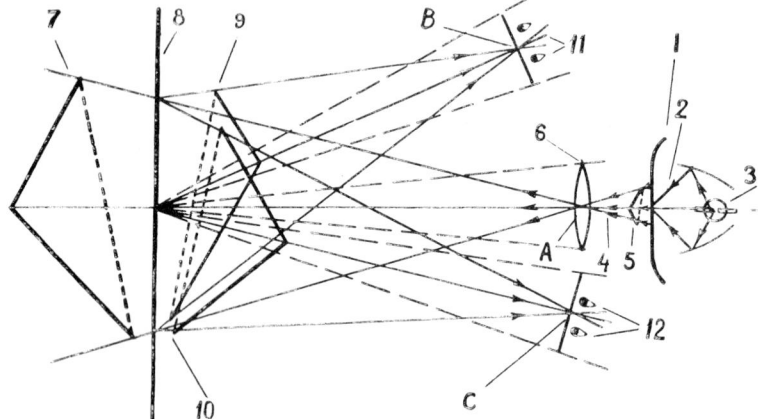
Fig. 7. Projection of hologram film.

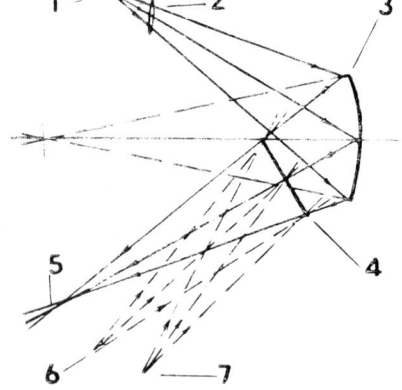
Fig. 8. Scheme of making the holographic screen.

Fig.7 illustrates a scheme which can be used to project a holographic film shot and printed by the methods described above. A holographic film (I) moves intermittently through a projector gate. In the projector gate a still hologram film frame is illuminated by a non-coherent light beam (2) from a light source (3) having a line spectrum. An object beam (4) forms a small space image (5) near the film (I). A projector lens (6) forms a large space image (7) which is virtual behind the screen (8) and real in front of it. Light beams reflected by the screen (8) form several images (9, 10, ...). Each viewing zone, for instance (II), has a single image (9). The perspectives of a picture (9) will be different for the two eyes of the spectator (II). Therefore, the spectator perceives this picture in three dimensions. A slight movement of the spectator's eyes (11) in the lateral direction causes changes in picture perspective like in case with real objects.

Our analysis of the projection hologram process shows that it is possible to reproduce the original sizes of the shot objects both in lateral and axial directions for the whole scene, but it will be exact only for the spectator sitting in the designed chair, for instance, in the centre of the auditorium. Spectators occupying other seats will see somewhat distorted pictures. But this distortion is not essential (4).

The statement has been made that it is impossible to project a space hologram image because the axial dimensions are changed as a square of the lateral ones. But it is possible to eliminate the distortion of this kind using proper optical decreasing during shooting. To get a proper ratio between the lateral and the axial dimensions of the space image, it is necessary that the magnification of the overall process beginning with the original object and finishing with the picture being viewed equals a unity. It is evident that deliberate disturbance of this condition may become a means of increasing the emotional effect.

Fig.8 illustrates the scheme of making holographic screens. A diverging beam passes from a coherent light source (I) through a compensating optical device (2) and falls upon a concave mirror (3) of large dimensions. A converging beam, reflected by the mirror (3), passes through a photoplate (4). This beam converges in a single point (5) which will be the primary centre of the holographic screen. Simultaneously, several diverging beams (6, 7) of the same coherent light source (I) pass through a photoplate (4) from the opposite side. The centres of the beams (6, 7) will be secondary focusing centres of the screen.

To provide the projection of colour pictures, it is necessary to have a thick emulsion layer. The photoplate (4) must be exposed with coherent light having different wavelengths in the blue, green and red parts of the spectrum.

Fig.9 illustrates a lamp housing scheme of a holographic movie projector. A mercury-cadmium lamp (I) having a short distance between the electrodes and line spectrum is used here.

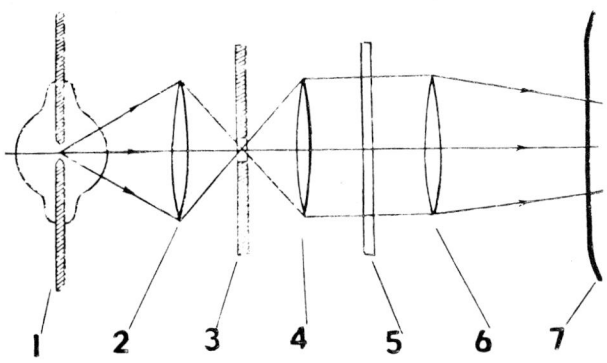

Fig. 9. Lamp housing for the hologram film projector.

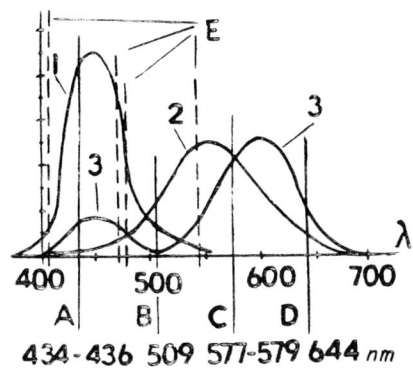

Fig. 10. Wavelengths of a mercury-cadmium lamp.

Rays from the lamp (I) pass through a lens (2), a diaphragm (3), collimator (4), an interference filter (5), a focusing lens (6) and a hologram film (7) at the projector gate. The reference beam falling on a hologram film (7) is homocentrical which is necessary for eliminating the angular aberration.

Fig.10 shows wavelengths of a mercury-cadmium lamp. In our example 4 parcels of wavelengths are used (A, B, C, D) and others (E) are eliminated by interference filters. Fig. 10 also shows the perception colour characteristics of an eye: blue (I), green (2) and red (3).

A light source and a lamp housing used in a conventional projector are not suitable for holographic movie projection. This is caused by chromatic and angular hologram aberrations (Figures 11, 12).

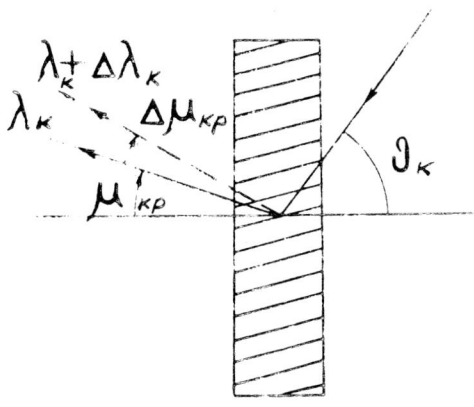

Fig. 11. Object ray direction versus wavelength (transmission hologram).

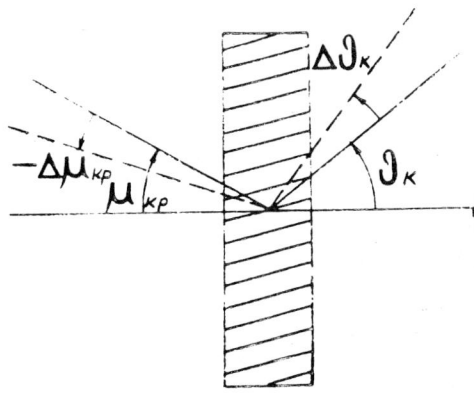

Fig. 12. Object ray direction versus the angle of reference ray incidence (transmission hologram).

Using non-coherent light for a hologram film production provides small graininess and speckle noise levels of the picture viewed by the observers. This picture may have high definition and sharpness in the entire space depth. This seems paradoxical as lenses, mirrors and the screen have large relative apertures in the overall holographic movie process. This paradox can be explained by the following. The image is viewed by a spectator in narrow beams limited by the eye pupil. With such narrow beams the space image is transmitted through the overall hologram movie process including the holograms and optics.

It is unnecessary that the film frame steadiness in the holographic projector gate is higher than that in conventional movie projectors. Such a possibility is ensured by using quasi-focused holograms in the secondary film, that is in a hologram film print. For the same reason, using gas discharge lamps instead of lasers for hologram film projection will not decrease the sharpness of the viewed pictures. But it is possible to achieve a higher projection light flux using laser instead of a discharge lamp. It is important for a large auditorium.

An opinion was described that it is impossible to achieve high quality of multicolour hologram image due to the interaction of the reproduction reference beam of one wavelength with gratings recorded with another wavelength.

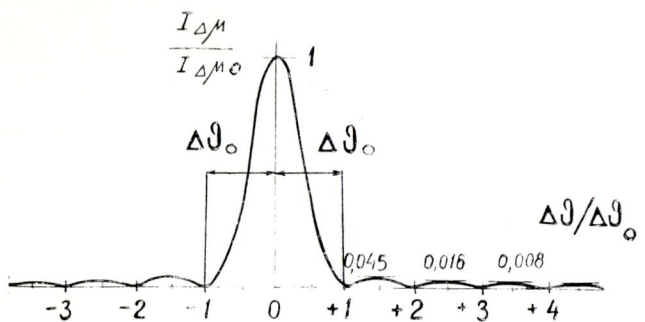

Fig. 13. Diffracted beam intensity versus the angle of reference ray incidence.

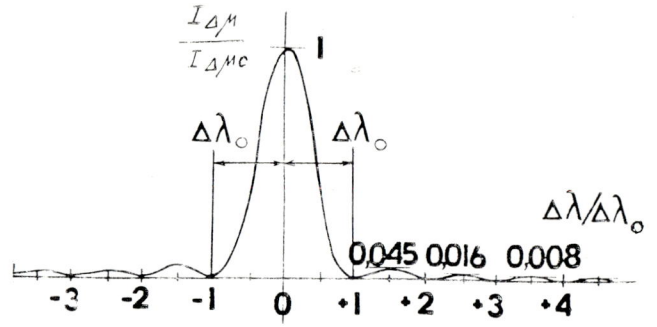

Fig. 14. Diffracted beam intensity versus the deviation of wavelength.

One can overcome this difficulty using thick holograms. It is possible to use several angles of incident reference rays of different wavelength during multicolour image reproduction. In this case we use the angular selectivity of the thick hologram (Fig.13). It is possible to use spectral selectivity of the thick reflection hologram during the multicolour image projection on the holographic screen (Fig.14).

Fig.15 illustrates the chromaticity coordinates (X, Y) of 12 typical movie objects. These are: white colour (I), snow (2), human face (3), paved road (4), yellow sand (5), blue sky (6), lake in fine weather (7), green grass (8), wheat (9), yellow dandelion (10), red poppy (II), blue corn-flower (12). Fig.15 shows a contour which has 4 angles corresponding to 4 wavelengths of a mercury-cadmium lamp: 435, 509, 578, 644 nanometres. All 12 coordinates of colour objects are inside the contour of lamp radiation.

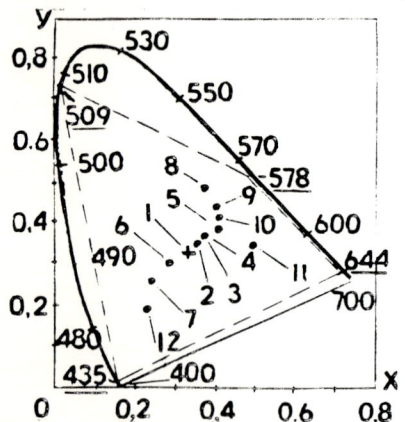

Fig. 15. Chromaticity coordinates of 12 typical movie objects.

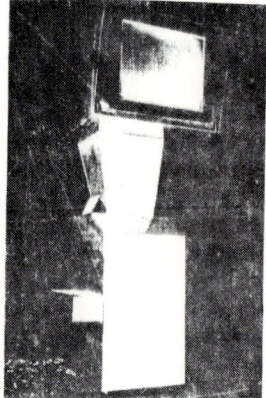

Fig. 16. Installation for the projection of small hologram still pictures on a holographic screen.

In the described process a multicolour image is transmitted by means of a line spectrum having only four parcels of wavelengths. It causes colour distortion. The analysis, conducted in NIKFI, shows that this colour distortion can be small and not exceed those encountered in the best modern motion picture process (5).

Compatibility with 3D television

The above principles of holographic cinematography correspond to those of 3D TV broadcasting. The most simple methods of 3D TV broadcasting are based on the following: two motion picture images corresponding to two angles of shooting are transmitted through television channels to two cathode-ray tubes, then the images are optically combined and viewed by spectators through lenticular plate screen or polarizing glasses. The complicated methods of 3D TV can provide (e.g. in cable television) image transmission at several perspectives in separate channels and wider viewing zone without glasses.

Hologram image quality

Theoretical investigation of the main optical properties of holograms were made by the author (6). The theoretical relations were verified using the experimental results obtained together with G.Sobolev and O.Serov.

As a result of our theoretical investigation we have obtained quantitative relations for the general values determining the quality hologram motion image: contrast, colour, sharpness, graininess, double contours. It has been shown that the holographic movie process, shooting - printing - projection of a multicolour hologram image, is nearly a linear process, but it is necessary to have the following conditions: 1) a reference beam intensity divided by an object beam intensity during recording is high enough; 2) an active thickness of hologram is large enough; 3) a reference beam source has a line spectrum during recording; the intervals between the spectrum lines are long enough; 4) an interval between the wavelengths, according to each other, during recording and reproduction of hologram is short enough; 5) difference between the angles of incident reference rays of different wavelength during recording is large enough; 6) reference beams are nearly homocentrical; 7) difference between the angles of incident reference rays during recording and reproduction are small enough.

For the above mentioned conditions, which is real in practical cases, the process of recording-reproduction of hologram is linear. It means the following: 1) brightness of each small element of object surface at the single wavelength is rendered independently of each other during recording and reproduction; 2) intensity of each object beam of small object surface element at the single wavelength during reconstruction is approximately proportional to the intensity corresponding to the object beam during recording.

This linear characteristic of the holographic process shows that it is possible to achieve in the ideal hologram recording-reproduction process the following results:
1) contrast of the hologram image can be the same as the contrast of the original object;
2) colour of the holographic image can be the same as the colour of the original object.

Our theory of the general optical properties of the hologram reproducing a colour space image shows that one can eliminate decreasing of a hologram diffraction efficiency due to the increase of the number of wavelengths. It is possible to eliminate decreasing of a diffraction efficiency due to the increase of the number of exposures. It can be obtained by an enlargement of emulsion thickness in phasor hologram of course in defined limits.

Since the quality of a hologram picture is very high, the problem of the holographic cinematography is so important. A well-made hologram picture is perceived as a real object, because this picture can have all the visual properties of the object. It is supposed that the main feature of the holographic cinematography will be a picture undistinguishable from the original objects. It is to be thought that holographic cinematography will have wider expressive possibilities than conventional motion pictures.

It will be not the replacement of the artistic means used but the increase in the artistic possibilities. Like silent scenes are specifically used in a sound film and black-and-white scenes are a particular case in a colour movie, it is supposed that a flat picture will be used as a special artistic device in holographic cinematography.

Experimental verification of the principles of holographic cinematography

The principles of shooting, printing and projection of hologram films described here have been verified experimentally in NIKFI (G.Sobolev, O.Serov, E.Suchman, I.Fedchuk, S.Papojan, L.Akimakina, I.Nalimov). The first step was the projection of a small hologram still picture on a holographic screen (Fig.16).

A short holographic 70 mm film loop have been shot with a pulse frequency ruby laser (Fig.17). This film loop has been projected (Fig.18) on the holographic screen having the size of 0,6 x 0,8 (Fig.19). The film loop runs two minutes. Four spectators could look at the screen simultaneously. The space picture was monochromatic (Fig.20).

The well known English specialist, Bernard Happe, he was in NIKFI on October 7, 1976, wrote: "... delegates who saw the actual results were left with the sense of having been present at an historical occasion, comparable with the classic demonstrations of past pioneers in film and television and having equally vast and perhaps unrecognised potentialities" (3).

Fig. 17. Installation for shooting 70 mm hologram film.

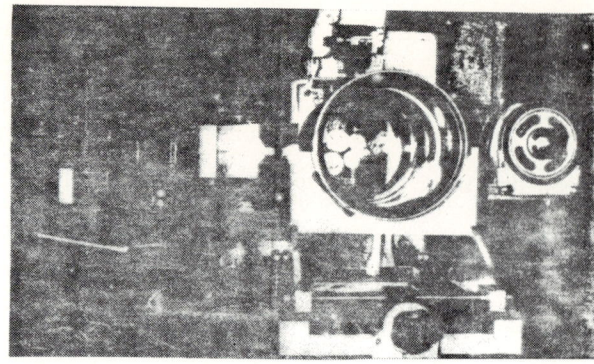

Fig. 18. Installation for 70 mm hologram film projection.

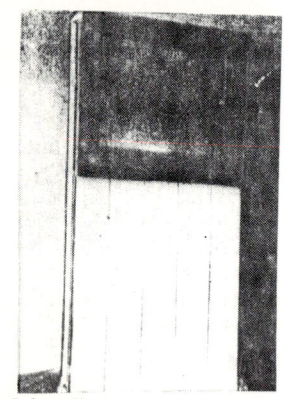

Fig. 19. Hologram screen for hologram film projection.

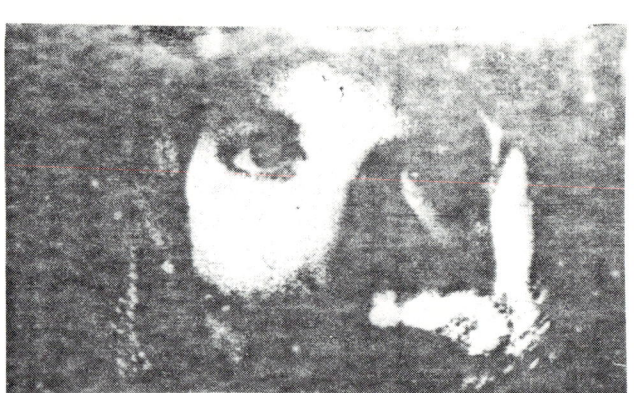

Fig. 20. Hologram movie space image, first projected on the holographic screen.

Application of holographic cinematography

Experimental verification of the holographic cinematography principles, developed in NIKFI, has shown that it is possible just now to construct the technical facilities of the hologram movie system with a small screen. This system, intended for shooting and projecting holographic films and slides with space monochromatic images on screens having the size of about I square metre can be used in practice as follows: Information (exhibition, advertisment), education, medicine and a means of a scientific research.

For the achievement of the hologram movie systems for the same purposes, having multicolour space image, it is necessary to carry out scientific researches to achieve colour hologram film stock, screen and equipment. It will demand several years.

It is difficult to solve the problem of the hologram movie system intended for shooting large scenes and for presentation of colour space pictures on a large screen for great auditorium having hundreds of spectators. For practical solution of this problem it is necessary to carry out experimental researches during several years. But the solution of this problem is available on the base of the just developed principles of holographic cinematography.

References

1. Komar V.G., About General Schemes of 3D Holographic Cinematography, Trudy NIKFI, N 78, 1975, p.131.
2. Komar V.G., About the Possibility of the Creation of Theatrical Holographic Cinematography Reproducing Space Colour Image, Technika Kino i Televidenia, 1975, NN 4, 5, pp.31, 34.
3. Happe B., The Most Extensive UNIATEC Paper Programme, The BKSTS Journal, 59, N 1, Jan.1977; Report on the 12-th UNIATEC Congress, SMPTE Journal, 86, N 3, March 1977, p.762.
4. Komar V.G., About the Projection Principles of Colour Space Hologram Motion Pictures, Trudy NIKFI, N 82, 1976, p.5.
5. Komar V.G., Ovechkis J.N., About Colour Reproduction of the Holographic Image, Technika Kino i Televidenia, N 9, 1976, p.18.
6. Komar V.G., About General Optical Properties of the Thick Holograms Reproducing Space Colour Images, Trudy NIKFl, N 82, 1976, p.60.

HOLOGRAPHIC ART WITH RECORDING IN THREE-DIMENSIONAL MEDIA ON THE BASIS OF LIPPMAN PHOTOGRAPHIC PLATES

Yu. N. Denisyuk

Optical Institute of Leningrad
USSR

The hologram technique initially proposed for application to a specific technical problem[1] in recent years has found increasingly wider use in the development of a new kind of art. The strange holographic pictures creating the full illusion of objects depicted are exhibited more and more frequently at the various exhibitions, museums, private collections and so on. The earliest idea of the possibility to utilize holography in this way was promoted in ref. (2) in connection with so-called holographic technique with recording in three-dimensional media. This idea was confirmed by the experiments carried out with the improved technology of Lippman colour photography [3-5].

The first hologram reconstructing three-dimensional image of a natural object was obtained in 1964 using the holographic method with recording in two-dimensional media [6]. This experiment is known to have been a turning point in the development of holography. The circumstances of publishing the experimental results in question have been indicative definitely enough of the fact that of all applications of holography the artistic holography is most interesting for the public in large.

Basically there has been little change in the technique of obtaining artistic holograms as compared to the earlier works [3,4,6]. In a number of cases such holograms are recorded in the passing beams, thus using the technique of registration in two dimensional media. In a number of cases such holograms are recorded in the passing beams, thus using the technique of registration in two dimensional media. In this case, the reconstruction is accomplished with the help of monochromatic laser or mercury arc lamp radiation.

The recording in the opposite beams, i.e. the holography in three-dimensional media is widely employed for the purposes of artistic holography too. The main feature of the three dimensional hologram is the ability for separating a definite monochromatic component from the reconstructing radiation. Accordingly such a hologram can be reconstructed with an ordinary incandescent lamp. This property is rather opportune for practical purposes.

The recording and the chemical treatment of three-dimensional holograms is mainly carried out with the help of the photographic plates and developers which were worked out by Lippman and other investigators of that period (e.g. see [7]). However, both the plates and the methods of their chemical treatment have undergone considerable changes which resulted in essential improvement of the quality of the image reconstructed. We shall discuss below the results of the development of artistic holography of this type in more detail.

As the whole bulk of photographic material does take part in the reconstruction of a three-dimensional hologram, the quality of the reconstructed image in this case especially depends upon the properties of the photographic plates used. The main directions of improving Lippman plates from the moment they first appeared till today have been : the search of methods to suppress the growth of haloid silver grains at the stage of obtaining emulsion and the search of hypersensitizators increasing the sensitivity of these photographic plates to the maximum.

Especially important are the methods to reduce the size of haloid silver crystals.

In the emulsion used for obtaining the first three-dimensional holograms the growth of emulsion grains was slowed down due to the excess concentration of KBr [4]. In this way, the emulsion containing haloid silver crystals of the order of 300 Å in size has been obtained. The initial sensitivity of such emulsion was very small. It was substantially raised after the hypersensitization in a solution of triethanolamine (TEA). This method of hypersensitization of Lippman photographic plates proved to be highly effective and is being widely used now. Diffraction efficiency of the holograms on such photographic plates came to about 5 %.

The next step to improve the emulsion in question was the development of so-called " Valenta " layer [8]. In this case due to the introduction of instantaneous emulsification at a low temperature and other measures the size of silver bromide grains has been reduced already at the emulsification stage. Reduction of grain size as well as higher concentration of silver in the emulsion allowed to increase the diffractional efficiency of holograms to the value of 20-24 % quite acceptable for practical purposes. In plate treatment a special pyrogallic developer was employed [9]. The energy required for recording a hologram was of the order of $20-30 \cdot 10^{-3}$ j/cm^2 [10].

Using the experience of obtaining holograms on " Valenta " layers, R.R. Protas, the chief author of the initial emulsion mentioned above [4], succeeded in creating highly efficient and technologically effective Lippman photographic plates LOI-2 [11].

In this particular case the growth of haloid silver crystals was further slowed down due to the increasing number of growth centers and also due to introduction of special inhibitors of growth. Chemical treatment of LOI-2 photographic plates is accomplished in the GP-2 developer, worked out specially for such plates[12].

The sensitivity of these plates on condition of pretreatment in TEA comes to $0,4.10^{-3}$ j/cm². Diffractional efficiency of three-dimensional reflection holograms, recorded on LOI-2 photographic plates reaches quite a high value of 50-55 %. Such plates have been manufactured in factory conditions for quite a long time.

The best available home-made though rather complicated in technology Lippman photographic plates are PE-1 photographic plates worked out by N.I. Kirillov [13,14]. In this case the growth of haloid silver crystals is hampered due to the fact that in emulsification process a highly diluted solution is used, and the concentration of emulsion is achieved by the method of gradual freezing-out and thawing-out. This method led to emulsion containing haloid silver crystals, not exceeding 100 Å in size [15]. The treatment of PE-I photographic plates is accomplished in a special method phenidone hydroquinoue developer [16,17]. The sensitivity of PE-I photographic plates amounts to about $0,05.10^{-3}$ j/cm² on condition that the plates are pre-treated in TEA [17]. Without such pre-treatment the sensitivity becomes about 10 times as low and thus approaches the sensitivity of LOI-2 photographic plates treated with TEA. Diffraction efficiency of the PE-I photographic plates exceeds 50 % [18]. Until recently the PE-I photographic plates have been manufactured in small-lot experimental production.

One of the most important prospects of the artistic holography is the transition from recording holograms on photographic plates to recording on the film. Obviously this will be fully realized only after the possibility is found to record high quality holograms in the light of pulsing lasers so far as in this case the rigid fixing of the photographic layer will be no longer necessary. At the present time three-dimensional holograms can be recorded on the special photographic film FP-GT [19]. This film is about twice as sensitive as LOI-2 photographic plates, but has somewhat lower diffraction efficiency of the order of 25 %. It is interesting to note that in case the hologram is being recorded in opposite beams the triacetate base is only suitable, because lavsan possesses a strong double-beam refraction. One should also take into account that if a gas discharge laser is to be employed for hologram recording, the film, while in storage, must not be rolled up in a roll to eliminate its deformation during exposure. It is desirable that the film be cut at once into separate sheets and packed like photographic plates.

Apart from the above mentioned photographic materials which have been quite widely employed for recording three-dimensional holograms or have been manufactured at an industrial scale, two experimental layers are worth mentioning, namely, the silver iodide-bromide emulsion somewhat resembling LOI-2 [20], and the sublimation layer representing a further development of the technique of synthetizing emulsions from diluted solutions [21]. The last emulsion is rather interesting due to the fact that the author concentrated it by way of sublimation, i.e. by evaporating water directly from ice, and is likely to have obtained haloid silver crystals of the smallest size. It is peculiar that in this case diffractional efficiency of holograms has not practically increased, and the sensitivity of photographic plates has considerably decreased. This experiment, however, was entirely of preliminary character. It would be premature to draw a final conclusion on the unsuitability of further decrease in the size of haloid silver grains.

Passing over to the methods of increasing the sensitivity of Lippman photographic plates it should be noted that for practical purposes apart from triethanolamine treatment [4] only hypersensitization by heating was proposed [22-23]. This method leads to nearly as great increase in sensitivity as does the hypersensitization in TEA, the shelf life of photographic plates being considerably longer. It is to be noted that the hypersensitization methods mentioned get sensitivity reserves from one and the same source and, therefore, cannot supplement each other.

Some scantiness of practical achievements in the sphere of increasing the sensitivity of photographic plates was compensated by the scientific success. In a number of papers some details of the nature of the reserves of sensitivity of Lippman photographic plates were made evident enough. The most interesting results were obtained by the so-called holographic method of latent image investigation [24-27]. This original method amounts to recording the interference picture of two plane waves upon a photographic plate without further development, the light scattered by this picture being registered simultaneously on this recording. In this particular case the light is being scattered by the latent image centers formed during exposure. Measuring the scattered light intensity one can investigate the behaviour of latent image in time. The results of the investigations by such method have shown that a strong regression of sensitivity is characteristic of superfinely dispersed photographic emulsions. During a period of about tens of seconds the intensity of the latent image contrives to decrease many times. The presence of such anomalous regression is also confirmed by the experiments aimed at reducing the time between the exposure and development [28]. It turned out that in this manner the light-sensitivity of Kodak 649 F photographic plates can be increased 25 times. Most likely, the same fact is indirectly confirmed by a considerable decrease in light-sensitivity of the emulsion obtained by sublimation method [21]. In all probability, when haloid silver grains are reduced in size below a certain limit (judging by PE-I emulsions this limit lies in the region of 100 Å) the duration of their memory rapidly reduces.

We now turn out attention to the problems of interaction of the light and a developed hologram. At present one may take it as a proven fact that after a strongly exposed Lippman photographic plate is developed a colloid silver forms in an emulsions layer. That means that the billions of minute compact grains, well reflecting the light are appearing in the volume of the layers. With lower exposures silver exists in the same form as on the ordinary photographic plates (29). To obtain a hologram of maximum diffraction efficiency it is to be exposed up to the exposures corresponding to the beginning of the colloid silver region (30). Most likely, this is due to the fact that with shorter exposures silver reflects light badly, and with longer exposures the structure of layers printed into the hologram is washed away.

The modulation mechanism of radiation from a three-dimensional hologram registered on a Lippman photographic plate is conveniently interpreted from the standpoint that the presence of silver particles in the emulsion layer does change the effective refraction index of the layer (31). These notions have been introduced as a result of an experimental work, in the course of which the anomalous values of the refractive index of the developed Lippman plate structure were actually discovered (32).

The estimation of the value of the light flux reflected by three-dimensional holograms of this type is complicated due to the fact that thanks to the harmonic nature of standing wave distribution in depth one and the same hologram has layers characterized by different exposure. In accordance with that, both small and large particles of colloid silver are represented in the volume of a hologram. There are two diametrically opposite viewpoints on the problem of what particular particles reflect maximum amount of light. There are arguments for the view that the main role in the formation of the image is played by the light, scattered on large-sized particles (33,34). There are arguments for the opposite viewpoint as well (32).

As far as the radiation sources are concerned which are used in the registration of the three-dimensional artistic holograms it should be noted that at present the most suitable for the purpose are helium-neon lasers with cold cathode. Argon and krypton lasers, though possess a considerably higher power, are less handy as they require water cooling. Among other things such a cooling is sometimes a source of vibrations due to water pulsations in the water supply and feed pipes.

Pulsed lasers are, no doubt, the most perspective radiation sources for registration of artistic holograms. While employing such lasers there is no necessity to have bulky vibration-proof installations, registration of any objects becomes possible including portraits, photographic films being employed as a photographic material. However, pulsed lasers are not sufficiently stable as yet, and, besides that some complications have risen here in providing the suitable photographic materials. It appeared, for example, that registration of holograms with diffraction efficiency exceeding 10 % is impossible if normal photographic plates and a mono-pulsed laser are employed. At the same time, a longer pulse in free generation mode makes it possible to achieve 30 % efficiency in hologram recording (35).

At the present time no special studies concerning the optimum choice of reconstructing source are available. It is known from the practice that for this purpose are used either mercury-arc lamps or incandescent lamps whose luminescence body is enlarged by means of optics up to the size sufficient to keep the reconstructing beam divergence within 30°. The sources of this type are relatively inexpensive and easy to handle, but their radiation is not sufficiently directive. Since this does cause washing away of the remote details of the image reconstructed, one should expect that in the near future the artistic holography will be restricted to the scenes of no more than tens of centimeters in depth. For the majority of art pieces and portraits this distance is sufficient.

The problem of colour reproduction is one of the most vital for the whole domain in question. In spite of a number of successful studies (e.g. see (36)) the solution of the problem is still long to come to. To some extent this drawback of present-day artistic holograms is compensated by the fact that individual monochromatic holograms, strange as it may be, reproduce the colour of objects to a certain extent (37). Most likely this striking effect is the consequence of exact reproduction of the object's microstructure and the experience -gained ability to correlate a certain microstructure with a certain colour.

The process of making artistic holograms is already placed to the category of art to a large extent. In this case very much depends on the choice of the object and the lighting. For example, the first and the simplest artistic three-dimensional holograms of a coin were obtained on the very same installation which was employed for the initial experiments and the substantiation of this method (38). At the remote time of the first experiments on 3-D holography it simply did not occur to the author of the paper mentioned (2,3) to make use of a relief coin as an object. The first large-sized artistic three-dimensional hologram of a complicated object was obtained in NIKFI by G.A. Sobolev in 1967. The studies carried out by those and other groups of investigators have found their reflection mainly as holograms presented at various exhibitions. The number of publications concerning this theme is rather limited (e.g. see (38-41)).

References

1. Gabor D., Proc. Roy. Soc. (London), A197, 454 (1949).
2. Denisyuk Yu.N., DAN (Academy of Sciences Reports), USSR, 144, n° 6, 1275 (1962).
3. Denisyuk Yu.N., Optica i Spectr. (Optics and Spectroscopy), XV, 522 (1963).
4. Denisyuk Yu.N., Protas R.R., Optica i Spectr. Optics and Spectroscopy, XIV, 721 (1963).
5. Lippman G., C.R. Acad. Sci., 112, 274 (1891).
6. Leith E.N., Upatnieks I., J. Opt. Soc. Am. 54, 1295 (1964).
7. Valenta E., Die Photographie in natürlichen Farben. Halle (1912).
8. Zagorskaya Z.A., Optico-mech. Promishlennost, Optical-mechanical Industry, n° 2, 72 (1973).
9. Andreyeva O.V., Sukhanov V.I., Optica i Spectr. XXX, 786 (1971).
10. Sukhanov V.I., Andreyeva O.V., Optico-mech. Promishlen. n° 3, 63 (1972).
11. Protas R.R., Krakau Yu.A., Mikhailova V.I. in " Recording Media for Holography " (Reguistriruyushchiye sredi dlya golografii) L. "Nauka", p. 41 (1975).
12. Usanov Yu.E. in " Recording Media for Holography ", L., "Nauka", p. 98 (1975).
13. Kirillov N.I., Vasilyeva N.V., Zelikman V.L., Journal of Scientific and Applied Photography and Cinematography (Zhurnal nauchnoi i prikladnoi fotografii i kinematografii) 15, n° , 441 (1970).
14. Kirillov N.I., Vasilyeva N.V., Zelikman V.L., Achievements of Scientific Photography (Uspekhi nauchnoi fotografii), 16, 204 (1972).
15. Kirillov N.I., Vasilyeva N.V., Senchenkov E.P., Feldsherov E.M., in " Recording Media for Holography ", p. 54 Z " Nauka " (1975).
16. Kirillov N.I., in " Recording Media for Holography ", p. 5, Z. " Nauka " (1975).
17. Kirillov N.I., Maslenkova N.G., Vasilyeva N.V., Petrenko A.S., Feldsherov E.M., Gulanyan E.Kh. in " the Problem of Holography " (Problema Golografii) iss. III, p. 112 Moscow (1973).
18. Sobolev G.A., Chursin V.N., Serov O.B. in Proceedings of the II-d All-Union Conference on Holography (Materiali II Vsesoyuznoi Konferentsii po golografii) v. I, p. 38 Kiev (1975).
19. Andreeva O.V., Borin A.V., Gafurova N.S., Mikheeva V.P., Prosalova N.A., Sukhanov V.I., in Proceedings of Seminar : Optical Holography and Its Application in Industry (Materiali Seminara : Opticheskaya golografia i yeyo primeneniye v promishlennosti) p. 35, Leningrad (1976).
20. Yaroslavskaya N.M. in " Recording Media for Holography " p. 48, Z. " Nauka " (1975).
21. Zagorskaya Z.A. in " Recording Media for Holography " p. 59, L. " Nauka " (1975).
22. Yaroskavskaya N.N., Andreyeva O.V., Sukhanov V.I. in " Recording Media for Holography " p. 106, L. " Nauka " (1975).
23. Yaroslavskaya N.N., Andreyeva O.V., Sukhanow V.I., Optico-mech. Promishl., n° 9, 55 (1975).
24. Sobolev G.A., Kirillov N.I., Proceedings of the Congress on Photographic Science (Trudi Kongressa po fotograficheskoi nauke) A23, Moscow (1970).
25. Girina M.G., Sobolev G.A., Rabinovitch Ts.M., Problems of Holography (Problemi Golografii) iss. III, 151, Moscow (1973).
26. Girina M.G., Sobolev G.A., Opt. and Spectrosc., 32, 216 (1972).
27. Sobolev G.A., Girina M.G. in " Recording Media for Holography ", p. 88, L. " Nauka " (1975).
28. Larionov N.P., Lukin A.V., Mustafin K.S. " Problems of Holography " iss. III, p. 90 Moscow (1973).
29. Andreyeva O.V., Veidenbakh V.A., Levina P.I., Sobol R.M., Sukhanov V.I., Journal of Scientific and Applied Photography and Cinematography, n° 1, p. 46 (1975).
30. Andreyeva O.V., Sukhanov V.I., Optics and Spectros.(Optica i Spectr.), XXX, p. 786 (1971).
31. Lord Rayleigh Phil. Mag., 47, 377 (1899).
32. Usanov Yu.E. in " Recording Media for Holography ", p. 98, L. " Nauka " (1975).
33. Kovachev M., Siinov V., Matveeva I., Quantum Electronics, 3, n° 11, 2399 (1976).
34. Andreyeva O.V., Sukhanov V.I., Veidenbakh V.A., Levina P.I., Sobol K.M., Proceedings of the II-d All-Union Conference on Holography. v.I, p. 76, Kiev (1975).
35. Varzobova N.D., Staselko D.I., in the Proceedings of Seminar : Optical Holography and its Application in Industry, p. 115, Leningrad (1976).
36. Zaitseva V.P., Tolchin V.G., Turukhano B.G., Proceedings of the II-d All-Union Conference on Holography, p. 39, Kiev (1975).
37. Sobolev G.A., Chursin V.N., Serov O.B., Proceedings of the II-d All-Union Conference on Holography, p. 38, Kiev (1975).
38. Denisyuk Yu.N., Sukhanov V.I., Opt. and Spectrum, XXV, 308 (1968).
39. Denisyuk Yu.N., Zemtsova E.G., Proceedings of the Conference on Implementation of Modern Optics; Achievements into the Practical Work of Museums (Materiali konferensii po vnedreniya v praktiku raboti muzeyev dostizhenii sovremennoi optiki). " Sov. Rossiya " pub. house., p. 31, Moscow (1975).
40. Papoyan S.M., Sobolev G.A., Technika kino i Televidehya (Cine and TV Technique), n° 8, p. 11 (1974).
41. Vanin V.A., Nazarova L.G., Problems of Holography, iss.I, p. 53, Moscow (1973).

AUTHOR INDEX

ARCHBOLD, E., Speckle Photography for Strain Measurement—A Critical Assessment, 258

ATKINSON, J. T., Holographic Interferometry Applied to Minimal Wear Measurement, 107

BLANDIN, M., Study by Holographic Interferometry of Dimensional Variability in Precision-Moulding Materials Used in Odontology, 130

BOCQUEMO, GERARD, Hologrammeteric Plotter (Apparatus for Three-Dimensional Measurement on an Image Reconstructed by Holographic Process), 76

BOUCHAREINE, P., Comparison of Wavelengths to the Primary Standard at the French National Institute of Technology, 38

BRUTTI, CARLO, Optical Differentiation of Moire—Holographic Fringes by Wavefront Reconstruction with White Light Sources, 296

CADORET, GAEL, Applications of Holography to the Study of Structures and Materials, 114

CAUSSIGNAC, J. M., Application of Holographic Interferometry to the Study of Structural Deformations in Civil Engineering, 136

CESARIO, DANIEL, Testing by Holographic Interferometry of Solid Propergol Engines, 181

CHRISTY, J. P., Minutes of the Round Table on Photo-Sensitive Surfaces, 355

CLIFTON, MICHAEL, Study by Holographic Interferometry of Mass Transfer During Electrochemical Processes at Solid-Liquid Interfaces, 143

CONSTANS, ALAIN, Optical Analyser of Vibrations, 29

DANDLIKER, RENE, Heterodyne Holographic Interferometry: A Review, 215

DEBRUS, S., Improvement of the Signal/Noise Ratio in the Subtraction of Images by Speckle Interferometry, 237

DELERY, JEAN, Holographic Interferometry Applied to the Metrology of Gaseous Flows, 192

DENISYUK, YU. N., Holographic Art with Recording in Three-Dimensional Media on the Basis of Lippman Photographic Plates, 365

DI CHIRICO, GIUSEPPE, Optical Differentiation of Moire—Holographic Fringes by Wavefront Reconstruction with White Light Sources, 296

DRAIN, L. E., Displacement and Vibration Measurement by Laser Interferometry, 52

DUBAS, M., Holographic Interferometry with the Possibility of Modifying the Fringes During Reconstruction, 174

DUBOURG, J. D., Application of Holographic Interferometry to Testing of Spun Structures, 186

DUROU, CHRISTIAN, Study by Holographic Interferometry of Dimensional Variability in Precision-Moulding Materials Used in Odontology, 130

———, Study by Holographic Interferometry of Mass Transfer During Electrochemical Processes at Solid-Liquid Interfaces, 143

DZIALOWSKI, YVES, Detection of Axial Displacements of a Diffusing Object, 245

EBBENI, JEAN P. L., Application in Civil Engineering of a Sandwich Speckle Method, 291

———, The Diffracting Gauge in Extensometry, 251

EBELING, KARL JOACHIM, Investigation of Cavitation Bubble Dynamics by High Speed Ruby Laser and Argon Ion Laser Holocinematography, 348

ELISSALDE, J. P., MODEX: Opto-Numerical Measurer of External Dimensions, 32

ENNOS, A. E., Speckle Photography for Strain Measurement—A Critical Assessment, 258

FELSKE, ARMIN, Holographic Analysis of Oscillations in Squealing Disk Brakes, 148

FLEURET, J., Analysis of the Diffraction Spectrum of a Population of Particles, 277

FONTAINE, J., Holographic Interferometry in Osteosynthesis, 202

FORTUNATO, G., Application of Interferential Correction of Spectra to the Detection of Pollutants in the Atmosphere, 14

GAGNAIRE, ALAIN, Interferometrical Setup for the Study of Thermic Turbulence in a Plane Airstream, 69

GARCIA, M., Determining the Inclination of a Diffusing Surface with Regard to Viewing Direction by Speckle Photography, 218

GOEDGEBUER, JEAN-PIERRE, Space-Time Optics in Shape and Surface Metrology, 3

GROSMANN, M., Dimensional Metrology of Length Standards by Holographic Interferometry with Phase Heterodynage, 92

GROUSSON, R., Study of the Distribution of Velocities in a Fluid by Speckle Photography, 266

HAPPE, ALFONS, Deformation Measurements on Connecting Rods by Speckle Photography, 270

HERRIAU, J. P., New Possibilities of Real-Time Interferometry with Photoconductive Electro-Optic Crystals $Bi_{12}Sio_{20}$, 226

HERNANDEZ, R., Determining the Inclination of a Diffusing Surface with Regard to Viewing Direction by Speckle Photography, 218

HOGMOEN, KARE, Holographic Methods Made Useful by Phase Modulated ESPI, 222

HUARD, ALINE, Measurement of Small Rotations, 19

HUIGNARD, J. P., New Possibilities of Real-Time Interferometry with Photoconductive Electro-Optic Crystals $Bi_{12}Sio_{20}$, 226

IMBERT, CHRISTIAN, Measurement of Small Rotations, 19

JANEST, A., Comparison of Wavelengths to the Primary Standard at the French National Institute of Metrology, 38

JEUDY, M., Autoprocessor Materials for the Recording of Phase Holograms: Photopolymers and Organo-Metallic Semiconductors, 229

JOLLY, NICOLE, Double Exposure Holographic Interferometry: Application to Nondestructive Testing and to Breaking Point Mechanics, 101

JOYEUX, D., Measurement in Real Time of Transversal Micro-Vibrations (Down to 1 Å) on Diffusing Objects: Random Moire Captors, 333

KOMAR, VICTOR G., Principle of the Holographic Cinematography, 358

KOWALIK, WALDEMAR, Interferometric Measurement of Heterogeneities in Semiconductors, 8

KRIENS, R.F.C., Some Considerations on the Quantitative Interpretation of Holographic Interferograms, 156.

LACHARME, JEAN-PAUL, Holographic Interferometry Applied to the Metrology of Gaseous Flows, 192

LACOURT, ALAIN, Image Spectrograms of Three-Dimensional Objects: Metrological Applications, 65

LALOR, M. J., Holographic Interferometry Applied to Minimal Wear Measurement, 107

LAMARE, MICHEL, Interferometer for Testing Infrared Materials and Optical Systems, 43

LANZL, FRANZ, Video-Electronic Analysis of Holographic Interferograms, 166

LAROCHE, DOMINIQUE, Hologrammetric Plotter (Apparatus for Three-Dimensional Measurement on an Image Reconstructed by Holographic Process), 76

LOKBERG, OLE J., Holographic Methods Made Useful by Phase Modulated ESPI, 222

MAITRE, H., Analysis of the Diffraction Spectrum of a Population of Particles, 277

MALLICK, S., Study of the Distribution of Velocities in a Fluid by Speckle Photography, 266

MARECHAL, A., Application of Interferential Correlation of Spectra to the Detection of Pollutants in the Atmosphere, 14

MAY, MARIE, Detection of Axial Displacements of a Diffusing Object, 245

MAYSTRE, DANIEL, Determination of the Index Profile of a Dielectric Plate by Optical Methods, 26

McKELVIE, J., The Measurement of Residual Stress by a Moire Fringe Method, 302

MENU, MICHEL, Decorrelation Produced in the Image of a Diffusing Object by a Rotation of the Incident Wave—Application to the Study of Surface States, 282

MERCIER, R., Holographic Testing of Aspherical Surfaces, 208

MEYRUEIS, P., Holographic Interferometry in Osteosynthesis, 202

———, Dimensional Metrology of Length Standards by Holographic Interferometry with Phase Heterodynage, 92

MONNERET, J., Photoelasticimetry and Holographic Interferometry: Applications to the Study of Stresses and Deformations, 82

MONTILLA, J., Determining the Inclination of a Diffusing Surface with Regard to Viewing Direction by Speckle Photography, 218

MOSS, B. C., Displacement and Vibration Measurement by Laser Interferometry, 52

PASTEUR, J., Contributions of Acoustical Holography to Mechanical Metrology, 340
PATANCHON, CLAUDE, Testing by Holographic Interferometry of Solid Propergol Engines, 181
PERRIN, J. C., Electronic Processing of Moire Fringes, Application to Moire Topography and Elements of Comparison with Photogrammetry, 311
PETTIGREW, R. M., Analysis of Grating Imaging and Its Application to Displacement Metrology, 325
PHAROK, M., Holographic Interferometry in Osteosynthesis, 202
PIGHINI, UMBERTO, Optical Differentiation of Moire—Holographic Fringes by Wavefront Reconstruction with White Light Sources, 296
PIRODDA, LUCIANO, Dimensional Metrology of Large Objects by Projection Moire Techniques, 318
POINTEAU, F., MODEX: Opto-Numerical Measurer of External Dimensions, 32
POIRIER, JACQUES, Double Exposure Holographic Interferometry: Application to Nondestructive Testing and to Breaking Point Mechanics, 101
RASTOGI, P., Photoelasticimetry and Holographic Interferometry: Applications to the Study of Stresses and Deformations, 82
RASUMOV, L. N., Holographical Disdrometry, 127
ROBLIN, G., Applications of Phase Modulation Interferometry to the Characterization of Materials and to Dimensional Metrology, 58
ROBLIN, MARIE-LOUISE, Decorrelation Produced in the Image of a Diffusing Object by a Rotation of the Incident Wave—Application to the Study of Surface States, 282
ROGER, ANDRE, Determination of the Index Profile of a Dielectric Plate by Optical Methods, 26
SAGAUT, J., Minutes of the Round Table on Photo-Sensitive Surfaces, 355
SANCHEZ, VICTOR, Study by Holographic Interferometry of Mass Transfer During Electrochemical Processes at Solid-Liquid Interfaces, 143
SCHLUTER, MICHAEL, Video-Electronic Analysis of Holographic Interferograms, 166
SCHUMANN, W., Holographic Interferometry with the Possibility of Modifying the Fringes During Reconstruction, 174
SEYZERIAT, Y., Contributions of Acoustical Holography to Mechanical Metrology, 340

SISAKYAN, I. N., Holographical Disdrometry, 127
SMIGIELSKI, PAUL, Testing by Holographic Interferometry of Solid Propergol Engines, 181
SOKOLOV, V., Improvement of the Signal/Noise Ratio in the Subtraction of Images by Speckle Interferometry, 237
SOULET, H., Study by Holographic Interferometry of Dimensional Variability in Precision-Moulding Materials Used in Odontology, 130
SPAJER, M., Photoelasticimetry and Holographic Interferometry: Applications to the Study of Stresses and Deformations, 82
SPEAKE, J. H., Displacement and Vibration Measurement by Laser Interferometry, 52
SURGET, JEAN, Holographic Interferometry Applied to the Metrology of Gaseous Flows, 192
TAILLAND, ALBERT, Interferometrical Setup for the Study of Thermic Turbulence in a Plane Airstream, 69
THERY, J. F., Analysis of the Diffraction Spectrum of a Population of Particles, 277
THOMAS, A., Electronic Processing of Moire Fringes. Application to Moire Topography and Elements of Comparison with Photogrammetry, 311
TRIBILLON, GILBERT, Two Wavelength Speckle Images Applied to Contouring, 286
TRIBILLON, J. L., Minutes of the Round Table on Photo-Sensitive Surfaces, 355
TURLIER, BERNARD, Hologrammetric Plotter (Apparatus for Three-Dimensional Measurement on an Image Reconstructed by Holographic Process), 76
———, MODEX: Opto-Numerical Measurer of External Dimensions, 32
VIENOT, JEAN-CHARLES, Space-Time Optics in Shape and Surface Metrology, 3
———, Optics in Europe, 2
VIRDEE, M. S., Speckle Photography for Strain Measurement—A Critical Assessment, 258
WALKER, C. A., The Measurement of Residual Stress by a Moire Fringe Method, 302
WOLFER, MELLE, Application of Interferential Correlation of Spectra to the Detection of Pollutants in the Atmosphere, 14
ZAKHAROV, V. M., Holographical Disdrometry, 127

SUBJECT INDEX

Acoustical Holography to Mechanical Metrology, Contributions of 340
Airstream, Interferometrical Setup for the Study of Thermic Turbulence in a Plane, 69
Analyser of Vibrations, Optical, 29
Analysis of the Diffraction Spectrum of a Population of Particles, 277
Analysis of Grating Imaging and Its Application to Displacement Metrology, 325
Analysis of Holographic Interferograms, Video-Electronic, 166
Analysis of Oscillations in Squealing Disk Brakes, Holographic, 148
(Apparatus for Three-Dimensional Measurement on an Image Reconstructed by Holographic Process), Hologrammetric Plotter, 76
Application in Civil Engineering of a Sandwich Speckle Method, 291
Application to Displacement Metrology, Analysis of Grating Imaging and Its, 325
Application of Holographic Interferometry to the Study of Structural Deformations in Civil Engineering, 136
Application of Holographic Interferometry to Testing of Spun Structures, 186
Application of Interferential Correlation of Spectra to the Detection of Pollutants in the Atmosphere, 14
Applications of Holography to the Study of Structures and Materials, 114
Applications of Phase Modulation Interferometry to the Characterization of Materials and to Dimensional Metrology, 58
Argon Ion Laser Holocinematography, Investigation of Cavitation Bubble Dynamics by High Speed Ruby Laser and, 348
Art with Recording in Three-Dimensional Media on the Basis of Lippman Photographic Plates, Holographic, 365
Aspherical Surfaces, Holographic Testing of, 208
Assessment, Speckle Photography for Strain Measurement—A Critical, 258
Atmosphere, Application of Interferential Correlation of Spectra to the Detection of Pollutants in the, 14
Autoprocessor Materials for the Recording of Phase Holograms: Photopolymers and Organo-Metallic Semiconductors, 229
Axial Displacements of a Diffusing Object, Detection of, 245

Basis of Lippman Photographic Plates, Holographic Art with Recording in Three-Dimensional Media on the, 365
$Bi_{12}SiO_{20}$, New Possibilities of Real-Time Interferometry with Photoconductive Electro-Optic Crystals, 226
Brakes, Holographic Analysis of Oscillations in Squealing Disk, 148
Breaking Point Mechanics, Double Exposure Holographic Interferometry: Application to Nondestructive Testing and to, 101
Bubble Dynamics by High Speed Ruby Laser and Argon Ion Laser Holocinematography, Investigation of Cavitation, 348

Captors, Measurement in Real Time of Transversal Micro-Vibrations (Down to 1 Å) on Diffusing Objects: Random Moire, 333
Cavitation Bubble Dynamics by High Speed Ruby Laser and Argon Ion Laser Holocinematography, Investigation of, 348
Cinematography, Principle of the Holographic, 358
Civil Engineering, Application of Holographic Interferometry to the Study of Structural Deformations in, 136
Civil Engineering of a Sandwich Speckle Method, Application in, 291
Comparison of Wavelengths to the Primary Standard at the French National Institute of Metrology, 38
Connecting Rods by Speckle Photography, Deformation Measurements on, 270
Contouring, Two Wavelength Speckle Images Applied to, 286
Contributions of Acoustical Holography to Mechanical Metrology, 340
Correlation of Spectra to the Detection of Pollutants in the Atmosphere, Application of Interferential, 14
Critical Assessment, Speckle Photography for Strain Measurement—A, 258
Crystals $Bi_{12}SiO_{20}$, New Possibilities of Real-Time Interferometry with Photoconductive Electro-Optic, 226

Decorrelation Produced in the Image of a Diffusing Object by a Rotation of the Incident Wave—Application to the Study of Surface States, 282
Deformation Measurements on Connecting Rods by Speckle Photography, 270

Deformations in Civil Engineering, Application of Holographic Interferometry to the Study of Structural, 136
Deformations, Photoelasticimetry and Holographic Interferometry: Applications to the Study of Stresses and, 82
Detection of Axial Displacements of a Diffusing Object, 245
Detection of Pollutants in the Atmosphere, Application of Interferential Correlation of Spectra to the, 14
Determination of the Index Profile of a Dielectric Plate by Optical Methods, 26
Determining the Inclination of a Diffusing Surface with Regard to Viewing Direction by Speckle Photography, 218
Dielectric Plate by Optical Methods, Determination of the Index Profile of a, 26
Differentiation of Moire—Holographic Fringes by Wavefront Reconstruction with White Light Sources, Optical, 296
(The) Diffracting Gauge in Extensometry, 251
Diffraction Spectrum of a Population of Particles, Analysis of the, 277
Diffusing Object, Detection of Axial Displacements of a, 245
Diffusing Object by a Rotation of the Incident Wave—Application to the Study of Surface States, Decorrelation Produced in the Image of a, 282
Diffusing Objects: Random Moire Captors, Measurement in Real Time of Transversal Micro-Vibrations (Down to 1 Å) on, 333
Diffusing Surface with Regard to Viewing Direction by Speckle Photography, Determining the Inclination of a, 218
Dimensional Metrology, Applications of Phase Modulation Interferometry to the Characterization of Materials and to, 58
Dimensional Metrology of Large Objects by Projection Moire Techniques, 318
Dimensional Metrology of Length Standards by Holographic Interferometry with Phase Heterodynage, 92
Dimensional Objects: Metrological Applications, Image Spectrograms of Three-, 65
Dimensional Variability in Precision-Moulding Materials Used in Odontology, Study by Holographic Interferometry of, 130
Dimensions, MODEX: Opto-Numerical Measurer of External, 32
Direction by Speckle Photography, Determining the Inclination of a Diffusing Surface with Regard to Viewing, 218
Disdrometry, Holographical, 127
Disk Brakes, Holographic Analysis of Oscillations in Squealing, 148
Displacement Metrology, Analysis of Grating Imaging and Its Application to, 325
Displacement and Vibration Measurement by Laser Interferometry, 52
Displacements of a Diffusing Object, Detection of Axial, 245
Distribution of Velocities in a Fluid by Speckle Photography, Study of the, 266
Double Exposure Holographic Interferometry: Application to Nondestructive Testing and to Breaking Point Mechanics, 101
Dynamics by High Speed Ruby Laser and Argon Ion Laser Holocinematography, Investigation of Cavitation Bubble, 348

Electrochemical Processes at Solid-Liquid Interfaces, Study by Holographic Interferometry of Mass Transfer During, 143
Electronic Analysis of Holographic Interferograms, Video-, 166
Electronic Processing of Moire Fringes. Application to Moire Topography and Elements of Comparison with Photogrammetry, 311
Electro-Optic Crystals in $Bi_{12}SiO_{20}$, New Possibilities of Real-Time Interferometry with Photoconductive, 226
Elements of Comparison with Photogrammetry, Electronic Processing of Moire Fringes. Application to Moire Topography and, 311
Engineering, Application of Holographic Interferometry to the Study of Structural Deformations in Civil, 136
Engineering of a Sandwich Speckle Method, Application in Civil, 291
Engines, Testing by Holographic Interferometry of Solid Propergol, 181
ESPI, Holographic Methods Made Useful by Phase Modulated, 222
Europe, Optics in, 2
Exposure Holographic Interferometry: Application to Nondestructive Testing and to Breaking Point Mechanics, Double, 101
Extensometry, The Diffracting Gauge in, 251
External Dimensions, MODEX: Opto-Numerical Measurer of, 32

Flows, Holographic Interferometry Applied to the Metrology of Gaseous, 192

Fluid by Speckle Photography, Study of the Distribution of Velocities in a, 266
French National Institute of Metrology, Comparison of Wavelengths to the Primary Standard at the, 38
Fringe Method, The Measurement of Residual Stress by a Moire, 302
Fringes. Application to Moire Topography and Elements of Comparison with Photogrammetry, Electronic Processing of Moire, 311
Fringes During Reconstruction, Holographic Interferometry with the Possibility of Modifying the, 174
Fringes by Wavefront Reconstruction with White Light Sources, Optical Differentiation of Moire—Holographic, 296

Gaseous Flows, Holographic Interferometry Applied to the Metrology of, 192
Gauge in Extensometry, The Diffracting, 251
Grating Imaging and Its Application to Displacement Metrology, Analysis of, 325

Heterodynage, Dimensional Metrology of Length Standards by Holographic Interferometry with Phase, 92
Heterodyne Holographic Interferometry: A Review, 215
Heterogeneities in Semiconductors, Interferometric Measurement of, 8
High Speed Ruby Laser and Argon Ion Laser Holocinematography, Investigation of Cavitation Bubble Dynamics by, 348
Holocinematography, Investigation of Cavitation Bubble Dynamics by High Speed Ruby Laser and Argon Ion Laser, 348
Hologrammetric Plotter (Apparatus for Three-Dimensional Measurement on an Image Reconstructed by Holographic Process), 76
Holograms: Photopolymers and Organo-Metallic Semiconductors, Autoprocessor Materials for the Recording of Phase, 229
Holographical Disdrometry, 127
Holographic Analysis of Oscillations in Squealing Disk Brakes, 148
Holographic Art with Recording in Three-Dimensional Media on the Basis of Lippman Photographic Plates, 365
Holographic Cinematography, Principle of the, 358
Holographic Fringes by Wavefront Reconstruction with White Light Sources, Optical Differentiation of Moire—, 296
Holographic Interferograms, Some Considerations on the Quantitative Interpretation of, 156
Holographic Interferograms, Video-Electronic Analysis of, 166
Holographic Interferometry: Application to Nondestructive Testing and to Breaking Point Mechanics, Double Exposure, 101
Holographic Interferometry: Applications to the Study of Stresses and Deformations, Photoelasticimetry and, 82
Holographic Interferometry Applied to the Metrology of Gaseous Flows, 192
Holographic Interferometry Applied to Minimal Wear Measurement, 107
Holographic Interferometry of Dimensional Variability in Precision-Moulding Materials Used in Odontology, Study by, 130
Holographic Interferometry of Mass Transfer During Electrochemical Processes at Solid-Liquid Interfaces, Study by, 143
Holographic Interferometry in Osteosynthesis, 202
Holographic Interferometry with Phase Heterodynage, Dimensional Metrology of Length Standards by, 92
Holographic Interferometry with the Possibility of Modifying the Fringes During Reconstruction, 174
Holographic Interferometry: A Review, Heterodyne, 215
Holographic Interferometry of Solid Propergol Engines, Testing by, 181
Holographic Interferometry to the Study of Structural Deformations in Civil Engineering, Application of, 136
Holographic Interferometry to Testing of Spun Structures, Application of, 186
Holographic Methods Made Useful by Phase Modulated ESPI, 222
Holographic Process), Hologrammetric Plotter (Apparatus for Three-Dimensional Measurement on an Image Reconstructed by, 76
Holographic Testing of Aspherical Surfaces, 208
Holography to Mechanical Metrology, Contributions of Acoustical, 340
Holography to the Study of Structures and Materials, Applications of, 114

Image of a Diffusing Object by a Rotation of the Incident Wave—Application to the Study of Surface States, Decorrelation Produced in the, 282
Image Reconstructed by Holographic Process), Hologrammetric Plotter (Apparatus for Three-Dimensional Measurement on an, 76
Image Spectrograms of Three-Dimensional Objects: Metrological Applications, 65
Images Applied to Contouring, Two Wavelength Speckle, 286
Images by Speckle Interferometry, Improvement of the Signal/Noise Ratio in the Subtraction of, 237
Imaging and Its Application to Displacement Metrology, Analysis of Grating, 325
Improvement of the Signal/Noise Ratio in the Method of Subtraction of Images by Speckle Interferometry, 237
Incident Wave—Application to the Study of Surface States, Decorrelation Produced in the Image of a Diffusing Object by a Rotation of the, 282
Inclination of a Diffusing Surface with Regard to Viewing Direction by Speckle Photography, Determining the, 218
Index Profile of a Dielectric Plate by Optical Methods, Determination of the, 26
Infrared Materials and Optical Systems, Interferometer for Testing, 43
Institute of Metrology, Comparison of Wavelengths to the Primary Standard at the French National, 38
Interfaces, Study by Holographic Interferometry of Mass Transfer During Electrochemical Processes at Solid-Liquid, 143
Interferential Correlation of Spectra to the Detection of Pollutants in the Atmosphere, Application of, 14
Interferograms, Some Considerations on the Quantitative Interpretation of Holographic, 156
Interferograms, Video-Electronic Analysis of Holographic, 166
Interferometer for Testing Infrared Materials and Optical Systems, 43
Interferometrical Setup for the Study of Thermic Turbulence in a Plane Airstream, 69
Interferometric Measurement of Heterogeneities in Semiconductors, 8
Interferometry: Application to Nondestructive Testing and to Breaking Point Mechanics, Double Exposure Holographic, 101
Interferometry: Applications to the Study of Stresses and Deformations, Photoelasticimetry and Holographic, 82
Interferometry Applied to the Metrology of Gaseous Flows, Holographic, 192
Interferometry Applied to Minimal Wear Measurement, Holographic, 107
Interferometry to the Characterization of Materials and to Dimensional Metrology, Applications of Phase Modulation, 58
Interferometry of Dimensional Variability in Precision-Moulding Materials Used in Odontology, Study by Holographic, 130
Interferometry, Displacement and Vibration Measurement by Laser, 52
Interferometry, Improvement of the Signal/Noise Ratio in the Subtraction of Images by Speckle, 237
Interferometry of Mass Transfer During Electrochemical Processes at Solid-Liquid Interfaces, Study by, 143
Interferometry in Osteosynthesis, Holographic, 202
Interferometry with Phase Heterodynage, Dimensional Metrology of Length Standards by Holographic, 92
Interferometry with Photoconductive Electro-Optic Crystals $Bi_{12}SiO_{20}$, New Possibilities of Real-Time, 226
Interferometry with the Possibility of Modifying the Fringes During Reconstruction, Holographic, 174
Interferometry: A Review, Heterodyne Holographic, 215
Interferometry of Solid Propergol Engines, Testing by Holographic, 181
Interferometry to the Study of Structural Deformations in Civil Engineering, Application of Holographic, 136
Interferometry to Testing of Spun Structures, Application of Holographic, 186
Interpretation of Holographic Interferograms, Some Considerations on the Quantitative, 156
Investigation of Cavitation Bubble Dynamics by High Speed Ruby Laser and Argon Ion Laser Holocinematography, 348
Ion Laser Holocinematography, Investigation of Cavitation Bubble Dynamics by High Speed Ruby Laser and Argon, 348

Large Objects by Projection Moire Techniques, Dimensional Metrology of, 318

Laser and Argon Ion Laser Holocinematography, Investigation of Cavitation Bubble Dynamics by High Speed Ruby, 348

Laser Holocinematography, Investigation of Cavitation Bubble Dynamics by High Speed Ruby Laser and Argon Ion, 348

Laser Interferometry, Displacement and Vibration Measurement by, 52

Length Standards by Holographic Interferometry with Phase Heterodynage, Dimensional Metrology of, 92

Light Sources, Optical Differentiation of Moire—Holographic Fringes by Wavefront Reconstruction with White, 296

Lippman Photographic Plates, Holographic Art with Recording in Three-Dimensional Media on the Basis of, 365

Mass Transfer During Electrochemical Processes at Solid-Liquid Interfaces, Study by Holographic Interferometry of, 143

Materials, Applications of Holography to the Study of Structures and, 114

Materials Used in Odontology, Study by Holographic Interferometry of Dimensional Variability in Precision-Moulding, 130

Materials and Optical Systems, Interferometer for Testing Infrared, 43

Materials for the Recording of Phase Holograms: Photopolymers and Organo-Metallic Semiconductors, Autoprocessor, 229

Measurement—A Critical Assessment, Speckle Photography for Strain, 258

Measurement of Heterogeneities in Semiconductors, Interferometric, 8

Measurement, Holographic Interferometry Applied to Minimal Wear, 107

Measurement of an Image Reconstructed by Holographic Process), Hologrammetric Plotter (Apparatus for Three-Dimensional, 76

Measurement by Laser Interferometry, Displacement and Vibration, 52

Measurement in Real Time of Transversal Micro-Vibrations (Down to 1 Å) on Diffusing Objects: Random Moire Captors, 333

The Measurement of Residual Stress by a Moire Fringe Method, 302

Measurement of Residual Stress by a Moire Fringe Method, The, 302

Measurements of Small Rotations, 19

Measurements on Connecting Rods by Speckle Photography, Deformation, 270

Measurer of External Dimensions, MODEX: Opto-Numerical, 32

Mechanical Metrology, Contributions of Acoustical Holography to, 340

Mechanics, Double Exposure Holographic Interferometry: Application to Nondestructive Testing and to Breaking Point, 101

Media on the Basis of Lippman Photographic Plates, Holographic Art with Recording in Three-Dimensional, 365

Metallic Semiconductors, Autoprocessor Materials for the Recording of Phase Holograms: Photopolymers and Organo-, 229

Method, Application in Civil Engineering of a Sandwich Speckle, 291

Method, The Measurement of Residual Stress by a Moire Fringe, 302

Methods, Determination of the Index Profile of a Dielectric Plate by Optical, 26

Methods Made Useful by Phase Modulated ESPI, Holographic, 222

Metrological Applications, Image Spectrograms of Three-Dimensional Objects:, 65

Metrology, Analysis of Grating Imaging and Its Application to Displacement, 325

Metrology, Applications of Phase Modulation Interferometry to the Characterization of Materials and to Dimensional, 58

Metrology, Comparison of Wavelengths to the Primary Standard at the French National Institute of, 38

Metrology, Contributions of Acoustical Holography to Mechanical, 340

Metrology of Gaseous Flows, Holographic Interferometry Applied to the, 192

Metrology of Large Objects by Projection Moire Techniques, Dimensional, 318

Metrology of Length Standards by Holographic Interferometry with Phase Heterodynage, Dimensional, 92

Metrology, Space-Time Optics in Shape and Surface, 3

Micro-Vibrations (Down to 1 Å) on Diffusing Objects: Random Moire Captors, Measurement in Real Time of Transversal, 333

Minimal Wear Measurement, Holographic Interferometry Applied to, 107

Minutes of the Round Table on Photo-Sensitive Surfaces, 355

MODEX: Opto-Numerical Measurer of External Dimensions, 32

Modifying the Fringes During Reconstruction, Holographic Interferometry with the Possibility of, 174

Modulated ESPI, Holographic Methods Made Useful by Phase, 222

Modulation Interferometry to the Characterization of Materials and to Dimensional Metrology, Applications of Phase, 58

Moire Captors, Measurement in Real Time of Transversal Micro-Vibrations (Down to 1 Å) on Diffusing Objects: Random, 333

Moire Fringe Method, The Measurement of Residual Stress by a, 302

Moire Fringes. Application to Moire Topography and Elements of Comparison with Photogrammetry, Electronic Processing of, 311

Moire—Holographic Fringes by Wavefront Reconstruction with White Light Sources, Optical Differentiation of, 296

Moire Techniques, Dimensional Metrology of Large Objects by Projection, 318

Moire Topography and Elements of Comparison with Photogrammetry, Electronic Processing of Moire Fringes. Application to, 311

Moulding Materials Used in Odontology, Study by Holographic Interferometry of Dimensional Variability in Precision-, 130

New Possibilities of Real-Time Interferometry with Photoconductive Electro-Optic Crystals $Bi_{12}SiO_{20}$, 226

Noise Ratio in the Subtraction of Images by Speckle Interferometry, Improvement of the Signal/, 237

Nondestructive Testing and to Breaking Point Mechanics, Double Exposure Holographic Interferometry: Application to, 101

Object, Detection of Axial Displacements of a Diffusing, 245

Object by a Rotation of the Incident Wave—Application to the Study of Surface States, Decorrelation Produced in the Image of a Diffusing, 282

Objects: Metrological Applications, Image Spectrograms of Three-Dimensional, 65

Objects by Projection Moire Techniques, Dimensional Metrology of Large, 318

Objects: Random Moire Captors, Measurement in Real Time of Transversal Micro-Vibrations (Down to 1 Å) on Diffusing, 333

Odontology, Study by Holographic Interferometry of Dimensional Variability in Precision-Moulding Materials Used in, 130

Optical Analyser of Vibrations, 29

Optical Differentiation of Moire—Holographic Fringes by Wavefront Reconstruction with White Light Sources, 296

Optical Methods, Determination of the Index Profile of a Dielectric Plate by, 26

Optical Systems, Interferometer for Testing Infrared Materials and, 43

Optic Crystals $Bi_{12}SiO_{20}$, New Possibilities of Real-Time Interferometry with Photoconductive Electro-, 226

Optics In Europe, 2

Optics in Shape and Surface Metrology, Space-Time, 3

Opto-Numerical Measurer of External Dimensions, MODEX: 32

Organo-Metallic Semiconductors, Autoprocessor Materials for the Recording of Phase Holograms: Photopolymers and, 229

Oscillations in Squealing Disk Brakes, Holographic Analysis of, 148

Osteosynthesis, Holographic Interferometry in, 202

Particles, Analysis of the Diffraction Spectrum of a Population of, 277

Phase Heterodynage, Dimensional Metrology of Length Standards by Holographic Interferometry with, 92

Phase Holograms: Photopolymers and Organo-Metallic Semiconductors, Autoprocessor Materials for the Recording of, 229

Phase Modulated ESPI, Holographic Methods Made Useful by, 222

Phase Modulation Interferometry to the Characterization of Materials and to Dimensional Metrology, Applications of, 58

Photoconductive Electro-Optic Crystals $Bi_{12}SiO_{20}$, New Possibilities of Real-Time Interferometry with, 226

Photoelasticimetry and Holographic Interferometry: Applications to the Study of Stresses and Deformations, 82

Photogrammetry, Electronic Processing of Moire Fringes. Application to Moire Topography and Elements of Comparison with, 311

Photographic Plates, Holographic Art with Recording in Three-Dimensional Media on the Basis of Lippman, 365

Photography, Deformation Measurements on Connecting Rods by Speckle, 270

Photography, Determining the Inclination of a Diffusing Surface with Regard to Viewing Direction by Speckle, 218

Photography for Strain Measurement—A Critical Assessment, Speckle, 258

Photography, Study of the Distribution of Velocities in a Fluid by Speckle, 266

Photopolymers and Organo-Metallic Semiconductors, Autoprocessor Materials for the Recording of Phase Holograms:, 229

Photo-Sensitive Surfaces, Minutes of the Round Table on, 355

Plane Airstream, Interferometrical Setup for the Study of Thermic Turbulence in a, 69

Plate by Optical Methods, Determination of the Index Profile of a Dielectric, 26

Plates, Holographic Art with Recording in Three-Dimensional Media on the Basis of Lippman Photographic, 365

Plotter (Apparatus for Three-Dimensional Measurement of an Image Reconstructed by Holographic Process), Hologrammetric, 76

Point Mechanics, Double Exposure Holographic Interferometry: Application of Nondestructive Testing and to Breaking, 101

Pollutants in the Atmosphere, Application of Interferential Correlation of Spectra to the Detection of, 14

Possibilities of Real-Time Interferometry with Photoconductive Electro-Optic Crystals $Bi_{12}SiO_{20}$, New, 226

Possibility of Modifying the Fringes During Reconstruction, Holographic Interferometry with the, 174

Precision-Moulding Materials Used in Odontology, Study by Holographic Interferometry of Dimensional Variability in, 130

Primary Standard at the French National Institute of Metrology, Comparison of Wavelengths to the, 38

Principle of the Holographic Cinematography, 358

Processes at Solid-Liquid Interfaces, Study by Holographic Interferometry of Mass Transfer During Electrochemical, 143

Process), Hologrammetric Plotter (Apparatus for Three-Dimensional Measurement on an Image Reconstructed by Holographic, 76

Processing of Moire Fringes. Application to Moire Topography and Elements of Comparison with Photogrammetry, Electronic, 311

Profile of a Dielectric Plate by Optical Methods, Determination of the index, 26

Projection Moire Techniques, Dimensional Metrology of Large Objects by, 318

Propergol Engines, Testing by Holographic Interferometry of Solid, 181

Quantitative Interpretation of Holographic Interferograms, Some Considerations on the, 156

Random Moire Captors, Measurement in Real Time of Transversal Micro-Vibrations (Down to 1 Å) on Diffusing Objects: 333

Ratio in the Subtraction of Images by Speckle Interferometry, Improvement of the Signal/Noise, 237

Real-Time Interferometry with Photoconductive Electro-Optic Crystals $Bi_{12}SiO_{20}$, New Possibilities of, 226

Real Time of Transversal Micro-Vibrations (Down to 1 Å) on Diffusing Objects: Random Moire Captors, Measurement in, 333

Reconstructed by Holographic Process), Hologrammetric Plotter (Apparatus for Three-Dimensional Measurement of an Image, 76

Reconstruction, Holographic Interferometry with the Possibility of Modifying the Fringes During, 174

Reconstruction with White Light Sources, Optical Differentiation of Moire—Holographic Fringes by Wavefront, 296

Recording of Phase Holograms: Photopolymers and Organo-Metallic Semiconductors, Autoprocessor Materials for the, 229

Recording in Three-Dimensional Media on the Basis of Lippman Photographic Plates, Holographic Art with, 365

Residual Stress by a Moire Fringe Method, The Measurement of, 302

Review, Heterodyne Holographic Interferometry: A, 215

Rods by Speckle Photography, Deformation Measurements on Connecting, 270

Rotation of the Incident Wave—Application to the Study of Surface States, Decorrelation Produced in the Image of a Diffusing Object by a, 282

Rotations, Measurement of Small, 19

Round Table on Photo-Sensitive Surfaces, Minutes of the, 355

Ruby Laser and Argon Ion Laser Holocinematography, Investigation of Cavitation Bubble Dynamics by High Speed, 348

Sandwich Speckle Method, Application in Civil Engineering of a, 291

Semiconductors, Autoprocessor Materials for the Recording of Phase Holograms: Photopolymers and Organo-Metallic, 229

Semiconductors, Interferometric Measurement of Heterogeneities in, 8

Sensitive Surfaces, Minutes of the Round Table on Photo-, 355

Setup for the Study of Thermic Turbulence in a Plane Airstream, Interferometrical, 69

Shape and Surface Metrology, Space-Time Optics in, 3

Signal/Noise Ratio in the Subtraction of Images by Speckle Interferometry, Improvement of the, 237

Small Rotations, Measurement of, 19

Solid-Liquid Interfaces, Study by Holographic Interferometry of Mass Transfer During Electrochemical Processes at, 143

Solid Propergol Engines, Testing by Holographic Interferometry of, 181

Some Considerations on the Quantitative Interpretation of Holographic Interferograms, 156

Sources, Optical Differentiation of Moire—Holographic Fringes by Wavefront Reconstruction with White Light, 296

Space-Time Optics in Shape and Surface Metrology, 3

Speckle Images Applied to Contouring, Two Wavelength, 286

Speckle Interferometry, Improvement of the Signal/Noise Ratio in the Subtraction of Images by, 237

Speckle Method, Application in Civil Engineering of a Sandwich, 291

Speckle Photography, Deformation Measurements on Connecting Rods by, 270

Speckle Photography, Determining the Inclination of a Diffusing Surface with Regard to Viewing Direction by, 218

Speckle Photography for Strain Measurement—A Critical Assessment, 258

Speckle Photography, Study of the Distribution of Velocities in a Fluid by, 266

Spectra to the Detection of Pollutants in the Atmosphere, Application of Interferential Correlation of, 14

Spectrograms of Three-Dimensional Objects: Metrologicql Applications, Image, 65

Spectrum of a Population of Particles, Analysis of the Diffraction, 277

Speed Ruby Laser and Argon Ion Laser Holocinematography, Investigation of Cavitation Bubble Dynamics by High, 348

Spun Structures, Application of Holographic Interferometry to Testing of, 186

Squealing Disk Brakes, Holographic Analysis of Oscillations in, 148

Standard at the French National Institute of Metrology, Comparison of Wavelengths to the Primary, 38

Standards by Holographic Interferometry with Phase Heterodynage, Dimensional Metrology of Length, 92

States, Decorrelation Produced in the Image of a Diffusing Object by a Rotation of the Incident Wave—Application to the Study of Surface, 282

Strain Measurement—A Critical Assessment, Speckle Photography for, 258

Stress by a Moire Fringe Method, The Measurement of Residual, 302

Stresses and Deformations, Photoelasticimetry and Holographic Interferometry: Applications to the Study of, 82

Structural Deformations in Civil Engineering, Application of Holographic Interferometry to the Study of, 136

Structures, Application of Holographic Interferometry to Testing of Spun, 186

Structures and Materials, Applications of Holography to the Study of, 114

Study by Holographic Interferometry of Dimensional Variability in Precision-Moulding Materials Used in Odontology, 130

Study of the Distribution of Velocities in a Fluid by Speckle Photography, 266

Study by Holographic Interferometry of Mass Transfer During Electrochemical Processes at Solid-Liquid Interfaces, 143

Subtraction of Images by Speckle Interferometry, Improvement of the Signal/Noise Ratio in the, 237

Surface Metrology, Space-Time Optics in Shape and, 3

Surface with Regard to Viewing Direction by Speckle Photography, Determining the Inclination of a Diffusing, 218

Surface States, Decorrelation Produced in the Image of a Diffusing Object by a Rotation of the Incident Wave—Application to the Study of, 282

Surfaces, Holographic Testing of Aspherical, 208

Surfaces, Minutes of the Round Table on Photo-Sensitive, 355
Systems, Interferometer for Testing Infrared Materials and Optical, 43

Table on Photo-Sensitive Surfaces, Minutes of the Round, 355
Techniques, Dimensional Metrology of Large Objects by Projection Moire, 318
Testing of Aspherical Surfaces, Holographic, 208
Testing and to Breaking Point Mechanics, Double Exposure Holographic interferometry: Application to Nondestructive, 101
Testing by Holographic Interferometry of Solid Propergol Engines, 181
Testing Infrared Materials and Optical Systems, Interferometer for, 43
Testing of Spun Structures, Application of Holographic Interferometry to, 186
Thermic Turbulence in a Plane Airstream, Interferometrical Setup for the Study of, 69
Three-Dimensional Measurement on an Image Reconstructed by Holographic Process), Hologrammetric Plotter (Apparatus for, 76
Three-Dimensional Media on the Basis of Lippman Photographic Plates, Holographic Art with Recording in, 365
Three-Dimensional Objects: Metrological Applications, Image Spectrograms of, 65
-Time Interferometry with Photoconductive Electro-Optic Crystals $Bi_{12}SiO_{20}$, New Possibilities of Real-, 226
-Time Optics in Shape and Surface Metrology, Space, 3
Topography and Elements of Comparison with Photogrammetry, Electronic Processing of Moire Fringes. Application to Moire, 311
Transfer During Electrochemical Processes at Solid-Liquid Interfaces, Study by Holographic Interferometry of Mass, 143

Transversal Micro-Vibrations (Down to 1 Å) on Diffusing Objects: Random Moire Captors, Measurement in Real Time of, 333
Turbulence in a Plane Airstream, Interferometrical Setup for the Study of Thermic, 69
Two Wavelength Speckle Images Applied to Contouring, 286

Variability in Precision-Moulding Materials Used in Odontology, Study by Holographic Interferometry of Dimensional, 130
Velocities in a Fluid by Speckle Photography, Study of the Distribution of, 266
Vibration Measurement by Laser Interferometry, Displacement and, 52
Vibrations, Optical Analyser of, 29
Video-Electronic Analysis of Holographic Interferograms, 166
Viewing Direction by Speckle Photography, Determining the Inclination of a Diffusing Surface with Regard to, 218

Wave—Application to the Study of Surface States, Decorrelation Produced in the Image of a Diffusing Object by a Rotation of the Incident, 282
Wavefront Reconstruction with White Light Sources, Optical Differentiation of Moire—Holographic Fringes by, 296
Wavelength Speckle Images Applied to Contouring, Two, 286
Wavelengths to the Primary Standard at the French National Institute of Metrology, Comparison of, 38
Wear Measurement, Holographic Interferometry Applied to Minimal, 107
White Light Sources, Optical Differentiation of Moire—Holographic Fringes by Wavefront Reconstruction with, 296

T
50
E97
1977

FEB 1 9 1980